Diagnosis, Management and Rehabilitation of Neurological Disorders

Diagnosis, Management and Rehabilitation of Neurological Disorders

Editor: Billy Acord

AMERICAN
MEDICAL PUBLISHERS
www.americanmedicalpublishers.com

AMERICAN
MEDICAL PUBLISHERS
www.americanmedicalpublishers.com

Cataloging-in-Publication Data

Diagnosis, management and rehabilitation of neurological disorders / edited by Billy Acord.
p. cm.
Includes bibliographical references and index.
ISBN 978-1-63927-290-7
1. Nervous system--Diseases. 2. Nervous system--Diseases--Diagnosis. 3. Nervous system--Diseases--Prevention.
4. Nervous system--Diseases--Treatment. 5. Neurology. I. Acord, Billy.
RC347 .D53 2021
616.8--dc23

American Medical Publishers,
41 Flatbush Avenue,
1st Floor, New York,
NY 11217, USA

ISBN 978-1-63927-290-7 (Hardback)

Contents

Preface

The world is advancing at a fast pace like never before. Therefore, the need is to keep up with the latest developments. This book was an idea that came to fruition when the specialists in the area realized the need to coordinate together and document essential themes in the subject. That's when I was requested to be the editor. Editing this book has been an honour as it brings together diverse authors researching on different streams of the field. The book collates essential materials contributed by veterans in the area which can be utilized by students and researchers alike.

The disorders of the nervous system are termed as neurological disorders. These include epilepsy, Parkinson's diseases, multiple sclerosis, peripheral neuropathy, Tourette's syndrome, narcolepsy, migraine, central neuropathy, autism, obsessive compulsive disorder, etc. A neurological examination and neuropsychological assessment can assess the impact of a neurological condition on memory, behavior or cognition, and identify problems in brain functioning. When a severe spine or brain injury results in a serious disability, neurorehabilitation offers innovative solutions to enable the individual to adapt to a new way of life. Therapies like occupational therapy, vision therapy, language therapy, physiotherapy, rehabilitation psychology, etc. are certain holistic practices of neurorehabilitation. Certain conditions, including neurotrauma and neuro-emergencies such as intracranial hemorrhage require neurosurgical interventions. Neuroradiology methods comprising of computer-assisted imaging, CT scan, MRI, PET, etc. have notable neurosurgical applications. This book includes some of the vital pieces of work being conducted across the world, on various topics related to the diagnosis, management and rehabilitation of neurological disorders. It will also provide interesting topics for research which interested readers can take up. The topics covered in this book offer the readers new insights in the field of neurology.

Each chapter is a sole-standing publication that reflects each author´s interpretation. Thus, the book displays a multi-facetted picture of our current understanding of application, resources and aspects of the field. I would like to thank the contributors of this book and my family for their endless support.

Editor

1

RNA binding proteins co-localize with small tau inclusions in tauopathy

Brandon F. Maziuk[1], Daniel J. Apicco[1], Anna Lourdes Cruz[1], Lulu Jiang[1], Peter E. A. Ash[1], Edroaldo Lummertz da Rocha[2], Cheng Zhang[2], Wai Haung Yu[3], John Leszyk[4], Jose F. Abisambra[5], Hu Li[2] and Benjamin Wolozin[1,6,7*]

Abstract

The development of insoluble, intracellular neurofibrillary tangles composed of the microtubule-associated protein tau is a defining feature of tauopathies, including Alzheimer's disease (AD). Accumulating evidence suggests that tau pathology co-localizes with RNA binding proteins (RBPs) that are known markers for stress granules (SGs). Here we used proteomics to determine how the network of tau binding proteins changes with disease in the rTg4510 mouse, and then followed up with immunohistochemistry to identify RNA binding proteins that co-localize with tau pathology. The tau interactome networks revealed striking disease-related changes in interactions between tau and a multiple RBPs, and biochemical fractionation studies demonstrated that many of these proteins including hnRNPA0, EWSR1, PABP and RPL7 form insoluble aggregates as tau pathology develops. Immunohistochemical analysis of mouse and human brain tissues suggest a model of evolving pathological interaction, in which RBPs co-localize with pathological phospho-tau but occur adjacent to larger pathological tau inclusions. We suggest a model in which tau initially interacts with RBPs in small complexes, but evolves into isolated aggregated inclusions as tau pathology matures.

Keywords: Alzheimer's disease, Neurofibrillary tangle, Immunohistochemistry, Mass spectrometry, Protein interactome, Protein aggregation

Introduction

RNA binding proteins (RBPs) have emerged as major factors in a number of neurodegenerative disorders, including tauopathies [31, 32]. RBPs represent a large class of proteins responsible for RNA metabolism, regulating key events including mRNA maturation, trafficking, and eventual translation. Importantly, many RBPs have been shown to have strong genetic links to neurodegenerative disease; for example, mutations in TDP-43, FUS, hnRNPA1, and ATXN2 have all been shown to cause amyotrophic lateral sclerosis (ALS), frontotemporal lobar degeneration (FTLD) and/or spinocerebellar ataxia [24]. Other RBPs have also been implicated in AD, Huntington's disease, and Creutzfeldt-Jakob disease [24].

* Correspondence: bwolozin@bu.edu
[1]Department of Pharmacology and Experimental Therapeutics, Boston University School of Medicine, Boston, MA, USA
[6]Department of Neurology, Boston University School of Medicine, Boston, MA, USA
Full list of author information is available at the end of the article

Following these initial discoveries, evidence now suggests that transient aggregation mediated by low-complexity domains within RBPs is crucial for the involvement of these proteins in disease. RBPs typically contain a "prion-like", low complexity glycine-rich domain in their sequences, which regulate their self-aggregation [21, 35]. This regulated aggregation is important for RBP function, as many RBPs regulate the formation of a variety of ribonucleoprotein (RNP) granules. Such granules, which include stress granules (SGs) and P-bodies, are responsible for the transient recruitment and aggregation of mRNA transcripts for varying purposes [29, 32]. SGs recruit mRNA into granules for storage and protection following translational stress, while P-bodies recruit mRNA for eventual degradation [29, 32]. Many disease-linked RBPs are proteins involved in the formation and maintenance of SGs.

In particular, recent findings have shown that T-cell intracellular antigen 1 (TIA1) and other RBPs interact with hyper-phosphorylated tau and accumulate in tandem with tau pathology in diseases such as Alzheimer's disease (AD) and frontotemporal dementia (FTD) [37, 38]. RBPs also regulate splicing of MAPT exon 10, which provides an additional mechanism for control of tau aggregation by determining whether the tau transcript codes for 3 or 4 microtubule repeats [9, 18, 40]. The aggregation of tau is a major pathological hallmark of these diseases, where the microtubule associated protein tau becomes hyperphosphorylated, and mislocalizes in the soma [41]. Once in the soma, tau begins to aggregate, eventually forming large neurofibrillary tangles and resulting in neurodegeneration. Recently, our lab demonstrated that tau interacts with TIA1 in SGs, which increases in sarkosyl insoluble aggregates. The association of tau with TIA1 and SGs also was shown to promote tau-mediated degeneration of primary hippocampal neurons [37]. This phenotype can be rescued by TIA1 knockdown in primary neurons [37]. We proceeded to show that TIA1 reduction (haploinsufficiency) is also protective in vivo, reducing neurodegeneration, rescuing cognition and increasing survival in the PS19 mouse model of tauopathy [1]. These results point to an essential role for TIA1 and SG biology in the pathophysiology of tauopathy.

We now present an enhanced view of the interaction of RBPs in the pathophysiology of tauopathy. We have used a proteomic approach to identify multiple RBPs that have significantly altered associations with tau as pathology develops in the rTg4510 mouse model of tauopathy including multiple splicing factors. The rTg4510 mouse model of tauopathy inducibly expresses P301L 4R0N tau under control of a tetracycline promoter, and when aged in the absence of doxycycline develop pathology by 4–5 months [30]. We also present optimized immunohistochemical methods for detecting RBPs in pathological tissues. We combine the optimized immunohistochemical methods with biochemical methods to validate the proteomic findings. We identify specific RBPs that associate with pathological phospho-tau early in disease and become increasingly insoluble as disease progresses. Together our results provide important evidence indicating that RNA binding protein pathologies are a major feature of tauopathy.

Results

TIA1 colocalizes with phosphorylated tau in tauopathy

Recent work from our laboratory demonstrates that TIA1 and other SG markers colocalize with pathological tau in Tg4510 and PS19 mouse brains as well as human post-mortem AD and FTDP-17 brain samples [1, 37, 38]. Immunohistochemical identification of TIA1 inclusions in pathological samples has proven to be particularly difficult, with some observing TIA1 inclusions, and others unable to observe inclusions [8, 11, 15]. While this could represent differences in the types of cases examined, we have also observed TIA1 reactivity to be highly sensitive to methodological conditions.

To improve reproducibility among laboratories, we set out to define optimal conditions for TIA1 immunohistochemistry. We began by determining optimal conditions for fixation of mouse tissue. rTg4510 tissues were drop-fixed in 4% buffered paraformaldehyde for 24 or 48 h, then transferred to 30% sucrose, incubated at 4 °C for 48 h, sectioned and subjected to immunohistochemistry. Shorter fixation times produced much stronger reactivity for both TIA1 and NeuN, which is also a RNA binding protein (Additional file 1: Figure S4). The 24 h fixation protocol was used for all subsequent experiments using animal tissues. We also optimized conditions to reduce background. Our previous studies used Sudan black to quench background fluorescence, however such quenching also has the drawback of reducing antibody signals [23, 38]. Photobleaching proved to provide much more effective quenching of autofluorescence without significant loss of antibody reactivity (Additional file 1: Figure S5). Application of shorter fixation times and photobleaching significantly improved the sensitivity for detecting RBPs in tissue samples.

Next we compared commercially available antibodies for immunohistochemical reactivity. We observed that the anti-TIA1 antibody from Abcam (Abcam cat#40693) gave the strongest results (Fig. 1, Additional file 1: Figure S1). Using the Abcam antibody, we consistently observed TIA1 colocalization with CP13 positive phospho-tau in the rTg4510 and the PS19 mouse models of tauopathy (Fig. 1a-c). The strength of reactivity varied with lot number (Fig. 1e), and thus the work in this manuscript used lot GR151575 (Fig. 1e). Reactivity with antibodies from other vendors did not work as well as the Abcam antibodies (Additional file 1: Figure S1). In all cases, specificity of the TIA1 reactivity was demonstrated by the absence of anti-TIA1 reactivity observed following immuno-adsorption or staining of TIA1 knockout brain tissue (Fig. 1d). These results demonstrate that immunohistochemical reactivity with the Abcam anti-TIA1 antibody is bona-fide TIA1 reactivity.

We proceeded to characterize how the co-localization of TIA1 and tau varied with the type of tau pathology. Analysis of patterns of co-localization in 6 month-old rTg4510 brain tissue demonstrated a distinct correlation of TIA1 co-localization with the size of CP13-positive tau inclusions (Fig. 1e, f). Abundant co-localization was observed with small CP13 reactive puncta, while little co-localization was observed with large fibrillar CP13 positive tau inclusions (Fig. 1e). Previous results from our laboratory indicate that TIA1 selectively interacts with oligomeric tau [1]. To test whether the small TIA1 reactive inclusions contained tau oligomers, we probed the tissues with the anti-tau oligomer antibody TOC1 [19].

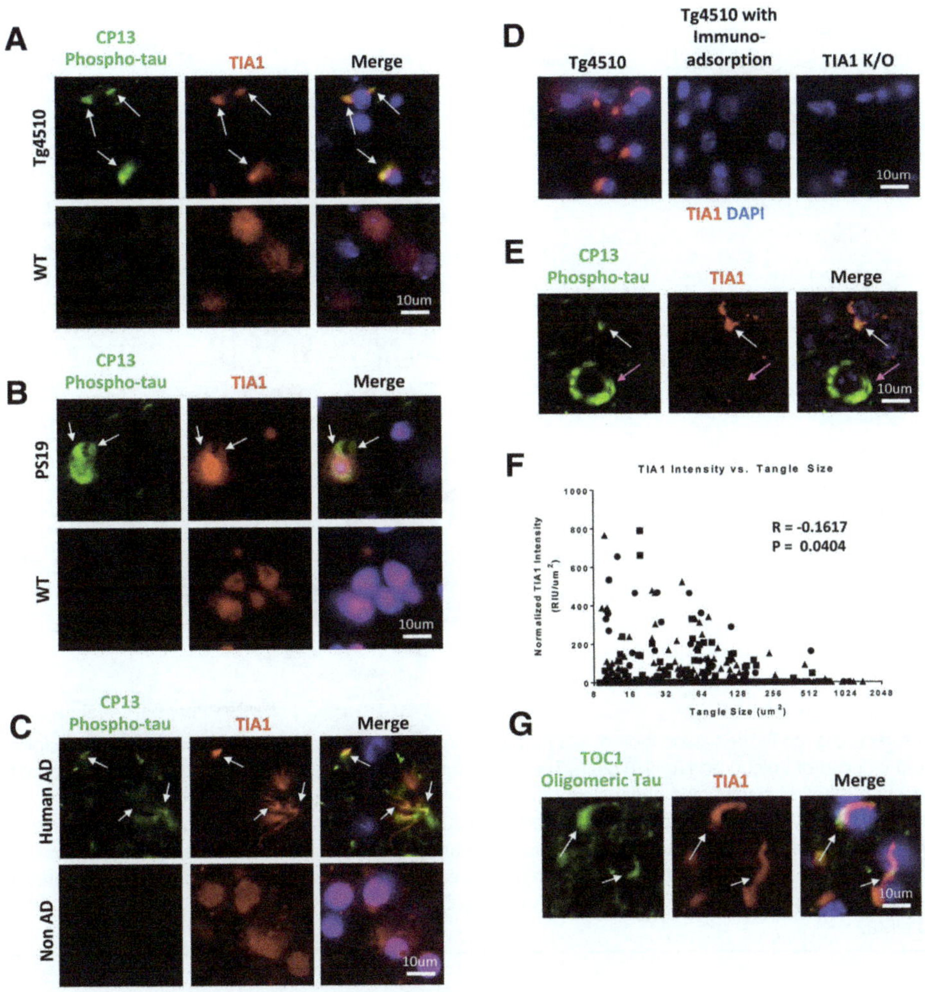

Fig. 1 TIA1 preferentially colocalizes with phosphorylated tau that is not in NFTs. Immunohistochemical analysis of the stress granule nucleating protein TIA1 (red) shows colocalization with phosphorylated tau (CP13 antibody against pSer202, green) in rTg4510 (**a**), PS19 (**b**) and human AD tissues (**c**) with DAPI in blue. **d** Immunohistochemistry of TIA1 on rTg4510 with and without immunosorbtion using TIA1 peptide, as well as TIA1 staining of TIA1 knockout mouse tissue, also show that the TIA1 antibody used for this analysis is specific for TIA1 and does not have off-target staining. **e** Further analysis of TIA1-tau colocalization in rTg4510 tissues indicates that TIA1 primarily has cytoplasmic colocalization with smaller tau aggregates (white arrow) over larger NFT-like tau aggregates (pink arrow). This is quantified in (**f**) using Imaris Bitplane software, where cytoplasmic TIA1 intensity negatively correlates with the size of tau tangles in neurons ($R = -0.1617$ with a two tailed p value of 0.0404). **g** TIA1 also colocalizes with oligomeric tau stained using the TOC1 antibody specific for oligomeric tau

Strong co-localization was observed between anti-TIA1 and the TOC1 tau oligomer specific antibodies (Fig. 1f, R = − 0.1617, $p = 0.0404$). From this we conclude that TIA1 preferentially associates with small tau inclusions containing oligomeric tau.

Other RNA binding proteins change in their association to tau in tauopathy

The association of phosphorylated tau with the stress granule protein TIA1 raises the possibility that other RBPs and ribosomal proteins might also interact with tau [1, 25, 26, 37]. To explore the tau interactome, we immunoprecipitated (IP) total human tau from 2.5 month old Tg4510 mouse frontal cortex expressing low or high amounts of human tau protein due to inclusion or exclusion of doxycycline from the mouse chow, respectively; IPs were performed using the Tau13 antibody, which is a high affinity antibody that selectively recognizes total human tau (independent of tau phosphorylation state) [5]. We also did IPs from MAPT −/− mice to provide negative controls for the IPs. This design allowed exploration of neurons exposed to relatively low or high amounts of stress, but preceding a phase with extensive neurodegeneration. The tau complexes were then analyzed using orbitrap mass spectrometry to identify proteins in the tau interactome (Fig. 2); any proteins pulled down by the Tau13 antibody from the MAPT−/− mice were removed from tau interactome observed in the rTg4510 IPs. This analysis identified a large

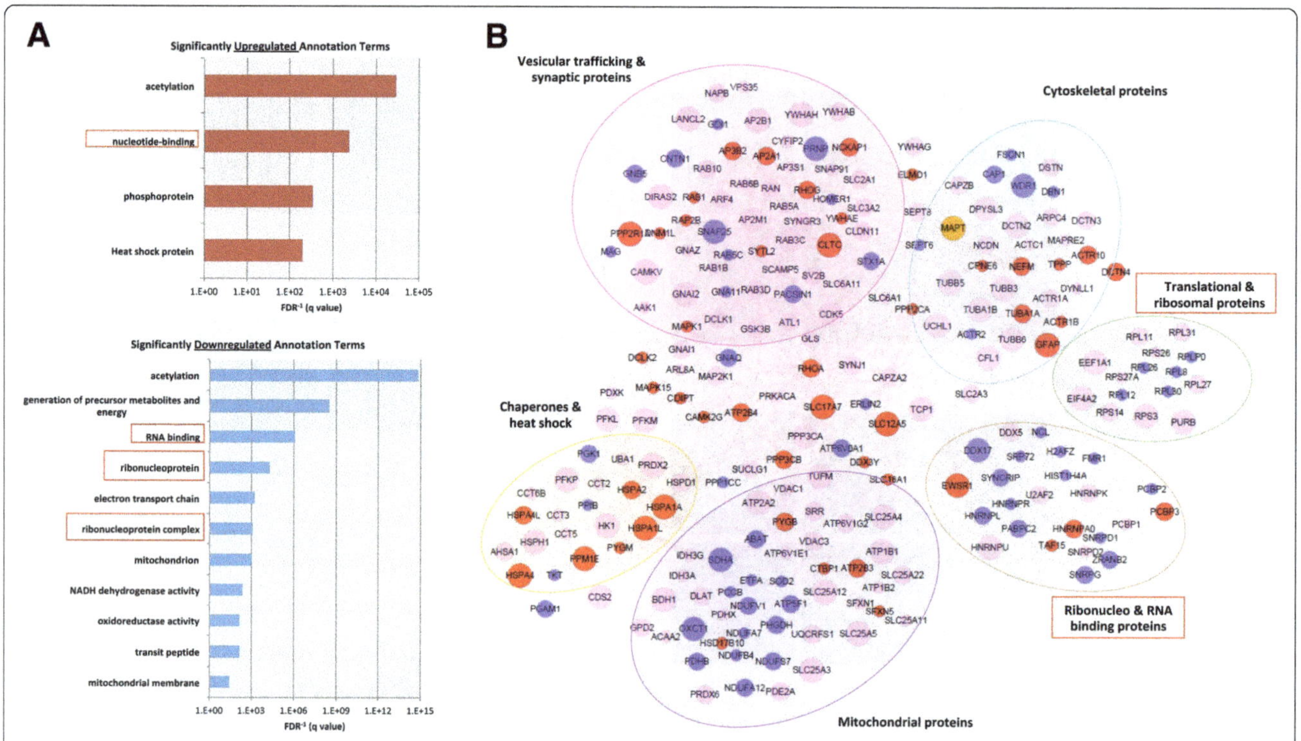

Fig. 2 RNA binding proteins change in their association to tau in tauopathy. Using Tau13 antibody, tau was immunoprecipitated from the cortex of either 2.5 month old induced or uninduced rTg4510 brains.The co-immunoprecipitating proteins were subsequently identified by mass spectrometry. A variety of nucleotide binding and ribonucleoproteins show differential association to tau in rTg4510 tissue compared to WT tissue as indicated by significantly upregulated and downregulated annotation terms (**a**). Part (**b**) represents a network diagram of proteins enriched in the tau interactome. Red nodes denote proteins whose association with tau is ≥5-fold increased in Tg4510 vs uninduced control. Blue nodes denote proteins whose association with tau is ≥5-fold decreased in Tg4510 vs control. Node size is proportional to degree of replication (*N* = 4), non-replicating proteins excluded. Key proteins include EWSR1, HNRNPA0, PABP, PCBP2, DDX5, TAF15, FMR1, RPL and RPS family members, and eIF4a2

number of proteins associated with tau, including RBPs, translational and ribosomal proteins, cytoskeletal proteins, chaperones, heat shock proteins and synaptic proteins (Fig. 2 and Additional file 2: Table S1).

Comparing the tau networks from the induced and un-induced rTg4510 mice revealed striking changes. Multiple proteins showed decreased association with tau in the 2.5 m induced rTg4510 mice. Proteins linked to RNA metabolism comprised 3 of the top 6 GO annotations categories of decreased proteins (Fig. 2a, bottom panel). Interaction of tau with some of these proteins, including HNRNPA0, HNRNPK, eIF4a, and RPL11, was validated by immunoprecipitating tau from 6 month old PS19 P301S tau mouse brains; these mice overexpress mutant P301S tau at 5 times endogenous levels (Additional file 1: Figure S2). Confirmation is also provided by comparison to our prior study showing the requirement of tau for the presence of many proteins in the TIA1 interactome network [37]. These changes corresponded to large-scale changes in the interaction of tau with these particular groups of proteins. For instance, 65% of proteins in the RBP group showed decreased association with disease; in contrast, only 16% of proteins in the synaptic and vesicular protein group showed decreased association with disease (Fig. 2a).

The groups of proteins whose interaction with tau increased with disease were equally striking. Heat shock proteins/chaperones and phospho-proteins were 2 of the 3 groups of proteins upregulated in the 2.5 m induced rTg4510 mouse cortex (Fig. 2a, top panel), with 37% of heat shock or chaperone proteins showing increased association with tau with disease (Fig. 2a). This upregulation is consistent with the stress associated with pre-tangle stage that predominates in the 2.5 m rTg4510 mouse [30]. Interestingly, the third group of proteins whose association with tau increased in the 2.5 m mice was nucleotide binding proteins; these include RBPs such as EWSR1, TAF15 and HNRNPA0, which we previously reported to co-localize with tau pathology in the rTg4510 model of tauopathy [37] (Fig. 2a, b). In addition, aggregation-enhancing mutations in EWSR1 and TAF15 are associated with ALS, suggesting that changes in the biology of these proteins are sufficient to drive neurodegenerative disease [6, 7]. Taken together, these results highlight that complexes containing tau and RBPs are a major feature of the pathophysiology of tauopathy.

RNA binding proteins become insoluble in tauopathy

A key feature of RBP biology is the formation of transient, dynamic complexes, such as stress granules, P-bodies, transport granules, and even splicing complexes. With time or in disease, many RBPs exhibit a tendency to develop into insoluble aggregates. To explore whether RBPs become insoluble in mouse models of tauopathy, we fractionated frontal cortex from 8 m old rTg4510 mice aged with low or high levels of human tau (i.e., ± dox). The samples were then biochemically separated into sarkosyl soluble or insoluble fractions and immunoblotted. Levels of RBPs in the sarkosyl soluble group did not change with disease state (Fig. 3a, b). In contrast, the amount of RBP present in the insoluble fraction was elevated in diseased tissues for most of the RBPs examined (Fig. 3a, b). Analysis of TDP-43 levels showed no change with disease state, consistent with the lack of association of this protein with tau aggregation processes. (Fig. 3a, b). These results suggest that a range of RBPs form insoluble aggregates with disease in tauopathy.

RNA binding proteins colocalize with phospho-tau but not mature neurofibrillary tangles in tauopathy

Using immunohistochemistry with confocal microscopy, we proceeded to examine how deposition of RBPs in disease correlated with tau pathology. We examined proteins identified by mass spectrometry in the tau network (Fig. 2), proteins that we had previously shown to associate with tau, as well as proteins thought to be involved in the RNA translational stress response and stress granule formation. Frontal cortex from 6 month old rTg4510 mice was used for the study because of the presence of both mature and evolving tau pathology at this age. Imaging of RBPs using fluorescence immunohistochemistry showed the accumulation of RBPs co-localized with or near diffuse deposits of phosphorylated tau identified with CP13, an antibody recognizing pS202 (Fig. 4,

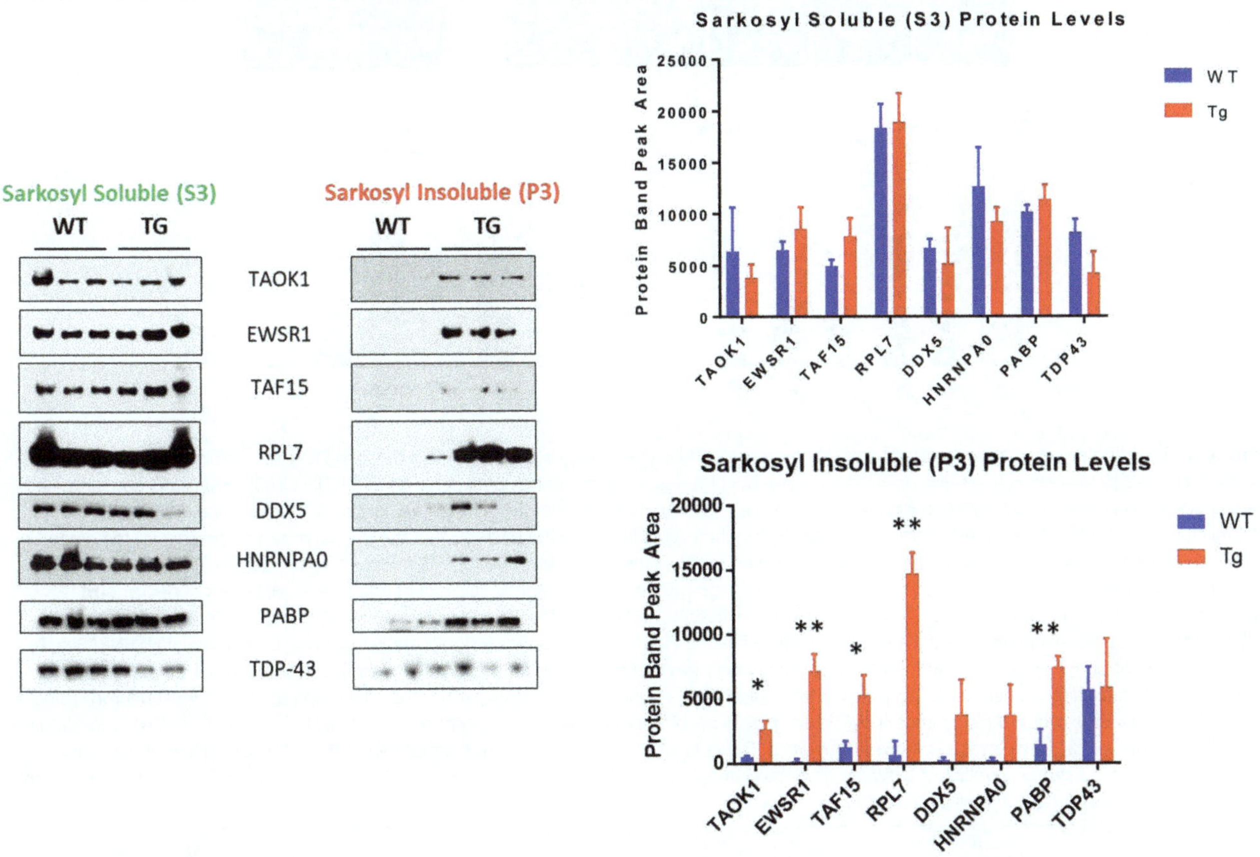

Fig. 3 RNA binding proteins become insoluble in the cortex of rTg4510 mice. (**a**, **b**) Immunoblots of the sarkosyl soluble (S3) and insoluble (P3) fractions isolated from rTg4510 cortical tissues indicate that many RBPs become insoluble as tau pathology develops. The fractions were also probed for TDP-43, which is not associated with tau aggregation. Quantification of these immunoblots (**c**, **d**) shows statistically significant RBP accumulation in the P3 fraction of induced rTg4510 mouse cortex using a two-tailed t-test ($p = 0.00599$ for TAOK1; $p = 0.0007599$ for EWSR1; $p = 0.0122$ for TAF15; $p = 0.000252$ for RPL7; $p = 0.00195$ for PABP; $p = 0.0926$ for DDX5; $p = 0.0638$ for HNRNPA0)

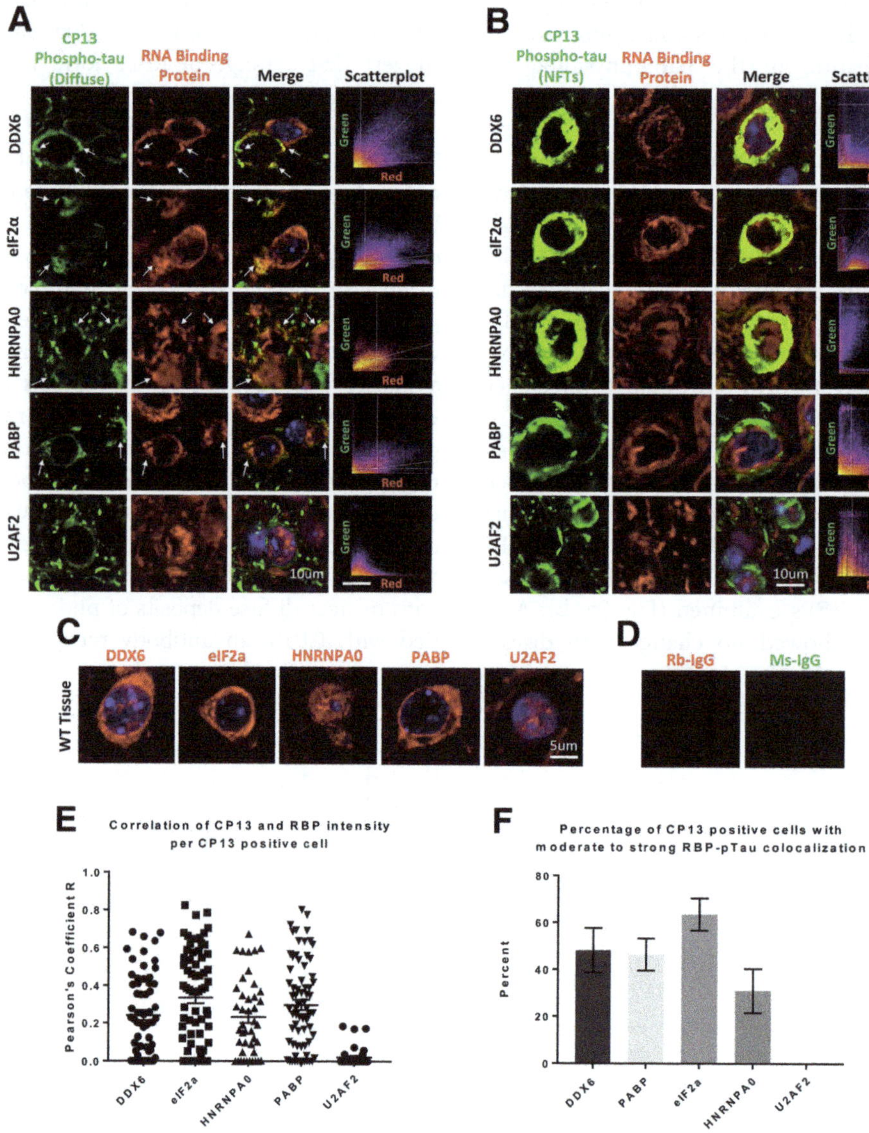

Fig. 4 RNA binding proteins show significant colocalization with diffuse phospho-tau but not NFTs in the rTg4510 cortex. (**a**) Immunohistochemical analysis of rTg4510 tissue ($n = 3$) has also revealed a significant colocalization in the cortex between the RBPs DDX6, PABP, HNRNPA0, and eIF2a (red) with pathological phospho-tau stained using the CP13 antibody (green). However, the RBP and splicing factor U2AF2 does not show significant correlation. To the right of each merged image is a scatterplot of the pixel intensities for each pixel of the image in the red channel vs. the green (Pearson correlation coefficients $r = 0.773$ for DDX6, 0.791 for eIF2α, 0.325 for HNRNPA0, 0.798 for PABP, and − 0.14 for U2AF2). This colocalization is greatly reduced and/or completely lost as tau aggregates into large NFTs which are brightly fluorescent and fill the cell bodies of neurons (**b**) ($r = 0.069$ for DDX6, 0.372 for eIF2α, 0.481 for PABP, − 0.03 for HNRNPA0, and − 0.009 for U2AF2). **c** Staining of wild-type C57Bl/6 mice also indicates that HNRNPA0 is predominantly nuclear in healthy animals, while the rTg4510 staining shows significant cytoplasmic localization of HNRNPA0 (**a**, **b**). (**d**) Negative controls IHC using rabbit and mouse normal IgG indicates that there is no off target staining or fluorescence in our tissues. **e** Pearson coefficients of correlation between CP13 positive tau with RBPs DDX6, eIF2α, HNRNPA0, PABP, and U2AF2 are graphed for individual neurons using ImageJ. For all cases except U2AF2, neurons show heterogeneity in colocalization between phospho-tau and the RBPs stained, from no colocalization to fully overlapping reactivity patterns in individual neurons. The percent of neurons with $r > 0.3$ is graphed in (**f**) as the percentage of neurons showing moderate to strong correlations between green:red intensity (DDX6 = 36% of neurons; eIF2α = 54% of neurons; HNRNPA0 = 35% of neurons; PABP = 33% of neurons; U2AF2 = 0% of neurons)

Additional file 1: Figure S3). The RBPs and proteins linked to RNA metabolism primarily colocalized with phosphorylated tau present in neuronal somas (Fig. 4a); scatterplots done on the images demonstrated that when overlap was present there was strong co-localization with tau pathology (Fig. 4a, e). We quantified the fraction of neurons exhibiting CP13 reactivity that also exhibited RBP reactivity (Fig. 4f, g). Robust correlation for CP13/RBP co-localization was observed for DDX6, eIF2α, hnRNPA0 and PABP, but not for U2AF2 (Fig. 4f, g); robust correlation was also observed for TIA1 (Fig. 1f). Interestingly, little colocalization was observed with mature NFTs showing bright condensed CP13 reactivity (Fig. 4b), consistent with the studies of TIA1 in Fig. 1e, f. Some of the RBPs (e.g., DDX6 and hnRNPA0) appear to remain as aggregates but accumulate adjacent to the tau NFTs, as if the RBPs are excluded from large, consolidated tau tangles, and pushed to the periphery of such tangles (Fig. 4b).

Finally, we examined the pattern of reactivity of RBPs in human tissue. We examined temporal cortex of late Braak stage V and VI human AD patients (Fig. 5). At this stage of disease, CP13 positive phospho-tau exists predominantly as NFTs (showing bright condensed CP13 reactivity), with little diffuse phospho-tau, such as is seen at 6–8 months of age in the rTg4510 mouse model (Fig. S6). RBP inclusions were readily apparent in the human tissues using antibodies against DDX6 and hnRNPA0 (Fig. 5). Comparison of the distribution of the RBP and tau deposits in individual neurons suggests an inverse correlation between RBP localization and mature NFTs similar to that observed in the rTg4510 mouse tissue (Fig. 5, intensity plots). In addition, we observed that some RBPs (e.g., DDX6 and hnRNP0) accumulated as pathological inclusions, but accumulated adjacent to the mature tangles, much like what was observed in rTg4510 mice (Fig. 5). These results demonstrate that deposits of RBPs occur near deposits of pathological tau, but suggest that RBPs are excluded from the aggregated tau as the deposits consolidate.

Discussion

Characterization of the tau interactome in the rTg4510 mouse model of tauopathy reveals striking disease-related changes in interactions between tau and multiple RBPs. About 65% of proteins in the RBP group show decreased tau association with disease. A smaller number of RBPs increase the association of tau with disease, including EWSR1, TAF15 and hnRNPA0, each of which have been linked to ALS (e.g., EWSR1 or TAF15) or have related family members linked to ALS (hnRNPA0 shares homology with hnRNPA2B1). Formation of aggregates was evident biochemically and by immuhistochemisty which is

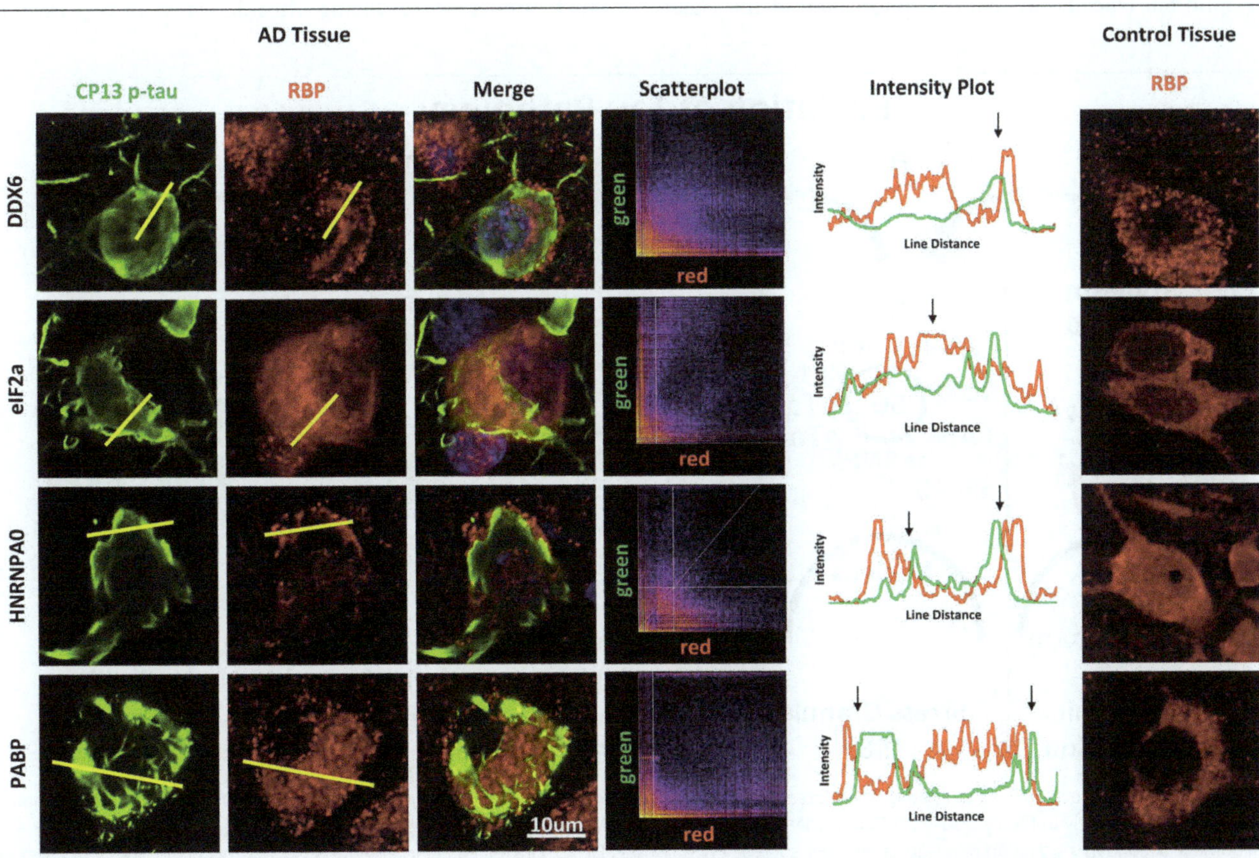

Fig. 5 RBPs show granularization and interfacing with NFTs in late stage human AD tissue. Immunohistochemical analysis of human AD frontal cortex tissue ($n = 6$) shows that RBPs (red) and NFTs stained with CP13 tau (green) do not colocalize ($r = 0.001$ for DDX6; 0.176 for eIF2α; 0.031 for HNRNPA0; and 0.222 for PABP). However, trace analyses across lines within the image (yellow bars) indicate that peak fluorescent intensities (reported as fluorescence intensity plots over the distance of the line for tau in green and the RBP in red) between phospho-tau and multiple RBPs are immediately adjacent to each other, indicating protein interfacing or interaction between the edges of NFTs and RBPs

consistent with prior studies showing multiple U1 spliceosome components that accumulate in the sarkosyl-insoluble fraction of AD tissues [2].

Imaging studies suggest that the pattern of co-localization varied depending on the type of RBP, and the size of the inclusion. For the RBPs examined in this manuscript, TIA1 showed the strongest co-localization with tau pathology. Co-localization for TIA1 with tau could be observed for phospho-tau (CP12, p-S202) and oligomeric tau (TOC1). Other RBPs, including hnRNPA0, DDX6, eIF2α and (to a lesser extent) PABP also co-localized with phospho-tau pathology. Direct co-localization was observed with small tau puncta, but for large tau inclusions these RBPs tended to accumulate immediately adjacent to the inclusion. The observation of RBP inclusions adjacent to tau tangles has also been observed for small nuclear ribonuclear proteins, such as U2AF2 [12, 13]. The differential localization of RBP inclusion with tau aggregate size suggests a model in which these proteins initially co-localize with tau but become excluded as the tau aggregates consolidate into mature neurofibrillary tangles. This work adds key information to our prior data indicating that phospho-tau (specifically, tau phosphorylated at the classic proline directed serine/threonine sites associated with pathology) promotes SG formation [37].

These results suggest the hypothesis that tau and RBPs exhibit progressive maturation (Fig. 6). This process appears to begin with a stress response that includes the translational stress response, which leads to cytoplasmic translocation of nuclear RBPs (Fig. 6a), followed by formation of functional SGs in part through interaction with phosphorylated, oligomeric tau (Fig. 6b) [1, 37]. The association of tau with SGs also produces aggregation, much as occurs with disease-prone RBPs during extended or repetitive periods of liquid liquid phase separation (LLPS) [22, 27, 28, 37]. Tau also undergoes LLPS, and does so in a manner that is accelerated by the presence of mRNA [14, 39, 42]. The similar biophysical behavior of tau and RBPs, the association of tau with SGs, the ability of tau to promote SG formation, and the tendency of tau to aggregate in the presence of SGs, provide a strong basis for a model based on the interaction of tau with SGs. Co-localization of tau with SG-associated RBPs suggest that the tau complexes with RBPs complexes that either are SGs or resemble SGs, which then mature during chronic stress into persistent

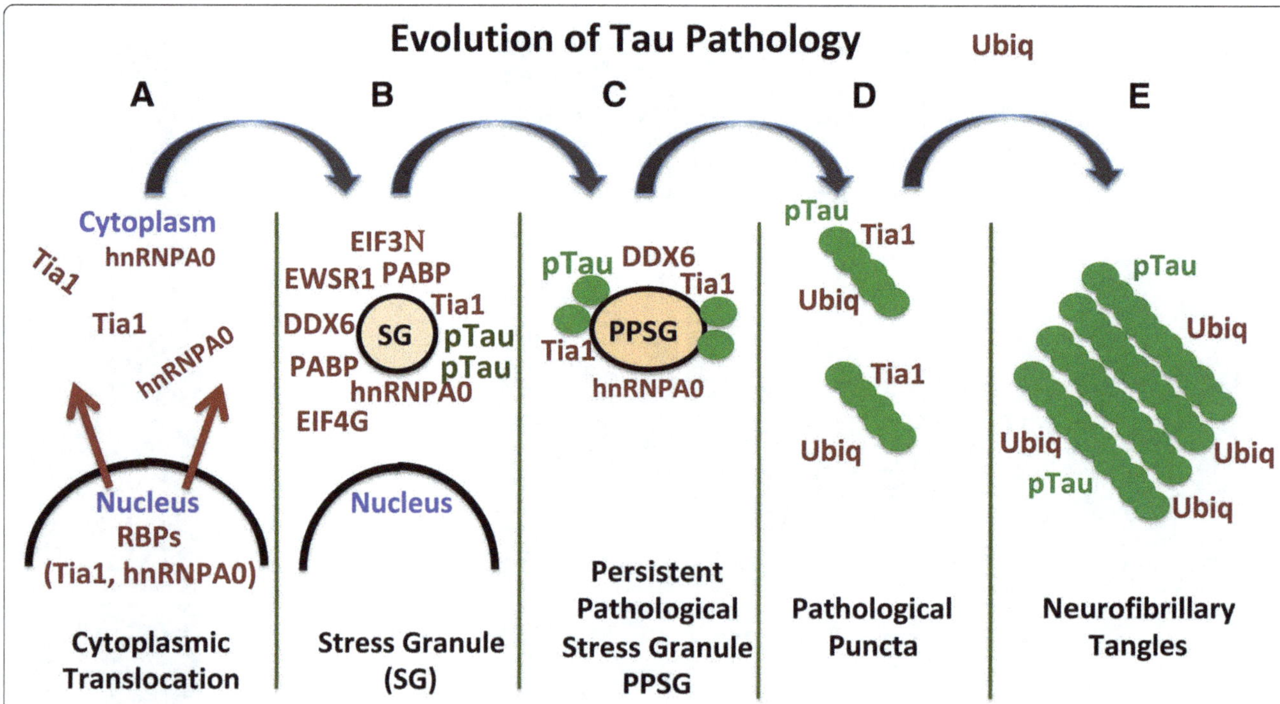

Fig. 6 Model for the evolution of tau pathology **a** Stress elicits translocation of nuclear RBPs, such as TIA1 and HNRNPA0, to the cytoplasm where they distribute diffusely in the cytoplasm. **b** Core nucleating RBPs, such as TIA1, coalesce to form SGs, in a process that is stimulated by tau oligomers. Secondary nucleation brings in other RBPs, such as EWSR1, PABP, DDX6, EIFs, etc. **c** Persistent stress (such as occurs in disease) causes particularly insoluble RBPs (and tau) to consolidate to form persistent pathological stress granules. More soluble RBPs separate from the persistent pathological SG, and disperse without aggregating. **d** The tau oligomers evolve into fibrils forming pathological puncta, which contain classic markers of pathology such as hyper-phosphorylation and ubiquitination [43]. Other RBPs that remain associated with the persistent pathological SG, begin to form aggregated puncta around the tau puncta. **e** Tau fibrillizes forming neurofibrillary tangles, which are relatively inert, do not act in the translational stress response and contain few RBSs

pathological SGs (Fig. 6c). Based on our data, we hypothesize that persistent SGs lose many types of RBPs and perhaps reducing their physiological activity (Fig. 6d); ubiquitin is added into the model at this point to reflect the well-documented association of ubiquitin with mature tangle pathology [43]. As the aggregated tau in these pathological SGs consolidates they appear to exclude many RBPs, and ultimately become the inert pathological structures that are referred to as neurofibrillary tangles (Fig. 6e). Meanwhile, the excluded RBPs form inclusions that accumulate adjacent to the neurofibrillary tangles. The latter stages of this hypothesis (steps 6c - e) remain to be explicitly tested beyond the correlative evidence presented above, but the model sets up a paradigm to guide future studies.

This model is further supported by accumulating evidence pointing to a persistent translational stress response as a key pathway leading to the accumulation of SGs. Chronic disease could produce a chronic stress, which leads to persistence of SGs. The high concentration of RBPs in SGs (100–400-fold higher than dispersed RBP levels) creates conditions that also promote aggregation of RBPs into insoluble amyloids, which over time accumulate [16, 24, 31]. Support for the persistent SG hypothesis comes from protection experiments in cell culture and in transgenic models. Bonini and colleagues demonstrated that chemical inhibition of the SG pathway rescues the ALS phenotype in drosophila [20]. Ataxin-2 deletion, which also inhibits the SG/translational stress response pathway, also delayed disease progression in a mouse model of ALS [3]. The relevance of the SG pathway to tauopathy was recently demonstrated by our observation that TIA1 reduction protects against disease progression in a mouse model of tauopathy [1]. Studies using primary neurons support these results by demonstrating that both RNAi knockdown of TIA1 and chemical inhibition of the SG pathway are able to prevent tau-mediated toxicity [37]. Thus, multiple independent lines of evidence demonstrate that RBPs, SGs and the translational stress response contribute to the pathophysiology of tauopathy and other neurodegenerative diseases.

The putative role for tau in regulating the RNA metabolism is supported by proteomic studies of tau interactomes from multiple different groups, which also identify similar classes of proteins that associate with tau. Immunoprecipitation and mass spectrometry of tau (both WT and P301L) binding proteins from SH-SY5Y neuroblastoma cells showed strong overlap of RBPs and ribosomal protein with the proteins identified in our study, including EWSR1, DDX5 & 17, hnRNPK, L, R and U, as well as ribosomal proteins RPL7, 8, 27 and 30 [10]. Proteomic studies of complexes containing tau from the rTg4510 mouse model and human AD tissues also report RNA binding and nucleotide binding proteins in their results, as well as multiple heat shock proteins and chaperones [4, 25]; work from the Abrisambra laboratory also directly demonstrates that tau over-expression inhibits RNA translation [25, 26]. Studies using HeLa cells report multiple RBPs in the tau interactome, including TIA1, hnRNP family members, and many ribosomal subunit proteins, and also demonstrated the presence of aggregated RBPs in the AD brain [36]. A recent study of tau-associated proteins in lymphoblastoid cell lines containing AKAP9 mutations linked to AD show enrichment of RNA binding and spliceosomal proteins in the tau proteome [17]. These proteins again include EWSR1, TAF15, DDX family members, and RPL family members, which parallels our findings. Finally, our own work previously demonstrated the association of TIA1 with tau by both proteomic analysis as well as immunoprecipitation [37]. These data demonstrate that identification of complexes containing tau and RBPs is a reproducible observation. The nature of the tau species responsible for binding to each RBP remains to be determined, and might differ between tau monomers, oligomers and fibrils; for instance, our recent study suggest that TIA1 exhibits preference for tau oligomers [1]. The studies also provide strong support for an emerging consensus that tau functions in stress to regulate the translation stress response through interaction with RBPs and SGs.

Our current study combines with accumulating prior studies to suggest that tauopathies (including AD) exist in the spectrum of neurodegenerative diseases and myopathies that are associated with dysfunction of RBPs, which already includes ALS and FTD. The current work advances the field by identifying a number of RBPs that can be reliably associated by immunohistochemistry with tau pathology, and providing optimized methods for detecting the association. These advances will facilitate rigor and reproducibility for this emerging field. Finally, the consistent role of RBPs and SGs in the mechanisms of multiple neurodegenerative diseases suggests that dysfunction of RBPs, SGs and translational stress response pathways plays a fundamental role in the pathophysiology of neurodegenerative diseases, and that seemingly disparate diseases might converge on common downstream mechanisms for neurodegeneration.

Materials and Methods

Animal husbandry and tissue collection

Animal husbandry for the rTg4510, PS19 and TIA1−/− mice was approved and performed as previously described in each indicated reference. For all brain harvesting, mice were anesthetized in an isoflurane chamber and perfused with ice cold PBS. Brains were then harvested and processed according to each subsequent experiment recorded below.

Human brain samples

Temporal cortex tissue from human brain was used for the immunohistochemical studies. The samples were de-identified. The cases are listed in Table 1 below:

Immunohistochemistry and quantification

All mouse ($n = 8$ rTg4510 and n = 8 wild type C57BL6/J) and human ($n = 7$ AD cases and $n = 8$ aged control) tissue was sectioned at 20um on a cryostat and stained as free floating sections using Netwell baskets (VWR Cat#29442–132) in 12 well Falcon plates (Cat#353043). Extracted mouse tissues were drop fixed in15mL 4% PFA for 24 h, transferred to 15 mL 30% sucrose in PBS for 48 h, then sectioned and stored in cryoprotectant (30% ethylene glycol; 30% glycerol; 40% PBS) at – 20 °C. We note that a recent study of stress granule pathology showed excellent labeling of TIA1 and other RBPs in a mouse model of tauopathy using perfusion with cold paraformaldehyde (4%), followed by drop fixation in cold paraformaldehyde for 2 h, transfer to 30% sucrose in PBS for at least 48 h [33]. Human tissues were fixed and stored in periodate-lysine-paraformaldehyde (PLP) until sectioned using a Leica VT1200S vibratome. To quench lipofuscin autofluorescence, sections were photobleached under a 1500 lm white LED bulb for a minimum of 72 h at 4 °C while suspended in PBS with no lid or covering [34]. Note that the photobleaching is time-limited; sections examined were observed to recover lipofuscin autofluorescence after approximately 1 week following the photobleaching. Sections were then washed 3× in PBS for 30 s each wash followed by a 5-min incubation in detergent media (TBS with 0.25% Triton-X). Human tissue, but not mouse tissue, was then incubated in 1%*w*/*v* sodium borohydride ($NaBH_4$; Sigma-Aldrich Cat#452882-25G) in PBS for 45 min to quench aldehyde autofluorescence which results from the over-fixation of tissue. All tissue was then incubated for 1 h in citrate based antigen unmasking solution (Vector Cat#H-3300) at 95 °C, with the exception of human TIA1 staining (Fig. 1), which was done using 0.05% citraconic anhydride (Sigma-Aldrich Cat#125318-25G) for 1 h at 95 °C. Tissue was blocked with gentle rotating for 2 h in detergent media with 5% donkey serum (Sigma-Aldrich D9663-10 mL) in PBS. After blocking, tissue was washed once in PBS for 30 s to remove excess detergent, then moved into 24 well plates with 200uL of the appropriate primary antibody dilutions and no basket overnight at 4 ° C. To prevent evaporation, the lids of each plate were lined with a damp Kimwipe. Antibody catalog numbers and dilutions are recorded in Additional file 3: Table S2.

Table 1 AD cases used for immunohistochemical studies

Age	Gender	Braak Stage
74	M	V
96	M	V
57	F	VI
80	F	VI
97	F	V
87	F	VI
87	F	II
67	F	II
84	F	I
70	M	I
90	F	II
93	F	II
91	M	II
79	M	VI
95	M	II

Following primary antibody incubation, tissue sections were washed 4× in PBS using baskets in 12 well plates for 5-min each wash. Tissues were then incubated in 200uL of secondary antibody solution containing a 1:500 dilution of the appropriate secondary antibody (Jackson Immunoresearch Alexafluor-488 and 549 secondaries) for 1.5 h in the dark at room temperature. If needed, after 1.5 h 300uL of a 1μg/mL DAPI solution was added to each tissue and incubated in the dark for 5 min at room temperature. Tissues were then washed 4× in PBS for 5 min each and carefully placed on Millenia 2.0 slides (StatLab Medical Products Cat#318) using a histology brush (Fisher Scientific Cat#NC0344756). Excess water was allowed to evaporate for 15 min, then each slide was mounted with 100uL of Prolong Gold Antifade Reagent (Thermo Fisher Cat#P36930) and a 1 mm cover slip.

All imaging was done using a Zeiss LSM 700 confocal microscope at either 40× or 63× magnification with a 1 AU pinhole. Following imaging, images were processed using either Imaris BitPlane software or the Fiji distribution of ImageJ. An Imaris BitPlane surfaces algorithm (Bitplane AG, Zurich, Switzerland) was used to quantify the tangle size vs. TIA1 intensity recorded in Fig. 1. The Coloc2 plugin of Fiji was used to determine colocalization parameters recorded in Fig. 4 following background subtraction with a rolling ball radius of 50. Intensity plots for Figs. 4 and 5 were also measured and generated using ImageJ-Fiji; regions of interest were set across a line using the line tool, and then measured using analyze- > plot profile.

Tau immunoprecipitation for mass spectrometry analysis

2.5 mo rTg4510 and uninduced transgenic mouse brains ($n = 4$/group) were extracted, slowly frozen by submersion in methanol on dry ice, and homogenized in RIPA lysis buffer (50 mM Tris-HCl, 150 mM NaCl, 1% Triton X-100, 0.1% SDS, 0.5% sodium deoxycholate, pH 6.8) supplemented with protease (Roche Cat#04693159001) and phosphatase inhibitor (Roche Cat#04906837001) cocktails. Tau was immunoprecipitated from 1 mg cortex lysate using 10 μg of the mouse monoclonal Tau13 antibody immobilized on Pierce Direct IP columns according to manufacturer's instructions (Thermo Scientific, Cat#26148). The Tau13 IP eluates were then separated on a Novex 4–12% Bis-Tris polyacrylamide gel (Life Technologies, Cat#NP0323) and stained with Simply Blue Coomassie G-250 SafeStain (Life Technologies, Cat#LC6060). Whole gel lanes were then excised and shipped to the UMass Mass Spectrometry and Proteomics facility for analysis by LC-MS/MS.

Proteomics, in gel digestion

Gel slices were cut into 1 × 1 mm pieces and placed in 1.5 ml eppendorf tubes with 1 ml of water for 30 min. The water was removed and 50ul of 250 mM ammonium bicarbonate was added. For reduction 20 μl of a 45 mM solution of 1, 4 dithiothreitol (DTT) was added and the samples were incubated at 50 C for 30 min. The samples were cooled to room temperature and then for alkylation 20 μl of a 100 mM iodoacetamide solution was added and allowed to react for 30 min. The gel slices were washed 2 X with 1 ml water aliquots. The water was removed and 1 ml of 50:50 (50 mM Ammonium Bicarbonate: Acetonitrile) was placed in each tube and samples were incubated at room temperature for 1 h. The solution was then removed and 200 ul of acetonitrile was added to each tube at which point the gels slices turned opaque white. The acetonitrile was removed and gel slices were further dried in a Speed Vac. Gel slices were rehydrated in 75 μl of 2 ng/μl trypsin (Sigma) in 0.01% ProteaseMAX Surfactant (Promega): 50 mM Ammonium Bicarbonate. Additional bicarbonate buffer was added to ensure complete submersion of the gel slices. Samples were incubated at 37C for 21 h. The supernatant of each sample was then removed and placed in a separate 1.5 ml eppendorf tube. Gel slices were further dehydrated with 100 ul of 80:20 (Acetonitrile: 1% formic acid). The extract was combined with the supernatants of each sample. The samples were then dried down in a Speed Vac. Samples were dissolved in 25 μl of 5% Acetonitrile in 0.1% trifluroacetic acid prior to injection on LC/MS/MS.

LC/MS/MS on Q Exactive

A 3.0 μl aliquot was directly injected onto a custom packed 2 cm × 100 μm C18 Magic 5 μm particle trap column. Labeled peptides were then eluted and sprayed from a custom packed emitter (75 μm × 25 cm C18 Magic 3 μm particle) with a linear gradient from 95% solvent A (0.1% formic acid in water) to 35% solvent B (0.1% formic acid in Acetonitrile) in 90 min at a flow rate of 300 nanoliters per minute on a Waters Nano Acquity UPLC system. Data dependent acquisitions were performed on a Q Exactive mass spectrometer (Thermo Scientific) according to an experiment where full MS scans from 300 to 1750 m/z were acquired at a resolution of 70,000 followed by 10 MS/MS scans acquired under HCD fragmentation at a resolution of 17,500 with an isolation width of 1.6 Da. Raw data files were processed with Proteome Discoverer (version 1.4) prior to searching with Mascot Server (version 2.5) against the Uniprot database. Search parameters utilized were fully tryptic with 2 missed cleavages, parent mass tolerances of 10 ppm and fragment mass tolerances of 0.05 Da. A fixed modification of carbamidomethyl cysteine and variable modifications of acetyl (protein N-term), pyro glutamic for N-term glutamine, oxidation of methionine were considered. Search results were loaded into the Scaffold Viewer (Proteome Software, Inc.).

Proteomic analysis

Quantitative proteomic analysis was performed using the total ion current (TIC) for proteins identified by LC-MS/MS normalized to the TIC level of TIA1 detected in each sample. Proteins identified in the TIA1 −/− samples were considered to be nonspecific binding proteins to the IP antibody and excluded from all subsequent analyses. Gene lists of detected proteins were then uploaded into the Database for Annotation, Visualization and Integrated Discovery (DAVID) resource available via the NIH website.

The proteins identified in the tau binding proteome in WT C57BL/6 J cortex resulted in 6 identified clusters with enrichment FDR < 0.05, and each of the proteins was associated to the cluster(s) based on its membership in the clustered gene sets. A network was created by adding a connection between protein pairs sharing annotation clusters. Edge weights were determined as the number of shared annotation clusters between protein pairs with thicker edges representing stronger functional associations between proteins (the smallest number of clusters shared between any two proteins was 1, and the largest was 8). The resulting network was visualized using the software Gephi 0.8.2 and arranged using the Force Atlas 2 layout algorithm. The network was generated using the python programming language (Python Software Foundation), and the networkx, numpy, and pandas python packages.

CoIP validation of mass spectrometry

Brains from 6 month old PS19 mice expressing P301S tau were extracted and freshly homogenized using 400uL IP lysis buffer (0.025 M Tris; 0.15 M NaCl; 0.001 M EDTA; 1% NP-40, 5% glycerol; pH 7.4) supplemented with protease inhibitor cocktail (Roche Cat#04693159001), phosphatase inhibitor cocktail (Roche Cat#04906837001), and 20 units RNase inhibitor (ThermoFisher Scientific #AM2964). 1 mg of each lysate was pre-cleared with 80uL Protein G Dynabeads, and 50uL of Protein G Dynabeads (Invitrogen #10004D) per lysate were prepared with either 10μg Tau13 according to manufacturer instructions. Lysates and beads were equivalently pre-cleared and prepared with 10μg normal mouse IgG. Pre-cleared lysates were immunoprecipitated with the antibody conjugated beads using gentle rotation overnight at 4 °C; protein was eluted by boiling in Bolt SDS-PAGE sample buffer for 10 min and analyzed using SDS-PAGE. Antibodies used in the western blot analysis are recorded in supplemental Table 2.

Sarkosyl fractionation and immunoblotting

Extracted brain tissue from ($n = 3$) rTg4510 and ($n = 3$) uninduced Tg4510 control mice was weighed and placed in a Beckman centrifuge tube, polycarbonate thick wall (Cat#362305). Tissue was homogenized in 4× weight/volume of homogenization buffer (50 mM Tris; 275 mM NaCl; 5 mM KCl; 1 mM PMSF; pH = 8.0 with protease inhibitors, phosphatase inhibitors and PMSF added immediately before use) and ultracentrifuged at 28 k rpm (29,800 g) in a TLA-55 rotor for 20 min at 4 °C using a Beckman Optima-TLX 120,000 ultracentrifuge. The supernatant was removed and stored at – 80 °C as the TBS soluble supernatant (supernatant S1); excess supernatant was then vacuumed off the pellet, and the pellet was suspended in sucrose buffer (10 mM Tris, pH = 7.4; 0.8 M NaCl; 10% sucrose; 1 mM EGTA; 1 mM PMSF). The suspension was ultracentrifuged at 22 k rpm (26,300 g) for 20 min at 4 °C. 450uL of the supernatant was transferred to a new tube with the pellet stored at – 80 °C (pellet P2). This supernatant was incubated with 1% Sarkosyl for 5 min with gentle rotation at room temperature then 1 h at 37 °C. After Sarkosyl incubation the solutions were ultracentrifuged at 55 k rpm (150,000 g) for 1 h at 4 °C. The supernatant was stored at – 80 °C (supernatant S3) and the pellet was resuspended in 60uL TE buffer (10 mM Tris, pH = 8.0, 1 mM EDTA) and also kept at – 80 °C as the P3 pellet.

The Sarkosyl soluble S3 supernatant and insoluble P3 pellet were analyzed for RBP content using SDS-PAGE and immunoblotting. As the protein concentration of all tissues were normalized at the start of fractionation, 20uL of the P3 pellet and 5uL of the S3 supernatant were run on a Bolt 4–12% Bis-Tris Plus polyacrylamide gel (Life Technologies Cat#NW04127BOX) at 120 V for 1 h and 15 min. Proteins were transferred onto nitrocellulose at 12 V for 1 h, and membranes were blocked in Tris-buffered saline (24 g Tris base; 88 g NaCl; 1 L water; pH = 7.6) with 0.1% Tween (TBS-T); with 5% non-fat milk for 2 h. After blocking membranes were incubated with 10 mL of the appropriate primary antibody in TBS-T overnight at 4 °C; primary antibodies and dilutions are recorded in supplementary Table 2. Membranes were washed 4× for 5 min each was in TBS-T, then incubated with the appropriate HRP-conjugated secondary antibody at a 1:15,000 dilution in TBS-T (Jackson Immunoresearch) for 1.5 h. Membranes were again washed 4× for 5 min each in TBS-T then developed using Pierce SuperSignal West Pico Chemiluminescent Substrate (Thermo Fisher Scientific Cat#34080). Blots were quantified using ImageJ peak analysis.

Additional files

Additional file 1: Figure S1. TIA1 antibodies demonstrate significant variability. Four commercial TIA1 antibodies were screened for performance in fluorescent immunohistochemical assays: Abcam 40,693 (A), Santa Cruz 1751 (B), Abcam 140,595 (C), and Cell Signaling 1398S (D). Of these, only Abcam 40,693 demonstrated affinity for both cytoplasmic and nuclear TIA1 with minimal background reactivity, but also shows performance variability between lots (E). **Figure S2.** Co-immunoprecipitation (co-IP) and immunoblotting (IB) of tau and RBPs from the brains of P301S tau mice ($n = 2$). HNRNPA0 (35kD), eIF4a2 (47kD), HNRNPK (55kd), and RPL11 (23kD) co-IPd with Tau13 (left) but not normal mouse IgG (right). **Figure S3.** Immunhistochemical analysis of rTg4510 tissue ($n = 3$) revealed a significant co-localization in the cortex between the PCBP2 ($r = 0.724$), RPL11 ($r = 0.728$), and eIF3h ($r = 0.315$) (red) with pathological phospho-tau stained with CP13 antibody (green). **Figure S4.** Duration of fixation affects sensitivity of RBP detection. Samples were fixed for 24 h (top row) or 48 h (bottom row) with 4%, and imaged for NeuN or TIA1; DAPI identifies nuclei. **Figure S5.** Photobleaching of tissue removes autofluorescence from lipofuscin and the extracellular matrix. Human AD tissue was treated with white light from an LED bulb for 72 h and then imaged. Untreated tissue shows significant autofluoresence in the red and green channels (top), which was removed with photobleaching (bottom). **Figure S6.** Consolidated but not diffuse phospho-tau is present in late stage tissue. Tangle morphology and intensity were compared in 6-month rTg4510 mouse tissue (left) and human AD tissue (right). In the human tissue, CP13 positive tau presents entirely as consolidated NFTs, which extend into the processes. The mouse tissue showed a continuum of pathological tau including diffuse cytoplasmic phospho-tau (white arrows), CP13 positive puncta, and intense, consolidated NFTs. (PDF 956 kb)

Additional file 2: Table S1. Mass spectometry data. This table provides quantification of the proteins identified by mass spectrometry, and shows # peptides identified, fold changes and *P*-values for each protein identified. (XLSX 65 kb)

Additional file 3: Table S2. List of antibodies used in the study. This table provides source information for each antibody, as well as the diltuion at which each antibody was used in the experiments. (XLSX 9 kb)

Acknowledgements

Human brain tissue was generously provided by the National Institute of Aging Boston University AD Center (P30AG13846). We would like to thank the following funding agencies for their support: BW: NIH (AG050471, NS089544, ES020395, AG056318) BrightFocus Foundation, Alzheimer Association, Cure Alzheimer's Fund and the Thome Medical Foundation; BM: NS106751. JA: NIH (NS091329, AG028383, MD009205), Alzheimer's Association NIRG-14-322441, Department of Defense AZ140097.

Authors' contributions

BFM designed experiments, carried out immunochemical and immunohistochemical experiments, and drafted the manuscript. DJA and LJ designed experiments, carried out immunochemical and immunohistochemical experiments. ALC carried out immunohistochemical experiments. ELdR, CZ and HL performed bio-informatics studies and designed related figures, JL performed mass spectroscopy, WHY and JFA provided tissues and helped to edit the manuscript. BW conceived of the study, participated in the studies coordination, designed experiments, drafted the manuscript. All authors read and approved the final manuscript.

Competing interests
BW is co-founder and Chief Scientific Officer of Aquinnah Phamraceuticals Inc.

Author details
[1]Department of Pharmacology and Experimental Therapeutics, Boston University School of Medicine, Boston, MA, USA. [2]Mayo Clinic, Rochester, MN, USA. [3]Department of Pathology and Cell Biology, Taub Institute for Alzheimer's Disease Research, Columbia University Medical Center, New York, NY, USA. [4]University of Massachusetts Medical Center, Worcester, MA, USA. [5]Sanders-Brown Center on Aging, Department of Physiology, Spinal Cord and Brain Injury Research Center, and Epilepsy Center, University of Kentucky, Lexington, KY, USA. [6]Department of Neurology, Boston University School of Medicine, Boston, MA, USA. [7]Department of Pharmacology and Neurology Program in Neuroscience, Boston University School of Medicine, 72 East Concord St., R614, Boston, MA 02118-2526, USA.

References
1. Apicco DJ, Ash PEA, Maziuk B, LeBlang C, Medalla M, Al Abdullatif A, Ferragud A, Botelho E, Ballance HI, Dhawan U et al (2018) Reducing the RNA binding protein TIA1 protects against tau-mediated neurodegeneration in vivo. Nat Neurosci 21:72–80. https://doi.org/10.1038/s41593-017-0022-z
2. Bai B, Hales CM, Chen PC, Gozal Y, Dammer EB, Fritz JJ, Wang X, Xia Q, Duong DM, Street C et al (2013) U1 small nuclear ribonucleoprotein complex and RNA splicing alterations in Alzheimer's disease. Proc Natl Acad Sci U S A 110:16562–16567. https://doi.org/10.1073/pnas.1310249110
3. Becker LA, Huang B, Bieri G, Ma R, Knowles DA, Jafar-Nejad P, Messing J, Kim HJ, Soriano A, Auburger G et al (2017) Therapeutic reduction of ataxin-2 extends lifespan and reduces pathology in TDP-43 mice. Nature 544:367–371. https://doi.org/10.1038/nature22038
4. Begcevic I, Kosanam H, Martinez-Morillo E, Dimitromanolakis A, Diamandis P, Kuzmanov U, Hazrati LN, Diamandis EP (2013) Semiquantitative proteomic analysis of human hippocampal tissues from Alzheimer's disease and age-matched control brains. Clin Proteomics 10:5. https://doi.org/10.1186/1559-0275-10-5
5. Combs B, Kanaan NM (2017) Exposure of the amino terminus of tau is a pathological event in multiple Tauopathies. Am J Pathol 187:1222–1229. https://doi.org/10.1016/j.ajpath.2017.01.019
6. Couthouis J, Hart MP, Erion R, King OD, Diaz Z, Nakaya T, Ibrahim F, Kim HJ, Mojsilovic-Petrovic J, Panossian S et al (2012) Evaluating the role of the FUS/TLS-related gene EWSR1 in amyotrophic lateral sclerosis. Hum Mol Genet 21:2899–2911. https://doi.org/10.1093/hmg/dds116
7. Couthouis J, Hart MP, Shorter J, DeJesus-Hernandez M, Erion R, Oristano R, Liu AX, Ramos D, Jethava N, Hosangadi D et al (2011) A yeast functional screen predicts new candidate ALS disease genes. Proc Natl Acad Sci U S A 108:20881–20890
8. Fujita K, Ito H, Nakano S, Kinoshita Y, Wate R, Kusaka H (2008) Immunohistochemical identification of messenger RNA-related proteins in basophilic inclusions of adult-onset atypical motor neuron disease. Acta Neuropathol 116:439–445. https://doi.org/10.1007/s00401-008-0415-x
9. Gu J, Chen F, Iqbal K, Gong CX, Wang X, Liu F (2017) Transactive response DNA-binding protein 43 (TDP-43) regulates alternative splicing of tau exon 10: implications for the pathogenesis of tauopathies. J Biol Chem 292:10600–10612. https://doi.org/10.1074/jbc.M117.783498
10. Gunawardana CG, Mehrabian M, Wang X, Mueller I, Lubambo IB, Jonkman JE, Wang H, Schmitt-Ulms G (2015) The human tau Interactome: binding to the Ribonucleoproteome, and impaired binding of the proline-to-leucine mutant at position 301 (P301L) to chaperones and the proteasome. Mol Cell Proteomics 14:3000–3014. https://doi.org/10.1074/mcp.M115.050724
11. Hackman P, Sarparanta J, Lehtinen S, Vihola A, Evila A, Jonson PH, Luque H, Kere J, Screen M, Chinnery PF et al (2013) Welander distal myopathy is caused by a mutation in the RNA-binding protein TIA1. Ann Neurol
12. Hales CM, Dammer EB, Diner I, Yi H, Seyfried NT, Gearing M, Glass JD, Montine TJ, Levey AI, Lah JJ (2014) Aggregates of small nuclear ribonucleic acids (snRNAs) in Alzheimer's disease. Brain Pathol 24:344–351. https://doi.org/10.1111/bpa.12133
13. Hales CM, Seyfried NT, Dammer EB, Duong D, Yi H, Gearing M, Troncoso JC, Mufson EJ, Thambisetty M, Levey AI et al (2014) U1 small nuclear ribonucleoproteins (snRNPs) aggregate in Alzheimer's disease due to autosomal dominant genetic mutations and trisomy 21. Mol Neurodegener 9:15. https://doi.org/10.1186/1750-1326-9-15
14. Hernandez-Vega A, Braun M, Scharrel L, Jahnel M, Wegmann S, Hyman BT, Alberti S, Diez S, Hyman AA (2017) Local nucleation of microtubule bundles through tubulin concentration into a condensed tau phase. Cell Rep 20:2304–2312. https://doi.org/10.1016/j.celrep.2017.08.042
15. Hirsch-Reinshagen V, Pottier C, Nicholson AM, Baker M, Hsiung GR, Krieger C, Sengdy P, Boylan KB, Dickson DW, Mesulam M et al (2017) Clinical and neuropathological features of ALS/FTD with TIA1 mutations. Acta Neuropathol Commun 5:96. https://doi.org/10.1186/s40478-017-0493-x
16. Hughes MP, Sawaya MR, Boyer DR, Goldschmidt L, Rodriguez JA, Cascio D, Chong L, Gonen T, Eisenberg DS (2018) Atomic structures of low-complexity protein segments reveal kinked beta sheets that assemble networks. Science 359:698–701. https://doi.org/10.1126/science.aan6398
17. Ikezu T, Chen C, DeLeo AM, Zeldich E, Fallin MD, Kanaan NM, Lunetta KL, Abraham CR, Logue MW, Farrer LA (2018) Tau phosphorylation is impacted by rare AKAP9 mutations associated with Alzheimer disease in African Americans. J Neuroimmune Pharmacol. https://doi.org/10.1007/s11481-018-9781-x
18. Ishigaki S, Fujioka Y, Okada Y, Riku Y, Udagawa T, Honda D, Yokoi S, Endo K, Ikenaka K, Takagi S et al (2017) Altered tau isoform ratio caused by loss of FUS and SFPQ function leads to FTLD-like phenotypes. Cell Rep 18:1118–1131. https://doi.org/10.1016/j.celrep.2017.01.013
19. Kanaan NM, Pigino GF, Brady ST, Lazarov O, Binder LI, Morfini GA (2013) Axonal degeneration in Alzheimer's disease: when signaling abnormalities meet the axonal transport system. Exp Neurol 246:44–53. https://doi.org/10.1016/j.expneurol.2012.06.003
20. Kim HJ, Raphael AR, LaDow ES, McGurk L, Weber RA, Trojanowski JQ, Lee VM, Finkbeiner S, Gitler AD, Bonini NM (2014) Therapeutic modulation of eIF2alpha phosphorylation rescues TDP-43 toxicity in amyotrophic lateral sclerosis disease models. Nat Genet 46:152–160. https://doi.org/10.1038/ng.2853
21. Li P, Banjade S, Cheng HC, Kim S, Chen B, Guo L, Llaguno M, Hollingsworth JV, King DS, Banani SF et al (2012) Phase transitions in the assembly of multivalent signalling proteins. Nature 483:336–340. https://doi.org/10.1038/nature10879
22. Lin Y, Protter DS, Rosen MK, Parker R (2015) Formation and maturation of phase-separated liquid droplets by RNA-binding proteins. Mol Cell 60:208–219. https://doi.org/10.1016/j.molcel.2015.08.018
23. Liu-Yesucevitz L, Bilgutay A, Zhang YJ, Vanderweyde T, Citro A, Mehta T, Zaarur N, McKee A, Bowser R, Sherman M et al (2010) Tar DNA binding protein-43 (TDP-43) associates with stress granules: analysis of cultured cells and pathological brain tissue. PLoS One 5:e13250. https://doi.org/10.1371/journal.pone.0013250
24. Maziuk B, Ballance HI, Wolozin B (2017) Dysregulation of RNA binding protein aggregation in neurodegenerative disorders. Front Mol Neurosci 10:89. https://doi.org/10.3389/fnmol.2017.00089
25. Meier S, Bell M, Lyons DN, Ingram A, Chen J, Gensel JC, Zhu H, Nelson PT, Abisambra JF (2015) Identification of novel tau interactions with endoplasmic reticulum proteins in Alzheimer's disease brain. J Alzheimers Dis 48:687–702. https://doi.org/10.3233/JAD-150298
26. Meier S, Bell M, Lyons DN, Rodriguez-Rivera J, Ingram A, Fontaine SN, Mechas E, Chen J, Wolozin B, LeVine H 3rd et al (2016) Pathological tau promotes neuronal damage by impairing ribosomal function and decreasing protein synthesis. J Neurosci 36:1001–1007. https://doi.org/10.1523/JNEUROSCI.3029-15.2016
27. Molliex A, Temirov J, Lee J, Coughlin M, Kanagaraj AP, Kim HJ, Mittag T, Taylor JP (2015) Phase separation by low complexity domains promotes stress granule assembly and drives pathological fibrillization. Cell 163:123–133. https://doi.org/10.1016/j.cell.2015.09.015
28. Nott TJ, Petsalaki E, Farber P, Jervis D, Fussner E, Plochowietz A, Craggs TD, Bazett-Jones DP, Pawson T, Forman-Kay JD et al (2015) Phase transition of a disordered nuage protein generates environmentally responsive membraneless organelles. Mol Cell 57:936–947. https://doi.org/10.1016/j.molcel.2015.01.013
29. Panas MD, Ivanov P, Anderson P (2016) Mechanistic insights into mammalian stress granule dynamics. J Cell Biol 215:313–323. https://doi.org/10.1083/jcb.201609081

30. Santacruz K, Lewis J, Spires T, Paulson J, Kotilinek L, Ingelsson M, Guimaraes A, DeTure M, Ramsden M, McGowan E et al (2005) Tau suppression in a neurodegenerative mouse model improves memory function. Science 309: 476–481
31. Shin Y, Brangwynne CP (2017) Liquid phase condensation in cell physiology and disease. Science 357. https://doi.org/10.1126/science.aaf4382
32. Shukla S, Parker R (2016) Hypo- and hyper-assembly diseases of RNA-protein complexes. Trends Mol Med 22:615–628. https://doi.org/10.1016/j.molmed.2016.05.005
33. Silva JM, Rodrigues S, Sampaio-Marques B, Gomes P, Neves-Carvalho A, Dioli C, Soares-Cunha C, Takashima A, Ludovico P, Wolozin B et al (2018) Dysregulation of autophagy and stress-granule related proteins in stress-driven tau pathology. Cell Death Differ (in press)
34. Sun Y, Ip P, Chakrabartty A (2017) Simple elimination of background fluorescence in formalin-fixed human brain tissue for immunofluorescence microscopy. J Vis Exp: Doi. https://doi.org/10.3791/56188
35. Sun Z, Diaz Z, Fang X, Hart MP, Chesi A, Shorter J, Gitler AD (2011) Molecular determinants and genetic modifiers of aggregation and toxicity for the ALS disease protein FUS/TLS. PLoS Biol 9:e1000614. https://doi.org/10.1371/journal.pbio.1000614
36. Thompson AD, Scaglione KM, Prensner J, Gillies AT, Chinnaiyan A, Paulson HL, Jinwal UK, Dickey CA, Gestwicki JE (2012) Analysis of the tau-associated proteome reveals that exchange of Hsp70 for Hsp90 is involved in tau degradation. ACS Chem Biol 7:1677–1686
37. Vanderweyde T, Apicco DJ, Youmans-Kidder K, Ash PE, Cook C, Lummertz da Rocha E, Jansen-West K, Frame AA, Citro A, Leszyk JD et al (2016) Interaction of tau with the RNA-binding protein TIA1 regulates tau pathophysiology and toxicity. Cell Rep 15:1–12. https://doi.org/10.1016/j.celrep.2016.04.045
38. Vanderweyde T, Yu H, Varnum M, Liu-Yesucevitz L, Citro A, Ikezu T, Duff K, Wolozin B (2012) Contrasting pathology of stress granule proteins TIA-1 and G3BP in Tauopathies. J Neurosci 32:8270–8283
39. Wegmann S, Eftekharzadeh B, Tepper K, Zoltowska KM, Bennett RE, Dujardin S, Laskowski PR, MacKenzie D, Kamath T, Commins C et al (2018) Tau protein liquid-liquid phase separation can initiate tau aggregation. EMBO J. https://doi.org/10.15252/embj.201798049
40. White MA, Kim E, Duffy A, Adalbert R, Phillips BU, Peters OM, Stephenson J, Yang S, Massenzio F, Lin Z et al (2018) TDP-43 gains function due to perturbed autoregulation in a Tardbp knock-in mouse model of ALS-FTD. Nat Neurosci. https://doi.org/10.1038/s41593-018-0113-5
41. Zempel H, Mandelkow E (2014) Lost after translation: missorting of tau protein and consequences for Alzheimer disease. Trends Neurosci 37:721–732. https://doi.org/10.1016/j.tins.2014.08.004
42. Zhang X, Lin Y, Eschmann NA, Zhou H, Rauch JN, Hernandez I, Guzman E, Kosik KS, Han S (2017) RNA stores tau reversibly in complex coacervates. PLoS Biol 15:e2002183. https://doi.org/10.1371/journal.pbio.2002183
43. Zhang Y, Chen X, Zhao Y, Ponnusamy M, Liu Y (2017) The role of ubiquitin proteasomal system and autophagy-lysosome pathway in Alzheimer's disease. Rev Neurosci 28:861–868. https://doi.org/10.1515/revneuro-2017-0013

2

Compromised astrocyte function and survival negatively impact neurons in infantile neuronal ceroid lipofuscinosis

Jenny Lange[1], Luke J. Haslett[2], Emyr Lloyd-Evans[2], Jennifer M. Pocock[3], Mark S. Sands[4], Brenda P. Williams[1†] and Jonathan D. Cooper[1,5,6*†]

Abstract

The neuronal ceroid lipofuscinoses (NCLs) are the most common cause of childhood dementia and are invariably fatal. Early localized glial activation occurs in these disorders, and accurately predicts where neuronal loss is most pronounced. Recent evidence suggests that glial dysfunction may contribute to neuron loss, and we have now explored this possibility in infantile NCL (INCL, CLN1 disease). We grew primary cultures of astrocytes, microglia, and neurons derived from Ppt1 deficient mice (*Ppt1*$^{-/-}$) and assessed their properties compared to wildtype (WT) cultures, before co-culturing them in different combinations (astrocytes with microglia, astrocytes or microglia with neurons, all three cell types together). These studies revealed that both *Ppt1*$^{-/-}$ astrocytes and microglia exhibit a more activated phenotype under basal unstimulated conditions, as well as alterations to their protein expression profile following pharmacological stimulation. *Ppt1*$^{-/-}$ astrocytes also displayed abnormal calcium signalling and an elevated cytoplasmic Ca^{2+} level, and a profound defect in their survival. *Ppt1*$^{-/-}$ neurons displayed decreased neurite outgrowth, altered complexity, a reduction in cell body size, and impaired neuron survival with prolonged time in culture. In co-cultures, the presence of both astrocytes and microglia from *Ppt1*$^{-/-}$ mice further impaired the morphology of both wild type and *Ppt1*$^{-/-}$ neurons. This negative influence was more pronounced for *Ppt1*$^{-/-}$ microglia, which appeared to trigger increased *Ppt1*$^{-/-}$ neuronal death. In contrast, wild type glial cells, especially astrocytes, ameliorated some of the morphological defects observed in *Ppt1*$^{-/-}$ neurons. These findings suggest that both *Ppt1*$^{-/-}$ microglia and astrocytes are dysfunctional and may contribute to the neurodegeneration observed in CLN1 disease. However, the dysfunctional phenotypes of *Ppt1*$^{-/-}$ glia are different from those present in CLN3 disease, suggesting that the pathogenic role of glia may differ between NCLs.

Keywords: Infantile batten disease, CLN1 disease, neuronal ceroid lipofuscinosis, Neuron-glial interactions, Astrocyte and microglial dysfunction

* Correspondence: cooperjd@wustl.edu
†Brenda P. Williams and Jonathan D. Cooper contributed equally to this work.
[1]Department of Basic and Clinical Neuroscience, King's College London, Institute of Psychiatry, Psychology & Neuroscience, Maurice Wohl Clinical Neuroscience Institute, 5 Cutcombe Road, London SE5 9RX, UK
[5]Department of Pediatrics, Los Angeles Biomedical Research Institute at Harbor-UCLA Medical Center, and David Geffen School of Medicine UCLA, 1124 West Carson Street, Hanley Hardison Building, Torrance, CA 90502, USA
Full list of author information is available at the end of the article

Introduction

The neuronal ceroid lipofuscinoses (NCLs, or Batten disease) are a large group of inherited lysosomal storage disorders, each caused by mutations in an individual gene [8, 32]. The age of onset and rate of disease progression are determined by which gene is deficient, and which mutation is present within that gene. The NCLs are collectively the largest cause of childhood dementia and share a range of similar clinical features, including loss of vision, epileptic seizures, and declines in mental and motor abilities, all invariably leading to premature death [2, 9, 50].

Despite recent advances in the treatment of late infantile NCL (CLN2 disease) via enzyme replacement therapy [30], the other forms of NCL all remain untreatable. This includes infantile NCL (INCL or CLN1 disease), an earlier onset and more rapidly progressing form of NCL, which is caused by a mutation in the *CLN1/PPT1* gene (Vesa et al., 1995; [5, 21]). However, it remains unclear how deficiency in palmitoyl protein thioesterase-1 (PPT1), the lysosomal enzyme encoded by this gene results in the devastating neurodegenerative impact upon the brain that typifies CLN1 disease.

The generation of an INCL mouse model ($Ppt1^{-/-}$ mice) [17] has made it possible to study the consequences of PPT1 deficiency [3, 23, 29], and test a range of pre-clinical interventions [19]. These mice display a progressive CLN1 disease-like phenotype, with well-defined declines in behavioural and neurological performance, and degenerative changes that are most pronounced within the thalamocortical system and cerebellum. Characteristically these mice display a profound activation of both microglia and astrocytes, which precedes neuron loss [23, 29]. The regional localization of this glial activation correlates closely with distribution of subsequent neuron loss, leading to the suggestion that these events may be causally related.

Rather than their traditional supportive role, evidence is emerging that both astrocytes and microglia can directly contribute to neuron loss in a variety of disease states [25]. Amongst these are several lysosomal storage disorders [11, 41, 48], including the juvenile form of NCL, CLN3 disease or JNCL [35, 51]. We have recently shown via primary cultures derived from Cln3 deficient mice that astrocytes and microglia are both dysfunctional, and when co-cultured can harm healthy neurons and kill Cln3 deficient neurons, suggesting a direct influence of glia upon the pathogenesis of this form of NCL [35].

The extent of glial activation is much more pronounced in CLN1 disease, but as longitudinal studies show [23, 29], still accurately predicts the sites where neuron loss subsequently occurs. This raises the question of what the role of glia in this most profoundly neurodegenerative form of NCL may be. As a first step to resolving this question, we derived primary cultures of either astrocytes or microglia from $Ppt1^{-/-}$ mice, and compared their properties and response to stimulation to those derived from wild type controls.

This in vitro study revealed a range of abnormal phenotypes different to those seen in Cln3 disease glia [35]. $Ppt1^{-/-}$ astrocytes and microglia both appeared more activated than WT cells under basal unstimulated culture conditions. However, most notable was a profound defect in the survival of $Ppt1^{-/-}$ astrocytes, which also displayed marked dysregulation of intracellular calcium handling, and subsequent death by apoptosis. Primary cortical neuron cultures from $Ppt1^{-/-}$ mice displayed abnormal dendritic morphology and impaired survival, especially of interneurons. Using a co-culture system wild type glia ameliorated these neuronal phenotypes, whereas $Ppt1^{-/-}$ astrocytes predominantly impaired the morphology of WT neurons, and $Ppt1^{-/-}$ microglia appeared to trigger increased $Ppt1^{-/-}$ neuronal death via apoptosis. These effects were more pronounced in mixed glial co-cultures when both $Ppt1^{-/-}$ astrocytes and microglia were grown together with neurons. Taken together, these data suggest that the dysfunction of astrocytes and microglia in CLN1 disease is detrimental to neurons and may lead to their loss.

Materials and methods

Animals

Homozygous Ppt1 deficient ($Ppt1^{-/-}$) mice were originally created through the deletion of the last exon in the *Ppt1* coding sequence [17], and were subsequently bred for at least 10 generations onto a C57/BL/6 J background [16]. The colony was maintained together with a colony of C57/BL/6 J wild type (WT) control mice originally derived from the crossing of heterozygous $Ppt1^{+/-}$ mice. All animal maintenance and experimental procedures were carried out according to the UK Scientific Procedures (Animals) Act (1986).

Tissue culture

Preparation of glial cultures

Mixed glial cells were isolated from post-natal day 1–5 (P1-P5) $Ppt1^{-/-}$ or WT mouse cerebral cortices, as previously described [31, 35]. Cultures reached confluence after 10–14 days, at which point they were composed of a base layer of non-dividing astrocytes and an upper layer of microglia as well as a few oligodendrocytes. To generate astrocyte cultures, microglia were removed by shaking cultures at 180 rpm for 10–12 h in a humidified incubator (5% CO_2, 37C, ([31]). The remaining astrocyte monolayer was treated with cytosine arabinoside (Ara-C) for 7 days to abolish any remaining dividing cells. Astrocytes for immunofluorescence staining experiments were plated at 1.5×10^4 cells per well onto poly-D-lysine (PDL, 25μg/ml, Sigma) coated glass coverslips in 24 well plates. For all functional assays unless otherwise described, astrocytes were plated at 14.25×10^4 cells per well onto PDL coated 6 well plates and kept in phenol-red-free DMEM supplemented with 4500 mg/L glucose (Sigma, Poole, UK), to prevent the presence of phenol red from providing a background interference.

Microglial cultures

Microglial cultures were isolated from mixed glial cultures using the shaking method described above [31, 35]. The harvested cells were resuspended in RPMI 1640 supplemented with 5% FBS, P/S, 2 mM L-Glutamine, macrophage colony-stimulating factor and granulate macrophage colony-stimulating factor, and consequently plated on PDL coated (as above) 1.5 cm coverslips placed in 24 well plates. They were maintained for 5 days, before growth factors were removed for a further two days. At the start of all experiments, all glial cultures had been maintained for approximately 21 days, yielding cultures of > 99% purity.

Microglia for co-cultures

Due to the difficulty of detaching microglia from PDL coated plastic surfaces, a different method was used to generate microglia for co-culturing with either astrocytes or neurons. These microglia were prepared from P3-P5 *Ppt1*$^{-/-}$ or WT mouse cerebral cortices, initially generating mixed glial cultures as described above. Conditioned medium from these mixed cultures was collected and cultures subsequently incubated at 37 °C with 1 mL/well of Trypsin-EDTA (0.5mgml-1, Sigma). After 45 min the cell monolayer, consisting primarily of astrocytes, was removed along with the supernatant and discarded. The remaining microglial cells were washed before being incubated in a 1:1 mixture of conditioned and fresh medium. These microglial cultures were ready to be used in co-culture experiments after 48 h, when they were removed from wells using 0.5 ml Accutase (Sigma) and a cell scraper (Corning). Microglia were then added to co-cultures at a concentration of 50×10^4 cells/ml per well.

Neuronal cultures

Neuronal cultures were generated from P0 *Ppt1*$^{-/-}$ and WT mouse cortices as previously described [35] and plated on PDL-coated 13 mm coverslips placed in 24 well plates at a concentration of 15×10^4 cells/cm^2. These cultures were maintained for 7, 9 or 14 days.

Co-cultures

Neurons to be grown in co-cultures were plated onto 19 mm diameter coverslips in placed in 12 well plates at a concentration of 15×10^4 cells/cm^2. After 7 days, glial cells previously maintained in culture for 21 days were added to these neurons at a density of 100×10^4 cells/ml for astrocytes or mixed glia, and 50×10^4 cells/ml for microglia. Additional coverslips plated with just glia (either astrocytes, microglia or of mixed glia), or only neurons were set up as control conditions. Co-cultures were grown for an additional 2 or 7 days before commencing analysis, making the neuronal cultures used in these experiments 9 or 14 days old. The composition of 14 day old neuronal cultures was assessed using cell type-specific phenotypic markers, as described below.

Pharmacological activation of glial cells Astrocytes were stimulated with lipopolysaccharide (LPS, 1 μg/ml LPS, Sigma) and interferon-gamma (IFN-γ, 100 U/ml, Thermo Scientific) for 24 or 48 h, while microglia were stimulated with LPS alone for either 6 or 24 h [35]. Mixed glial cultures were stimulated for 24 h using both stimuli.

Immunofluorescence staining Cultures were fixed in 4% PFA and stained as previously described [35] and across all experiments, nuclei were counterstained with DAPI. Cell type-specific markers were used to assess culture composition, with glial fibrillary acidic protein (GFAP, Rabbit polyclonal, 1:500, Dako) being used to identify astrocytes, Cluster of Differentiation 68 (CD68, Rat monoclonal, 1:250, Serotec) to identify microglia, Map2 MAP2 (mouse monoclonal, 1:1000, Abcam) to identify neurons and O4 (Mouse monoclonal, 1:100, Covance) to identify oligodendrocytes. Cytoskeletal components were examined by staining cultures with Phalloidin, α- and β-tubulin (monoclonal and polyclonal antibodies respectively, both from Sigma and used at 1:1000). All secondary antibodies were used 1:1000 and obtained from Invitrogen. Coverslips were mounted using Fluoromount G and stained cells were visualized using a Zeiss *AxioImager Z1* microscope (Carl Zeiss, Ltd) and images captured with a monochrome *AxioCamMR3* camera, using *AxioVision* 4.8 Imaging software.

Morphological Assessment Images of 10 random fields were taken per coverslip of GFAP stained, non-overlapping astrocytes and soma size was measured using *ImageJ* (National Institutes of Health, Bethesda, MD). Neuron soma size and neurite length were measured using *ImageJ* on images of 10 random fields per coverslip in neuronal cultures or 20 random fields per coverslip in co-cultures. The length of the longest neurite was obtained per neuron; as well as mean neurite length per neuron. The morphological response of microglia to stimulation was assessed by assigning cells to 3 sub categories; Type 1 microglia which exhibit extended processes and an elongated cell body, Type 2 microglia with retracted processes and a rounded cell body and Type 3 microglia exhibiting a very small, rounded cell body. 5 random fields of at least 200 cells were counted per coverslip per condition and the percentage of each subtype was determined [35].

Live/Dead Assay Cell viability in WT and *Ppt1*$^{-/-}$ astrocyte cultures was assessed using a Live/Dead kit (Invitrogen) according to the manufacturer's instructions. Briefly, Live/Dead dye was diluted in DMSO and added 30 min prior to fixation of cultures 24, 48 or 72 h after plating astrocytes onto coverslips. Astrocyte cultures were then co-stained with GFAP and 10 random fields were counted per coverslip.

Calcium Measurements Astrocytes were plated in μ-Slide 8 well imaging dishes (Sigma) at 1.5cells/cm^2. Cells were loaded with 5 μM Fura2-AM (Teflabs) in DMEM with 1% BSA and 0.025% Pluronic acid F127 for 1 h at room temperature, before being washed in imaging buffer (HBSS, 1 mM HEPES and 1 mM $MgCl_2$) and left at room temperature for 10 min. Intracellular Ca^{2+} concentrations and responses were recorded with a Colibri LED microscope system, using an Axiocam Mrm CCD camera and *Axiovision* software (Version 4.7) with an additional physiology module for live cell Ca^{2+} imaging (Zeiss). Recordings of the Ca^{2+} probe baselines were taken prior to agonist addition in order to provide basal Ca^{2+} measurements of the cell, and all agonist-induced responses were compared at peak height.

Release of endoplasmic reticulum (ER) Ca^{2+} was measured after addition of 5 μM of thapsigargin (Sigma). Lysosomal Ca^{2+} release was measured using a method adapted from [26]. Following with 5 μM ionomycin (Calbiochem) to prevent release from intracellular Ca^{2+} stores, the addition of 10 μM nigericin (Sigma) was used to induce Ca^{2+} release from the lysosome. To examine store operated Ca^{2+} entry, ER Ca^{2+} release was triggered as described, and following return to the baseline, 1 mM Ca^{2+} was added to the imaging medium. In order to trigger Ca^{2+} plasma membrane influx, 100 μM ATP (Sigma) was added after astrocytes were loaded with Fura2-AM in imaging buffer and 1 mM $CaCl_2$.

Protein Secretion To compare pharmacologically stimulated cytokine secretion by WT and *Ppt1*$^{-/-}$ glia, supernatant was collected from mixed glial, astrocyte and microglial cultures and analysed with ELISA kits (Signosis, Santa Clara, CA). A mouse oxidative stress ELISA kit was selected to quantify microglial cytokine secretion based on previous evidence that *Ppt1*$^{-/-}$ neurons exhibit signs of oxidative stress [49], a custom mouse cytokine ELISA kit was used to examine cytokine secretion by astrocytes based on pilot cytokine data generated in our laboratory, and cytokine secretion in mixed glial cultures was assessed with a mouse cytokine ELISA kit (all kits from Signosis). All these experiments were carried out according to the manufacturer's instructions.

Statistical analysis All data were collected into Microsoft Excel spreadsheets and analysed in Graphpad Prism. All data are presented as mean ± SEM. A student's t-test was used to analyse data between two groups, data sets with more than two groups were analysed with a one-way ANOVA using Bonferroni's correction. All experiments were repeated in triplicate, using three technical replicates per experiment, unless otherwise stated. Data were considered statistically significant if $p \leq 0.05$, p values are marked * if $p \leq 0.05$, ** if $p \leq 0.01$ and *** if $p \leq 0.001$.

Results

Ppt1$^{-/-}$ astrocytes exhibit an activated phenotype under basal conditions

Astrocyte activation can be observed as early as 3 months of age in *Ppt1*$^{-/-}$ mice, and progressively becomes more pronounced and widespread towards the later stages of the disease [23, 29]. Before comparing the in vitro phenotypes of *Ppt1*$^{-/-}$ and WT astrocytes we first defined the cellular composition of our astrocyte cultures. One week after plating, over 99% of DAPI positive cells in astrocyte cultures of either genotype were positive for the astrocyte marker glutamine synthetase (GS) (WT: 99.2% GS + ve, 0.71% CD68 + ve, and 0% O4 + ve; *Ppt1*$^{-/-}$: 99.6% GS + ve, 0.35% CD68 + ve, and 0% O4 + ve).

Using GFAP as a marker whose increased expression is associated with astrocyte activation, we first compared the morphology of astrocytes in cultures derived from mice of both genotypes, under both basal and simulated conditions (Fig. 1A). Under basal conditions the morphology of cultured *Ppt1*$^{-/-}$ astrocytes was more heterogeneous than that seen in WT cultures, with many GFAP-positive astrocytes appearing larger in *Ppt1*$^{-/-}$ cultures (Fig. 1A). The proportion of GFAP-positive cells under basal unstimulated conditions was significantly higher in *Ppt1*$^{-/-}$ astrocyte cultures than in WT cultures (Fig. 1B). Cultures were then stimulated with LPS and IFNγ, to examine astrocyte response to a standardized pharmacological stimulus [35]. Although the proportion of GFAP-positive astrocytes was significantly upregulated in WT cultures after both 24 h and 48 h of stimulation, no further changes in the proportion of GFAP-positive astrocytes were observed in *Ppt1*$^{-/-}$ cultures upon stimulation (Fig. 1B). These data reveal that although more *Ppt1*$^{-/-}$ astrocytes express GFAP under basal culture conditions, this proportion does not increase further upon stimulation with LPS/ IFNγ, as it does with WT astrocytes.

To quantify the morphological responses of astrocytes to stimulation, we measured astrocyte soma size (Fig. 1C). Under basal conditions, *Ppt1*$^{-/-}$ astrocytes exhibited a

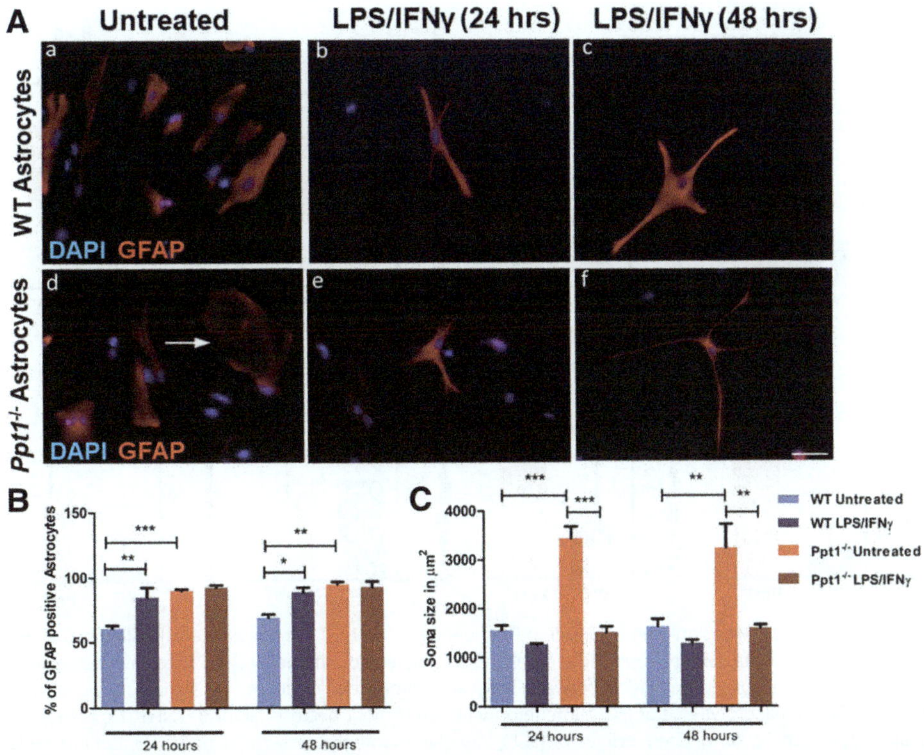

Fig. 1 Ppt1 deficient *(Ppt1*$^{-/-}$*)* astrocytes exhibit a reactive phenotype. **A** Wild type (WT) and *Ppt1*$^{-/-}$ Astrocytes were stained for glial fibrillary associated protein (GFAP) to assess GFAP expression and astrocyte morphology under basal and stimulated (LPS/IFNγ) conditions after 24 and 48 h. Immunostaining for GFAP (red) revealed the morphology of *Ppt1*$^{-/-}$ and WT astrocytes under basal and stimulated conditions, with *Ppt1*$^{-/-}$ astrocytes exhibiting a large flattened morphology (white arrow) under basal conditions. **B** A higher percentage of *Ppt1*$^{-/-}$ astrocytes stained positively for (GFAP) under basal conditions, with no significant upregulation following stimulation. **C** The soma size or spread of *Ppt1*$^{-/-}$astrocytes was much larger under basal conditions than in WT astrocytes, however was reduced following stimulation for 24 and 48 h. (Data shown as Mean ± SEM using a one way ANOVA, $n = 3$). Scale bar = 50 μm

markedly and significantly larger soma size than WT astrocytes. However, upon stimulation for 24 or 48 h *Ppt1*$^{-/-}$ astrocytes dramatically reduced their cell body size, a typical response of cultures astrocytes to pharmacological stimulation, so that they were now of a similar size to stimulated WT astrocytes at either time point (Fig. 1C). Taken together these data reveal that although appearing morphologically different to WT astrocytes under basal conditions, *Ppt1*$^{-/-}$ astrocytes still respond robustly to pharmacological stimulation by changing their shape.

Increased secretion of soluble factors by Ppt1$^{-/-}$ astrocytes

Increased secretion of specific soluble proteins is a hallmark of astrocyte activation (Lucas et al., 2006), and we used an ELISA kit to investigate the release of soluble factors following stimulation in our astrocyte cultures. No significant differences in the release of TNFα, IL-1α, IFNγ, FGFB, or EGF were observed between WT and *Ppt1*$^{-/-}$ astrocytes following 24 h exposure to LPS and IFNγ (Fig. 2). However, significantly higher amounts of IL-4, SCF and VEGF were released by *Ppt1*$^{-/-}$ astrocytes (35.98 ± 17.80%; 10.74 ± 3.98%; 5.90 ± 4.16% more, respectively) than by WT astrocytes (− 14.12 ± 9.39%; − 1.68 ± 3.47%; − 12.61 ± 3.02% less, respectively) (Fig. 2). Not only does this provide further evidence that *Ppt1*$^{-/-}$ astrocytes respond differently to stimulation than their WT counterparts, but is also in marked contrast to data obtained from *Cln3*$^{-/-}$ astrocytes which exhibit reduced release of soluble factors following stimulation [35]. Furthermore, lactate secretion from *Ppt1*$^{-/-}$ astrocytes was significantly increased 1.4 fold ($p < 0.05$, data not shown) under basal conditions, which may point towards the possibility of mitochondrial dysfunction as shown in aging and prematurely aging mice [39], with mitochondrial failure and a shift in transcriptional activities of lactate dehydrogenases resulted in increased brain lactate levels. Mitochondrial dysfunction is closely linked with oxidative stress and has previously been reported in other forms of the NCLs ([34]; Jalanko et al., 2009).

Ppt1$^{-/-}$ astrocyte survival is impaired in vitro

There are fewer cells that stain positively for GS or S100β in 6 month old *Ppt1*$^{-/-}$ mice [29], suggesting that

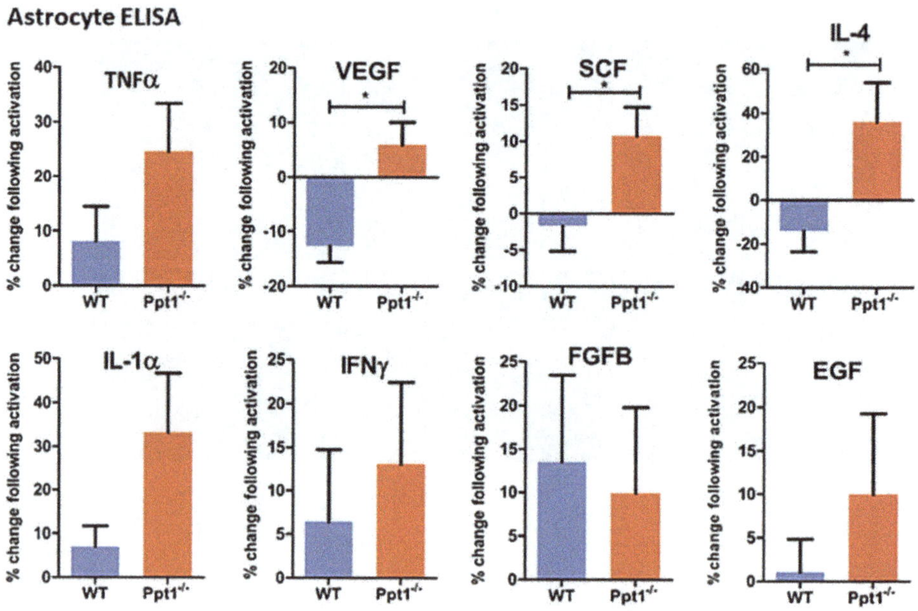

Fig. 2 Expression of SCF, VEGF and IL-4 was significantly higher in Ppt1 deficient (*Ppt1*$^{-/-}$) astrocytes after stimulation. Supernatant was collected from wild type (WT) and *Ppt1*$^{-/-}$ astrocyte cultures kept under basal and stimulated conditions for 24 h. Cytokine release was assessed using an ELISA kit (Signosis) and calculating changes in expression between basal and stimulated conditions. No statistically significant changes were observed in the secretion of Epidermal growth factor (EGF), Interferon γ (IFNγ), Tumour necrosis factor α (TNFα), Expression of Basic fibroblast growth factor (FGFb) and Interleukin 1-α (IL1α). Stem cell factor (SCF), Vascular endothelial growth factor (VEGF) and Interleukin-4 (IL-4) was significantly higher in *Ppt1*$^{-/-}$ astrocyte cultures following stimulation than in WT cultures. (Data shown as Mean ± SEM using a t-test, $n = 3$)

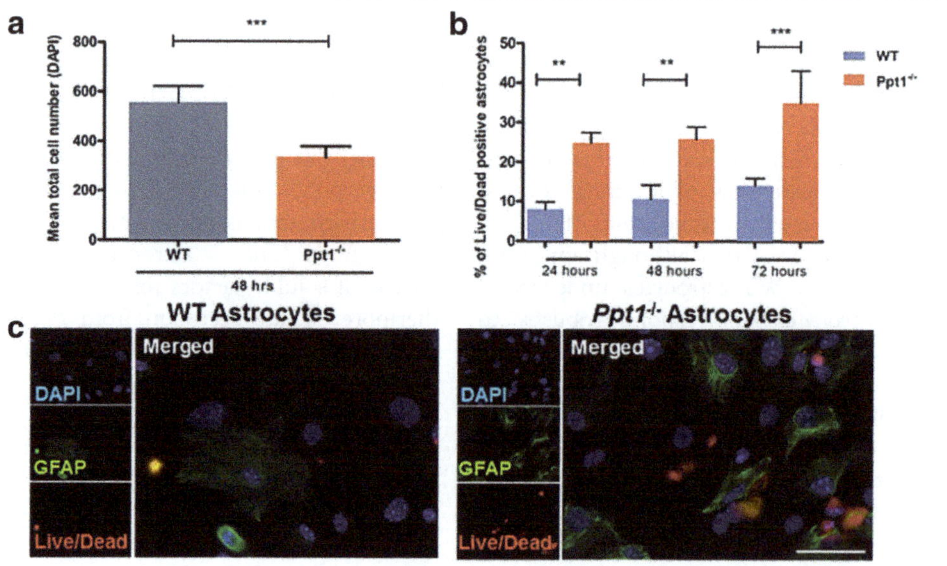

Fig. 3 Impaired Survival of Ppt1 deficient (*Ppt1*$^{-/-}$) astrocytes. A lower cell number was observed in *Ppt1*$^{-/-}$ astrocyte cultures than in wild type (WT) cultures, due to decreased cell survival. **a** Under basal and stimulated conditions the number of *Ppt1*$^{-/-}$astrocytes per coverslip was markedly lower than in WT cultures as indicated by significantly reduced counts of cells stained with the nuclear marker DAPI. **b** At 24, 48 and 72 h following seeding onto coverslips cell death was significantly higher in *Ppt1*$^{-/-}$ astrocyte cultures than in their WT counterparts. **c** A live/dead cell death marker (red) revealed the increased presence of dying cells in *Ppt1*$^{-/-}$astrocyte cultures stained for GFAP (green), compared to those derived from WT mice. (Data shown as Mean ± SEM using a one way ANOVA, n = 3). Scale bar = 50 μm

astrocyte survival may be impaired in CLN1 disease in the later stages of disease progression. These data prompted us to investigate *Ppt1*$^{-/-}$ astrocyte survival in our cultures, and during our immunofluorescence staining experiments, we noticed fewer DAPI-stained nuclei in *Ppt1*$^{-/-}$ astrocyte cultures under all basal and stimulated conditions. Counts of the number of DAPI-stained cells in these *Ppt1*$^{-/-}$ astrocyte cultures confirmed they were significantly reduced in number (Fig. 3a). However, no significant differences in the percentage of proliferating cells was observed in *Ppt1*$^{-/-}$ astrocyte cultures (data not shown), and as such could not account for significantly reduced number of astrocytes in these cultures. Consequently, a Live/Dead kit was used to investigate cell survival 24, 48 and 72 h after plating the cells onto coverslips (Fig. 3b, c). Across all time points, significantly more astrocytes were positive for both GFAP and for the Live/Dead marker in *Ppt1*$^{-/-}$ cultures, indicative of them undergoing cell death (24.83 ± 1.49%; 25.87 ± 1.78%; 34.99 ± 4.66%, at 24, 48 and 72 h respectively) compared to their WT counterparts (7.95 ± 1.08%; 10.57 ± 2.10%; 14.09 ± 1.08% at 24, 48 and 72 h respectively) (Fig. 3b, c). These data reveal that *Ppt1*$^{-/-}$ astrocyte survival is markedly impaired in vitro, suggesting an inherent vulnerability of *Ppt1*$^{-/-}$ astrocytes rather than any defect in their ability to proliferate. Although these tissue culture data for impaired astrocyte survival are consistent with our previous histological evidence from the *Ppt1*$^{-/-}$ brain [29], it remains unclear what the extent and nature of astrocyte death may be in vivo, and it will be important to resolve this issue.

Ppt1$^{-/-}$astrocytes display abnormal Ca^{2+} signalling with increased cytoplasmic Ca^{2+} levels

De-regulated Ca^{2+} signalling has been implicated in the pathogenesis of several neuro-degenerative disorders, including several lysosomal storage disorders (reviewed in Lloyd-Evans, 2016). After the endoplasmic reticulum (ER), the lysosome represents an important intracellular Ca^{2+} store [6, 26]. Thapsigargin is an inhibitor of the sarco/endoplasmic reticulum ATPase (SERCA) that at high concentrations induces Ca^{2+} release from the ER by blocking uptake and uncovering the ER leak channels (Chandrachud et al., 2015). Nigericin is a H^{+}/K^{+} ionophore that will release Ca^{2+} from both lysosomes and mitochondria by collapsing the H^{+} gradient [26]. However, prior pre-treatment with ionomycin results in Ca^{2+} release from all intracellular stores apart from lysosomes, where its action is blocked by the lysosomal glyocaylix, meaning that resultant effects of nigericin on Ca^{2+} release are solely from lysosomal stores [26].

The base-line level of cytoplasmic Ca^{2+} was significantly elevated in *Ppt1*$^{-/-}$ astrocytes (Fig. 4a), suggesting that they are more stressed than their WT counterparts, and this likely caused by the significantly higher number of spontaneous Ca^{2+} release events that we observed in the *Ppt1*$^{-/-}$ astrocytes (data not shown). Stimulation of *Ppt1*$^{-/-}$ and WT astrocytes with thapsigargin revealed no difference between genotypes in the level of Ca^{2+} released from the ER (thapsigargin, Fig. 4b) or from lysosomes (ionomycin/nigericin, Fig. 4c) or, suggesting Ca^{2+} release from these intracellular stores was unaffected in *Ppt1*$^{-/-}$ astrocytes. ATP mediated intracellular Ca^{2+} elevation was significantly lower in *Ppt1*$^{-/-}$ astrocytes (Fig. 4d), and following thapsigargin depletion of the ER there was a small, but significant, increase in store operated Ca^{2+} entry in *Ppt1*$^{-/-}$ astrocytes (Fig. 4e). Taken together these data suggest that *Ppt1*$^{-/-}$ astrocytes display marked abnormalities in Ca^{2+} signalling with elevated levels of both spontaneous Ca^{2+} release events and resting basal cytoplasmic Ca^{2+} that may be related to their pronounced survival defect.

Ppt1$^{-/-}$ microglia exhibit an activated morphological phenotype in vitro

Alongside activation of astrocytes, microglial activation is also evident in *Ppt1*$^{-/-}$ mice as early as three months of age with marked up-regulation of CD68 [27]. We first defined the cellular composition of our microglial cultures. One week after plating microglial cultures showed over 99% of DAPI stained cells expressed CD68 (99.1 and 99.8% CD68 + ve in WT and *Ppt1*$^{-/-}$, respectively; with only 0.89 and 0.14% being GFAP+ve). To assess the in vitro properties of *Ppt1*$^{-/-}$ microglia, we first assessed their ability to undergo morphological transformation following pharmacological stimulation with LPS, a response that is impaired in Cln3 deficient microglia [35]. Following stimulation with LPS, WT and *Ppt1*$^{-/-}$ microglial cultures were stained with CD68 and α-tubulin, and as described previously cells were classified into 3 morphological subcategories: type 1 cells – microglia with extended processes (non-activated); type 2 cells – microglia with retracted processes (partly activated); type 3 cells– rounded cells with a small soma (fully activated) [35].

Under basal conditions, the vast majority of CD68-positive microglia in WT cultures exhibited a bipolar or rod-like morphology (Fig. 5A). Quantitatively these Type 1 cells were the predominant microglial subtype in WT cultures (Fig. 5B). In contrast, far fewer Type 1 cells were present in *Ppt1*$^{-/-}$ microglial cultures under basal conditions (Fig. 5B), with the majority of CD68-positive microglia exhibiting a rounded Type 2 morphology (Fig. 5C). As expected, following LPS stimulation WT microglia rounded up, and Type 2 microglia became the prevailing cell type in these cultures (Fig. 5C). Little change in *Ppt1*$^{-/-}$

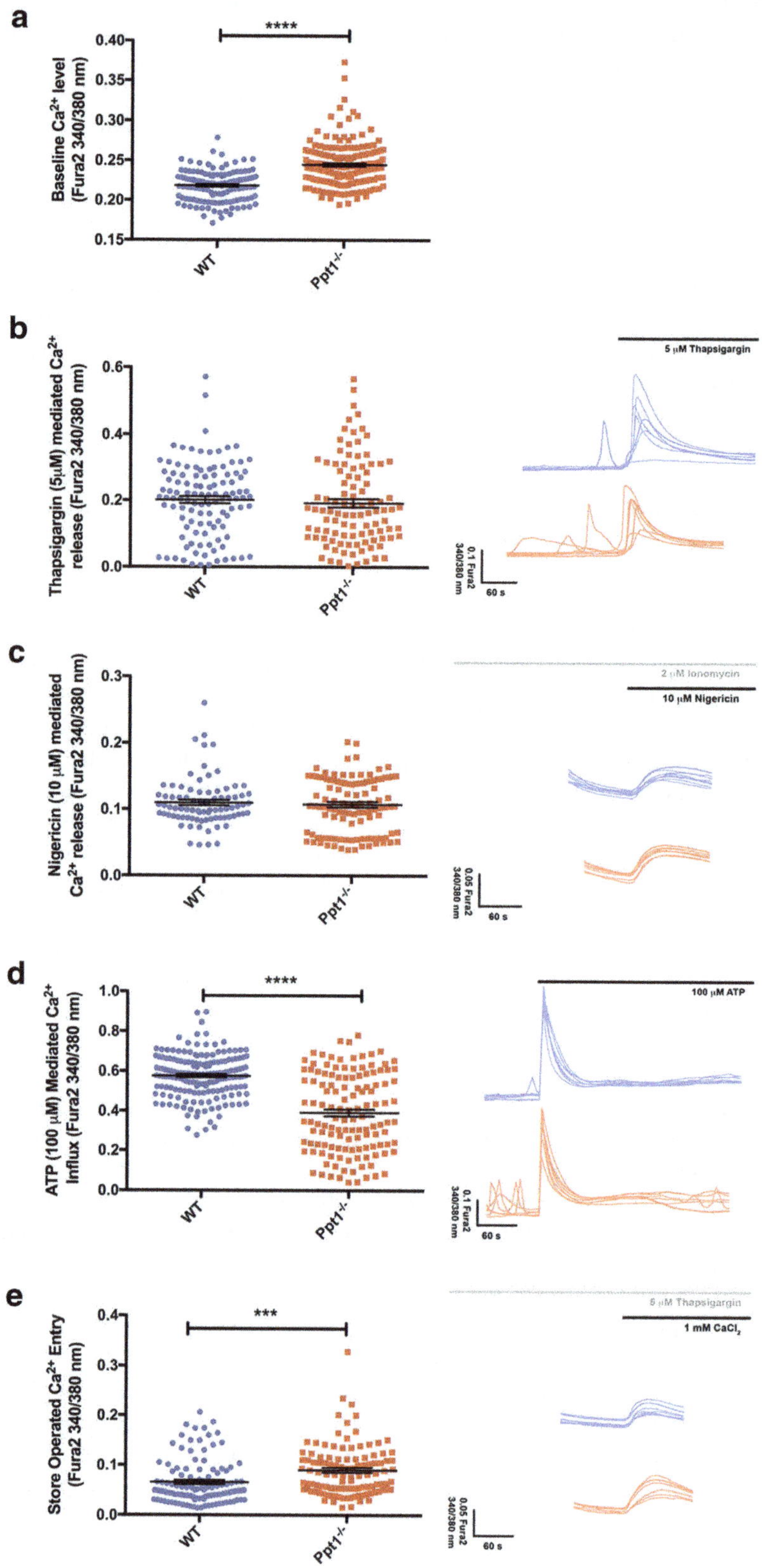

Fig. 4 (See legend on next page.)

(See figure on previous page.)
Fig. 4 Ppt1 deficient (*Ppt1*$^{-/-}$) astrocytes exhibit changes in Ca^{2+} homeostasis. Ca^{2+} measurements were completed 48 h after seeding of wild type (WT) and *Ppt1*$^{-/-}$ astrocytes under basal conditions to assess changes in Ca^{2+} homeostasis. Representative Ca^{2+} traces shown for WT (blue) and *Ppt1*$^{-/-}$ (red) astrocytes (**b-e**). (**a**) Baseline Ca^{2+} levels measured with Fura2 dye (340/380 nm) were significantly higher in *Ppt1*$^{-/-}$astrocytes. (**b**) No statistically significant differences were detected between WT and *Ppt1*$^{-/-}$ astrocytes in Ca^{2+} released from endoplasmic reticulum Ca^{2+} stores, mediated by 5 μM thapsigargin. **c** Lysosomal Ca^{2+} release, triggered by 10 μM nigericin was also not significantly altered in *Ppt1*$^{-/-}$astrocytes. **d** ATP (100 μM) mediated Ca^{2+} influx was significantly lower in *Ppt1*$^{-/-}$astrocytes than in their WT counterparts. **e** Store-operated Ca^{2+} entry, triggered by 5 μM thapsigargin followed by 1 mM $CaCl_2$, was significantly higher in *Ppt1*$^{-/-}$astrocytes than in WT astrocytes

microglial morphology was observed following stimulation for 6 or 24 h, potentially because these mutant microglia already appeared morphologically to be activated (Fig. 5A, C). Very few Type 3 microglia were observed in either WT or *Ppt1*$^{-/-}$ microglial cultures (data not shown), consistent with previous observations that prolonged stimulation is required to fully activate microglia to become small and rounded [35]. Taken together these data suggest that *Ppt1*$^{-/-}$ microglia appear morphologically to be already activated under basal conditions and do not dramatically alter their morphology any further when stimulated pharmacologically.

IL-1β secretion is increased in stimulated Ppt1$^{-/-}$ microglia

Oxidative stress has previously been suggested to play a role in poor cellular health in patient derived primary *Ppt1*$^{-/-}$ deficient fibroblasts [49]. Because microglia are implicated in mediating the effects of oxidative stress [7, 13],

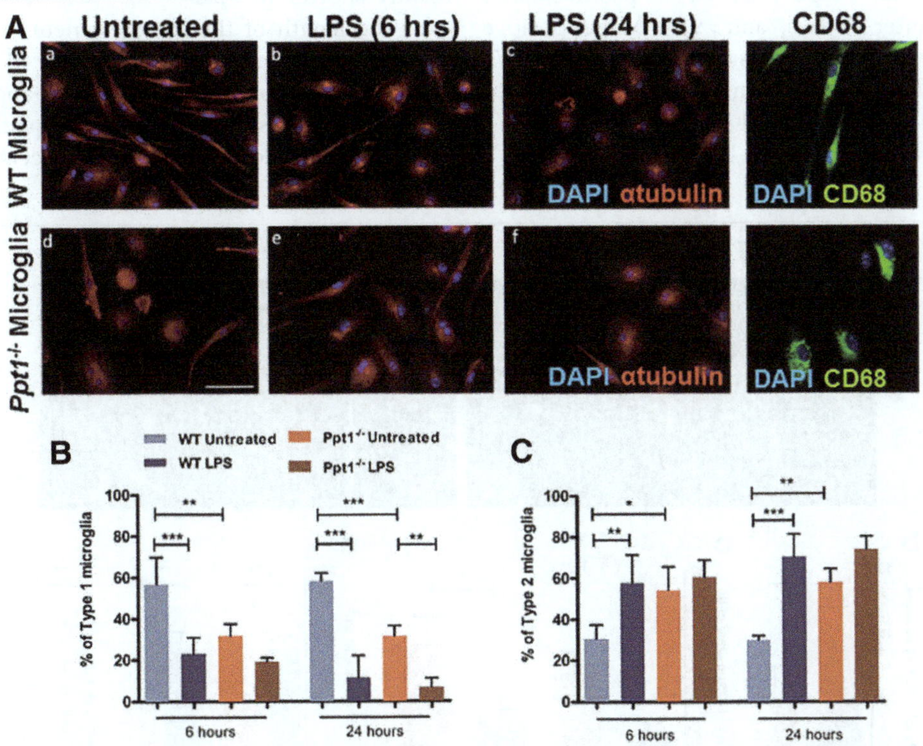

Fig. 5 Ppt1 deficient (*Ppt1*$^{-/-}$) microglia appear morphologically more activated under basal conditions. To assess microglial activation, we first needed to define their morphology. Type 1 or 'resting' microglia exhibit a rod shaped morphology. Type 2 or activated microglia exhibit a round soma and withdrawn processes. Microglial morphology was visualised by staining for CD68 and α-tubulin and 5 random fields were counted per coverslip. α-tubulin was used to assess microglial morphology since CD68 was observed to be more concentrated in the soma and processes were not well visualised using this antibody in these images microglial morphology was visualised with α-tubulin, as CD68 is more concentrated in the soma and processes were not well visualised with this antigen. Morphology was assessed under basal conditions and after stimulation with LPS for 6 and 24 h. **A** *Ppt1*$^{-/-}$ microglia appeared morphologically more activated under basal conditions, as more cells with rounded cell bodies were present. Following treatment with LPS, WT microglia assumed a more reactive morphology. Scale bar = 50 μm. **B** A higher percentage of Type 1 microglia were present in WT microglial cultures than in *Ppt1*$^{-/-}$microglial cultures, and the percentage of Type 1 microglia was reduced following stimulation. **C** More Type 2 microglia were present under basal conditions in *Ppt1*$^{-/-}$ cultures. The percentage of Type 2 microglia increased following stimulation with LPS after 6 and 24 h in WT microglial cultures. (Data shown as Mean ± SEM using a one way ANOVA, $n = 3$)

we investigated the release of soluble proteins typically associated with oxidative stress from *Ppt1*$^{-/-}$ microglia. ELISA assays of these proteins revealed no significant differences in the release of TNFα, TGFβ, IL-1α, IL-6, IL-10, IL-12 or MCP-1 from WT and *Ppt1*$^{-/-}$ microglial cultures (Additional file 1: Figure S1). However, the release of IL-1β was significantly higher in *Ppt1*$^{-/-}$ cultures (5.48 ± 4.32%) following 24 h stimulation than in WT cultures (− 8.41 ± 2.04%), suggesting that *Ppt1*$^{-/-}$ microglia may adapt a more inflammatory phenotype upon stimulation than WT microglia (Additional file 1: Figure S1). It will be important to provide *Ppt1*$^{-/-}$ microglia with an oxidative stress stimulus and determine if the expression of these factors is further altered.

Ppt1$^{-/-}$ neurons display impaired survival of interneurons and morphological defects

Before defining the impact of Ppt1 deficient astrocytes or microglia upon neurons, we first assessed neuronal culture composition. In neuronal cultures > 82% of cells stained positively for Map2 (16% GFAP positive, 0.3% CD68 positive) after 7 days, and > 93% Map2 positive (5% GFAP positive, 3% CD68 positive) after 14 days. We then compared the survival and morphology of cultured primary cortical neurons derived from WT and *Ppt1*$^{-/-}$ mice (Fig. 6a).

Inhibitory neurons are amongst the most vulnerable cells in *Ppt1*$^{-/-}$ mice [3], and although there was little difference in overall cell number between *Ppt1*$^{-/-}$ and WT neuronal cultures after 7 DIV (data not shown), the percentage of inhibitory neurons stained with a cocktail of antibodies against the interneuron markers parvalbumin, calbindin and calretinin was significantly lower in *Ppt1*$^{-/-}$ cortical neuron cultures (4.79 ± 1.41% vs 9.94 ± 1.10% WT) (Fig. 6b). Subsequently, after 14 DIV a significantly higher proportion of cells in *Ppt1*$^{-/-}$ neuronal cultures stained positive for cleaved caspase 3 (6.61 ± 1.13% vs 2.49 ± 0.45% WT), suggesting their impaired longer-term survival in vitro (Fig. 6c). We also quantified neuronal morphology as an indicator of in vitro development and neuronal health, and first revealed that the soma size of Map2-positive neurons was significantly smaller in *Ppt1*$^{-/-}$ cultures at all time points (Fig. 6d).

Assessing neurite complexity (Fig. 7), revealed that at 7, 9, and 14 DIV the mean neurite length was significantly shorter in *Ppt1*$^{-/-}$ vs. WT neurons (Fig. 7a), as was the length of their longest neurite, (Fig. 7b) which by 14 DIV is typically the axon [12, 14, 46]. Changes in neurite morphology were readily apparent in Map2 stained cultures (Fig. 7c), but quantitative analysis revealed specific arborization defects. Although the

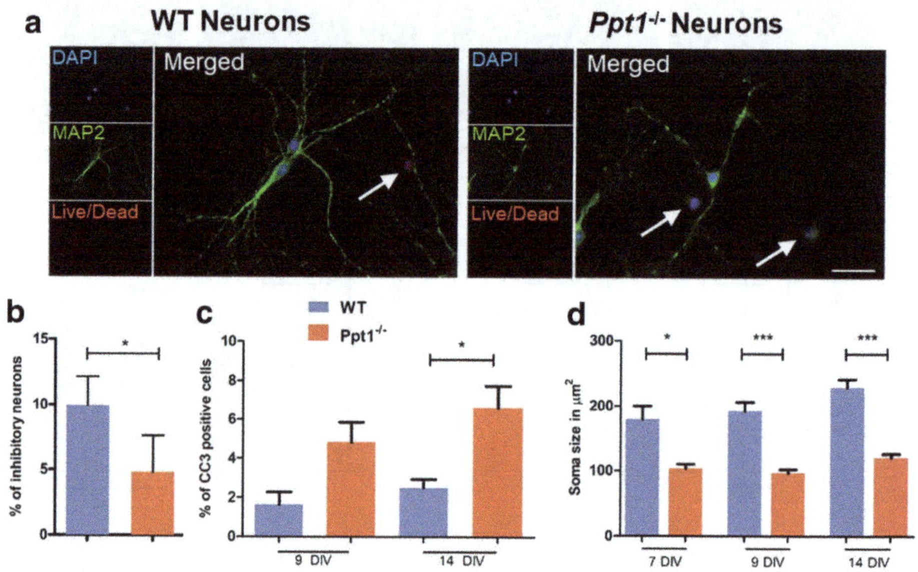

Fig. 6 Ppt1 deficient (*Ppt1*$^{-/-}$) neuronal cultures exhibit altered culture composition and impaired survival. **a** To analyse neuronal cell death and visualise neuronal morphology, WT and *Ppt1*$^{-/-}$neuronal cultures were stained for Map2 and CC3. Pictures of 14 day old cultures stained with nuclear marker DAPI (blue), showing more Map2 positive cells (green) in *Ppt1*$^{-/-}$neuronal cultures labelled with CC3 (red) that in WT cultures. Arrows indicate CC3/ DAPI positive cells in WT cultures and Map2/CC3/DAPI positive cells in *Ppt1*$^{-/-}$ neuronal cultures. Scale bar = 50 μm. **b** Significantly fewer neurons were positively stained for parvalbumin/calretinin/calbindin in *Ppt1*$^{-/-}$ neuronal cultures, indicating that fewer inhibitory neurons are present than in WT neuronal cultures after 7 days in vitro. **c** The percentage of CC3 positive cells did not differ between WT and *Ppt1*$^{-/-}$neuronal cultures after 9DIV, but was significantly higher in *Ppt1*$^{-/-}$cultures after 14DIV. **d** *Ppt1*$^{-/-}$neurons stained for Map2 had significantly smaller soma sizes after 7, 9 and 14 days in vitro (DIV) compared to WT neurons. (Data shown as Mean ± SEM using a one way ANOVA, *n* = 3)

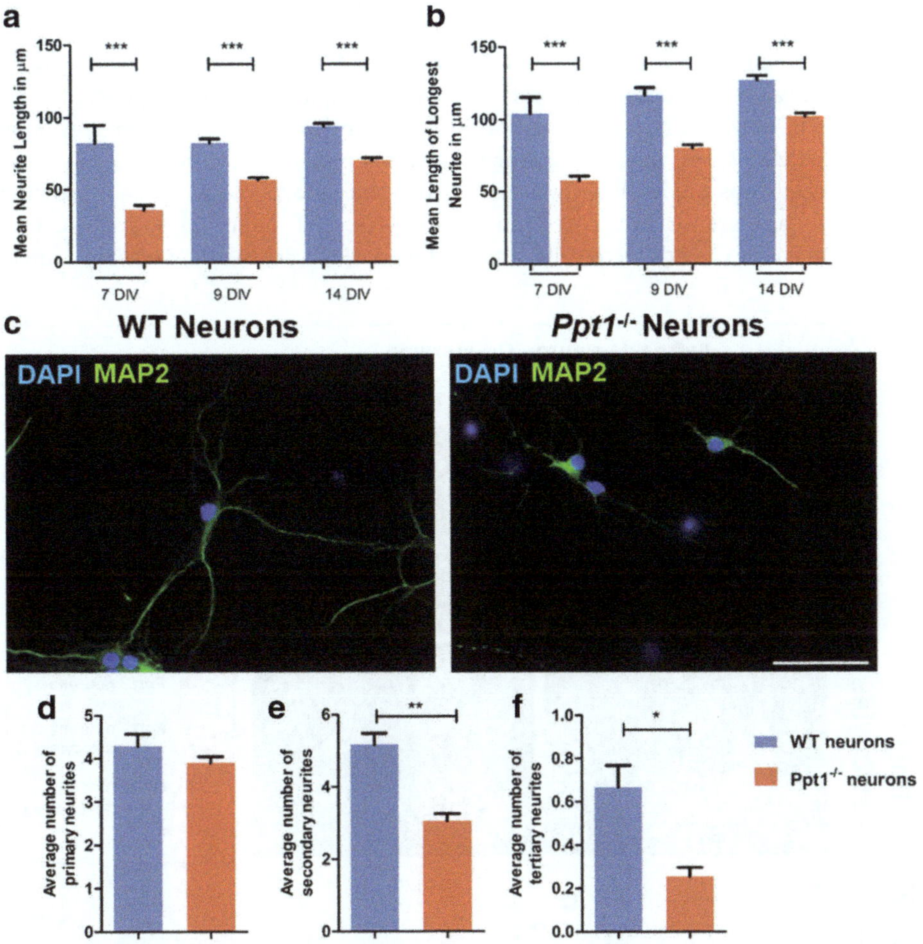

Fig. 7 Ppt1 deficient ($Ppt1^{-/-}$) neurite outgrowth is impaired. Wild type (WT) and $Ppt1^{-/-}$ neuronal cultures were stained with MAP2 to visualise neuronal processes and DAPI to label all nuclei. Neurite length was measured using ImageJ. **a** Mean neurite length is significantly shorter in $Ppt1^{-/-}$ neuronal cultures than in WT cultures after 7, 9 and 14 days in vitro (DIV). **b** The length of the longest neurites, assumed to be the axon is significantly reduced in $Ppt1^{-/-}$neurons compared to WT neurons after 7, 9 and 14 DIV. **c** $Ppt1^{-/-}$ neurons stained for Map2 appear to have shorter neurites than their WT counterparts and appear less complex. **d** The number of primary neurites did not differ between WT and $Ppt1^{-/-}$ neuronal cultures after 7 days in vitro (DIV). **e** $Ppt1^{-/-}$ neurons had significantly fewer secondary neurites than their WT counterparts after 7 DIV. **f** Significantly fewer tertiary neurites were present in $Ppt1^{-/-}$neuronal cultures. (Data shown as Mean ± SEM using a one way ANOVA, n = 3). Scale bar = 50 µm

average number of primary neurites was similar between neurons of both genotypes (Fig. 7d), *Ppt1*$^{-/-}$ neurons had significantly fewer secondary (Fig. 7e) or tertiary (Fig. 7f) neurites at 7 DIV. Taken together these data suggest that *Ppt1*$^{-/-}$ neurons not only show impaired morphology in vitro, suggesting they are in poor health, but also display a moderate impairment in their survival. It will now be important to study neuronal arborization in vivo, to see whether these findings are corroborated in Ppt1-deficient mice.

Co-cultures reveal the impact of Ppt1 deficiency on cellular interactions

To assess the impact of *Ppt1*$^{-/-}$ astrocytes or microglia on each other, or WT and *Ppt1*$^{-/-}$ neurons, we grew these cell types together in different combinations (e.g. astrocytes with microglia, neurons with astrocytes, neurons with microglia, neurons with both astrocytes and microglia), using the morphological phenotypes defined above and survival as outcome measures as outcome measures. As such, for co-cultures were stained with Map2 and CC3, soma size as well as neurite length and complexity were measured and the percentage of cells undergoing apoptosis was determined. Where appropriate, microglia were labelled with CD68 and astrocytes with GFAP.

Detrimental impact of Ppt1$^{-/-}$ astrocytes upon neuron morphology

We first assessed different co-culture combinations of astrocytes and neurons of different genotypes, which revealed obvious effects upon neuron survival and morphology (Fig. 8a). Quantifying these changes, cell death in neuron-astrocyte co-cultures, as revealed by CC3 immunostaining, was significantly greater when *Ppt1*$^{-/-}$ astrocytes were present (Fig. 8b), either in combination with WT neurons (13.41 ± 2.18%) or *Ppt1*$^{-/-}$ neurons (14.03 ± 2.61%). This cell death was predominantly of astrocytes rather than neurons, as there was little correlation between the overall percentage of CC3-positive cells, and those positive for both CC3 and the neuron marker Map2 (data not shown). Less cell death was evident in *Ppt1*$^{-/-}$ neuron/WT astrocyte co-cultures (Fig. 8b), but this difference was not statistically significant.

In contrast, *Ppt1*$^{-/-}$ astrocytes appeared to have a detrimental impact on neuronal health, as judged by neuronal morphology. When grown with *Ppt1*$^{-/-}$ astrocytes, WT neuronal soma size was significantly reduced (Fig. 8c), with decreased mean neurite length

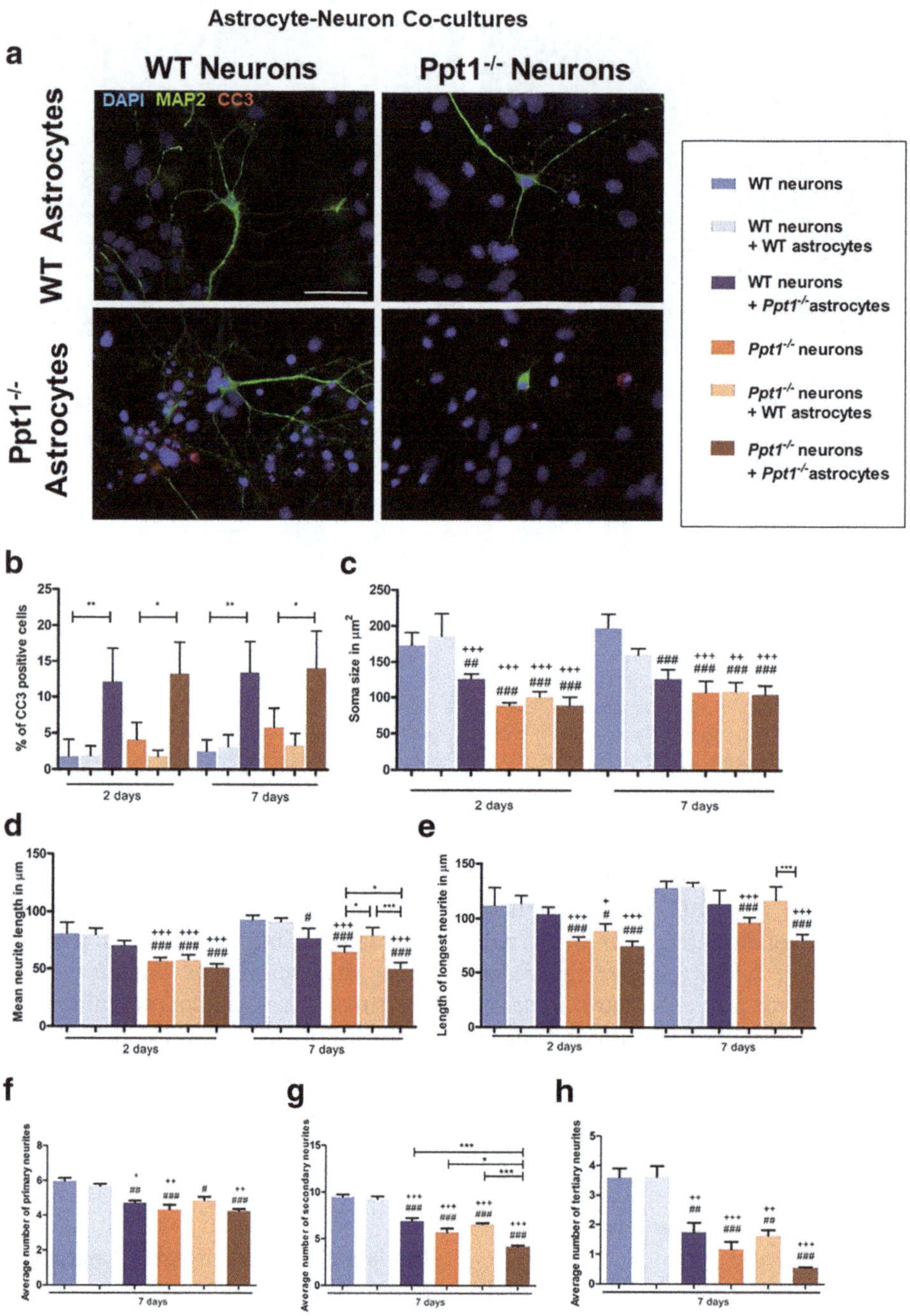

Fig. 8 (See legend on next page.)

(See figure on previous page.)
Fig. 8 Wild Type (WT) astrocytes ameliorate morphological defects in Ppt1 deficient (*Ppt1*$^{-/-}$) neurons. **a** WT and *Ppt1*$^{-/-}$ astrocytes and neurons were cultured together for 2 or 7 days, and stained with MAP2 (green) and cleaved caspase 3 (CC3, red) to examine cell survival and neuronal morphology. **b** After both 2 and 7 days in culture, the percentage of CC3 expressing cells was significantly higher in both WT and *Ppt1*$^{-/-}$ co-cultures when grown with *Ppt1*$^{-/-}$ astrocytes. **c** After both 2 and 7 days in culture, soma size in all *Ppt1*$^{-/-}$ culture conditions was significantly smaller than in WT monocultures, and WT neuron/WT astrocyte co-cultures. Although *Ppt1*$^{-/-}$ astrocytes had little effect upon *Ppt1*$^{-/-}$ neuronal soma size, WT neuron soma size was significantly reduced when grown with Ppt1−/−astrocytes after 2 and 7 days in culture. **d** After 2 days in co-culture, the mean neurite length was shorter in *Ppt1*$^{-/-}$ neurons under all conditions. Following 7 days in co-culture, *Ppt1*$^{-/-}$ astrocytes had a detrimental impact on both WT and *Ppt1*$^{-/-}$ neurite outgrowth, whereas WT astrocytes improved neurite outgrowth in *Ppt1*$^{-/-}$ neurons. **e** Axon length in all *Ppt1*$^{-/-}$ cultures was significantly shorter than in WT neuron and WT neuron/WT astrocyte co-cultures after 2 days. After 7 days, *Ppt1*$^{-/-}$ axon length was significantly reduced in the presence of *Ppt1*$^{-/-}$ astrocytes, whereas WT astrocytes significantly improved axon growth. **f** The number of primary neurites was significantly lower in in all *Ppt1*$^{-/-}$ cultures as well as WT neuron/ *Ppt1*$^{-/-}$ astrocyte cultures. **g** Secondary neurite number was significantly reduced in WT neuron/ *Ppt1*$^{-/-}$ astrocyte cultures, *Ppt1*$^{-/-}$ neuron cultures *Ppt1*$^{-/-}$ neuron/WT astrocyte and *Ppt1*$^{-/-}$ neuron/ *Ppt1*$^{-/-}$ astrocyte co-cultures compared to WT neuron cultures and WT neuron/WT astrocyte cultures. Secondary neurite number remained significantly lower in *Ppt1*$^{-/-}$ neuron/ *Ppt1*$^{-/-}$ astrocyte cultures than in WT neuron/ *Ppt1*$^{-/-}$ astrocyte cultures, *Ppt1*$^{-/-}$ astrocyte cultures and *Ppt1*$^{-/-}$ neuron/ *Ppt1*$^{-/-}$ astrocyte cultures. **h** The number of tertiary neurites was significantly lower in WT neuron/ *Ppt1*$^{-/-}$ astrocyte cultures, *Ppt1*$^{-/-}$ neuron cultures, *Ppt1*$^{-/-}$ neuron/WT astrocyte and *Ppt1*$^{-/-}$ neuron/ *Ppt1*$^{-/-}$astrocyte co cultures than in WT neuronal cultures and WT neuron/WT astrocyte cultures. (Data shown as Mean ± SEM using a one-way ANOVA, $n = 3$, # represents significant difference to WT neuron cultures, + represents significant difference to WT neuron/WT astrocyte cultures)

(Fig. 8d), but not of the longest neurite, probably the axon (Fig. 8e). These effects were more pronounced after 7 days in co-culture, when WT neurons started resembling *Ppt1*$^{-/-}$ neurons morphologically (Fig. 8a). Although *Ppt1*$^{-/-}$ astrocytes did not adversely impact *Ppt1*$^{-/-}$ neuron soma size after 7 days in co-culture (Fig. 8c), the neurites of these *Ppt1*$^{-/-}$ neurons were not only significantly shorter than in *Ppt1*$^{-/-}$ neuronal cultures (50.41 ± 2.71 μm vs 65.02 ± 2.51 μm) (Fig. 8d, e), but also had fewer secondary neurites (4.2 ± 0.11vs 5.69 ± 0.4 secondary neurites) (Fig. 8g). In contrast, WT astrocytes appeared to improve *Ppt1*$^{-/-}$ neurite outgrowth (80.02 ± 3.31 μm) (Fig. 8d-e), but had no discernible impact on *Ppt1*$^{-/-}$ neurite complexity (Fig. 8f-h).

The overall impact of *Ppt1*$^{-/-}$ astrocytes appeared to be primarily detrimental to neuronal health, adversely affecting their morphology rather than triggering their death. Although the morphological defects of *Ppt1*$^{-/-}$ neurons can be partially restored by WT astrocytes, their survival is not improved, which suggests an intrinsic defect in *Ppt1*$^{-/-}$ neuronal survival, despite any possible cross-correction effect of Ppt1 enzyme secreted from WT astrocytes.

Detrimental impact of Ppt1$^{-/-}$ microglia upon neuron survival

We next grew co-cultures of microglia with neurons, in order to assess the impact of these cell types from mice of different genotypes upon each other and revealed a profound impact of *Ppt1*$^{-/-}$ microglia upon neuron survival (Fig.9a). Cell death as indicated by CC3 staining appeared significantly higher in *Ppt1*$^{-/-}$ neuron/ *Ppt1*$^{-/-}$ microglial co-cultures than in WT neuron/ *Ppt1*$^{-/-}$ microglial co-cultures, *Ppt1*$^{-/-}$ neuronal cultures and *Ppt1*$^{-/-}$ neuron/ *WT* microglial co-cultures (Fig. 9a) There were significantly more double- labelled CC3- and Map2-positive cells in *Ppt1*$^{-/-}$ neuron/ *Ppt1*$^{-/-}$ microglial co-cultures (16.05 ± 1.57%) than in WT neuron/ *Ppt1*$^{-/-}$ microglial cultures (6.31 ± 1.19%) (Fig. 9b). While these data suggest a direct cytotoxic effect of *Ppt1*$^{-/-}$ microglia upon neurons, a combined effect of *Ppt1*$^{-/-}$ microglia with the small number of astrocytes also present in these cultures cannot be ruled out.

We next investigated neuronal morphology in these co-cultures and saw no significant effects of either WT or *Ppt1*$^{-/-}$ microglia upon the soma size of neurons of either genotype (Fig. 9c). No detrimental effects of *Ppt1*$^{-/-}$ microglia could be observed on neurite outgrowth (Fig. 9d-e), or neurite complexity (Fig. 9f-h). However, in *Ppt1*$^{-/-}$ neuron/ WT microglia co-cultures there were more secondary (8.1 ± 0.29 vs 5.8 ± 0.08 *Ppt1*$^{-/-}$ neurons) and tertiary neurites (2.33 ± 0.2 vs 1.41 ± 0.15 *Ppt1*$^{-/-}$ neurons) (Fig. 9g-h) than in *Ppt1*$^{-/-}$ neuron cultures, and these neurites were longer (74.96 ± 0.84 μm *Ppt1*$^{-/-}$ neuron vs. 87.37 ± 3.35 μm *Ppt1*$^{-/-}$/ WT co-culture) (Fig. 9d-e), suggesting a positive influence of WT microglia upon neurite outgrowth and complexity, perhaps by their secretion of Ppt1 enzyme.

Overall, these data reveal that, compared to *Ppt1*$^{-/-}$ astrocytes, *Ppt1*$^{-/-}$ microglia exert less of an influence upon neuronal morphology, but exert a significant negative effect upon the survival of *Ppt1*$^{-/-}$ neurons with much less of an impact upon WT neurons. As such, *Ppt1*$^{-/-}$ microglia do not appear to be intrinsically neurotoxic, but it appears that *Ppt1*$^{-/-}$ neurons may simply be more vulnerable to their influence.

Astrocytes drive microglial activation in Ppt1$^{-/-}$ mixed glial cultures

Mixed glial cultures containing both astrocytes and microglia of either genotype were grown to examine whether the presence of other glial cells may improve or exacerbate the phenotypes exhibited by either cell type in monocultures. The presence of microglia appeared to have little impact on *Ppt1*$^{-/-}$ astrocytes, as the percentage of GFAP-expressing cells remained significantly

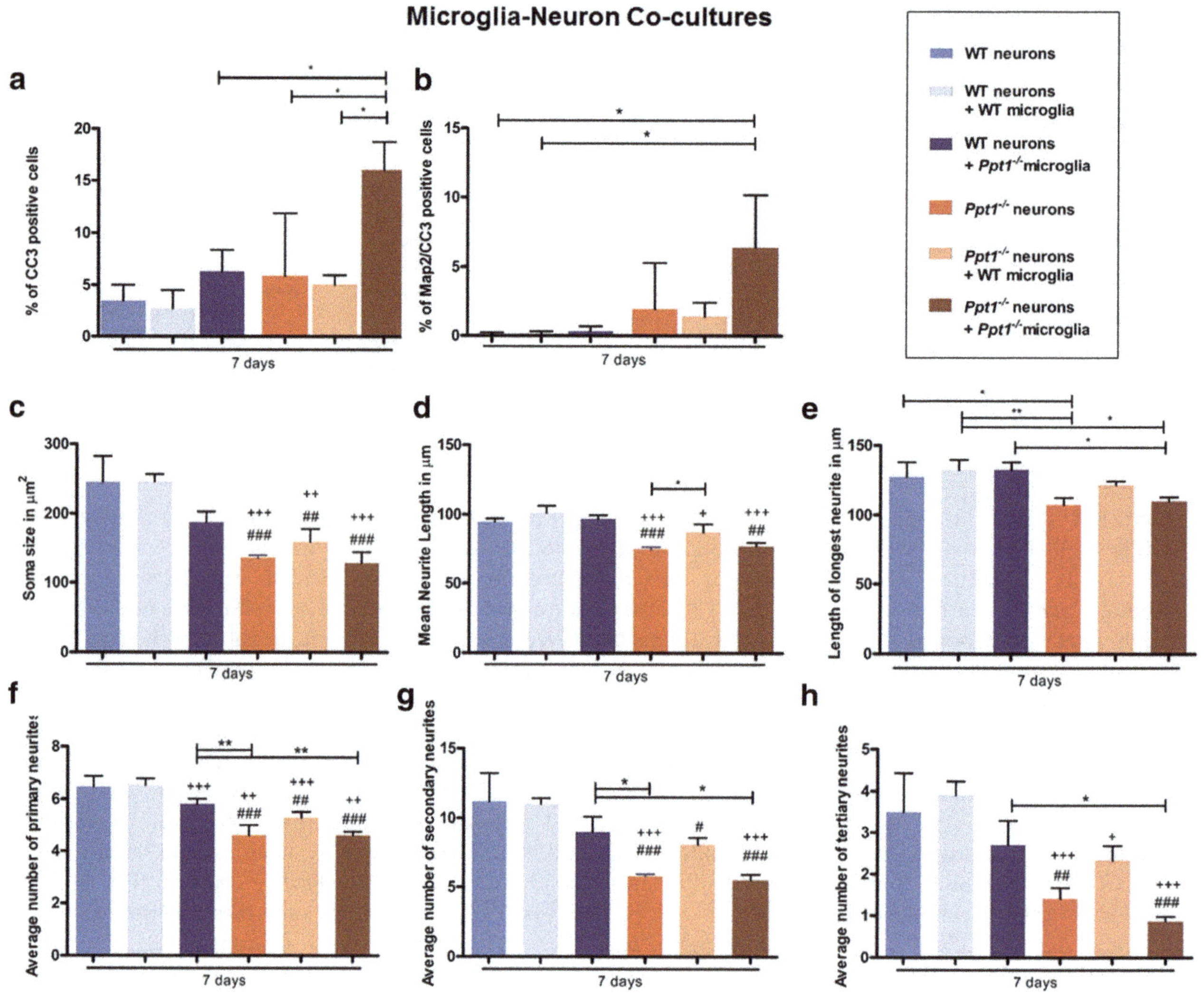

Fig. 9 Ppt1 deficient ($Ppt1^{-/-}$) microglia trigger $Ppt1^{-/-}$ neuronal death. Wild type (WT) and $Ppt1^{-/-}$ neurons and microglia were cultured together 7 days, and stained with MAP2 and cleaved caspase 3 (CC3) to examine cell survival and neuronal morphology. **a** The percentage of CC3 expressing cells was statistically significantly higher in $Ppt1^{-/-}$neuron/ $Ppt1^{-/-}$ microglial co-cultures than in all other cultures. In contrast $Ppt1^{-/-}$microglia had a smaller impact on WT neuronal survival. **b** The percentage of Map2/CC3 expressing cells was highly variable within all cultures, although was significantly higher in $Ppt1^{-/-}$ neuron/ $Ppt1^{-/-}$microglial co-cultures than in WT neuron and WT neuron/WT microglial cultures. **c** Neuronal soma size was significantly smaller for $Ppt1^{-/-}$ neurons across all co-cultures, compared to the soma size of WT neurons in either WT monocultures or WT neuron/WT microglial co-cultures. Neuronal soma size was somewhat reduced in WT neuron/ $Ppt1^{-/-}$ microglial co-cultures, but this difference was not statistically significant. **d** $Ppt1^{-/-}$ microglia do not have a detrimental effect on $Ppt1^{-/-}$or WT mean neurite length. However, the presence of WT microglia had a beneficial impact in increasing $Ppt1^{-/-}$neurite length. **e** Axon length in WT or $Ppt1^{-/-}$ cultures was unaffected by $Ppt1^{-/-}$microglia, and the apparently beneficial impact of WT microglia upon axon length was not as pronounced as on mean neurite length. **f** The number of primary neurites was significantly lower across all $Ppt1^{-/-}$ cultures. **g** Secondary neurite number was significantly reduced in $Ppt1^{-/-}$neuron cultures, $Ppt1^{-/-}$ neuron/WT microglia and $Ppt1^{-/-}$ neuron/ $Ppt1^{-/-}$ microglia co-cultures compared to WT neuron cultures and WT neuron/WT microglial cultures. A trend towards increased secondary neurite number was observed in $Ppt1^{-/-}$ neuron/ WT microglial cultures. **h** The number of tertiary neurites was significantly lower in $Ppt1^{-/-}$ neuron cultures, $Ppt1^{-/-}$neuron/WT microglial and $Ppt1^{-/-}$ neuron/ $Ppt1^{-/-}$microglial cultures than in WT neuronal cultures and WT neuron/WT microglial cultures. (Data shown as Mean ± SEM using a one way ANOVA, n = 3; # represents significant difference to WT neuron cultures, + represents significant difference to WT neuron/WT microglia cultures)

higher in $Ppt1^{-/-}$ cultures under basal conditions and after exposure to LPS alone than in WT cultures (Fig. 10a). Similarly, $Ppt1^{-/-}$ astrocyte soma size in these mixed glial cultures was consistently significantly larger than that of WT astrocytes, until stimulated with LPS and IFNγ, after which $Ppt1^{-/-}$ astrocyte cell body size decreased dramatically (Fig. 10b), as we had seen in astrocyte monocultures (Fig. 1b). Significantly fewer cell nuclei were counted in $Ppt1^{-/-}$ mixed glial cultures (408.1 ± 28.89 vs 720.7 ± 43.19 WT) (Fig. 10c), suggesting that the presence of WT microglia does not improve $Ppt1^{-/-}$ astrocyte survival.

The phenotypes seen in $Ppt1^{-/-}$ microglial monocultures also persisted in mixed astrocyte:microglial cultures. Under basal conditions, Type 1 microglia (63.99 ± 1.39%) were the predominant cell type in WT cultures (Fig. 10d), which transformed into rounded Type 2 activated microglia following stimulation (81.53 ± 4.03% LPS, 78.82 ± 6.74% LPS/IFNγ) (Fig. 10e). Under all conditions, activated Type 2 microglia (75.70 ± 3.17% basal; 96.02 ± 1.84% LPS; 95.31 ± 1.91% LPS/IFNγ) were the prevailing type of microglia present in $Ppt1^{-/-}$ microglial cultures (Fig. 8e), however the percentage of activated Type 2 microglia was significantly higher in mixed astrocyte:microglial cultures than in microglial monocultures (Fig. 10e). Furthermore, following stimulation with LPS only or LPS and IFNγ, the percentage of Type 2 microglia significantly increased in $Ppt1^{-/-}$ mixed astrocyte:-microglial cultures (Fig. 10f), which was not observed in cultures of microglia alone. These data suggest that $Ppt1^{-/-}$ astrocytes may drive microglial activation, and also induce a greater microglial response following pharmacological stimulation.

Neuron/astrocyte/microglial co-cultures

The presence of even some $Ppt1^{-/-}$ astrocytes appears to influence morphological measures of microglial activation (Fig. 10), and may also potentially influence any negative impact of microglia (Fig. 9). Therefore, we hypothesised that growing mixed cultures that contain

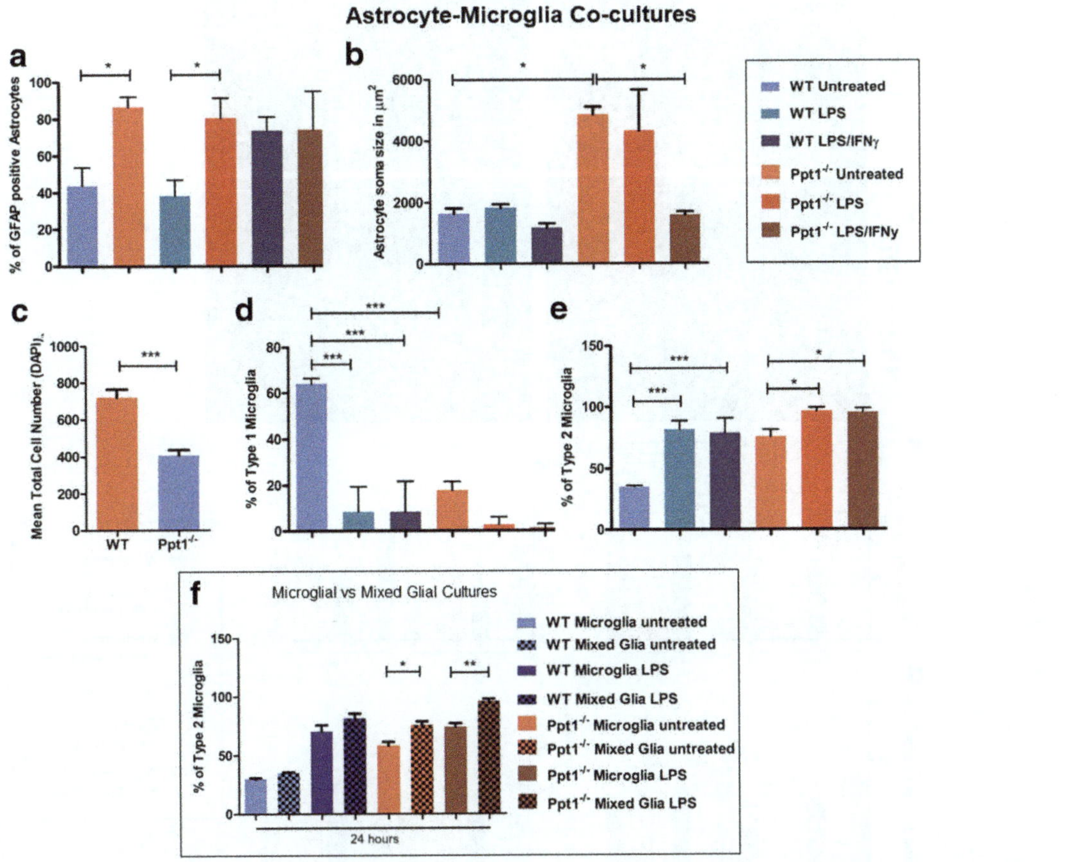

Fig. 10 Ppt1 deficient (*Ppt1*$^{-/-}$) astrocytes drive microglial activation. *Ppt1*$^{-/-}$ and wild type (WT) cultures were maintained as mixed glial (astrocyte and microglial) cultures to examine whether monoculture phenotypes were maintained under basal and stimulated (LPS or LPS/IFNγ) conditions. **a** Under basal conditions, the proportion of cells expressing the astrocyte marker glial fibrillary acidic protein (GFAP) expression was higher in *Ppt1*$^{-/-}$ mixed glial cultures than in their WT counterparts. GFAP expression did not increase in WT or *Ppt1*$^{-/-}$ mixed glial cultures following addition of LPS, but did increase in WT following stimulation with LPS/IFNγ. **b** Under basal conditions, compared to WT cultures, *Ppt1*$^{-/-}$ astrocytes exhibited a larger cell body size, which was decreased dramatically in response to exposure to LPS/IFNγ but not LPS only. **c** The mean total cell number as indicated by DAPI staining was statistically significantly lower in *Ppt1*$^{-/-}$ mixed glial cultures. **d** The percentage of Type 1 (resting, elongated) microglia was significantly higher in WT cultures under basal conditions than in *Ppt1*$^{-/-}$ mixed glial cultures. **e** The percentage of Type 2 (activated, rounded) microglia was higher in *Ppt1*$^{-/-}$ mixed glial cultures under basal conditions. In both WT and *Ppt1*$^{-/-}$ mixed glial cultures the percentage of Type 2 microglia was increased following stimulation with LPS, or with LPS/IFNγ. **f** *Ppt1*$^{-/-}$ microglia appear morphologically more activated in mixed glia cultures than in microglial cultures, as indicated by a significantly higher percentage of Type 2 microglia under both basal and stimulated (LPS) conditions in mixed glial cultures. (Data shown as Mean ± SEM using a one way ANOVA, n = 3)

proportionately more astrocytes than microglia, a situation that may more closely reflect the composition if the brain in vivo [20], might result in the most detrimental environment for neurons. This appeared to be the case with progressively more CC3-positive cells being observed in these co-cultures with time (Fig. 11), with a corresponding negative influence upon neuronal morphology (Fig. 11a).

After 7 days in co-culture, cell death levels were at their highest in *Ppt1*$^{-/-}$ neuron/ *Ppt1*$^{-/-}$ mixed glial cultures (22.67 ± 2.17% of cells CC3 + ve) (Fig. 11b), a significant increase compared to cell death after 2 days of co-culturing *Ppt1*$^{-/-}$ neurons with *Ppt1*$^{-/-}$ mixed glia (14.05 ± 2.33% of cells CC3 + ve), which was comparable to that in WT neuron/ *Ppt1*$^{-/-}$ mixed glial co-cultures. In contrast, there appeared to be a moderate protective effect of WT astrocytes resulting in decreased cell death in *Ppt1*$^{-/-}$ neuron/WT mixed glial cultures (5.58 ± 0.32% of cells CC3 + ve) (Fig. 11b), perhaps as a result of cross correction by Ppt1 enzyme secreted by these WT cells.

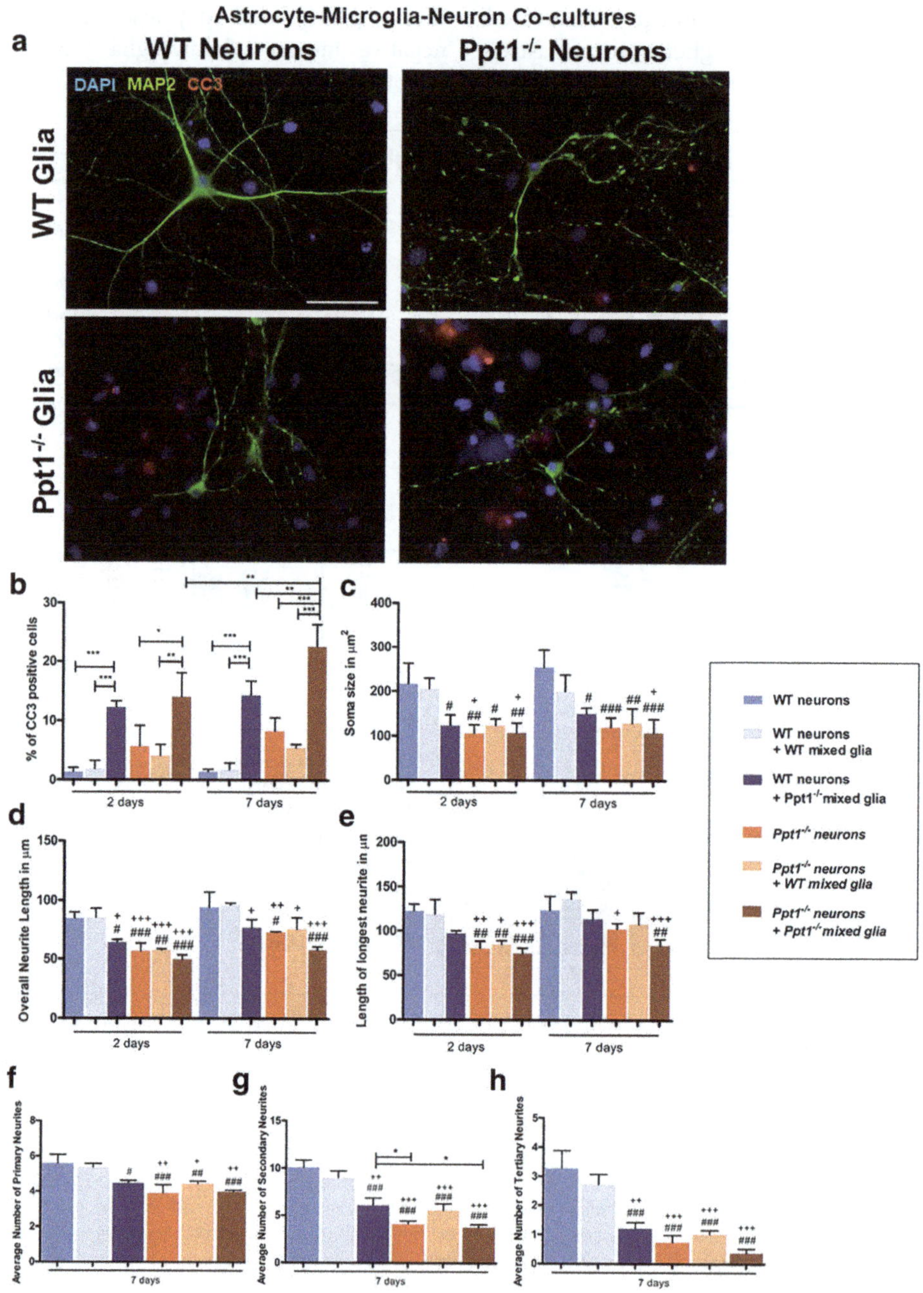

Fig. 11 (See legend on next page.)

(See figure on previous page.)

Fig. 11 Ppt1 deficient (*Ppt1*$^{-/-}$) astrocyte/microglia interactions escalate cell death in *Ppt1*$^{-/-}$ neuron/ *Ppt1*$^{-/-}$ mixed glial cultures after prolonged co-culturing. **a** Wild type (WT) and *Ppt1*$^{-/-}$ neurons and mixed glia (astrocytes and microglia) were cultured together for 2 and 7 days and stained with MAP2 and cleaved caspase 3 (CC3) to examine cell survival and neuronal morphology. Cells were stained with neuronal marker Map2 (green), cell death marker CC3 (red) and nuclear stain DAPI (blue). After 7 days in co-culture, more CC3 positive cells were observed in WT neuron/ *Ppt1*$^{-/-}$ mixed glial co-cultures and *Ppt1*$^{-/-}$ neuron/ *Ppt1*$^{-/-}$ mixed glial co-cultures. Scale bar = 50 μm. **b** The percentage of CC3 positive cells was significantly higher in WT neuron/ *Ppt1*$^{-/-}$ mixed glial co-cultures, and *Ppt1*$^{-/-}$ neuron/ *Ppt1*$^{-/-}$ mixed glial co-cultures after both 2 and 7 days. After 7 days, the percentage of CC3 positive cells had significantly increased within *Ppt1*$^{-/-}$ neuron/ *Ppt1*$^{-/-}$ mixed glial co-cultures. **c** Soma size across *Ppt1*$^{-/-}$ cultures was smaller than in WT and WT neuron/WT mixed glial co-cultures after both 2 and 7 days in culture. WT neuron size was significantly reduced following the addition of *Ppt1*$^{-/-}$ mixed glia at both time points. **d** After both 2 and 7 days in culture, mean neurite length was shorter across all *Ppt1*$^{-/-}$ cultures than in WT monocultures or WT neuron/WT mixed glial co-cultures. WT neurite length was significantly reduced in the presence of *Ppt1*$^{-/-}$ mixed glia after 2 and 7 days in culture. *Ppt1*$^{-/-}$ neurite length was also somewhat reduced after 7 days in co-culture with *Ppt1*$^{-/-}$ mixed glia. **e** *Ppt1*$^{-/-}$ axon length was consistently shorter than that of WT neurons, and although WT axon length was somewhat reduced in WT neuron/ *Ppt1*$^{-/-}$ mixed glial co-cultures, this was not statistically significant. **f** The number of primary neurites was significantly lower in WT neuron/ *Ppt1*$^{-/-}$ mixed glial cultures, as well as in all *Ppt1*$^{-/-}$ cultures. **g** Secondary neurite number was significantly reduced in WT neuron/ *Ppt1*$^{-/-}$ mixed glial cultures, *Ppt1*$^{-/-}$ neuron cultures, *Ppt1*$^{-/-}$ neuron/WT mixed glial and *Ppt1*$^{-/-}$ neuron/ *Ppt1*$^{-/-}$ mixed glial co-cultures compared to WT neuron cultures and WT neuron/WT mixed glial cultures. Secondary neurite number remained significantly lower in *Ppt1*$^{-/-}$ neuron and *Ppt1*$^{-/-}$ neuron/*Ppt1*$^{-/-}$ mixed glial cultures than in WT neuron/*Ppt1*$^{-/-}$ mixed glial cultures. **h** The number of tertiary neurites was significantly lower in WT neuron/*Ppt1*$^{-/-}$ mixed glial cultures, *Ppt1*$^{-/-}$ neuron cultures, *Ppt1*$^{-/-}$ neuron/WT mixed glial and *Ppt1*$^{-/-}$ neuron/*Ppt1*$^{-/-}$ mixed glial cultures than in WT neuronal cultures and WT neuron/WT mixed glial cultures. (Data shown as Mean ± SEM using a one way ANOVA, n = 3; # represents significant difference to WT neuron cultures, + represents significant difference to WT neuron/WT microglia cultures)

Exploring effects upon neuronal morphology revealed that the soma size of *Ppt1*$^{-/-}$ neurons was not affected by the presence of either WT or *Ppt1*$^{-/-}$ mixed glia, and remained significantly smaller than their WT counterparts (Fig. 11c). However, when WT neurons were grown with *Ppt1*$^{-/-}$ mixed glia, their neuronal soma size was significantly reduced, suggesting a detrimental impact of these Ppt1 deficient astrocytes and microglia upon neuronal health (Fig. 11c). Indeed, *Ppt1*$^{-/-}$ astrocytes and microglia also had a pronounced and rapid impact upon WT average neurite length (Fig. 11d-e), which was significantly shorter after only 2 days in co-culture (64.25 ± 1.24 μm vs WT neurons 84.72 ± 3.13 μm). Although some growth in WT neurite length was apparent with continued time in culture, average neurite length remained shorter after 7 days of co-culture in WT neuron/ *Ppt1*$^{-/-}$ mixed glial cultures (76.70 ± 4.01 μm) vs. WT/WT cultures (96.59 ± 0.80 μm). The combined presence of *Ppt1*$^{-/-}$ astrocytes and microglia also significantly impacted neurite complexity with significant reductions in the average number of primary (4.48 ± 0.09 vs 5.61 ± 0.28 WT neurons, Fig. 11f), secondary (6.07 ± 0.45 vs 10.08 ± 0.44 WT neurons, Fig. 11g) and tertiary neurites in WT neurons (1.21 ± 0.12 vs 3.29 ± 0.35 WT neurons, Fig. 11h). In contrast, the presence of WT astrocytes and microglia produced no significant improvements in *Ppt1*$^{-/-}$ neuronal soma size (Fig. 11c), neurite outgrowth (Fig. 11d-e) or complexity (Fig. 11f-h).

Taken together, these data suggest that the presence of both *Ppt1*$^{-/-}$ astrocytes and microglia has the most significant detrimental effects upon neurons, not just upon their morphology, but also upon the survival of both WT and *Ppt1*$^{-/-}$ neurons. WT astrocytes and microglia were not capable of rescuing the pronounced defects of *Ppt1*$^{-/-}$ neurons, providing further evidence that these neurons are intrinsically more vulnerable than WT neurons.

Discussion

The purpose of undertaking this tissue culture study was not to replicate disease progression in the Ppt1 deficient brain, but rather to assess intrinsic defects in *Ppt1*$^{-/-}$ glial biology. Indeed, in addition to detailing defects in the morphology and survival of neurons, this study is the first to characterize *Ppt1*$^{-/-}$ glia in vitro. Our data reveal that not only do *Ppt1*$^{-/-}$ microglia and astrocytes exhibit a range of abnormal phenotypes, they also appear capable of harming neurons, especially if these neurons are Ppt1 deficient. Whilst *Ppt1*$^{-/-}$ astrocytes predominantly affect neuronal morphology, the presence of *Ppt1*$^{-/-}$ microglia alone was enough to impact neuron survival, a negative influence that was greater if both *Ppt1*$^{-/-}$ astrocytes and *Ppt1*$^{-/-}$ microglia were present, resulting in pronounced neuron loss. These data suggest that glial cells are more affected by Ppt1 deficiency than previously anticipated, and this may directly influence neuron survival in CLN1 disease and it will be important to explore this also occurs in vivo. However, as discussed below, these defects in Ppt1 deficient glia are quite distinct from those we recently reported in similar experiments in CLN3 disease [35], another member of this group of disorders in which a negative influence of functionally compromised glia upon neuronal survival is evident. Taken together, these data provide further evidence that although these disorders broadly share similar features, they may differ markedly in how individual cell types are impacted by disease.

Effects of Ppt1 deficiency upon glial biology

Compared to the morphologically attenuated glial activation that is evident in CLN3 disease [35], mouse models of CLN1 disease exhibit much more pronounced and early onset activation of both astrocytes and microglia in the same CNS regions that later display the most neuron loss. This includes the thalamocortical system [23], cerebellum [29], and long before these events occur in the brain this glial activation is also evident at all levels of the spinal cord [43]. These reactive events are accompanied by a pronounced upregulation of chemokines and cytokines in vivo [29], that can be normalized by a combination of brain directed gene therapy and anti-inflammatory drugs [28]. Our data demonstrate that cultured *Ppt1*$^{-/-}$ astrocytes are very responsive to pharmacological stimulation (Figs. 1), and *Ppt1*$^{-/-}$ microglia appear morphologically more activated even under basal unstimulated conditions (Fig. 5). This is quite different to the properties of cultured *Cln3*$^{-/-}$ astrocytes and microglia that both respond slowly and incompletely [35], mirroring the relative extent of morphological transformation of these cell types with disease progression in murine CLN1 and CLN3 diseases in vivo [23, 35, 37, 38]. Another marked phenotypic difference between glia isolated from mouse models of these two forms of NCLs is the reduced secretion of chemokines and cytokines by *Cln3*$^{-/-}$ astrocytes and microglia in culture [35], compared to the elevated secretion of a subset of these factors by *Ppt1*$^{-/-}$ astrocytes (Fig. 2) and microglia (Additional file 1: Figure S1), data that mirrors in vivo findings for much broader elevation of such factors [29]. Any comparison between in vitro and in vivo studies must be made with caution. However, there are differences between these expression profiles in terms of which chemokines are upregulated. These may reflect the markedly different ages of the mice when these studies were performed, the inherent differences in a tissue culture environment and the living brain, as well as the pharmacological stimulation used in the current in vitro study.

Another fundamental difference between data from Cln1 and Cln3 glia, is that compared to the disruption of intermediate filaments seen in *Cln3*$^{-/-}$ astrocytes [35], the current study found no evidence for any similar cytoskeletal defects in *Ppt1*$^{-/-}$ astrocytes (data not shown). Taken together, these data emphasize that although glia from both of these mouse models of NCL display a variety of abnormal phenotypes in culture, and that these consistently affect astrocytes more than microglia, the nature and extent of these glial defects differ markedly between CLN1 and CLN3 disease.

One of the most striking findings of the current study is the compromised survival of *Ppt1*$^{-/-}$ astrocytes in all culture conditions, a phenotype not exhibited by *Cln3*$^{-/-}$ astrocytes in vitro [35]. It will be important to determine if this vulnerability of *Ppt1*$^{-/-}$ astrocytes is related to abnormal calcium signalling properties that they display (Fig.4), but it might be expected that these raised cytoplasmic calcium levels would contribute to their death. It will be crucial to determine if *Ppt1*$^{-/-}$ astrocyte survival is also impaired in vivo to a similar extent that we have observed in vitro. Although there is already evidence for fewer astrocytes expressing S100β and glutamine synthetase in *Ppt1*$^{-/-}$ mice during the later stages of the disease [29], which points towards a loss of these astrocyte populations in vivo, this issue is yet to be fully resolved. Our in vitro findings do provide further corroborative evidence for an inherent vulnerability of *Ppt1*$^{-/-}$astrocytes, and this would be expected to adversely impact neurons and their survival during CLN1 disease progression. However, the exact mechanisms underlying such cell-specific vulnerability of astrocytes remain unclear and will need to be investigated in subsequent in vivo studies.

Effects of Ppt1 deficiency upon neurons

CLN1 disease is perhaps the most profoundly neurodegenerative form of NCL, with near total cortical neuron loss at autopsy [18]. However, the mechanisms by which a deficiency in a de-palmitoylating lysosomal enzyme leads to this remarkably dramatic neuron loss remain unclear. Previous studies culturing neurons derived from *Ppt1*$^{-/-}$ mice surprisingly did not reveal any overt defect in their survival, or firing properties, but did find alterations in the synaptic vesicle pool size [47]. These are suggestive of an early stage synaptic pathology, and *Ppt1*$^{-/-}$ mice do go on to display progressive changes in synaptic organization within the thalamocortical system [24]. Our data in this study provide new evidence that *Ppt1*$^{-/-}$ cortical neurons show compromised survival in our culture conditions (Fig. 6c), display a markedly smaller cell soma (Fig. 6d) and compromised neurite organization in vitro, that involves secondary and tertiary neurites rather than those arising directly from the soma (Fig. 7d-f). This neuron loss was most pronounced for GABAergic interneurons (Fig. 6b), consistent with in vivo data from these *Ppt1*$^{-/-}$ mice [3, 23]. Taken together, these data reveal that Ppt1 deficiency does not by itself result in the same profound extent of neuron loss seen in vivo, suggesting that other non-neuron-intrinsic mechanisms may operate to influence neurodegeneration in this disorder.

Assessing the glial contribution to neuron loss

There is an increasing body of evidence that glial activation and/or dysfunction have an active role in the pathogenesis of several neurodegenerative disorders [1, 36, 40, 44]. Although not necessarily themselves instigators of neurodegeneration, any disruption of the normal support functions served by astrocytes or microglia may plausibly lead to neuron dysfunction or loss. The concept that glial cells can contribute to disease severity and accelerate its progression is perhaps most advanced in ALS and Parkinson disease, but it is becoming apparent this principle also extends to several lysosomal storage disorders. Lysosomal dysfunction and disrupted autophagy were thought to lead to the toxic role of astrocytes in multiple sulfatase deficiency, with the suggestion that a genetic defect in astrocytes alone was sufficient to drive the disease [11]. Microglial activation has also been suggested to contribute adversely to the pathogenesis in various LSDs including MPSI, MPSIIIB, Sandhoff disease and GM2 gangliosidosis [22, 33, 48], but the extent to which this happens in the NCLs is less clear.

The close correlation between the sites where early localized glial activation occurs in every form of NCL, and the extent of subsequent neuron loss [9, 34], has always begged the question of whether these events may be mechanistically related to one another. So far, the role of glia in these disorders has been investigated in most depth in CLN3 disease, and there is recent evidence for a negative influence of Cln3 deficient microglia [51] and astrocytes [4] upon neurons. We have recently extended these observations, using very similar primary culture and co-culture systems used in the present study [35]. This approach revealed a variety of defects in the biology of *Cln3*$^{-/-}$ microglia and astrocytes, and when grown in co-cultures, these dysfunctional glia harmed healthy neurons and resulted in the death of *Cln3*$^{-/-}$ neurons [35]. These data prompted us to investigate whether there was any influence of *Ppt1*$^{-/-}$ astrocytes and microglia upon neuron health in CLN1 disease.

Despite the marked differences in the phenotypes displayed in vitro by *Ppt1*$^{-/-}$ glia (this study) and *Cln3*$^{-/-}$ glia [35], a consistent feature is that astrocytes and microglia deficient in either gene exert a negative influence of upon neuronal morphology and survival. The relative contributions of either *Cln3*$^{-/-}$ astrocytes or microglia to these processes are yet to be defined, as are the underlying mechanisms, but the current study has provided the first insights into the negative influence of glia in CLN1 disease. Our data reveals that this adverse impact varies between cell types, with *Ppt1*$^{-/-}$ astrocytes only exerting a moderate influence on neuronal morphology (Fig. 8), whereas *Ppt1*$^{-/-}$ microglia are capable of inducing the death of neurons (Fig. 9), especially if these neurons are themselves Ppt1 deficient. These differences likely reflect how each of these cell-types are compromised by Ppt1 deficiency, and it will be important to investigate these mechanisms by transcriptional profiling. Whatever the underlying mechanisms may be, this adverse influence appeared to be exacerbated when both *Ppt1*$^{-/-}$ astrocytes and *Ppt1*$^{-/-}$ microglia were present (Fig. 11). This is consistent with recent data suggesting that under certain conditions microglia may prime astrocytes to become neurotoxic [25]. However, the specific factors implicated in this priming mechanism were not all upregulated by *Ppt1*$^{-/-}$ microglia, and it may be that different mechanisms operate in CLN1 disease. However, data from our co-cultures of *Ppt1*$^{-/-}$ astrocytes and *Ppt1*$^{-/-}$ microglia together (without any neurons present) suggest that these mutant astrocytes appear to drive these *Ppt1*$^{-/-}$ microglia to become further activated (Fig. 10). Indeed, it seems plausible that the toxic effect of *Ppt1*$^{-/-}$ microglia (Fig. 9) may be exacerbated by the consequence of impaired *Ppt1*$^{-/-}$ astrocyte function and/or survival to create conditions that promote neuron loss, perhaps via the increased release of IL-1β from microglia (Additional file 1: Figure S1), but this awaits experimental verification. Previous data had suggested astrocyte activation was beneficial in this disorder (Macauley et al., 2011), since disease progression was accelerated in *GFAP*$^{-/-}$/*Vimentin*$^{-/-}$/*Ppt1*$^{-/-}$ mice in which glial activation is genetically suppressed. This is in marked contrast to data from our co-culture systems that suggest a negative influence of *Ppt1*$^{-/-}$ astrocytes upon neuron health (Figs. 8, 11), and it will be essential to validate these effects in vivo.

In order to address such issues, we are already generating mice in which Ppt1 can be inactivated in a cell type-specific manner. Such approaches would normally be complicated by the release of Ppt1 enzyme from genetically unmodified cells to cross correct the cell type in which Ppt1 has been inactivated. However, the generation of mice expressing a biologically active membrane tethered Ppt1 enzyme that is not secreted [42] provides important proof of principle for this strategy, which will be employed in our future studies. Nevertheless, this in vitro study has provided novel evidence for a greater extent of dysfunction in *Ppt1*$^{-/-}$ glial cells than was previously anticipated and highlights the need for targeting glial cells in developing therapies for CLN1 disease. Gene therapy using newer generations of adeno-associated viruses has shown increasing efficacy, especially if these are targeted simultaneous to the brain and spinal cord [43]. These vectors predominantly transduce neurons, and these would secret Ppt1 enzyme to cross correct Ppt1 deficient glia within the CNS and correct their defects [10, 15, 45]. Regardless of such cross-correction, the administration of anti-inflammatory compounds to *Ppt1*$^{-/-}$ mice provides additional benefit above gene therapy alone [28], and greater therapeutic efficacy may yet be provided by testing other drugs of this type. In a clinical setting, such neuroimmunomodulatory or anti-inflammatory approaches may be of use either before gene therapy can be administered or as an adjunct to gene therapy.

The NCLs have been grouped together traditionally on the basis of certain broadly similar clinical and pathological themes, but as new data emerge it is becoming evident that this is an oversimplification. It is clear that cultured astrocytes and microglia generated from CLN1 (this study) and CLN3 mice [35] are dysfunctional and exert a negative influence upon neuron health and may ultimately contribute to neurodegeneration. However, as the CLN1 data in this study reveal, the nature of these events differs markedly between these two major forms of NCL. Perhaps this is not surprising given that these two forms of NCL are caused by mutations in genes that encode such different types of proteins, one a lysosomal hydrolase (CLN1/PPT1) and the other a transmembrane protein whose function remains obscure (CLN3). Nevertheless, these findings have direct implications for our understanding of the cellular pathogenesis of these disorders and in this respect, it will be important to extend these studies to other forms of NCL.

Additional file

Additional file 1: Figure S1. Expression of Interleukin-1β is increased following stimulation of Ppt1 deficient (*Ppt1*$^{-/-}$) microglia. Supernatant was collected from wild type (WT) and *Ppt1*$^{-/-}$ microglial cultures kept under basal and stimulated conditions for 24 h. Release of cytokines linked to oxidative stress was assessed using an ELISA kit (Signosis) and calculating changes in expression between basal and stimulated conditions. Secretion of Tumor necrosis factor α (TNFα), Transforming Growth Factor β (TGFβ), Monocyte Chemoattractant Protein-1 (MCP-1), Interleukin (IL)-1α, IL-6, IL-10 and IL-12 was not statistically significant between WT and *Ppt1*$^{-/-}$ microglia, however secretion of IL-1β was significantly higher in *Ppt1*$^{-/-}$ cultures. (Data shown as Mean ± SEM using a t-test, $n = 3$). (TIF 2444 kb)

Acknowledgements
Prof. Tammy Kielian, Dr. Jill Weimer, Dr. Alison Barnwell, Dr. Allison Najafi and Dr. Hemanth Nelvagal are thanked for their very useful comments upon the manuscript.

Funding
This study was supported by NIH NINDS 043205 to MSS and JDC, Saoirse Foundation and Irish Health Research Board to JDC and BPW, a Medical Research Council DTA studentship to JL, the Batten Disease Support and Research Association (USA), the Batten Disease Family Association (UK), The Natalie Fund and the Bletsoe Family. Prof. Tammy Kielian, Dr. Jill Weimer, Dr. Alison Barnwell, Dr. Allison Najafi and Dr. Hemanth Nelvagal are thanked for their very useful comments upon the manuscript.

Authors' contributions
The study was designed and supervised by JDC and BPW, with the input of all the authors. JL performed all aspects of the tissue culture experiments, and analysed these data; the microglial cytokine assays were performed with and supervised by JMP; the calcium imaging experiments were performed by JL and LJH and supervised by EL-E; MSS provided the Ppt1 deficient mice, and advice on their maintainence and breeding. The manuscript was written by JDC, BPW, and JL with input from all the authors, who approved the final version of the manuscript.

Competing interests
The authors declare that they have no competing interests.

Author details
[1]Department of Basic and Clinical Neuroscience, King's College London, Institute of Psychiatry, Psychology & Neuroscience, Maurice Wohl Clinical Neuroscience Institute, 5 Cutcombe Road, London SE5 9RX, UK. [2]School of Biosciences, Cardiff University, Museum Avenue, Cardiff CF10 3AX, UK. [3]Department of Neuroinflammation, Institute of Neurology, University College London, 1 Wakefield Street, London WC1N 1PK, UK. [4]Departments of Medicine and Genetics, Washington University School of Medicine, 660 South Euclid Avenue, St. Louis, MO 63110, USA. [5]Department of Pediatrics, Los Angeles Biomedical Research Institute at Harbor-UCLA Medical Center, and David Geffen School of Medicine UCLA, 1124 West Carson Street, Hanley Hardison Building, Torrance, CA 90502, USA. [6]Department of Pediatrics, Washington University School of Medicine, Campus Box 8208, 660 S Euclid Avenue, St. Louis, MO 63110, USA.

References

1. Benarroch EE (2013) Microglia. Neurology 81(1079):LP-1088
2. Bennett MJ, Rakheja D (2013) The neuronal ceroid-lipofuscinoses. Dev Disabil Res Rev 17:254–259
3. Bible E, Gupta P, Hofmann SL, Cooper JD (2004) Regional and cellular neuropathology in the palmitoyl protein thioesterase-1 null mutant mouse model of infantile neuronal ceroid lipofuscinosis. Neurobiol Dis 16:346–359
4. Burkovetskaya M, Karpuk N, Xiong J, Bosch M, Boska MD, Takeuchi H, Suzumura A, Kielian T (2014) Evidence for aberrant astrocyte Hemichannel activity in juvenile neuronal ceroid Lipofuscinosis (JNCL). PLoS One 9:e95023
5. Cárcel-Trullols J, Kovács AD, Pearce DA (2015) Cell biology of the NCL proteins: what they do and don't do. Biochim Biophys Acta - Mol Basis Dis 1852:2242–2255
6. Christensen KA, Myers JT, Swanson JA (2002) pH-dependent regulation of lysosomal calcium in macrophages. J Cell Sci 115:599–607
7. Colton CA, Chernyshev ON, Gilbert DL, Vitek MP (2000) Microglial contribution to oxidative stress in Alzheimer's disease. Ann N Y Acad Sci 899:292–307
8. Cooper JD (2010) The neuronal ceroid lipofuscinoses: the same, but different? Biochem Soc Trans 38:1448–1452
9. Cooper JD, Tarczyluk MA, Nelvagal HR (2015) Towards a new understanding of NCL pathogenesis. Biochim Biophys Acta - Mol Basis Dis 1852:2256–2261
10. Dayton RD, Wang DB, Klein RL (2012) The advent of AAV9 expands applications for brain and spinal cord gene delivery. Expert Opin Biol Ther 12:757–766
11. Di Malta C, Fryer JD, Settembre C, Ballabio A (2012) PNAS plus: astrocyte dysfunction triggers neurodegeneration in a lysosomal storage disorder. Proc Natl Acad Sci 109:E2334–E2342
12. Dotti G, Sullivan A, Biology C, College AM, York N (1988) The establishment of polarity by hippocampal
13. Fan R, Pan T, Zhu A-L, Zhang M-H (2017) Anti-inflammatory and anti-arthritic properties of naringenin via attenuation of NF-κB and activation of the heme oxygenase HO -1/related factor 2 pathway. Pharmacol Reports 69:1021–1029
14. Fletcher TL, Cameron P, De Camilli P, Banker G (1991) The distribution of synapsin I and synaptophysin in hippocampal neurons developing in culture. J Neurosci 11:1617–1626
15. Foust KD, Nurre E, Montgomery CL, Hernandez A, Chan CM, Kaspar BK (2009) Intravascular AAV9 preferentially targets neonatal-neurons and adult-astrocytes in CNS. Nat Biotechnol 27:59–65
16. Griffey MA, Wozniak D, Wong M, Bible E, Johnson K, Rothman SM, Wentz AE, Cooper JD, Sands MS (2006) CNS-directed AAV2-mediated gene therapy ameliorates functional deficits in a murine model of infantile neuronal ceroid lipofuscinosis. Mol Ther 13:538–547
17. Gupta P, Soyombo AA, Atashband A, Wisniewski KE, Shelton JM, Richardson JA, Hammer RE, Hofmann SL (2001) Disruption of PPT1 or PPT2 causes neuronal ceroid lipofuscinosis in knockout mice. Proc Natl Acad Sci U S A 98:13566–13571
18. Haltia M, Rapola J, Santavuori P, Keranen A (1973) Infantile type of so-called neuronal ceroid-Lipofuscinosis part 2. Morphological and Biochemical Studies 18:269–285
19. Hawkins-Salsbury JA, Cooper JD, Sands MS (2013) Pathogenesis and therapies for infantile neuronal ceroid lipofuscinosis (infantile CLN1 disease). Biochim Biophys Acta - Mol Basis Dis 1832:1906–1909
20. Herculano-Houzel S (2014) The glia/neuron ratio: how it varies uniformly across brain structures and species and what that means for brain physiology and evolution. Glia 62:1377–1391

21. Hofmann S, Atashband A, Cho S, Das A, Gupta P, Lu J (2002) Neuronal ceroid Lipofuscinoses caused by defects in soluble lysosomal enzymes (CLN1 and CLN2). Curr Mol Med 2:423–437
22. Jeyakumar M, Thomas R, Elliot-Smith E, Smith DA, Van der Spoel AC, D'Azzo A, Perry VH, Butters TD, Dwek RA, Platt FM (2003) Central nervous system inflammation is a hallmark of pathogenesis in mouse models of GM1 and GM2 gangliosidosis. Brain 126:974–987
23. Kielar C, Maddox L, Bible E, Pontikis CC, Macauley SL, Griffey MA, Wong M, Sands MS, Cooper JD (2007) Successive neuron loss in the thalamus and cortex in a mouse model of infantile neuronal ceroid lipofuscinosis. Neurobiol Dis 25:150–162
24. Kielar C, Wishart TM, Palmer A, Dihanich S, Wong AM, Macauley SL, Chan CH, Sands MS, Pearce DA, Cooper JD, Gillingwater TH (2009) Molecular correlates of axonal and synaptic pathology in mouse models of batten disease. Hum Mol Genet 18:4066–4080
25. Liddelow SA, Guttenplan KA, Clarke LE, Bennett FC, Bohlen CJ, Schirmer L, Bennett ML, Münch AE, Chung S, Peterson TC, Wilton DK, Frouin A, Napier BA (2017) Neurotoxic reactive astrocytes are induced by activated microglia. Nature 541:481–487
26. Lloyd-Evans E, Morgan AJ, He X, Smith DA, Elliot-Smith E, Sillence DJ, Churchill GC, Schuchman EH, Galione A, Platt FM (2008) Niemann-pick disease type C1 is a sphingosine storage disease that causes deregulation of lysosomal calcium. Nat Med 14:1247–1255
27. Macauley SL, Roberts MS, Wong AM, McSloy F, Reddy AS, Cooper JD, Sands MS (2012) Synergistic effects of central nervous system-directed gene therapy and bone marrow transplantation in the murine model of infantile neuronal ceroid lipofuscinosis. Ann Neurol 71:797–804
28. Macauley SL, Wong AMS, Shyng C, Augner DP, Dearborn JT, Pearse Y, Roberts MS, Fowler SC, Cooper JD, Watterson DM, Sands MS (2014) An anti-neuroinflammatory that targets dysregulated glia enhances the efficacy of CNS-directed gene therapy in murine infantile neuronal ceroid lipofuscinosis. J Neurosci 34:13077–13082
29. Macauley SL, Wozniak DF, Kielar C, Tan Y, Cooper JD, Sands MS (2009) Cerebellar pathology and motor deficits in the palmitoyl protein thioesterase 1-deficient mouse. Exp Neurol 217:124–135
30. Markham D (2017) Cerliponase alfa: first global approval. Drugs 77:1247–1249
31. McCarthy KD, de Vellis J (1980) Preparation of separate astroglial and oligodendroglial cell cultures from rat cerebral tissue. J Cell Biol 85:890–902
32. Mole SE, Cotman SL (2015) Genetics of the neuronal ceroid lipofuscinoses (batten disease). Biochim Biophys Acta - Mol Basis Dis 1852:2237–2241
33. Ohmi K, Greenberg DS, Rajavel KS, Ryazantsev S, Li HH, Neufeld EF (2003) Activated microglia in cortex of mouse models of mucopolysaccharidoses I and IIIB. Proc Natl Acad Sci U S A 100:1902–1907
34. Palmer DN, Barry LA, Tyynelä J, Cooper JD (2013) NCL disease mechanisms. Biochim Biophys Acta - Mol Basis Dis 1832:1882–1893
35. Parviainen L, Dihanich S, Anderson GW, Wong AM, Brooks HR, Abeti R, Rezaie P, Lalli G, Pope S, Heales SJ, Mitchison HM, Williams BP, Cooper JD (2017) Glial cells are functionally impaired in juvenile neuronal ceroid lipofuscinosis and detrimental to neurons. Acta Neuropathol Commun 5:74
36. Pekny M, Pekna M, Messing A, Steinhäuser C, Lee J-M, Parpura V, Hol EM, Sofroniew MV, Verkhratsky A (2016) Astrocytes: a central element in neurological diseases. Acta Neuropathol 131:323–345
37. Pontikis CC, Cella CV, Parihar N, Lim MJ, Chakrabarti S, Mitchison HM, Mobley WC, Rezaie P, Pearce DA, Cooper JD (2004) Late onset neurodegeneration in the Cln3 −/− mouse model of juvenile neuronal ceroid lipofuscinosis is preceded by low level glial activation. Brain Res 1023:231–242
38. Pontikis CC, Cotman SL, MacDonald ME, Cooper JD (2005) Thalamocortical neuron loss and localized astrocytosis in the Cln3Deltaex7/8 knock-in mouse model of batten disease. Neurobiol Dis 20:823–836
39. Ross JM, Öberg J, Brené S, Coppotelli G, Terzioglu M, Pernold K, Goiny M, Sitnikov R, Kehr J, Trifunovic A, Larsson N-G, Hoffer BJ, Olson L (2010) High brain lactate is a hallmark of aging and caused by a shift in the lactate dehydrogenase a/B ratio. Proc Natl Acad Sci U S A 107:20087–20092
40. Rossi D, Volterra A (2009) Astrocytic dysfunction: insights on the role in neurodegeneration. Brain Res Bull 80:224–232
41. Shen LH, Li Y, Chen J, Zacharek A, Gao Q, Kapke A, Lu M, Raginski K, Vanguri P, Smith A, Chopp M (2007) Therapeutic benefit of bone marrow stromal cells administered 1 month after stroke. J Cereb Blood Flow Metab 27:6–13
42. Shyng C, Macauley SL, Dearborn JT, Sands MS (2017) Widespread Expression of a Membrane-Tethered Version of the Soluble Lysosomal Enzyme Palmitoyl Protein Thioesterase-1. JIMD Rep. 36:85–92.
43. Shyng C, Nelvagal HR, Dearborn JT, Tyynelä J, Schmidt RE, Sands MS, Cooper JD (2017) Synergistic effects of treating the spinal cord and brain in CLN1 disease. Proc Natl Acad Sci 114(29): E5920–E5929
44. Sofroniew MV, Vinters HV (2010) Astrocytes: biology and pathology. Acta Neuropathol 119:7–35
45. Swain GP, Prociuk M, Bagel JH, O'Donnell P, Berger K, Drobatz K, Gurda BL, Haskins ME, Sands MS, Vite CH (2014) Adeno-associated virus serotypes 9 and rh10 mediate strong neuronal transduction of the dog brain. Gene Ther 21:28–36
46. Threadgill R, Bobb K, Ghosh A (1997) Regulation of dendritic growth and remodeling by rho, Rac, and Cdc42. Neuron 19:625–624
47. Virmani T, Gupta P, Liu X, Kavalali ET, Hofmann SL (2005) Progressively reduced synaptic vesicle pool size in cultured neurons derived from neuronal ceroid lipofuscinosis-1 knockout mice. Neurobiol Dis 20:314–323
48. Wada R, Tifft CJ, Proia RL (2000) Microglial activation precedes acute neurodegeneration in Sandhoff disease and is suppressed by bone marrow transplantation. Proc Natl Acad Sci U S A 97:10954–10959
49. Wei H, Kim SJ, Zhang Z, Tsai PC, Wisniewski KR, Mukherjee AB (2008) ER and oxidative stresses are common mediators of apoptosis in both neurodegenerative and non-neurodegenerative lysosomal storage disorders and are alleviated by chemical chaperones. Hum Mol Genet 17:469–477
50. Weimer J, Kriscenski-Perry E, Elshatory Y, Pearce D (2002) The neuronal ceroid lipofuscinoses: mutations in different proteins result in similar disease. NeuroMolecular Med 1:111–124
51. Xiong J, Kielian T (2013) Microglia in juvenile neuronal ceroid lipofuscinosis are primed toward a pro-inflammatory phenotype. J Neurochem 127:245–258

Neutrophil elastase inhibition effectively rescued angiopoietin-1 decrease and inhibits glial scar after spinal cord injury

Hemant Kumar[1], Hyemin Choi[1], Min-Jae Jo[1], Hari Prasad Joshi[1], Manjunatha Muttigi[2], Dario Bonanomi[3], Sung Bum Kim[4], Eunmi Ban[5], Aeri Kim[5], Soo-Hong Lee[2], Kyoung-Tae Kim[6,7], Seil Sohn[1], Xiang Zeng[8*] and Inbo Han[1*]

Abstract

After spinal cord injury (SCI), neutrophil elastase (NE) released at injury site disrupts vascular endothelium integrity and stabilization. Angiopoietins (ANGPTs) are vascular growth factors that play an important role in vascular stabilization. We hypothesized that neutrophil elastase is one of the key determinants of vascular endothelium disruption/destabilization and affects angiopoietins expression after spinal cord injury. To test this, tubule formation and angiopoietins expression were assessed in endothelial cells exposed to different concentrations of recombinant neutropil elastase. Then, the expression of angiopoietin-1, angiopoietin-2, and neutrophil elastase was determined at 3 h and at 1, 3, 5, 7, 14, 21, and 28 days in a clinically relevant model of moderate compression (35 g for 5 min at T10) spinal cord injury. A dichotomy between the levels of angiopoietin-1 and angiopoietin-2 was observed; thus, we utilized a specific neutrophil elastase inhibitor (sivelestat sodium; 30 mg/kg, i.p., b.i.d.) after spinal cord injury. The expression levels of neutropil elastase and angiopoietin-2 increased, and that of angiopoietin-1 decreased after spinal cord injury in rats. The sivelestat regimen, optimized via a pharmacokinetics study, had potent effects on vascular stabilization by upregulating angiopoietin-1 via the AKT pathway and preventing tight junction protein degradation. Moreover, sivelestat attenuated the levels of inflammatory cytokines and chemokines after spinal cord injury and hence subsequently alleviated secondary damage observed as a reduction in glial scar formation and the promotion of blood vessel formation and stabilization. As a result, hindlimb locomotor function significantly recovered in the sivelestat-treated animals as determined by the Basso, Beattie, and Bresnahan scale and footprint analyses. Furthermore, sivelestat treatment attenuated neuropathic pain as assessed by responses to von Frey filaments after spinal cord injury. Thus, our result suggests that inhibiting neutropil elastase by administration of sivelestat is a promising therapeutic strategy to inhibit glial scar and promote functional recovery by upregulating angiopoietin-1 after spinal cord injury.

Keywords: Neutrophil elastase, Spinal cord injury, Glial scar, Angiopoietins, Functional recovery, Neuropathic pain

Introduction

Spinal cord injury (SCI) is a clinically devastating condition that can cause either temporary or permanent disability in young adults [49, 76]. SCI pathology is multifaceted. It involves several major biological cascades [48, 54, 55] and results in expeditious and enduring changes to the structure and function of microvessels [27, 56, 85], such as a loss of structural organization and microcirculation, a disruption of the blood-spinal cord barrier (BSCB), and endothelial cell (EC) and vascular remodelling [49, 85]. Vascular damage following SCI augments secondary damage, and vascular protection or the maintenance of vascular integrity mitigates this damage [30, 35, 49]. ECs participate in all facets of vascular homeostasis and play a variety of critical roles in the control of vascular functions, including in thrombosis, inflammation, and vascular wall

* Correspondence: zengx33@mail.sysu.edu.cn; hanib@cha.ac.kr
[8]Department of Histology and Embryology, Zhongshan School of Medicine, Sun Yat-sen University, Guangzhou 510080, Guangdong Province, China
[1]Department of Neurosurgery, CHA University School of Medicine, CHA Bundang Medical Center, Seongnam-si, Gyeonggi-do 13496, Republic of Korea
Full list of author information is available at the end of the article

remodeling. The death of ECs disengages the vascular network, and ischemia results in apoptosis and cell death of central nervous system (CNS) cells due to the lack of blood supply [23, 54, 55].

Recovery from SCI is preceded by angiogenesis, the extent of which correlates with neural regeneration, suggesting that angiogenesis may play a significant role in repair. Angiopoietins (ANGPTs) are vascular growth factors involved in blood vessel formation and maturation, as well as in EC survival [31, 86], and are critical regulators of vascular functions in the brain [46, 98] and spinal cord [30, 35, 74]. There are four members of the ANGPT family: ANGPT-1, – 2, and – 4 are expressed in humans, and ANGPT-3, an ortholog of ANGPT-4, is expressed in mice [31, 92]. ANGPT-1 and ANGPT-2 are released from ECs, with ANGPT-1 constitutively expressed in normal CNS vasculature [61] and ANGPT-2 weakly expressed under homeostatic conditions and increased during hypoxia, inflammation, and vascular remodeling [86]. ANGPT-1 exerts anti-inflammatory effects, reduces vessel permeability, and protects against plasma leakage in the adult vasculature [28, 87]. These effects are opposed by the actions of ANGPT-2. Therefore, it is not surprising that ANGPT-2 can antagonize the benefits to vascular integrity from endogenous or exogenous ANGPT-1 after SCI.

Neuropathological changes in spinal cord tissue result from acute inflammatory reactions that can involve elastases derived from neutrophils. Neutrophils also play a critical role in the initial events in demyelinating neuroinflammatory diseases and are intimately linked with the status of the blood-brain barrier/BSCB [7]. For example, neutrophils release a destructive proteolytic enzyme called neutrophil elastase (NE) [44]. At the vascular interface, NE induces cellular damage and dysfunction, degradation of the extracellular matrix, and pathways leading to cell death [44]. In response to proinflammatory stimuli, NE regulates the adhesion of leukocytes, clears their path for diapedesis/transmigration [62, 83, 97], and mediates the degradation of endothelial junction proteins [34, 39]. Furthermore, NE can induce apoptosis of ECs [96] and has a broad substrate specificity [69, 89]. These complex secondary pathomechanisms are responsible for extending spinal cord damage into previously uncompromised segments [66, 67, 73]. NE can induce vascular damage leading to spinal cord ischemia [84] and is also a determinant of long-term functional recovery after traumatic brain injury [81].

We hypothesized that NE might be a key determinant for the disruption/destabilization of the vascular endothelium and alter ANGPT expression after SCI. To test this, we utilized a selective NE inhibitor (sivelestat sodium; 30 mg/kg, i.p.,b.i.d.) in a rat model of moderate compression (35 g for 5 min at T10) SCI. Sivelestat attenuates NE-induced pathologies and is approved for use in patients with acute lung injury in Japan and the Republic of Korea [5, 90], and attenuates the perioperative inflammatory response in pediatric patients undergoing cardiopulmonary bypass surgery [38]. Moreover, administration of sivelestat attenuated the ischemia [41], and the chemo-attractant mRNA and protein [88] in an experimental model of SCI. However, the effect of NE inhibition on the glial scar, secondary damage, vascular stabilization, ANGPTs, ECs survival and angiogenesis after SCI remains to be determined. In the current study, we ascertain the role of NE with ANGPTs after SCI and suggest that NE inhibition endows multidimensional therapeutic strategy in tissue protection and glial scar inhibition in treating SCI.

Material and methods

Cell culture and treatment

In an attempt to understand the biological role of NE in ECs, we used HUVEC (ATCC) cells. HUVECs were cultured in fully supplemented endothelial growth medium as per the manufacturer instructions. Recombinant human NE protein (R&D Systems, Minneapolis, USA) was activated with 50 μg/ml Cathepsin C in assay buffer before use as per manufacturer instruction and was used at a functional concentration of 100 ng/ml, 250 ng/ml and 500 ng/ml and 1000 ng/ml, in ECs. Corning matrigel matrix was used for the tubule formation assay as per the manufacturer recommendations. Briefly, matrigel matrix was polymerized at 37 °C in a 24 well plate and HUVEC cells (passage 3) at a seeding density of 1.2×10^5. The EGM-2 bullet kit medium were supplemented with human NE at a concentration of 100 ng/ml (group 2), 250 ng/ml (group 3), 500 ng/ml (group 4), and 1000 ng/ml (group 5). HUVEC supplemented with the only medium served as control (group 1). After 18 h, capillary-like tubules was stained with calcein AM fluorescent dye on the matrilgel. Images were randomly acquired using Cytation 3 Cell Imaging Multi-Mode Reader (Biotek Instruments,Inc., Winooski, VT, USA).

Subjects and surgical procedures

Total 146 adult female Sprague-Dawley (SD) rats were used in the study. Rats (220–240 g) for this study were purchased from Orient Bio Inc. (Seongnam, Korea), housed in a facility at 55–65% humidity and controlled temperature of 24 ± 3 °C with light / dark cycle of 12 h, and had free access to food and water. All animal procedures were performed according to the approved protocol by the Institutional Animal Care and Use Committee (IACUC) of CHA University (IACUC160076) and Principles of laboratory animal care [63]. The animals were anesthetized with Zoletil® (50 mg/kg, Virbac Laboratories, France) / Rompun® (10 mg/kg, Bayer, Korea) solution administered intraperitoneally. Complete anesthesia was assessed using hindlimb withdrawal in response to a noxious foot pinch.

After skin preparation and precise positioning of anesthetized rats, a laminectomy was performed to expose T10 spinal cord. The vertebral column was supported and stabilized by Allis clamps at T8 and T12 spinous processes as described previously [48, 75]. A metal impounder (35 g × 5 min) was then gently applied on T10 dura, resulting in moderate standing weight compression. Following compression injury, the surgical site was closed by suturing the muscle and fascia and suturing the skin; followed by external povidone-iodine application. Animals were kept on a heating pad to maintain body temperature, and then 5 mL of 0.9% sterile saline injected subcutaneously. Manual bladder expression of urine was performed twice daily until a bladder reflex was established.

Drugs and treatments

Sivelestat sodium (Dong-A-pharma, Seoul, Korea) was dissolved in distilled water and administered twice daily intraperitoneally at the dose of 30 mg/kg 1 h after SCI; Animals were sacrificed at different time point's day post-injury (DPI) after vehicle or sivelestat treatment. The total number of injection(s) varied based on the efficacy parameter and time points. DPI-1 (2 injections of sivelestat were given), DPI-7 (14 injections of sivelestat were given), DPI-14 (28 injections of sivelestat were given), DPI-28 (28 injections of sivelestat were given till day 14 and animals were observed till day 28 for behavioral studies and efficacy experiments).

Pharmacokinetic study

We used SCI-injured animals for a pharmacokinetic study to correlate PK-PD. Sivelestat was treated at 1 h after SCI, The blood, brain and spinal cord samples were collected at 15 min, 30 min, 60 min, 90 min, 120 min, 150 min, 180 min after a single dose of sivelestat. For 1D, 7D, and 14 D samples, the blood, brain and spinal cord were collected after two, 14 and 28 doses of sivelestat respectively. Plasma was separated from blood using centrifugation and stored at − 80 °C until analysis. Spinal cord tissues were collected and washed with PBS and homogenized in PBS. Analysis of sivelestat in samples were performed as described in the Additional file 1.

Behavioral assessment

Footprint analysis

Gait behavior and motor coordination were evaluated on 1, 7, 14, 21 and 28 days following injury, and after sivelestat treatment using the manual method as described previously with some modifications [80]. Right fore and hind paws, left fore and hind limb was painted with dyes of different colors and animals were placed over an absorbent paper surrounded by cage border. The animals were encouraged to walk in a straight line by putting a clue at the finish line. The footprint pattern was then digitalized and representative pictures were shown to assess coordination.

Hindlimb locomotor score

Hindlimb motor function was evaluated using the open-field Basso, Beattie, and Bresnahan (BBB) [10] locomotor test on 1, 7, 14, 21 and 28 days following injury, and after sivelestat treatment. The animals hindlimb locomotor score were evaluated by two experienced investigators who were blinded to treatment group.

Test for nociception

Nociception was checked using Von Frey filaments (Bioseb) as per the reported method [17]. Animals were acclimatized in rat chambers (Ugo Basile), and filaments probing were done when the animals were calm and not moving. The simplified up-down method (started with 2 g) was used to determine the mechanosensitivity/paw withdrawal threshold (PWT) with Von Frey filaments on 1, 7, 14, 21 and 28 days following injury, and after sivelestat treatment. A positive response includes flinching, licking, vocalization, or overt behavioral cue corresponding to discomfort. A total of five stimuli per test were recorded and average of three readings (lowest and highest removed) was used to determine the average PWT in Sham and after injury or treatment with sivelestat.

qRT-PCR

Quantitative real-time PCR was carried out using an SYBR Green Master Mix and the mRNA detection was analyzed using an ABI StepOne Real-time PCR System (Applied Biosystems, Foster City, CA, USA). Primer sequences for the genes of interest and the reference gene 18S or GAPDH were as given in Additional file 2: Table S1: Typical profile times were the initial step, 95 °C for 10 min followed by a second phase at 95 °C for 15 s and 60 °C for 30 s for 40 cycles with a melting curve analysis. The target mRNA level was normalized with the level of the 18S or GAPDH and compared with the control. Data were analyzed using the ΔΔCT method.

Western blot analysis

Spinal cord tissues were collected and washed with PBS, placed at 4° C, and homogenized using T 25 digital homogenizer (IKA, Seoul, Korea) in lysis buffer (1× RIPA lysis buffer) and then finally passed through a $31^{1/2}$ gauge syringe needle and centrifuged at 14,000 rpm at 4°C for 15 min. Protein concentration was determined in supernatants using Bio-Rad DC Protein Assay (Hercules, CA, USA). Equal amounts (40 μg) of protein were separated electrophoretically by 10% SDS-PAGE electrophoresis, and the resolved proteins were transferred to PVDF membranes (Millipore, Bedford, MA, USA). The membranes were incubated for 1 h with 5% non-fat skim milk prepared in TBS buffer to block nonspecific binding. The membranes were then incubated overnight with primary antibodies to ANGPT-1 (1:10000, Abcam, Cambridge, UK), ANGPT-2 (1:1000, Abcam), NE (1;1000), ZO-1 (1:500, Invitrogen, California, USA), AKT (1:1000, Cell Signaling Technology, Danvers, MA, USA), pAKT (1:1000, Cell Signaling Technology,) LC3B (1:1000, Cell Signaling Technology,), and Actin (1:10000, ABM). After 1-h incubation with corresponding secondary antibodies, The blots were visualized with a PowerOpti-ECL (Animal Genetics Inc., Gainesville, FL, USA) detection system, according to the recommended procedure. Immunoreactivity was detected using the BIORAD ChemiDoc™ XRS.

Immunohistochemistry and immunofluorescence

At 1, 7, 14 and 28 days after compression of the spinal cord at T10, animals were anesthetized with mixture of Zoletil® (50 mg/kg, Virbac Laboratories, France) / Rompun® (10 mg/kg, Bayer, Korea) solution administered intraperitoneally and perfused with 0.9% saline followed by 4% paraformaldehyde for tissue fixation. The spinal cord at the compression site was removed, and immersed in 4% paraformaldehyde for 1 day, and then embedded in paraffin, sectioned at 5 or 10 μm, dewaxed, and stained with antibodies against ANGPT-1 (1:500, Abcam, Cambridge, UK), ANGPT-2 (1:200, Abcam), N.E. (1;50, Abcam), GFAP (1:1000, Abcam), GFAP (1:200, Sigma), Iba-1 (1:200, Abcam) ZO-1 (1:50, Invitrogen, California, USA), Occludin (1:50, Invitrogen), RECA (1:100, Abcam), TGF-β1 (1:100, Abcam), NG-2 (1:500, Millipore), α-SMA (1:500, Abcam)-PECAM(1:100, Abcam), Caspase-3 (1:50, Abcam), CD68 (1:200, Abcam), Arginase (1:100, Abcam), Laminin (1:50, Abcam), BDNF (1:200, Alomone Lab), NT-3 (1:200, Alomone Lab), NT-4 (1:200, Alomone Lab), vWF (1:50, Abcam), NF-200 (1:1000, Abcam), Tuj-1 (1:200, Abcam), VEGF (1:100, Abcam) TNF-α (1:50, Abcam), Fibronectin (1:50, Abcam), RECA-1 (1:100, Abcam), IL-6 (1:50, Abcam). Secondary antibodies (1:200 or 1:500) were goat anti-rabbit Alexa Fluor® 488 (Abcam), goat anti-rabbit Alexa Fluor® 647 (Abcam), goat anti-rabbit Alexa Fluor® 568 (Abcam), goat anti-rabbit Alexa Fluor® 555 (Abcam), goat anti-mouse Alexa Fluor® 488 (Abcam), goat anti-mouse Alexa Fluor® 568 (Abcam), chicken-anti-goat 647 (Invitrogen).Following washing after secondary antibody incubation DAPI (1:500) was incubated for 10 min. Sections were mounted and examined using a fluorescence microscope (Zeiss, Oberkochen, Germany or Leica, Germany).

Statistical analysis

All data were analyzed using Graph Pad Prism ver. 5.01 (Graph Pad, Inc., La Jolla, CA, USA). All data are expressed as mean ± SEM. One-way ANOVA was performed and Dunnett's posthoc test was used to analyze the in-vitro data. The in-vivo PCR data was analyzed using One-way ANOVA followed by Tukey's test. The BBB scores were analyzed statistically with Kruskal-Wallis test at each time point. The nociception data were analyzed using Two-way ANOVA followed by Bonferroni test. P-values < 0.05 were considered statistically significant.

Results

NE suppresses capillary-like tubule formation and ANGPT expression in ECs, whereas proinflammatory factors differentially modulate ANGPT expression

We used various strategies to elucidate the effects of NE and inflammation on ANGPT expression in ECs. First, we determined the effects of recombinant human NE protein (100, 250, 500, and 1000 ng/ml) on human umbilical vein ECs (HUVECs); HUVECs treated with medium only served as the control. The results of a tubule formation assay showed a dose-dependent decrease in the tubule-covered area, total tube length, and total numbers of tubes, with significant effects observed at NE concentrations of 500 ng/ml and 1000 ng/ml (Fig. 1a–d).The expression of ANGPT-1 and ANGPT-2in HUVECs treated with NE for 24 h was determined by RT-PCR and immunocytochemistry. The expression of ANGPT-1 was dose-dependently decreased following the addition of NE (Fig. 1e and f), and ANGPT-2 expression decreased at higher doses (500 ng/ml and 1000 ng/ml) (Fig. 1g and h). Additionally, the effect of inflammation on ANGPT expression was assessed by treating HUVECs with lipopolysaccharide ([LPS] 2 μg/ml) and tumor necrosis factor alpha ([TNF-α] 100 ng/ml). ANGPT-1 expression increased at 0.5 and 3 h and decreased at 6, 9, and 12 h after the addition of LPS to the medium (Fig. 1i), whereas expression was significantly decreased by TNF-α at 3–12 h (Fig. 1j). By contrast, ANGPT-2 expression increased at 3, 6, and 9 h after the addition of LPS (Fig. 1k) and increased at 3 and 9 h after TNF-α was added (Fig. 1l). These results suggest that NE and inflammation differentially modulate the expression of ANGPTs in ECs.

SCI disrupts vascular endothelial integrity and alters NE and ANGPT expression

We characterized the time course of NE, ANGPT-1, and ANGPT-2 mRNA and protein expression at the epicenter of the damaged spinal cords in rats at 3 h and 1, 3, 5, 7, 14, 21, and 28 days after moderate compression injury (35 g for 5 min) (Fig. 2a).There was a significant increase in NE expression from 3 h to 5 days after SCI, with maximum expression observed at 1 day after SCI (Fig. 2b and ei). A dichotomy was observed between ANGPT-1 and ANGPT-2 expression patterns. The expression of ANGPT-1 was initially drastically reduced (the maximum decrease was observed 1 day after SCI when NE expression was maximal) and then increased at 7 days after SCI and remained elevated (Fig. 2c and eii), whereas the expression of ANGPT-2

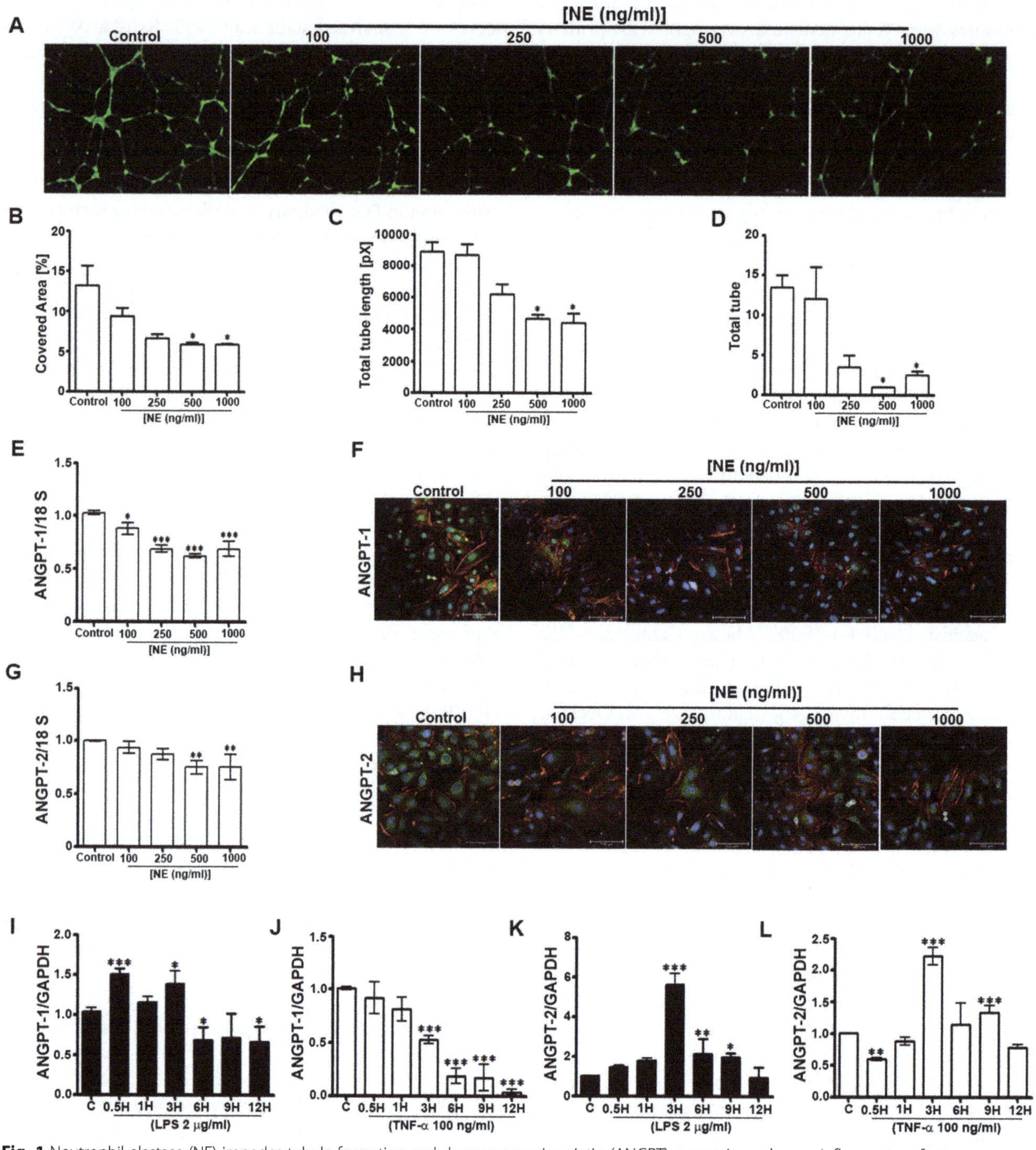

Fig. 1 Neutrophil elastase (NE) impedes tubule formation and decreases angiopoietin (ANGPT) expression, whereas inflammatory factors differentially modulate ANGPT expression, in human umbilical vein endothelial cells (HUVECs). A tubule formation assay was performed as described in the Methods section. Recombinant human NE was added at concentrations of 100, 250, 500, and 1000 ng/ml (HUVECs exposed only to medium served as the control) to determine tubule formation (**a**), the percentage of covered area (**b**), total tube length (**c**), and total numbers of tubes (**d**). **e** and **g** Total RNA was prepared from HUVECs exposed to various concentrations of NE for 24 h to determine the expression of *ANGPT1* and *ANGPT2*. **f** and **h** ANGPT-1 and ANGPT-2 immunocytochemistry was performed on fixed HUVECs as described in the Methods sections. **i–l** *ANGPT1* and *ANGPT2* mRNA expression was determined by real-time quantitative reverse transcription-polymerase chain reaction in HUVECs collected 0.5, 1, 3, 6, 9, and 12 h after treatment with lipopolysaccharide ([LPS] 2 μg/ml) and tumor necrosis factor alpha ([TNF-α] 100 ng/ml). 18S was used as the internal control. Data represent means ± SEMs (n = 2–3/group performed in triplicate). $^{*}p < 0.05$, $^{**}p < 0.01$, $^{***}p < 0.001$ vs. control

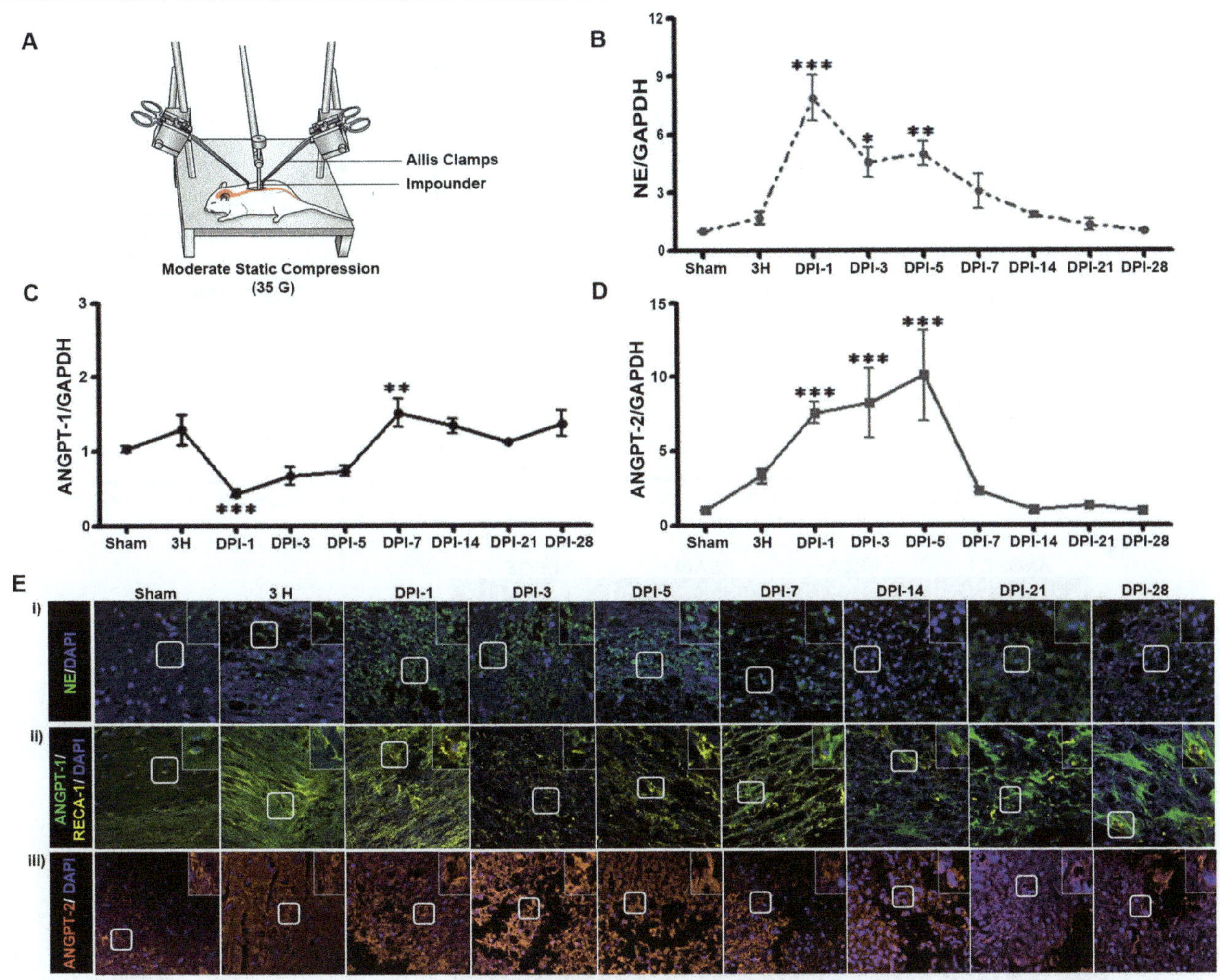

Fig. 2 Neutrophil elastase (NE) and angiopoietin-2 (ANGPT-2) expression increases and angiopoietin-1 (ANGPT-1) and rat endothelial cell antigen (RECA-1) expression decreases after spinal cord injury (SCI) at the epicenter of the damage. **a** Schematic showing SCI method. Total RNA was prepared from spinal cord tissues at the epicenter of the damage collected 3 h and 1, 3, 5, 7, 14, 21, and 28 days after SCI to determine the expression of NE (**b**), ANGPT-1 (**c**), and ANGPT-2 (**d**). **e** Representative images of immunohistochemistry performed on longitudinal sections for NE (i), ANGPT-1 and RECA-1(ii), and ANGPT-2 (iii) at different time points after SCI [3 fields/slide, $n = 2$–3/group (sham = 2, and injury = 3)]. GAPDH was used as internal controls for real-time quantitative reverse transcription–polymerase chain reaction. Data represent means ± S.E.M. [$n = 2$–3/group (sham = 2, and injury = 3) performed in triplicates]. *$p < 0.05$, **$p < 0.01$, ***$p < 0.001$ compared with Sham group

continuously increased through 5 days after SCI and unexpectedly decreased at 7 days, when the ANGPT-1 expression was increased, and consequently returned to normal (Fig. 2d and eiii). As ANGPTs are expressed primarily by ECs, we determined the integrity of the vascular endothelium using immunohistochemistry (IHC) with rat EC antigen ([RECA-1] Fig. 2eii), which revealed progressive damage after SCI. RECA-1-stained vessels were readily identified within the injured spinal cord.

Sivelestat increases ANGPT-1 and decreases ANGPT-2 and NE expression after SCI

The data above indicated that peak expression of NE occurred 1 day after SCI, accompanied by increased ANGPT-2 and decreased ANGPT-1 expression. Therefore, we determined the effect of inhibiting NE on ANGPT expression at DPI-1. One group of animals was treated with sivelestat (30 mg/kg, i.p., b.i.d.), a specific inhibitor of NE, and the concentrations in plasma, brain, and spinal cord were monitored over 14 days (Fig. 3a). Samples from sham, injured untreated, and injured sivelestat-treated (two doses) animals were prepared on DPI-1. Interestingly, sivelestat treatment prevented the SCI-induced decrease in ANGPT-1 expression (Fig. 3b–d) and attenuated the SCI-induced increase in ANGPT-2 at DPI-1 (Fig. 3e and f). Treatment with sivelestat also significantly reduced NE expression at DPI-1 (Fig. 3e and g). Additionally, the phosphorylation of

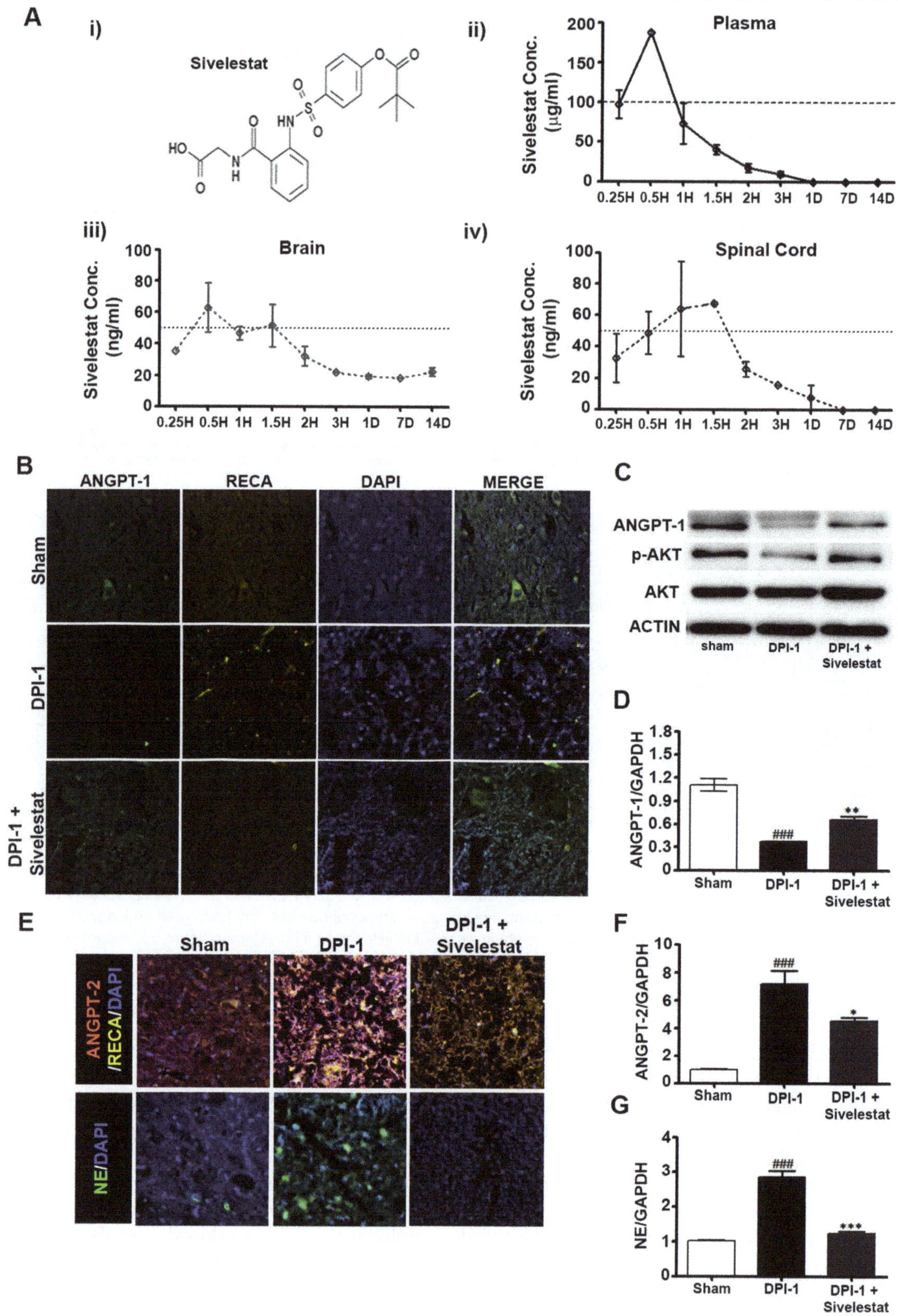

Fig. 3 (See legend on next page.)

(See figure on previous page.)
Fig. 3 Neutrophil elastase (NE) inhibition via sivelestat prevented the spinal cord injury (SCI)-induced modulation of angiopoietins (ANGPTs) and inhibited the expression of NE. **a** Molecular structure of sivelestat (i) and concentrations determined in plasma (ii), brain (iii), and spinal cord (iv) at several time points as described in the Methods section ($n = 2$/timepoint). Samples from sham, injured untreated, and injured sivelestat-treated animals were prepared DPI-1 as described in the Methods section. Representative images of immunohistochemistry for ANGPT-1 and rat endothelial cell antigen (RECA-1) (**b**) and ANGPT-2 and NE (**e**) at 1 day after SCI [3 fields/slide, $n = 2–3$/group (sham = 2, injury = 3 and sivelestat = 3)]. **c** Western blots of ANGPT-1, p-AKT, and AKT expression at 1 day after injury. Actin was used as internal controls for western blot [$n = 2–3$/group (sham = 2, injury = 3 and sivelestat = 3)]. Total RNA and spinal extracts from sham or injured untreated (Injury) or after sivelestat treatment were prepared 1 day after the injury. RT-PCR results of Ang-1 (**d**), Ang-2 (**f**) and N.E (**g**) expression 1 day after injury [$n = 2–3$/group (sham = 2, injury = 3 and sivelestat = 3) performed in triplicates]. GAPDH was used as internal controls for real-time quantitative reverse transcription–polymerase chain reaction. Data represent means ± S.E.M. $^{###}p < 0.001$ vs. sham group. $^{*}p < 0.05$, $^{**}p < 0.01$, $^{***}p < 0.001$ vs. Injury group

AKT (p-AKT) was reduced in the spinal cord following SCI, which was prevented by treatment with sivelestat (Fig. 3c).

NE inhibition attenuates the expression of inflammatory cytokines and chemokines after SCI

We and others have shown that the peak expression of inflammatory markers (cytokines and chemokines) occurs in the acute phase of SCI [48]. Therefore, we determined the effect of NE inhibition on inflammatory parameters using spinal cord samples from sham, injured untreated (Injury), and injured sivelestat-treated (two doses) animals prepared on DPI-1. Sivelestat treatment significantly attenuated the SCI-induced expression of TNF-α (Fig. 4a and c) and interleukin (IL)-6 (Fig. 4b and d).Similarly, the induction of the inflammatory mediators inducible nitric oxide synthase (iNOS) and IL-1β was also significantly decreased with sivelestat (Fig. 4e and f, respectively). Interestingly, sivelestat reversed the suppression of the anti-inflammatory cytokine IL-10 (Fig. 4g) and significantly reduced the SCI induction of C-C motif chemokine ligand (CCL)-2 and CCL-3 (Fig. 4h and i, respectively) and TGF-β (Fig. 4j).

NE inhibition attenuates TJ disruption and blood-spinal cord permeability after SCI

The disruption of the BSCB and subsequent blood infiltration after SCI initiates a secondary injury cascade via the production of inflammatory mediators, such as IL-6, TNF-α, and iNOS. To establish the role of NE in BSCB disruption, we examined samples from sham, injured untreated, and injured sivelestat-treated (two doses) animals prepared DPI-1. NE inhibition via sivelestat administration significantly inhibited the loss of the tight junction (TJ) proteins occludin and zonula occludens-1 (ZO-1) after injury (Fig. 5a–d). SCI induced haemorrhage which was reduced by sivelestat treatment, and expression of cleaved PARP and LC3B was also decreased following treatment (Fig. 5b).

NE inhibition attenuates secondary damage and prevents glial scar formation after SCI

Fibrotic scar tissue is rich in microglia, astroglia, and laminin, and fibronectin forms at the lesion site after SCI in rodents and humans [77]. Several axon-inhibitory molecules present at this scar tissue facade hinder axon regeneration. To evaluate the role of NE in secondary damage and glial scar formation, we examined samples from sham, injured untreated, and injured sivelestat-treated (for 14 days) animals prepared 28 days after SCI (DPI-28). We observed decreases in secondary damage and glial scar formation in animals treated with sivelestat, which were associated with substantial inhibitory effects observed via IHC for microglia (Iba-1), astroglia (GFAP), and fibronectin (Fig. 6a) at the injury site. The decreases in astroglial and microglial activation, as well as macrophage activation, were confirmed at the transcriptional level via quantitative reverse transcription-PCR analysis (Fig. 6bi–iii). IHC for laminin showed a J- or T-shaped morphology under normal conditions, whereas fibrotic scars rich in laminin formed at the epicenter after SCI (Fig. 6c). Treatment with sivelestat attenuated the laminin expression at DPI-28 as well as SCI-induced transforming growth factor beta (TGF-β) expression (Fig. 6d). IHC for the neuronal marker TuJ1 and the axonal marker neurofilament (NF) revealed significant regeneration in the lesions of sivelestat-treated rats (Fig. 6e).

NE inhibition stabilizes vascular endothelium formation after SCI

SCI substantially disrupts the vasculature, especially at the epicenter of the injury, where blood vessels undergo remodeling and structural changes before they become functional. Therefore, we delineated the changes associated with vascular-related proteins and factors after SCI in untreated and sivelestat-treated animals. SCI-induced increases in TGF-β, platelet-derived growth factor beta (PDGF-β), neuropilin1, and platelet-endothelial cell adhesion molecule (PECAM), indicative of vascular changes/damage and vascular remodeling. In contrast to those of untreated (vehicle-injected) animals, the spinal cords of sivelestat-treated animals at 7 and 14 days after SCI exhibited an attenuation of these increases in vascular-related factors and a stabilization of the vasculature (Fig. 7a and b). IHC of spinal cord microvessels also revealed a change in the expression of neural/glial

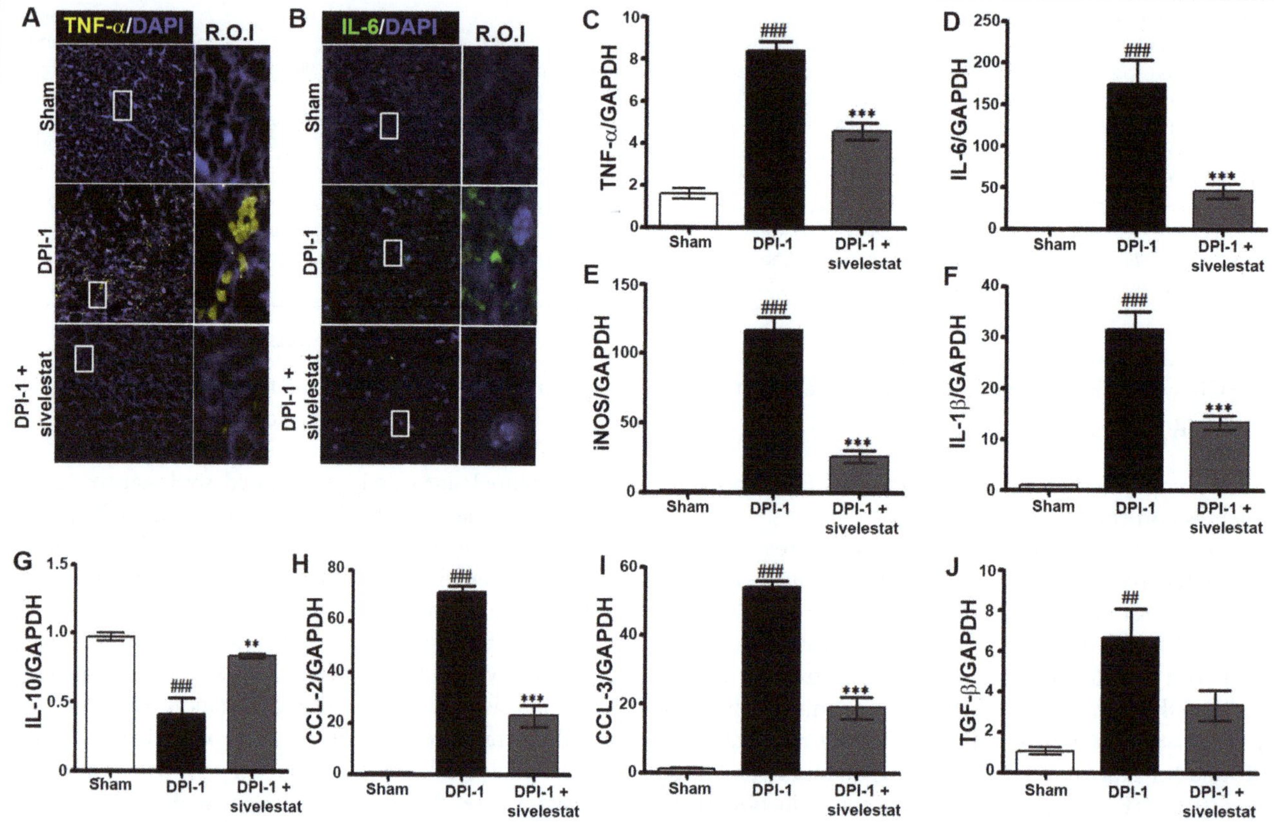

Fig. 4 Neutrophil elastase (NE) inhibition reduces the expression and production of inflammatory mediators (cytokines and chemokines) after spinal cord injury (SCI). Samples from sham or injured untreated (Injury) or after sivelestat treatment were prepared 1 day after injury as described in the Methods section. Representative section of tumor necrosis factor-alpha [TNF-α (**a**)], Interleukin [IL-6 (**b**)] immunofluorescence 1 day after SCI [3 fields/slide, $n = 2–3$/group (sham = 2, injury = 3 and sivelestat = 3)]. Total RNA from sham, vehicle (injury)- or sivelestat-treated samples were prepared DPI-1 after injury as described in the Methods section. RT-PCR results showing relative expression levels of TNF-α (**c**), IL-6 (**d**), inducible nitric oxide synthase [iNOS (**e**)], IL-1β (**f**), IL-10 (**g**), chemokine (C-C motif) ligand 2 [CCL-2 (**h**)], chemokine (C-C motif) ligand 3 [CCL-3 (**i**)] and TGF-β (**j**) after injury/treatment [$n = 2–3$/group (sham = 2, injury = 3 and sivelestat = 3) performed in triplicates]. GAPDH was used as internal controls for real-time quantitative reverse transcription–polymerase chain reaction. Data represent means ± S.E.M. $^{##}p < 0.01$, $^{###}p < 0.001$ vs. sham group. $^{*}p < 0.05$, $^{**}p < 0.01$, $^{***}p < 0.001$ vs. Injury group

antigen2 (NG-2), alpha-smooth muscle actin (α-SMA) (Fig. 7c), and von Willebrand factor (vWF) (Fig. 7d). Treatment with sivelestat increases the expression of NG-2, α-SMA and vWF.

NE inhibition improves functional recovery, attenuates nociception, and provides neuroprotection after SCI

We performed behavioral tests at 1, 7, 14, 21, and 28 days following SCI on untreated animals and animals treated for 14 days with sivelestat. Specifically, walking patterns (gait) were evaluated via manual analyses of footprints. After SCI, all animals showed distinct reductions in motor coordination in forepaw-hindpaw stepping. Those treated with sivelestat showed a marked recovery of gait and improved motor coordination in comparison with untreated animals (Fig. 8a). The performance of sham rats remained unchanged throughout the testing period. Functional recovery was also assessed in open-field testing using the 21-point Basso, Beattie, and Bresnahan (BBB) locomotor test [10]. Animals developed paraplegia after SCI, corresponding to a low BBB score, but showed evidence of modest improvements of motor function as early as 3–4 days after injury (Fig. 8b). Recovery then continued reasonably faster for another 1–2 weeks and then subsequently proceeded at a slower rate. The recovery in motor function was significantly faster in animals treated with sivelestat than in untreated animals. Interestingly, BBB scores were significantly higher in the sivelestat group even after the treatment had stopped (i.e., after 14 days). As nociception is one of the most common devastating conditions after SCI, we evaluated the effect of sivelestat on SCI-induced pain using vonFrey filaments. After SCI, animals exhibited tactile hypersensitivity (decrease in withdrawal threshold) on days 7, 14, 21, and 28, but not at day 1 (Fig. 8c). Treatment with sivelestat significantly attenuated the SCI-induced hypersensitivity.

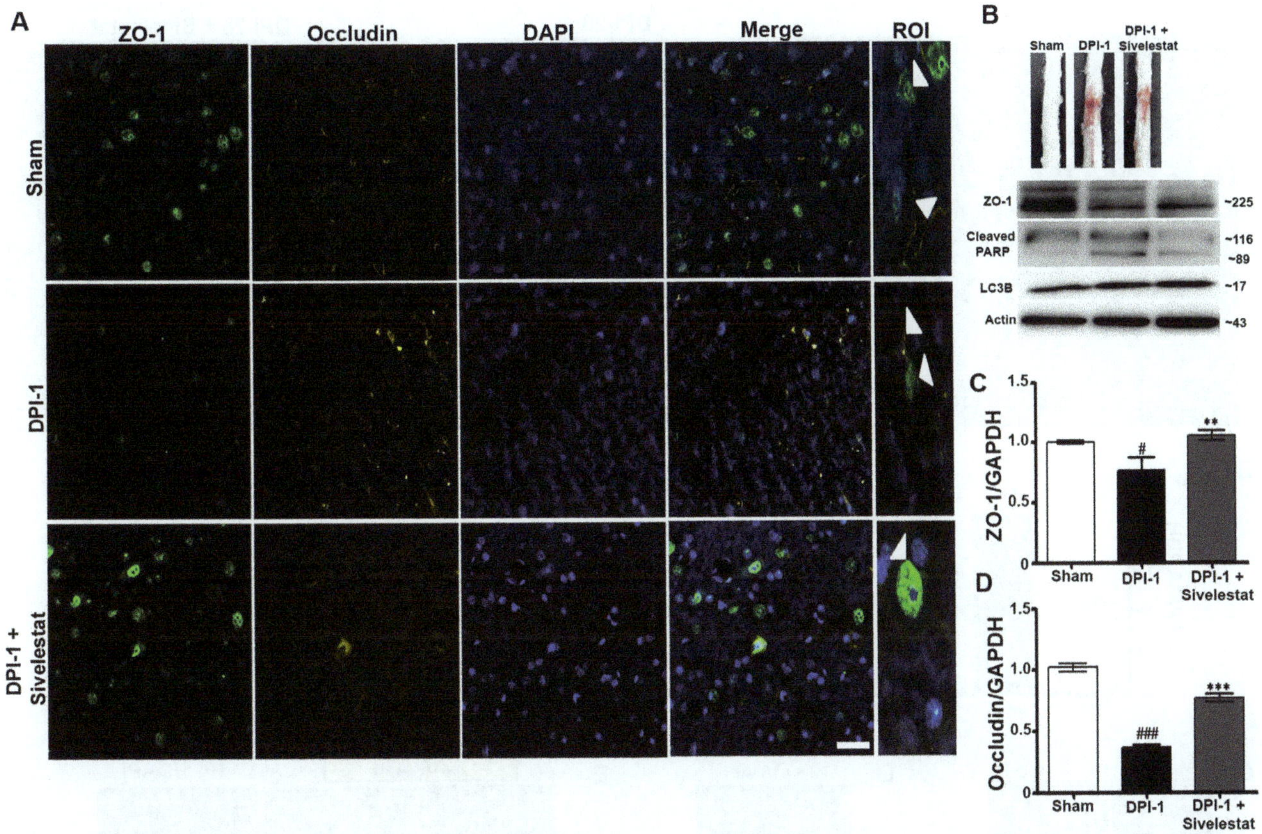

Fig. 5 Neutrophil elastase inhibition prevented the spinal cord injury (SCI)-induced disruption of tight junction protein (ZO-1 and Occludin) in a rat model. Samples from sham, injured untreated, and injured sivelestat-treated animals were prepared DPI-1 as described in the Methods section. **a** Representative section of zonula occludens (ZO)-1 and occludin immunofluorescence, at DPI-1[3 fields/slide, $n = 2–3$/group (sham = 2, injury = 3 and sivelestat = 3)]. **b** Western blots of ZO-1, cleaved PARP, and LC3B expression at 1 day after injury. Actin was used as internal controls for western blot. Total RNA and spinal extracts from sham or injured untreated (Injury) or after sivelestat treatment were prepared DPI-1 as described in the Methods section. RT-PCR results of occludin and zonula occludens (ZO)-1 expression 1 day after injury [$n = 2–3$/group (sham = 2, injury = 3 and sivelestat = 3) performed in triplicates] (**c** and **d**). GAPDH was used as internal controls for real-time quantitative reverse transcription–polymerase chain reaction. Data represent means ± S.E.M. $^{\#}p < 0.05$, $^{\#\#\#}p < 0.001$ vs. sham group. $^{**}p < 0.01$, $^{***}p < 0.001$ vs. Injury group

Finally, the expression of neurotrophic factors, which are mediators of neuronal and axonal plasticity and regeneration after SCI, was examined by IHC. The expression of brain-derived neurotrophic factor (BDNF), ANGPT-1, neurotrophin3 (NT-3), and neurotrophin4 (NT-4) declined, whereas the expression of GFAP increased, after SCI (Fig. 8d, g and f). Consistent with the other findings in this study, sivelestat treatment prevented the decrease in BDNF, ANGPT-1, NT-3, and NT-4 and attenuated the increase in GFAP expression.

Discussion

In the present study, we demonstrate that NE is a key determinant of vascular endothelium disruption/destabilization and affects ANGPT expression in vitro (ECs) and in a rat model of SCI. First, the effects of various concentrations of recombinant NE were assessed on the most-characterized type of ECs, HUVECs. ECs form capillary-like structures in response to angiogenic factors present in the growth medium; however, the addition of NE recombinant protein dose-dependently prevented the formation of capillary-like tubules, reducing the total length and numbers of tubes. These findings suggest that NE influences a required step in the process of angiogenesis. NE also suppressed the expression of ANGPT-1 and ANGPT-2 in ECs. The reduced expression of these factors may explain the decreased tubule formation in ECs. We also found that factors known to induce inflammation (i.e., LPS and TNF-α), which impact SCI and ANGPT in ECs [26, 48, 49], decreased the expression of ANGPT-1 and increased that of ANGPT-2, suggesting that NE and inflammation differentially modulate the expression of ANGPTs in ECs.

The expression patterns of NE and ANGPTs and the roles of NE inhibition in neuroinflammation, BSCB disruption, vascular damage, functional recovery, and neuroprotection after SCI were also examined. The compression injury used in this study results in damage to both the

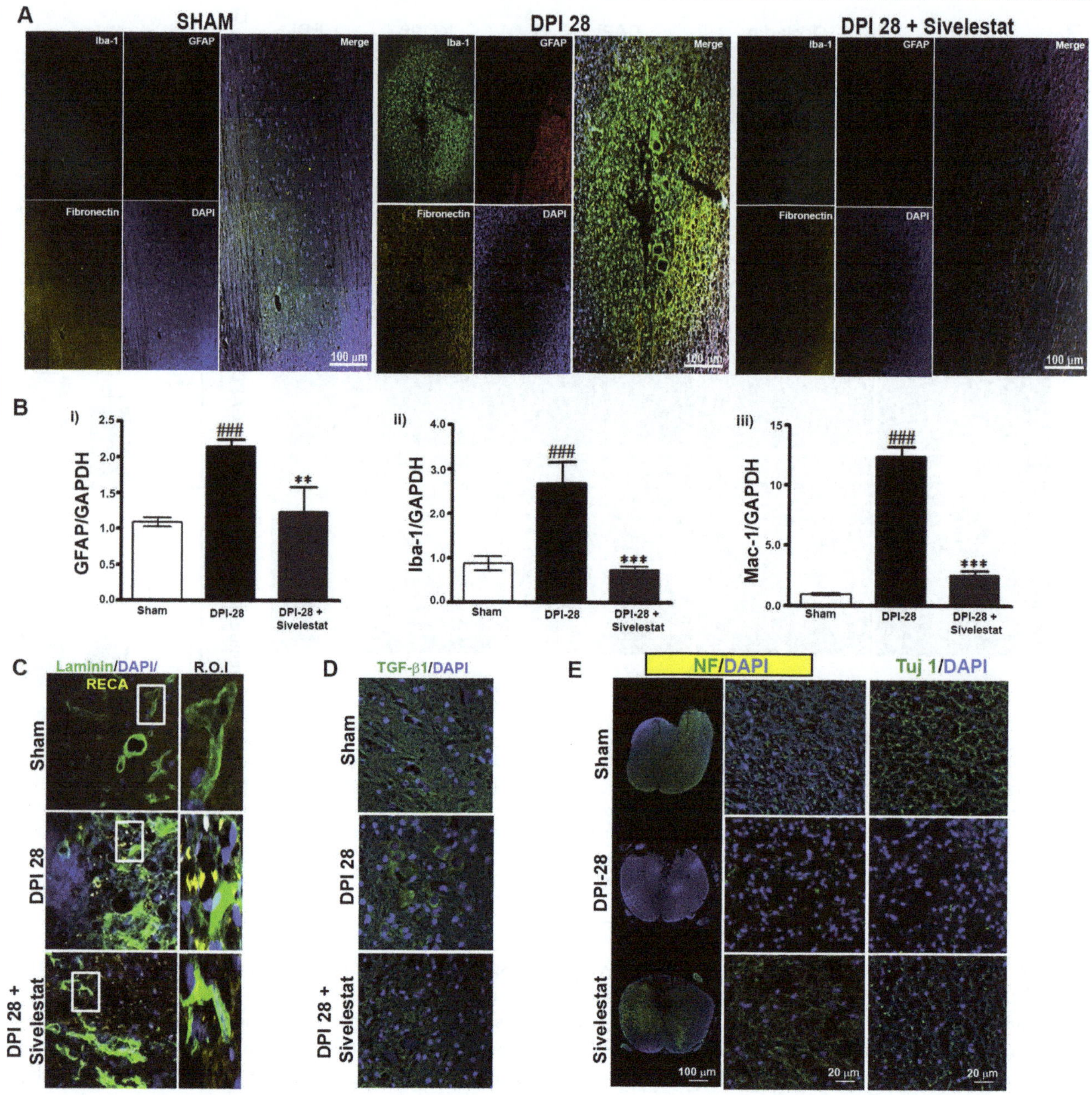

Fig. 6 Neutrophil elastase inhibition attenuates inhibitory glial/fibrotic scarring and secondary damage after spinal cord injury (SCI). Samples from sham or injured untreated (Injury) or after sivelestat treatment were prepared at DPI-28 as described in the Methods section. Sivelestat (30 mg/kg, i.p.) was given for 14 days (28 doses) and animals were sacrificed at day 28. Zeiss confocal microscope was used to evaluate the Immunofluorescence after IHC. The merge function of Zeiss microscope was used and 16 areas were evaluated. **a** Representative merges images (longitudinal) for Iba-1 (green), GFAP (red), fibronectin (yellow) at DPI-28. RT-PCR results of GFAP (**b**-i), Iba-1 (**b**-ii), and Mac-1 (**b**-iii), expression at DPI-28 ($n = 3$/group). Representative images of laminin (green), rat endothelial cell antigen (RECA-1; yellow) (**c**), TGF-β1 (green) (**d**) and neurofilaments (N.F) and Tuj-1 (**e**), at DPI-28 (3 fields/slide, $n = 3$/group). GAPDH was used as internal controls for real-time quantitative reverse transcription–polymerase chain reaction. Data represent means ± S.E.M. $^{###}p < 0.001$ vs. sham group. $^{**}p < 0.01$, $^{***}p < 0.001$ vs. Injury group

dorsal and ventral spinal cord, transient ischemia, and impaired blood flow, which act synergistically to promote secondary pathology as typically observed in SCI in humans. Neutrophil infiltration is linked to progressive damage to the BSCB after SCI, and the prevention of this promotes functional recovery [52, 64]. In the present study, the expression of NE increased significantly from 3 h to 5 days, with maximal expression observed at DPI-1. We also observed damage to the vascular endothelium at this time point via IHC for RECA-1. Since the release of NE from

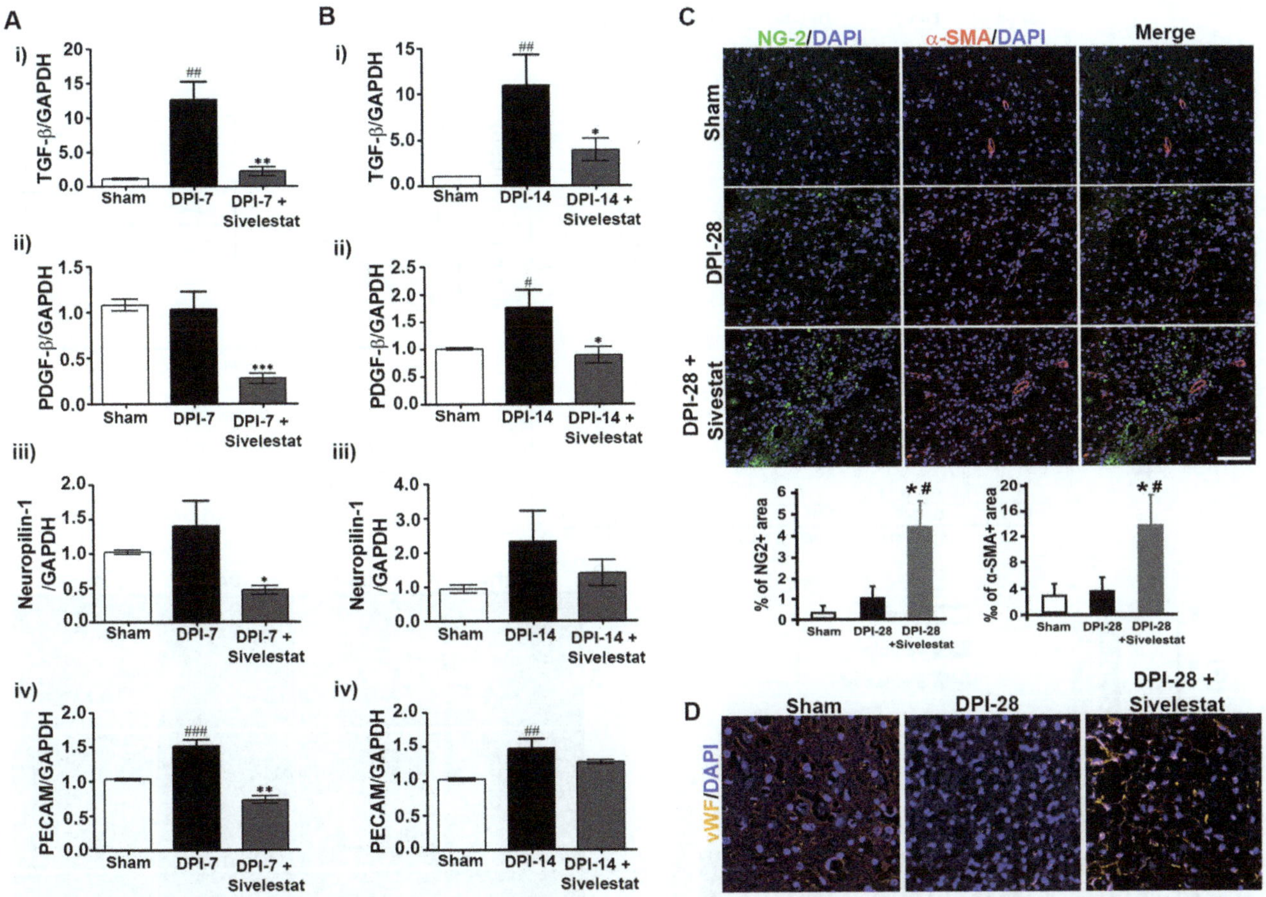

Fig. 7 Neutrophil elastase inhibition stabilizes vascular endothelium after spinal cord injury (SCI). Samples from sham or injured untreated (Injury) or after sivelestat treatment were prepared 7 or 14 or 28 day after injury as described in the Methods section. Total RNA from sham, vehicle (injury) - or sivelestat-treated samples were prepared 7 and 14 day after injury as described in the Methods section. RT-PCR results showing relative expression levels of Transforming growth factor-β1 [TGF-β1; **a** and **b** (i)], Platelet-derived growth factor [PDGF; **a** and **b** (ii)], Neuropilin-1 [**a** and **B** (iii)], and platelet endothelial cell adhesion molecule [PECAM; **a** and **b** (iv)] [$n = 2–3$/group (sham = 2, injury = 3 and sivelestat = 3) performed in triplicates]. Representative section of neural/glial antigen [NG-2 (**c**); green], alpha-smooth muscle actin [α- SMA(**c**) red]; Von Willebrand factor [vWF (**d**) orange] after SCI. Bar charts show the percentage of NG2 or α-SMA positive area per randomly selected field (3 fields/slide, $n = 3$/group) (**c**). GAPDH was used as internal controls for real-time quantitative reverse transcription–polymerase chain reaction. Data represent means ± S.E.M. $^{\#}p < 0.05$, $^{\#\#}p < 0.01$, $^{\#\#\#}p < 0.001$ vs. sham group. $^{**}p < 0.01$, $^{***}p < 0.001$ vs. Injury group

neutrophils can damage ECs, we surmised that NE can also affect the proteins released from ECs. Interestingly, ANGPT-1 expression was downregulated when the expression of NE was maximal and corresponded to the initial progressive damage to the endothelium. From day 7 on, ANGPT-1 expression increased, which may be due to expression in a subset of glial and neuronal cells [13]. ANGPT-2 sensitizes ECs to inflammation, enhancing vascular responsiveness to proinflammatory cytokines [26]. ANGPT-1 and ANGPT-2 have similar affinities for the same receptor but produce opposing effects on blood vessels. SCI induces a lasting decrease in ANGPT-1 levels [14, 35, 58, 74] but a persistent upregulation of ANGPT-2 [22]. Indeed, we observed an increase in ANGPT-2 expression that may counterbalance the ANGPT-1 decrease after SCI. These findings suggest that SCI-induced NE expression can cause endothelium damage and modulate the activity of ANGPT-1 activity after SCI.

To examine this, we used sivelestat, a specific inhibitor of NE. The pharmacokinetic results in the present study show that the concentration of sivelestat in blood peaks at around 30 min, with no accumulation, observed even after 28 doses, indicating that the drug is cleared from the body. Sivelestat treatment significantly attenuated the SCI-induced NE and ANGPT-2 expression, reduction in ANGPT-1, and endothelial damage. These results suggest that specific inhibition of NE prevents the modulation of ANGPT, restores SCI-induced endothelial damage, and prevents vascular disruptions.

SCI is accompanied by a series of intense immune responses, including inflammation, the synthesis of chemokines and cytokines, and the infiltration of peripheral

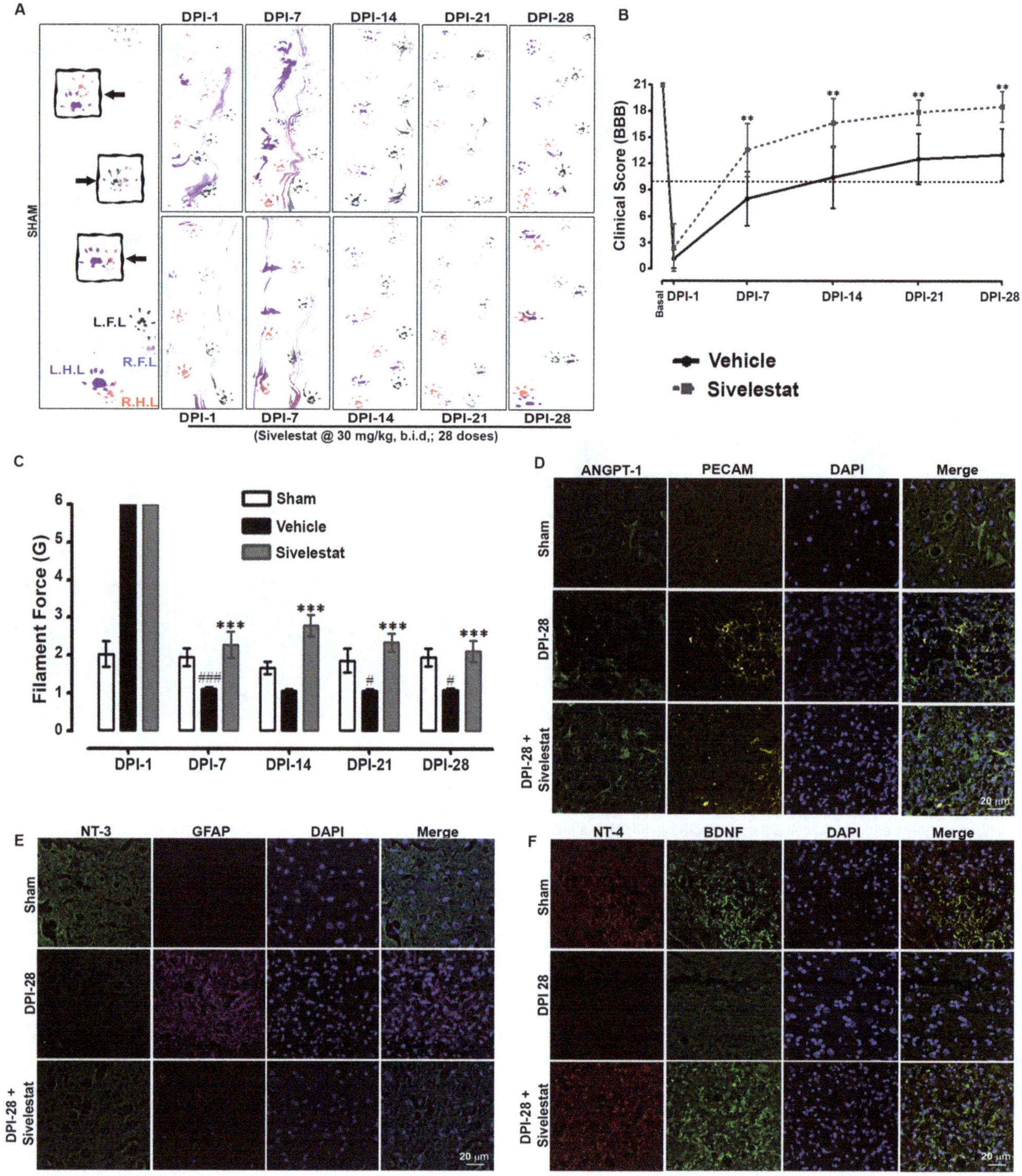

Fig. 8 Sivelestat treatment improved motor functions and attenuated neuropathic pain after spinal cord injury (SCI). **a** Representative footprints of animal walking 1, 7, 14, 21 and 28 days after SCI. Black: Left forepaw footprints; Blue: Right forepaw footprints; Purple: left hind paw footprints; Red: Right hind paw footprints. **b** Functional recovery was then assessed in open-field testing by using the 21-point Basso, Beattie, and Bresnahan (BBB) locomotor test at 1, 7, 14, 21 and 28 days after SCI. **c** Nociception was evaluated using Von Frey-filaments; SCI-induced hypersensitivity (decrease in withdrawal threshold) was assessed at 1, 7, 14, 21 and 28 days after SCI (sham = 6, injury = 16 and sivelestat = 12 for behavioral experiments). Representative sections of ANGPT-1 and PECAM (**d**), neurotrophin-3 (NT-3) and Glial fibrillary acidic protein (GFAP) (**e**), and neurotrophin-4 (NT-4) and Brain-derived neurotrophic factor (BDNF) (**f**), at DPI-28. Data represent means ± S.E.M. $^{\#}p < 0.05$, $^{\#\#\#}p < 0.001$ vs. sham group. $^{***}p < 0.001$ vs. Injury group

leukocytes to the site of damage. Sivelestat was shown to prevent neutrophil infiltration in a rat model of SCI [88]. In addition to their production of NE [6, 48], infiltrating neutrophils contribute to inflammation via their production of IL-6 and TNF-α [42]. Systemic inflammation promotes vascular endothelial injury and results in organ dysfunction [60, 91]. Inflammatory cells and ECs express ANGPTs [4, 25, 53], suggesting that ANGPT signaling plays a central role in commencing and continuing the inflammatory response. Of note, ANGPT-1 and ANGPT-2 show dichotomous pro- and anti-inflammatory properties, with ANGPT-1 primarily considered anti-inflammatory and ANGPT-2 proinflammatory, which are influenced by other inflammatory mediators and proteolytic enzymes, such as NE [2, 9, 33, 44]. NE can regulate acute as well as chronic inflammation [21]. The data presented here demonstrate that SCI induces the expression of cytokines (TNF-α, IL-6, iNOS, and IL-1β) and chemokines (CCL-2 and CCL-3); the expression of the anti-inflammatory cytokine IL-10 decreased. The inhibition of NE with sivelestat treatment decreased these proinflammatory factors, including the expression of TNF-α, suggesting the beneficial role of NE inhibition on the ANGPT pathway after SCI. A previous report suggested that ANGPT-1 inhibits TNF-α-stimulated leukocyte transmigration [28], and ANGPT-2 sensitizes ECs to TNF-α and plays a crucial role in the induction of inflammation [26]. IL-6 is a proinflammatory cytokine that sharply increases in the acute phase after SCI and not only downregulates ANGPT-1 signaling [45] but also stimulates defective angiogenesis [29]. In the present study, treatment with sivelestat attenuated the SCI-induced IL-6 expression as well as that of iNOS, which is implicated in immune responses, inflammation, and apoptosis following SCI [79]. ANGPTs regulate vascular reactivity after hemorrhagic shock in rats through the Tie-2-nitric oxide pathway [95]. ANGPT-1 neutralizes the activity of proinflammatory factors on ECs, suppressing EC permeability induced by vascular endothelial growth factor, thrombin, bradykinin, and histamine [70, 72]. In addition, we observed that sivelestat augmented anti-inflammatory IL-10 while attenuating SCI-induced proinflammatory IL-1β, CCL-2, and CCL-3. Altogether, the evidence from this study supports the notion that acute sivelestat inhibition prevents the decrease in ANGPT-1 and acts as an anti-inflammatory in an experimental model of SCI.

NE is downstream in the humoral mediator network and is essential in vascular endothelial injury and increased permeability [78]. By contrast, ANGPT-1 suppresses vascular leakage/inflammation and expedites angiogenesis [51]. Previous studies showed that ANGPT-1 lowers vascular leakage by strengthening related endothelial molecules and regulating interendothelial adhesion. However, NE can protect the blood-brain barrier by increasing ANGPT-1 expression and EC survival in an animal model of focal ischemia [37]. BSCB disruption following SCI enables leukocytes, including neutrophils, to infiltrate the injured parenchyma, contributing to secondary injury [1, 32, 99]. As TJs provide a "barrier" or "fence" to regulate permeability and endothelial dysfunction [12], we speculated that NE is involved in the degradation of TJ proteins after SCI. We previously showed that the expression of the TJ proteins occludin and ZO-1 is decreased after moderate compression injury [48], suggesting a disruption of the BSCB. The decline in ANGPT-1 and increase in ANGPT-2 correspond with marked blood-brain barrier breakdown after brain injury [68]. Similarly, ANGPT-1 treatment was shown to attenuate BSCB permeability in an animal model of SCI [30, 35]. In the present study, the expression of the TJ proteins occludin and ZO-1 was reduced after SCI, suggesting BSCB disruption, which was prevented by treatment with sivelestat. Thus, specific inhibition of NE effectively prevented ANGPT-1 disruption and increased TJ protein expression after SCI.

After SCI, fibrotic scar tissue at the lesion site becomes rich in microglia, astroglia, and laminin and fibronectin forms in rodents and humans [77], which impedes axonal regeneration [50, 82]. NE damages fibronectins, laminins, and other matrix proteins, resulting in increased vascular permeability and haemorrhaging in tissues [36, 40]. Damaged ECs and the basal lamina deposited at the epicenter of the lesion diminish angiogenesis and are concurrent with cystic cavity formation. Sivelestat treatment reduced glial scar formation and secondary damage, facilitating neuronal regeneration at DPI-28. This was associated with a strong reduction in the amounts of microglia and astroglia at the injury site, as observed by IHC. However, SCI also induces inflammation, which contributes to fibrosis scarring in part via TGF-β signaling. Indeed, we observed a significant increase in TGF-β expression after SCI, which was attenuated in animals treated with sivelestat in accordance with the reduction in scar tissue formation.

Blood vessel density correlates with improved functional outcomes; hence sparing or regenerating the vasculature postinjury is desirable [43]. Although blood vessels tend to grow rapidly into the lesion site after SCI, there is substantial regression around 14 days postinjury [18, 56]. After nerve injury, PDGF-β expression increases [71], which leads to increased EC proliferation [11]. In the present study, sivelestat attenuated the increase in PDGF-β expression at the transcriptional level, as well as that for neuropilin1, whose expression in the spinal cord is normally low but is also upregulated after hemisection and dorsal column crush where it acts as an inhibitory molecule in regulating the organization of the sensory network [3]. Similarly, sivelestat attenuated the SCI-induced increase in PECAM-1, which acts as a mechanosensor in ECs [65] and whose localization to cell junctions is regulated by ANGPT-1 to maintain cellular integrity [28]. Another study found a significant loss of PECAM-1-positive cells at early time points (until day 3) after SCI, with a significant increase from 7 days onwards [94]. SCI also causes a robust decrease in vWF [59], which was observed in the present study. However, RECA-1 stained vessels could be readily identified within the injured spinal cord, suggesting that, although ECs are present, there is altered angiogenesis, which is prevented with NE inhibition.

As SCI can cause paralysis, altered motor coordination, and even neuropathic pain, we assessed these via behavioral tests in the animal model. A footprint analysis revealed reduced motor coordination in forepaw-hindpaw stepping after SCI. The functional impairment is likely influenced by the decrease in ANGPT-1 and related vascular dysfunction [15, 49, 73, 93], as exogenous administration of ANGPT-1 has shown favorable effects on both functional and vascular recovery [30, 35]. Accordingly, animals treated with sivelestat showed an increased ANGPT-1, marked recovery of gait and improved motor coordination compared with that of untreated injured animals. This functional recovery was also reflected by the increase in the BBB score in sivelestat-treated animals. The functional improvements were also accompanied by a reduction in SCI-induced hypersensitivity, an indicator of neuropathy, as assessed by hindpaw responses to stimulation with von Frey filaments. Thus functional recovery is a reflection of the increase in the regenerated area of the lesion. Previous reports suggest that enhanced axon regeneration correlates with functional recovery after SCI [19, 24]. Interestingly, intravenous injection of ANGPT-1 and αvβ3 integrin peptide results in almost complete recovery after SCI [30]. In the current study, the regeneration may have been facilitated by the increase in ANGPT-1, which promotes neurite outgrowth [47] and supports the differentiation of neural progenitor cells via the AKT pathway [8], as evidenced in the present study by the increase in AKT phosphorylation in sivelestat-treated animals. Furthermore, sivelestat treatment also maintained the levels of several neurotrophins (BDNF, NT-3, and NT-4) that are linked with EC survival [20] and are dramatically reduced in the adult spinal cord [16, 57]. In the current study, we treated sivelestat 1 h after injury. However additional studies are required to see whether sivelestat also work if treatment is delayed till 3~4 h after injury simulating the clinical settings. Secondly, owing to the short half-life of the sivelestat we treated it twice a day; it would be interesting to observe the effect of sivelestat as continuous infusion with a lower dose or increase the half-life or directly delivering the sivelestat into the spinal cord by several available approaches.

Conclusions

In conclusion, our results indicate that NE expression is increased after SCI, resulting in a dissociation of ECs from microvessels, reduced ANGPT-1 expression, decreased angiogenesis, tissue damage, vascular destabilization, BSCB breakdown, and cell injury. The inhibition of NE via treatment with sivelestat significantly attenuated SCI-induced inflammation, prevented the decrease in ANGPT-1 expression, and attenuated the increase in ANGPT-2, BSCB breakdown, and cell injury. As a result, secondary damage, functional impairment, and neuropathic pain were reduced while vascular stabilization was promoted. Thus, NE inhibition could serve as a promising therapeutic strategy after SCI.

Acknowledgments

This work was supported by a grant of the National Research Foundation of Korea (NRF) (NRF-2015H1D3A1066543 and NRF-2017R1C1B2011772), Korea Healthcare Technology Research & Development Project, Ministry for Health & Welfare Affairs, Republic of Korea (HI16C1559) and the National Natural Science Foundation of China (31600780).

Authors' contributions

HK and KK carried out the molecular genetic studies, participated in the sequence alignment and drafted the manuscript. HC,MJ, HJ,MM, and SK carried out the immunoassays. EB and AK participated in pharmacokinetic study. IH, XZ, and SL participated in the design of the study and performed the statistical analysis. DB conceived of the study and participated in its design and coordination. All authors read and approved the final manuscript.

Competing interests

The authors declare that they have no competing interests.

Author details

[1]Department of Neurosurgery, CHA University School of Medicine, CHA Bundang Medical Center, Seongnam-si, Gyeonggi-do 13496, Republic of Korea. [2]Department of Biomedical Science, CHA University, Seongnam-si, Gyeonggi-do, Republic of Korea. [3]Molecular Neurobiology Laboratory, Division of Neuroscience, San Raffaele Scientific Institute, Milan, Italy. [4]Department of Neurosurgery, Kyung Hee University, Dongdaemun-gu, Seoul 02447, Republic of Korea. [5]College of Pharmacy, CHA University, Seongnam-si, Gyeonggi-do, Republic of Korea. [6]Department of Neurosurgery, Kyungpook National University Hospital, Kyungpook National University, 130, Dongdeok-ro, Jung-gu, Daegu 41944, Republic of Korea. [7]Department of Neurosurgery, School of Medicine,Kyungpook National University, 130, Dongdeok-ro, Jung-gu, Daegu 41944, Republic of Korea. [8]Department of Histology and Embryology, Zhongshan School of Medicine, Sun Yat-sen University, Guangzhou 510080, Guangdong Province, China.

References

1. Abbott NJ, Ronnback L, Hansson E (2006) Astrocyte-endothelial interactions at the blood-brain barrier. Nat Rev Neurosci 7:41–53. https://doi.org/10.1038/nrn1824
2. Acarin L, Gonzalez B, Castellano B (2000) Neuronal, astroglial and microglial cytokine expression after an excitotoxic lesion in the immature rat brain. Eur J Neurosci 12:3505–3520
3. Agudo M, Robinson M, Cafferty W, Bradbury EJ, Kilkenny C, Hunt SP, McMahon SB (2005) Regulation of neuropilin 1 by spinal cord injury in adult rats. Mol Cell Neurosci 28:475–484. https://doi.org/10.1016/j.mcn.2004.10.008
4. Ahmad S, Cudmore MJ, Wang K, Hewett P, Potluri R, Fujisawa T, Ahmed A (2010) Angiopoietin-1 induces migration of monocytes in a tie-2 and

integrin-independent manner. Hypertension 56:477–483. https://doi.org/10.1161/HYPERTENSIONAHA.110.155556

5. Aikawa N, Kawasaki Y (2014) Clinical utility of the neutrophil elastase inhibitor sivelestat for the treatment of acute respiratory distress syndrome. Ther Clin Risk Manag 10:621–629. https://doi.org/10.2147/TCRM.S65066
6. Ankeny DP, Popovich PG (2009) Mechanisms and implications of adaptive immune responses after traumatic spinal cord injury. Neuroscience 158:1112–1121. https://doi.org/10.1016/j.neuroscience.2008.07.001
7. Aube B, Levesque SA, Pare A, Chamma E, Kebir H, Gorina R, Lecuyer MA, Alvarez JI, De Koninck Y, Engelhardt B, Prat A, Cote D, Lacroix S (2014) Neutrophils mediate blood-spinal cord barrier disruption in demyelinating neuroinflammatory diseases. J Immunol 193:2438–2454. https://doi.org/10.4049/jimmunol.1400401
8. Bai Y, Cui M, Meng Z, Shen L, He Q, Zhang X, Chen F, Xiao J (2009) Ectopic expression of angiopoietin-1 promotes neuronal differentiation in neural progenitor cells through the Akt pathway. Biochem Biophys Res Commun 378:296–301
9. Bartholdi D, Schwab ME (1997) Expression of pro-inflammatory cytokine and chemokine mRNA upon experimental spinal cord injury in mouse: an in situ hybridization study. Eur J Neurosci 9:1422–1438
10. Basso DM, Beattie MS, Bresnahan JC (1995) A sensitive and reliable locomotor rating scale for open field testing in rats. J Neurotrauma 12:1–21. https://doi.org/10.1089/neu.1995.12.1
11. Battegay EJ, Rupp J, Iruela-Arispe L, Sage EH, Pech M (1994) PDGF-BB modulates endothelial proliferation and angiogenesis in vitro via PDGF beta-receptors. J Cell Biol 125:917–928
12. Bazzoni G, Dejana E (2004) Endothelial cell-to-cell junctions: molecular organization and role in vascular homeostasis. Physiol Rev 84:869–901. https://doi.org/10.1152/physrev.00035.2003
13. Beck H, Acker T, Wiessner C, Allegrini PR, Plate KH (2000) Expression of angiopoietin-1, angiopoietin-2, and tie receptors after middle cerebral artery occlusion in the rat. Am J Pathol 157:1473–1483
14. Belaaouaj A, McCarthy R, Baumann M, Gao Z, Ley TJ, Abraham SN, Shapiro SD (1998) Mice lacking neutrophil elastase reveal impaired host defense against gram negative bacterial sepsis. Nat Med 4:615–618
15. Benton RL, Maddie MA, Worth CA, Mahoney ET, Hagg T, Whittemore SR (2008) Transcriptomic screening of microvascular endothelial cells implicates novel molecular regulators of vascular dysfunction after spinal cord injury. J Cereb Blood Flow Metab 28:1771–1785. https://doi.org/10.1038/jcbfm.2008.76
16. Blesch A, Yang H, Weidner N, Hoang A, Otero D (2004) Axonal responses to cellularly delivered NT-4/5 after spinal cord injury. Mol Cell Neurosci 27:190–201
17. Bonin RP, Bories C, De Koninck Y (2014) A simplified up-down method (SUDO) for measuring mechanical nociception in rodents using von Frey filaments. Mol Pain 10:26. https://doi.org/10.1186/1744-8069-10-26
18. Casella GT, Marcillo A, Bunge MB, Wood PM (2002) New vascular tissue rapidly replaces neural parenchyma and vessels destroyed by a contusion injury to the rat spinal cord. Exp Neurol 173:63–76
19. Coumans JV, Lin TT-S, Dai HN, MacArthur L, McAtee M, Nash C, Bregman BS (2001) Axonal regeneration and functional recovery after complete spinal cord transection in rats by delayed treatment with transplants and neurotrophins. J Neurosci 21:9334–9344
20. Donovan MJ, Lin MI, Wiegn P, Ringstedt T, Kraemer R, Hahn R, Wang S, Ibañez CF, Rafii S, Hempstead BL (2000) Brain derived neurotrophic factor is an endothelial cell survival factor required for intramyocardial vessel stabilization. Development 127:4531–4540
21. Doring G (1994) The role of neutrophil elastase in chronic inflammation. Am J Respir Crit Care Med 150:S114–S117. https://doi.org/10.1164/ajrccm/150.6_Pt_2.S114
22. Durham-Lee JC, Wu Y, Mokkapati VU, Paulucci-Holthauzen AA, Nesic O (2012) Induction of angiopoietin-2 after spinal cord injury. Neuroscience 202:454–464. https://doi.org/10.1016/j.neuroscience.2011.09.058
23. Engelhardt B, Coisne C (2011) Fluids and barriers of the CNS establish immune privilege by confining immune surveillance to a two-walled castle moat surrounding the CNS castle. Fluids Barriers CNS 8:4. https://doi.org/10.1186/2045-8118-8-4
24. Fawcett JW (2009) Recovery from spinal cord injury: regeneration, plasticity and rehabilitation. Brain 132:1417–1418
25. Feistritzer C, Mosheimer BA, Sturn DH, Bijuklic K, Patsch JR, Wiedermann CJ (2004) Expression and function of the angiopoietin receptor Tie-2 in human eosinophils. J Allergy Clin Immunol 114:1077–1084. https://doi.org/10.1016/j.jaci.2004.06.045
26. Fiedler U, Reiss Y, Scharpfenecker M, Grunow V, Koidl S, Thurston G, Gale NW, Witzenrath M, Rosseau S, Suttorp N (2006) Angiopoietin-2 sensitizes endothelial cells to TNF-α and has a crucial role in the induction of inflammation. Nat Med 12:235
27. Figley SA, Khosravi R, Legasto JM, Tseng YF, Fehlings MG (2014) Characterization of vascular disruption and blood-spinal cord barrier permeability following traumatic spinal cord injury. J Neurotrauma 31:541–552. https://doi.org/10.1089/neu.2013.3034
28. Gamble JR, Drew J, Trezise L, Underwood A, Parsons M, Kasminkas L, Rudge J, Yancopoulos G, Vadas MA (2000) Angiopoietin-1 is an antipermeability and anti-inflammatory agent in vitro and targets cell junctions. Circ Res 87:603–607
29. Gopinathan G, Milagre C, Pearce OM, Reynolds LE, Hodivala-Dilke K, Leinster DA, Zhong H, Hollingsworth RE, Thompson R, Whiteford JR (2015) Interleukin-6 stimulates defective angiogenesis. Cancer Res 75:3098–3107
30. Han S, Arnold SA, Sithu SD, Mahoney ET, Geralds JT, Tran P, Benton RL, Maddie MA, D'Souza SE, Whittemore SR, Hagg T (2010) Rescuing vasculature with intravenous angiopoietin-1 and alpha v beta 3 integrin peptide is protective after spinal cord injury. Brain 133:1026–1042. https://doi.org/10.1093/brain/awq034
31. Hansen TM, Moss AJ, Brindle NP (2008) Vascular endothelial growth factor and angiopoietins in neurovascular regeneration and protection following stroke. Curr Neurovasc Res 5:236–245
32. Hawkins BT, Davis TP (2005) The blood-brain barrier/neurovascular unit in health and disease. Pharmacol Rev 57:173–185. https://doi.org/10.1124/pr.57.2.4
33. Hayashi M, Ueyama T, Nemoto K, Tamaki T, Senba E (2000) Sequential mRNA expression for immediate early genes, cytokines, and neurotrophins in spinal cord injury. J Neurotrauma 17:203–218
34. Hermant B, Bibert S, Concord E, Dublet B, Weidenhaupt M, Vernet T, Gulino-Debrac D (2003) Identification of proteases involved in the proteolysis of vascular endothelium cadherin during neutrophil transmigration. J Biol Chem 278:14002–14012. https://doi.org/10.1074/jbc.M300351200
35. Herrera JJ, Sundberg LM, Zentilin L, Giacca M, Narayana PA (2010) Sustained expression of vascular endothelial growth factor and angiopoietin-1 improves blood-spinal cord barrier integrity and functional recovery after spinal cord injury. J Neurotrauma 27:2067–2076. https://doi.org/10.1089/neu.2010.1403
36. Houtz PK, Jones PD, Aronson NE, Richardson LM, Lai-Fook SJ (2004) Effect of pancreatic and leukocyte elastase on hydraulic conductivity in lung interstitial segments. J Appl Physiol (1985) 97:2139–2147. https://doi.org/10.1152/japplphysiol.00567.2004
37. Ikegame Y, Yamashita K, Hayashi S, Yoshimura S, Nakashima S, Iwama T (2010) Neutrophil elastase inhibitor prevents ischemic brain damage via reduction of vasogenic edema. Hypertens Res 33:703–707. https://doi.org/10.1038/hr.2010.58
38. Inoue N, Oka N, Kitamura T, Shibata K, Itatani K, Tomoyasu T, Miyaji K (2013) Neutrophil elastase inhibitor sivelestat attenuates perioperative inflammatory response in pediatric heart surgery with cardiopulmonary bypass. Int Heart J 54:149–153
39. Ionescu CV, Cepinskas G, Savickiene J, Sandig M, Kvietys PR (2003) Neutrophils induce sequential focal changes in endothelial adherens junction components: role of elastase. Microcirculation 10:205–220. https://doi.org/10.1038/sj.mn.7800185
40. Ishikawa N, Oda M, Kawaguchi M, Tsunezuka Y, Watanabe G (2003) The effects of a specific neutrophil elastase inhibitor (ONO-5046) in pulmonary ischemia-reperfusion injury. Transpl Int 16:341–346. https://doi.org/10.1007/s00147-003-0556-8
41. Iwamoto S, Higashi A, Ueno T, Goto M, Iguro Y, Sakata R (2009) Protective effect of sivelestat sodium hydrate (ONO-5046) on ischemic spinal cord injury. Interact Cardiovasc Thorac Surg 8:606–609
42. Jablonska E, Kiluk M, Markiewicz W, Piotrowski L, Grabowska Z, Jablonski J (2001) TNF-alpha, IL-6 and their soluble receptor serum levels and secretion by neutrophils in cancer patients. Arch Immunol Ther Exp (Warsz) 49:63–69
43. Kaneko S, Iwanami A, Nakamura M, Kishino A, Kikuchi K, Shibata S, Okano HJ, Ikegami T, Moriya A, Konishi O (2006) A selective Sema3A inhibitor enhances regenerative responses and functional recovery of the injured spinal cord. Nat Med 12:1380
44. Kawabata K, Hagio T, Matsuoka S (2002) The role of neutrophil elastase in acute lung injury. Eur J Pharmacol 451:1–10
45. Kayakabe K, Kuroiwa T, Sakurai N, Ikeuchi H, Kadiombo AT, Sakairi T, Matsumoto T, Maeshima A, Hiromura K, Nojima Y (2012) Interleukin-6

promotes destabilized angiogenesis by modulating angiopoietin expression in rheumatoid arthritis. Rheumatology. https://doi.org/10.1093/rheumatology/kes093.
46. Kim H, Lee JM, Park JS, Jo SA, Kim YO, Kim CW, Jo I (2008) Dexamethasone coordinately regulates angiopoietin-1 and VEGF: a mechanism of glucocorticoid-induced stabilization of blood-brain barrier. Biochem Biophys Res Commun 372:243–248. https://doi.org/10.1016/j.bbrc.2008.05.025
47. Kosacka J, Figiel M, Engele J, Hilbig H, Majewski M, Spanel-Borowski K (2005) Angiopoietin-1 promotes neurite outgrowth from dorsal root ganglion cells positive for Tie-2 receptor. Cell Tissue Res 320:11–19
48. Kumar H, Jo MJ, Choi H, Muttigi MS, Shon S, Kim BJ, Lee SH, Han IB (2017) Matrix Metalloproteinase-8 Inhibition Prevents Disruption of Blood-Spinal Cord Barrier and Attenuates Inflammation in Rat Model of Spinal Cord Injury. Mol Neurobiol. https://doi.org/10.1007/s12035-017-0509-3
49. Kumar H, Ropper AE, Lee S-H, Han I (2017) Propitious Therapeutic Modulators to Prevent Blood-Spinal Cord Barrier Disruption in Spinal Cord Injury. Mol Neurobiol 54(5):3578–3590. https://doi.org/10.1007/s12035-016-9910-6
50. Leal-Filho MB (2011) Spinal cord injury: From inflammation to glial scar. Surg Neurol Int 2:112. https://doi.org/10.4103/2152-7806.83732
51. Lee HS, Han J, Bai HJ, Kim KW (2009) Brain angiogenesis in developmental and pathological processes: regulation, molecular and cellular communication at the neurovascular interface. FEBS J 276:4622–4635. https://doi.org/10.1111/j.1742-4658.2009.07174.x
52. Lee SM, Rosen S, Weinstein P, van Rooijen N, Noble-Haeusslein LJ (2011) Prevention of both neutrophil and monocyte recruitment promotes recovery after spinal cord injury. J Neurotrauma 28:1893–1907. https://doi.org/10.1089/neu.2011.1860
53. Lemieux C, Maliba R, Favier J, Theoret JF, Merhi Y, Sirois MG (2005) Angiopoietins can directly activate endothelial cells and neutrophils to promote proinflammatory responses. Blood 105:1523–1530. https://doi.org/10.1182/blood-2004-09-3531
54. Ling X, Liu D (2007) Temporal and spatial profiles of cell loss after spinal cord injury: Reduction by a metalloporphyrin. J Neurosci Res 85:2175–2185. https://doi.org/10.1002/jnr.21362
55. Liu XZ, Xu XM, Hu R, Du C, Zhang SX, McDonald JW, Dong HX, Wu YJ, Fan GS, Jacquin MF, Hsu CY, Choi DW (1997) Neuronal and glial apoptosis after traumatic spinal cord injury. J Neurosci 17:5395–5406
56. Loy DN, Crawford CH, Darnall JB, Burke DA, Onifer SM, Whittemore SR (2002) Temporal progression of angiogenesis and basal lamina deposition after contusive spinal cord injury in the adult rat. J Comp Neurol 445:308–324
57. Maisonpierre PC, Belluscio L, Friedman B, Alderson RF, Wiegand SJ, Furth ME, Lindsay RM, Yancopoulos GD (1990) NT-3, BDNF, and NGF in the developing rat nervous system: parallel as well as reciprocal patterns of expression. Neuron 5:501–509
58. Man S, Ubogu EE, Ransohoff RM (2007) Inflammatory cell migration into the central nervous system: a few new twists on an old tale. Brain Pathol 17: 243–250. https://doi.org/10.1111/j.1750-3639.2007.00067.x
59. Matsushita T, Lankford KL, Arroyo EJ, Sasaki M, Neyazi M, Radtke C, Kocsis JD (2015) Diffuse and persistent blood–spinal cord barrier disruption after contusive spinal cord injury rapidly recovers following intravenous infusion of bone marrow mesenchymal stem cells. Exp Neurol 267:152–164
60. McGill SN, Ahmed NA, Christou NV (1998) Endothelial cells: role in infection and inflammation. World J Surg 22:171–178
61. Nag S, Papneja T, Venugopalan R, Stewart DJ (2005) Increased angiopoietin2 expression is associated with endothelial apoptosis and blood-brain barrier breakdown. Lab Invest 85:1189–1198. https://doi.org/10.1038/labinvest.3700325
62. Nakatani K, Takeshita S, Tsujimoto H, Kawamura Y, Sekine I (2001) Inhibitory effect of serine protease inhibitors on neutrophil-mediated endothelial cell injury. J Leukoc Biol 69:241–247
63. National Research Council (2011) Guide for the care and use of laboratory animals. Eighth Edition. Washington, DC: National Academies Press.
64. Neirinckx V, Coste C, Franzen R, Gothot A, Rogister B, Wislet S (2014) Neutrophil contribution to spinal cord injury and repair. J Neuroinflammation 11:150
65. Newman PJ (1999) Switched at birth: a new family for PECAM-1. J Clin Invest 103:5–9
66. Noble LJ, Mautes AE, Hall JJ (1996) Characterization of the microvascular glycocalyx in normal and injured spinal cord in the rat. J Comp Neurol 376: 542–556. https://doi.org/10.1002/(SICI)1096-9861(19961223)376:4<542::AID-CNE4>3.0.CO;2-1
67. Noble LJ, Wrathall JR (1989) Distribution and time course of protein extravasation in the rat spinal cord after contusive injury. Brain Res 482:57–66
68. Nourhaghighi N, Teichert-Kuliszewska K, Davis J, Stewart DJ, Nag S (2003) Altered expression of angiopoietins during blood-brain barrier breakdown and angiogenesis. Lab Invest 83:1211–1222
69. Okajima K, Harada N, Uchiba M, Mori M (2004) Neutrophil elastase contributes to the development of ischemia-reperfusion-induced liver injury by decreasing endothelial production of prostacyclin in rats. Am J Physiol Gastrointest Liver Physiol 287:G1116–G1123. https://doi.org/10.1152/ajpgi.00061.2004
70. Oubaha M, Gratton JP (2009) Phosphorylation of endothelial nitric oxide synthase by atypical PKC zeta contributes to angiopoietin-1-dependent inhibition of VEGF-induced endothelial permeability in vitro. Blood 114: 3343–3351. https://doi.org/10.1182/blood-2008-12-196584
71. Oya T, Zhao YL, Takagawa K, Kawaguchi M, Shirakawa K, Yamauchi T, Sasahara M (2002) Platelet-derived growth factor-b expression induced after rat peripheral nerve injuries. Glia 38:303–312
72. Pizurki L, Zhou Z, Glynos K, Roussos C, Papapetropoulos A (2003) Angiopoietin-1 inhibits endothelial permeability, neutrophil adherence and IL-8 production. Br J Pharmacol 139:329–336. https://doi.org/10.1038/sj.bjp.0705259
73. Popovich PG, Horner PJ, Mullin BB, Stokes BT (1996) A quantitative spatial analysis of the blood-spinal cord barrier. I. Permeability changes after experimental spinal contusion injury. Exp Neurol 142:258–275. https://doi.org/10.1006/exnr.1996.0196
74. Ritz MF, Graumann U, Gutierrez B, Hausmann O (2010) Traumatic spinal cord injury alters angiogenic factors and TGF-beta1 that may affect vascular recovery. Curr Neurovasc Res 7:301–310
75. Ropper AE, Zeng X, Anderson JE, Yu D, Han I, Haragopal H, Teng YD (2015) An efficient device to experimentally model compression injury of mammalian spinal cord. Exp Neurol 271:515–523. https://doi.org/10.1016/j.expneurol.2015.07.012
76. Rossignol S, Schwab M, Schwartz M, Fehlings MG (2007) Spinal cord injury: time to move? J Neurosci 27:11782–11792. https://doi.org/10.1523/JNEUROSCI.3444-07.2007
77. Ruschel J, Hellal F, Flynn KC, Dupraz S, Elliott DA, Tedeschi A, Bates M, Sliwinski C, Brook G, Dobrindt K, Peitz M, Brustle O, Norenberg MD, Blesch A, Weidner N, Bunge MB, Bixby JL, Bradke F (2015) Axonal regeneration. Systemic administration of epothilone B promotes axon regeneration after spinal cord injury. Science 348:347–352. https://doi.org/10.1126/science.aaa2958
78. Russell JA (2006) Management of sepsis. N Engl J Med 355:1699–1713. https://doi.org/10.1056/NEJMra043632
79. Satake K, Matsuyama Y, Kamiya M, Kawakami H, Iwata H, Adachi K, Kiuchi K (2000) Nitric oxide via macrophage iNOS induces apoptosis following traumatic spinal cord injury. Brain Res Mol Brain Res 85:114–122
80. Scali M, Begenisic T, Mainardi M, Milanese M, Bonifacino T, Bonanno G, Sale A, Maffei L (2013) Fluoxetine treatment promotes functional recovery in a rat model of cervical spinal cord injury. Sci Rep 3:2217. https://doi.org/10.1038/srep02217
81. Semple BD, Trivedi A, Gimlin K, Noble-Haeusslein LJ (2015) Neutrophil elastase mediates acute pathogenesis and is a determinant of long-term behavioral recovery after traumatic injury to the immature brain. Neurobiol Dis 74:263–280. https://doi.org/10.1016/j.nbd.2014.12.003
82. Silver J, Miller JH (2004) Regeneration beyond the glial scar. Nat Rev Neurosci 5:146–156. https://doi.org/10.1038/nrn1326
83. Smedly LA, Tonnesen MG, Sandhaus RA, Haslett C, Guthrie LA, Johnston RB Jr, Henson PM, Worthen GS (1986) Neutrophil-mediated injury to endothelial cells. Enhancement by endotoxin and essential role of neutrophil elastase. J Clin Invest 77:1233–1243. https://doi.org/10.1172/JCI112426
84. Taoka Y, Okajima K, Murakami K, Johno M, Naruo M (1998) Role of neutrophil elastase in compression-induced spinal cord injury in rats. Brain Res 799:264–269
85. Tator CH, Koyanagi I (1997) Vascular mechanisms in the pathophysiology of human spinal cord injury. J Neurosurg 86:483–492. https://doi.org/10.3171/jns.1997.86.3.0483
86. Thomas M, Augustin HG (2009) The role of the Angiopoietins in vascular

morphogenesis. Angiogenesis 12:125–137. https://doi.org/10.1007/s10456-009-9147-3
87. Thurston G, Rudge JS, Ioffe E, Zhou H, Ross L, Croll SD, Glazer N, Holash J, McDonald DM, Yancopoulos GD (2000) Angiopoietin-1 protects the adult vasculature against plasma leakage. Nat Med 6:460–463. https://doi.org/10.1038/74725
88. Tonai T, Shiba K, Taketani Y, Ohmoto Y, Murata K, Muraguchi M, Ohsaki H, Takeda E, Nishisho T (2001) A neutrophil elastase inhibitor (ONO-5046) reduces neurologic damage after spinal cord injury in rats. J Neurochem 78: 1064–1072
89. Travis J (1988) Structure, function, and control of neutrophil proteinases. Am J Med 84:37–42
90. Tsuboko Y, Takeda S, Mii S, Nakazato K, Tanaka K, Uchida E, Sakamoto A (2012) Clinical evaluation of sivelestat for acute lung injury/acute respiratory distress syndrome following surgery for abdominal sepsis. Drug Des Devel Ther 6:273
91. Ueno H, Hirasawa H, Oda S, Shiga H, Nakanishi K, Matsuda K (2002) Coagulation/fibrinolysis abnormality and vascular endothelial damage in the pathogenesis of thrombocytopenic multiple organ failure. Crit Care Med 30: 2242–2248. https://doi.org/10.1097/01.CCM.0000030445.64104.E0
92. Valenzuela DM, Griffiths JA, Rojas J, Aldrich TH, Jones PF, Zhou H, McClain J, Copeland NG, Gilbert DJ, Jenkins NA, Huang T, Papadopoulos N, Maisonpierre PC, Davis S, Yancopoulos GD (1999) Angiopoietins 3 and 4: diverging gene counterparts in mice and humans. Proc Natl Acad Sci U S A 96:1904–1909
93. Whetstone WD, Hsu JY, Eisenberg M, Werb Z, Noble-Haeusslein LJ (2003) Blood-spinal cord barrier after spinal cord injury: relation to revascularization and wound healing. J Neurosci Res 74:227–239. https://doi.org/10.1002/jnr.10759
94. Whetstone WD, Hsu JYC, Eisenberg M, Werb Z, Noble-Haeusslein LJ (2003) Blood-spinal cord barrier after spinal cord injury: Relation to revascularization and wound healing. J Neurosci Res 74:227–239
95. Xu J, Lan D, Li T, Yang G, Liu L (2012) Angiopoietins regulate vascular reactivity after haemorrhagic shock in rats through the Tie2-nitric oxide pathway. Cardiovasc Res 96:308–319. https://doi.org/10.1093/cvr/cvs254
96. Yang JJ, Kettritz R, Falk RJ, Jennette JC, Gaido ML (1996) Apoptosis of endothelial cells induced by the neutrophil serine proteases proteinase 3 and elastase. Am J Pathol 149:1617–1626
97. Young RE, Thompson RD, Larbi KY, La M, Roberts CE, Shapiro SD, Perretti M, Nourshargh S (2004) Neutrophil elastase (NE)-deficient mice demonstrate a nonredundant role for NE in neutrophil migration, generation of proinflammatory mediators, and phagocytosis in response to zymosan particles in vivo. J Immunol 172:4493–4502
98. Zacharek A, Chen J, Cui X, Li A, Li Y, Roberts C, Feng Y, Gao Q, Chopp M (2007) Angiopoietin1/Tie2 and VEGF/Flk1 induced by MSC treatment amplifies angiogenesis and vascular stabilization after stroke. J Cereb Blood Flow Metab 27:1684–1691. https://doi.org/10.1038/sj.jcbfm.9600475
99. Zlokovic BV (2008) The blood-brain barrier in health and chronic neurodegenerative disorders. Neuron 57:178–201. https://doi.org/10.1016/j.neuron.2008.01.003

4

Membrane binding, internalization, and sorting of alpha-synuclein in the cell

Caterina Masaracchia[1], Marilena Hnida[1], Ellen Gerhardt[1], Tomás Lopes da Fonseca[1], Anna Villar-Pique[1], Tiago Branco[3], Markus A. Stahlberg[4], Camin Dean[4], Claudio O. Fernández[2], Ira Milosevic[3] and Tiago F. Outeiro[1,5,6*]

Abstract

Alpha-synuclein (aSyn) plays a crucial role in Parkinson's disease (PD) and other synucleinopathies, since it misfolds and accumulates in typical proteinaceous inclusions. While the function of aSyn is thought to be related to vesicle binding and trafficking, the precise molecular mechanisms linking aSyn with synucleinopathies are still obscure. aSyn can spread in a prion-like manner between interconnected neurons, contributing to the propagation of the pathology and to the progressive nature of synucleinopathies. Here, we investigated the interaction of aSyn with membranes and trafficking machinery pathways using cellular models of PD that are amenable to detailed molecular analyses. We found that different species of aSyn can enter cells and form high molecular weight species, and that membrane binding properties are important for the internalization of aSyn. Once internalized, aSyn accumulates in intracellular inclusions. Interestingly, we found that internalization is blocked in the presence of dynamin inhibitors (blocked membrane scission), suggesting the involvement of the endocytic pathway in the internalization of aSyn. By screening a pool of small Rab-GTPase proteins (Rabs) which regulate membrane trafficking, we found that internalized aSyn partially colocalized with Rab5A and Rab7. Initially, aSyn accumulated in Rab4A-labelled vesicles and, at later stages, it reached the autophagy-lysosomal pathway (ALP) where it gets degraded. In total, our study emphasizes the importance of membrane binding, not only as part of the normal function but also as an important step in the internalization and subsequent accumulation of aSyn. Importantly, we identified a fundamental role for Rab proteins in the modulation of aSyn processing, clearance and spreading, suggesting that targeting Rab proteins may hold important therapeutic value in PD and other synucleinopathies.

Keywords: Alpha-synuclein, Parkinson's disease, Uptake, Spreading, Rab proteins

Introduction

The aggregation and accumulation of proteins in the brain is a common feature among several neurodegenerative disorders such as Parkinson's disease (PD) and dementia with Lewy bodies (DLB) [24]. These diseases are part of a group known as synucleinopathies, characterized by the accumulation of proteinaceous inclusions enriched in alpha-synuclein (aSyn) [55, 56, 59, 60, 64], an abundant protein in the brain that is found in presynaptic terminals and also in other subcellular compartments. The precise physiological function of aSyn remains elusive, but it is thought to be involved in synaptic vesicle trafficking and biology [47]. aSyn can be divided into three distinct regions based on the amino acid composition: the N-terminal region (residues 1–60) adopts amphipathic α-helical structure when associated with membranes ([13, 15]; the central region (residues 61–95) is highly hydrophobic and essential for aggregation ([23]; and the C-terminal region (residues 96–140) is enriched in acidic residues and is involved in several protein-protein interactions [26], conferring to the protein a chaperone-like function [31, 43, 54]. aSyn is an intrinsically disordered protein (IDP), characterized by the lack of defined secondary structure under physiological conditions. However, aSyn adopts helical structure in the first 100 residues upon interaction with membranes [21, 28, 51].

* Correspondence: touteir@gwdg.de
[1]Department of Experimental Neurodegeneration, Center for Biostructural Imaging of Neurodegeneration, Center for Nanoscale Microscopy and Molecular Physiology of the Brain, University Medical Center Goettingen, 37073 Göttingen, Germany
[5]Max Planck Institute for Experimental Medicine, Göttingen, Germany
Full list of author information is available at the end of the article

The precise physiological form of aSyn is still a matter of debate. Initially, the protein was thought to be monomeric, existing in a balanced equilibrium between a cytosolic, free state, and a state bound to the plasma membrane and vesicles. Recently, multimeric forms of aSyn, mainly tetrameric, have been isolated from various cell types [3, 9, 65], initiating a debate about its natural state.

Several mutations in the *SNCA* gene have been identified in familial forms of PD (A53T [45], A30P [32], E46K [67], H50Q [2], G51D [35] and A53E [44]). In addition, overexpression of wild-type aSyn (aSyn WT) due to duplication [16] or triplication [49] of the *SNCA* gene are also associated with autosomal dominant forms of PD.

Intense efforts have focused on the study of the molecular mechanisms underlying aSyn misfolding and aggregation. Recently, cell-to-cell spreading of aSyn has become an attractive model to explain the progressive nature of these diseases and the typical patterns of pathology deposition in neuroanatomically connected regions of the diseased brain. Multiple studies demonstrated that aSyn oligomers and pre-formed fibrils (PFFs) enter cultured cells and accumulate in the cytoplasm [37, 38, 63]. However, it is still unclear how aSyn enters cells and where aggregation starts. The hypothesis that aSyn multimerizes upon interacting with lipid membranes [9] raised the question of whether α-helical aSyn multimers directly transition into β-strand-rich cytotoxic forms, or whether it is the unstructured, monomeric form that transitions to aggregates inside cells, during the processing and compartmentalization in different organelles and the interaction with effector proteins.

We have previously shown that small Ras-like GTPases (Rabs) proteins, key mediators of the membrane trafficking and vesicle recycling, can also modulate aSyn oligomerization and aggregation [5, 17, 25]. Rabs act as molecular "switches" that alternate between two conformational states: the GTP-bound 'on' form, and the GDP-bound 'off' form [57]. Notably, mutations in RAB genes (e.g. *RAB7L1 and RAB39B*) and in their regulators or effectors have been implicated in several neurological and neurodevelopmental disorders, suggesting that impairment in the function of these proteins might be linked to familial forms of PD [36, 48, 66].

Here, we investigated the internalization of aSyn and the role of Rab proteins in mediating this process. We found that membrane interactions by aSyn enable internalization and that Rab proteins mediate the intracellular distribution of the protein. In total, our study provides novel insight into the molecular mechanisms associated with aSyn internalization and aggregation, paving the way for future intervention strategies aimed at interfering with the spreading of aSyn pathology.

Materials and methods

aSyn purification

aSyn WT, aSyn A30P and aSyn A11P/V70P were obtained by transforming *E.coli* BL21-DE3 competent cells with plasmids encoding corresponding cDNA sequences (pET21-aSyn, pET21-A30P, pET21-A11P/V70P).

Purification was performed as previously reported [26] with minor modifications. Briefly, BL21-DE3 cells were grown in LB medium in the presence of ampicillin (100 μg/ml). Protein expression was induced with 1 mM IPTG for 4 h at 37 °C. Afterwards, cultures were harvested and the cell pellet was resuspended in Lysis Buffer (50 mM Tris HCL, 150 mM NaCl, 1 mM EDTA and Inhibitor Protease cocktail) at pH 8.0. Cells were recovered, sonicated on ice, boiled for 20 min at 95 °C, and cell debris were discarded by centrifugation. Subsequent precipitation first with streptomycin sulphate (10 mg/ml) and later with ammonium sulphate (361 mg/ml) was used to obtain aSyn-enriched precipitate.

Anion exchange high-performance liquid-chromatography (AEC) was carried out on an Äkta-HPLC Purifier (GE Healthcare). The pellet was resuspended then in 25 mM Tris-HCl (pH 7.7), and loaded onto a Mono Q column or bounded to a Hi-Trap column (GE Healthcare). The monomeric proteins were eluted at ~300 mM NaCl with a linear salt gradient of elution buffer from 0 mM to 1 M NaCl. The pure proteins (judged by PAGE) were dialyzed overnight against the appropriate buffer and further size exclusion chromatography (SEC) purification step using a Superdex 75 column (GE Healthcare) was performed.

Protein concentration was estimated from the absorbance at 274 nm using an extinction coefficient of 5600 M^{-1} cm^{-1}. The protein stocks were frozen in single aliquots at – 80 °C.

Fibril formation

Three aliquots of 300 μL of aSyn WT were prepared from the protein stocks, and diluted in phosphate saline buffer (PBS) to reach a final concentration of 60 μM. Samples were incubated in an Eppendorf Thermomixer Comfort (Eppendorf, USA) with 0.02% sodium azide at 600 rpm and 37 °C.

The transition of aSyn from initial soluble monomeric form to aggregated state was determined by measuring light scattering in a Jasco FP-8200 spectrofluorometer (Jasco Inc., MD, USA) with an excitation wavelength of 330 nm and emission range from 320 to 340 nm at 25 ° C. Solutions without protein were used as negative controls. All experiments were carried out in triplicates.

Cell line cultures and treatments with aSyn

Human neuroglioma H4 cells were maintained at 37 °C and 5% CO_2 environment, in Opti-MEM medium (PAN,

Germany) supplemented with 10% fetal calf serum (ThermoFisher) and 1% penicillin-streptomycin (ThermoFisher).

Cells were seeded in different well-plate formats, one day prior to transfection, at a density of 8.5 to $1*10^5$ cells/ml. The day after, cells were treated with different concentrations of aSyn monomers or aSyn fibrils for 24 h. At the end of the treatment, cells were extensively washed with PBS and then briefly treated with trypsin in order to remove the residual proteins still outside of the cells or bound to the dish (for a maximum time of 30 s), incubated again with medium (in order to stop the trypsinization) and then washed one last time with PBS.

Transfections were performed with calcium phosphate following the procedure from www.flemingtonlab.com. Shortly, 3 h prior to transfection, fresh cell medium was added to the cells. DNA was diluted in 1× HBS buffer with 25 mM 4-(2-hydroxyethyl)-1-piperazineethanesulfonic acid, 140 mM NaCl, 5 mM KCl, 0.75 mM $Na_2HPO_4 \cdot 2H_2O$, 6 mM Dextrose, pH 7.1. After mixing, 2.5 M $CaCl_2$ was added dropwise and vigorously mixed. Followed 20 min of incubation, the mixture was added dropwise to the cells. In the next morning cells were fed with fresh medium.

Immunoblot analysis

Cells were solubilised with RIPA buffer (50 mM Tris pH 8.0, 150 mM NaCl, 0.1% Sodium-Dodecyl-Sulphate (SDS), 1% Nonidet P40, 0.5% Sodium-Deoxycholate, protease inhibitors) and protein quantification was done using the Bradford assay (BioRad). All samples were measured in triplicate.

Cell lysates were separated by SDS-PAGE under reducing conditions in 12% separating gels with 7% stacking gels. After electrophoresis, proteins were transferred onto 0.45 μm nitrocellulose membranes for 20 min per membrane at constant 25 mA in a semi-dry transfer chamber Trans-Blot® Turbo™ Transfer Solution from Bio-Rad (Bio-Rad Laboratories, Inc., Hercules, CA, USA). Free binding sites were blocked with 10% (*w*/*v*) skim milk dissolved in TBS-T (50 mM Tris (hydroxymethyl)-aminomethane (TRIS) supplemented with 0.05% (*v*/v) Tween-20) for 1 h at Room Temperature (RT).

For detection, the primary antibodies were dissolved in TBS and incubated over night at 4 °C. Detection of proteins on immunoblots was performed to detect aSyn, (aSyn C-20, 1:1000, Santa-Cruz Biotechnology) β-actin (Actin beta, 1:10.000, Sigma Aldrich), tubulin (tubulin β-III, 1:10.000, Santa-Cruz Biotechnology), transferrin receptor (Tfr Receptor, 1:1000, Invitrogen-Life Technologies) and Rab GTPases fused to GFP (GFP Ab, 1:1000, Roche). After washing with TBS-T, secondary HRP-conjugated antibodies (GE Healthcare) were diluted 10,000-fold in TBS-T and incubated with the membrane for 2 h at RT.

Membranes were visualized using Fusion Fx (Vilber Lourmat, Marne-la-Vallée, France) with Immobilon Western Chemiluminescent HRP Substrate (Merck Millipore, Billerica, MA, USA).

Dot blot analysis

All HPLC samples were boiled for 10 min at 95 °C at 650 rpm and then spun down briefly at 10,000 g and 4 ° C. Samples were loaded fully onto a 96 well custom-manufactured Dot Blot machine. A vacuum pump was used to suck sample through a 0.2 μm pore size Protean nitrocellulose membrane (Schleicher & Schuell Bioscience GmbH, Dassel, Germany). The membrane was subsequently blocked with 5% skim milk in TBS (to prevent unspecific staining) for 1 h.

Membranes were incubated with primary antibody (aSyn BD transduction, 1:2000; BD Biosciences) diluted in 1% skim milk in TBS, or 5% bovine serum albumin (BSA) in TBS, overnight at 4 °C. At the end of the incubation, membranes were washed three times with TBS-T for 10 min.

Subsequently, membranes were incubated with HRP-conjugated secondary antibody (GE Healthcare) diluted 1:10,000 in TBS. Afterwards, membranes were visualized using Fusion Fx (Vilber Lourmat, Marne-la-Vallée, France) with Immobilon Western Chemiluminescent HRP Substrate (Merck Millipore, Billerica, MA, USA).

Triton X-100 fractionation assay

Cells were plated and treated as described above. At the end of the treatment, cells were lysed in Lysis Buffer I (25 mM Tris pH 7.5, 150 mM NaCl, 1 mM EDTA and cocktail of protease inhibitors) and centrifuged at 100.000 g for 30 min at 4 °C. Supernatants were collected (soluble fraction) and the pellets (insoluble fraction) were washed with cold PBS and transferred to new tubes. Samples were centrifuged once again 14.000 rpm for 10 min at 4 °C and the resulting pellet, corresponding to the insoluble fraction, was subsequently resuspended in Lysis Buffer II (75 mM Tris, pH 6.8, 3% SDS, 15% Glycerin, 3.75 mM EDTA pH 7.4 and a cocktail of protease inhibitors). Finally, samples were sonicated (10 pulse/second) and immunoblotting analysis were performed as described above.

Membrane biotinylation assay

The day before the experiment, cells were plated in 100 mm Petri dishes at a density of $4*10^6$ cells, and grown until 60–70% confluence. Thereafter, cells were treated with 1 μM of aSyn recombinant monomers or fibrils of different aSyn variants (aSyn WT, aSyn A30P or aSyn A11P/V70P) for 24 h.

At the end of the treatment, cells were treated as previously described [20, 46]. Briefly, cells were rinsed 3 times in ice-cold PBS and further incubated in PBS containing 1.5 mg/ml of EZ-Link Sulfo-NHS-SS-Biotin (ThermoFisher) with gentle rocking, for 30 min, at 4 °C.

The non-bound biotin was removed by incubating cells with 100 mM solution of Glycine for 15 min at 4 °C.

To remove the excess of glycine, cells were briefly washed with PBS and thereafter cell lysate was prepared in PBS containing Protease Inhibitor α-complete (La Roche, Basel, Switzerland), 0.1% SDS and 1% Triton X-100. The lysates were sonicated for 30 s and centrifuged for 5 min at 17000 × g. The supernatant was further incubated with 100 μL of NeutrAvidin Agarose Resin (ThermoFisher) for 2 h, in a rotatory shaker with gentle agitation, at 4 °C. After the incubation with the resin, the supernatant (corresponding to the Cytoplasmic cell lysate fraction) was collected, and a Bradford assay was performed to evaluate the amount of total protein concentration in each of the samples.

Biotinylated proteins were then washed 3 times with PBS and then eluted with 2× Laemmli Buffer by boiling the samples at 95 °C for 5 min.

Samples were then processed by western blotting. Transferrin receptor was used as a positive control of the biotinylated fraction, whereas tubulin was used as a positive control for the cytoplasmic cell lysate fraction.

Immunocytochemistry (ICC)

For ICC analysis, cells were plated in multi-well plates with different formats, previously coated with coverslips. 24 or 48 h after transfection, H4 cells were washed with PBS and fixed with 4% paraformaldehyde (PFA) for 10 min at RT, followed by a permeabilisation step with 0.5% Triton X-100 (Sigma Aldrich, St. Louis, MO, USA) for 10 min at RT.

After blocking in 10% normal goat serum (PAA, Cölbe, Germany)/DPBS for 1 h, cells were incubated with primary antibody. Primary antibodies used were: rabbit anti-aSyn (Human aSyn (C-20): sc-7011-R, 1:1000, Santa Cruz Biotechnology, Dallas, USA) or mouse anti-aSyn (Human aSyn, 610,787, 1:2000, BD Transduction), for 3 h or overnight and secondary antibody (Alexa Fluor 555 donkey anti-mouse IgG and/or Alexa Fluor 555 goat anti rabbit IgG, (Life Technologies- Invitrogen, Carlsbad, CA, USA) for 2 h at RT. In some of the experiments cells were incubated with Phalloidin (Phalloidin 488, A12379, Phalloidin 594, A12381, 1:50 in PBS, ThermoFisher Scientific, Massachusetts, USA), in order to stain acting filaments. Phalloidin was added to the samples after the secondary antibody, for 1–2 h at RT. Finally, cells were stained with Hoechst 33,258 (Life Technologies- Invitrogen, Carlsbad, CA,USA) (1:5000 in DPBS) for 5 min, washed again and then fixed with Mowiol (Sigma Aldrich, St. Louis, MO, USA) for epifluorescence and confocal microscopy.

Microscopy and imaging

Images in Fig. 1 and in Fig. 5 were acquired using a Leica Inverted Microscope DMI 6000 B (Leica, Wetzlar, Germany), using a 40× objective (HCX Pl Fluotar) or a 63× objective (HCX Pl Fluotar).

Total internal reflection (TIRF) microscopy was conducted on a Zeiss Axio Observer.Z1 microscope equipped with 405, 488, 561 and 639 nm lasers and an Evolve 512 EMCCD camera (Photometrics) and Zen Blue software, using 63× EC Plan-NEOFLUAR and 100× α Plan-APOCHROMAT objectives.

For the screening of Rab proteins, images were taken using an Olympus IX81-ZDC microscope system, with a 40× objective (FN 26,5). Sixty four images were randomly taken using the Olympus Scan^R Image Analysis Software in three independent experiments. Analysis of the aggregation patterns were performed qualitatively, by the experimenters, and the percentage of cells with inclusions was determined.

All the other images were acquired using a Airyscan Confocal Zeiss LSM800 with 40× or 63× oil immersion objectives.

For the colocalization analysis in Fig. 3, the Pearson's R value was calculated via the use of Coloc2 plugin from ImageJ.

For the mean fluorescence intensity values in Figs. 8 was used Fiji (ImageJ) Software and GraphPad Prism for the statistical analyses and graph generation.

NMR experiments

All NMR spectra were recorded on Bruker 600 MHz Avance III spectrometer, equipped with a cryogenically cooled triple resonance $^{1}H(^{13}C/^{15}N)$ TCI probe. Experiments were recorded at 15 °C using protein samples dissolved in Buffer B supplemented with 10% D_2O. For the ^{1}H-^{15}N HSQC experiments we used 16 scans, 1024 complex points (sweep-width of 16 ppm in the ^{1}H dimension) and 256 complex points (sweep-width of 26 ppm in the ^{15}N dimension). Sequence-specific assignments for the backbone of aSyn WT and aSyn A11P/V70P were transferred from previously published studies [39, 42]. Only unambiguously assigned, well resolved peaks were included in the analysis. The I/I_0 ratios obtained for aSyn WT and aSyn A11P/V70P, in absence and presence of SUVs were plotted as a function of the protein sequence to obtain the intensity perturbation profiles [33]. Mean weighted chemical shifts displacements (MWΔCS) for ^{1}H-^{15}N were calculated as $[(\Delta\delta^{1}H)^2 + (\Delta\delta^{15}N/10)^2]^{1/2}$.

Acquisition and processing of NMR spectra were performed using TOPSPIN 3.2 (Bruker Biospin). 2D spectra analyses were performed with CCPN.

SUV preparation

SUVs were prepared from a molar ratio of 1:1 of Coagulation Reagent I containing DOPE:DOPS:DOPC (5:3:2 *w*/w) and DOPC (both Avanti Polar Lipids Inc., USA)

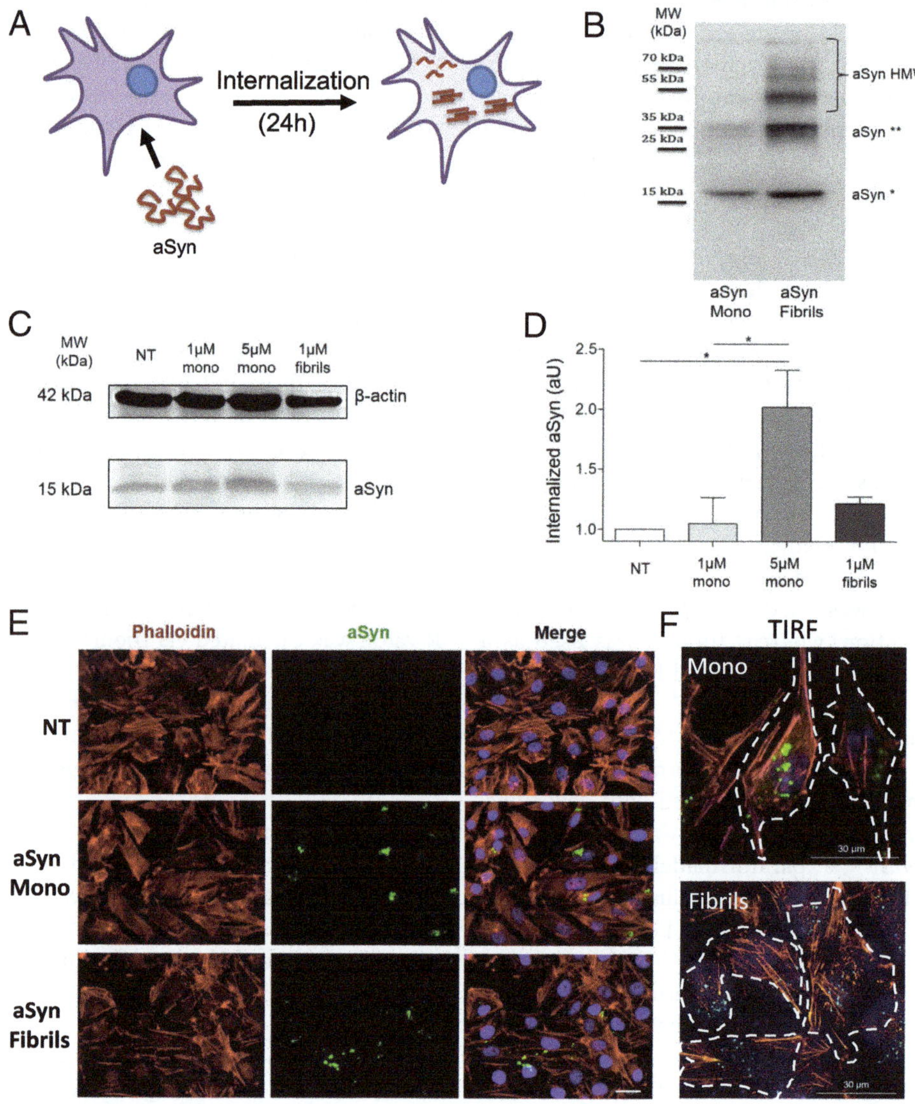

Fig. 1 Recombinant aSyn monomers and fibrils are internalized by H4 cells. **a** Recombinant aSyn monomers (aSyn Mono) or fibrils (aSyn Fibrils) were added to the cell culture medium and incubated for 24 h. **b** SDS-PAGE and immunoblot analysis of the recombinant monomeric or fibrillar species of aSyn used in the experiments. The monomers show also the presence of a small fraction of dimers, as displayed by the faint band at 35 kDa. In the fibril preparation, one can observe the presence of higher molecular weight (HMW) species that are stable even on an SDS-PAGE. **c** Western blot (WB) of H4 cells after treatment with aSyn, confirming the internalization of aSyn monomers or fibrils as seen by the increase in the levels of aSyn in cells that were treated with monomers or fibrils (actin is used as a loading control). **d** Quantification of the immunoblots. Statistical test was performed using one-way ANOVA followed by Tukey's post-hoc tests, $*p < 0.01$. **e** Epifluorescence microscopy of cells treated as indicated. Scale bar: 30 μm. **f** TIRF microscopy of cells treated with aSyn monomers or fibrils (green) and stained with Phalloidin (red) confirm intracellular localization of aSyn. Scale bar: 30 μm

dissolved in chloroform yielding a final molar ratio of DOPE:DOPS:DOPC (5:3:12 w/w). The lipid solution formed a thin film under evaporation of the solvent with nitrogen gas and was further dried by lyophilization under vacuum. The dried phospholipids were dissolved in MES buffer (20 mM MES, 100 mM NaCl, pH 6.5) and underwent several cycles of freeze-thawing and water bath sonication until the solution became clear. The size distribution was also checked by DLS. For the NMR experiments a SUV stock solution of 85 mM (6.6% *w*/*v*) in respect to the monomers was used.

Software and statistical analyses

Colocalization in ICC samples was measured by using ImageJ software and Pierson's Coefficient was calculated and detailed colocalization analysis were performed with the use of Coloc2 Plugin from Fiji (ImageJ software). Figures were composed with CorelDRAW X8 (Corel Corporation, Ottawa, Canada) or with Microsoft Power Point (Microsoft Corporation).

Statistical analysis was performed using Microsoft Excel (Microsoft Corporation) and GraphPad PRISM 5 (GraphPad Software, San Diego, CA, USA). Images were processed with ImageJ V1.41, NIH, USA and/or CorelDRAW X8 (Corel Corporation, Ottawa, Canada).

Statistical tests performed were Student's-two-tailed t-test, one-way-Analysis of Variance (ANOVA) and repeated-measures ANOVA for grouped analysis, followed by Tukey's post-hoc tests for multiple comparison.

Data were expressed as mean ± SEM and a 0.5% general significance level was defined, with significance levels as follows: *: $p < 0.05$; **: $p < 0.01$; ***: $p < 0.001$.

Results

aSyn is internalized and forms intracellular inclusions

In order to investigate the molecular determinants of aSyn internalization, we compared the behaviour of two distinct forms of recombinant aSyn (monomers or fibrillar aggregates, hereafter named fibrils) (Fig. 1a). aSyn monomers and fibrils were generated as described in Methods, and the species were characterized by Transmission Electron Microscopy (TEM) [62] and by SDS-PAGE (Fig. 1b). The preparation of aSyn monomers showed also the presence of a smaller amount of dimers, as illustrated by the band at ~ 35 kDa. Due to contrasting studies reporting the ability of monomeric aSyn to passively enter cells, we tested two different concentrations of this form (1 μM and 5 μM), and 1 μM of fibrils of aSyn (calculated based on the initial concentration of monomeric aSyn). The internalization of aSyn was analysed by immunoblotting and by immunocytochemistry (ICC) after 24 h of incubation with cells. Immunoblot analysis revealed that both aSyn monomers and fibrils were internalized, given the increase in the levels of aSyn in treated cells (Fig. 1c and d). Furthermore, microscopy analysis (confocal and TIRF microscopy) demonstrated that, in cells exposed to monomeric aSyn, the protein accumulated in distinct perinuclear puncta, whereas in cells exposed to fibrils aSyn accumulated in larger cytosolic inclusions (Fig. 1e-f).

aSyn interacts with the plasma membrane and accumulates as high molecular weight species

Based on previously reported aSyn binding to membranes, we hypothesized that the internalization of aSyn might involve an interaction with the plasma membrane. In order to test this, we performed a cell surface biotinylation assay in cells treated with aSyn. aSyn was found in the biotinylated fraction treated with either monomers or fibrils, indicating that extracellular aSyn interacted with the plasma membrane (Fig. 2a and b), and then accumulated within the cell in punctae (Fig. 1e-f).

To further investigate the biochemical nature of the intracellular aSyn, we performed size exclusion chromatography (SEC) of cell lysates, and then dot blot analysis of the various fractions collected. Recombinant aSyn monomers were assessed by SEC in parallel, in order to establish the elution profile of this form of aSyn (Additional file 1: Figure S1A and S1B).

As expected, dot blot analysis of non-treated cells (NT) showed low signal, given the lower levels of aSyn (Fig. 2c). In cells treated with aSyn monomers, we detected the protein in fractions C1 to C4 (black box, Fig. 2c) as reported in the chromatogram (Additional file 1: Figure S1B). We also detected aSyn in the final B fractions (from B6 to B15, red box), indicating the presence of higher molecular weight aSyn species as well (Fig. 2c). Such species were also found in cells treated with aSyn fibrils (fractions B5 to B11, red box), but not monomeric species of aSyn (Fig. 2c), suggesting that the species accumulating in the cells were biochemically distinct depending on the species of aSyn added to the cells.

To further confirm the biochemical differences observed, we performed differential fractionation of the cell lysates using Triton X-100. Immunoblot analysis showed higher levels of Triton X-100-soluble aSyn in cells treated with monomers, and higher levels of Triton X-100-insoluble aSyn in cells treated with fibrils, consistent with the results of the SEC analysis (Fig. 2d, left side). Interestingly, we also detected the formation of high molecular weight aSyn species in cells treated with monomeric aSyn, suggesting that, upon internalization, aSyn monomers start to aggregate (Fig. 2d, right side).

Taken together, these results suggest that both monomeric and fibrillar aSyn enter cells and accumulate in aggregated, high molecular weight species.

aSyn partially colocalizes with Rab5A and Rab7

Next, we performed a microscopy-based screen of mammalian Rab proteins in order to identify the interplay between aSyn and the trafficking pathways. The experiment consisted of treating cells overexpressing each individual mammalian Rab protein (fused to EGFP: Rab-GFP) with aSyn monomers or fibrils in order to assess (i) the effect of aSyn on the subcellular distribution of the Rab proteins, and (ii) the effect of each Rab protein on the subcellular distribution of aSyn. From the screen, we selected a set of Rab proteins whose localization was altered, or that colocalized with aSyn (Additional file 2: Table S1). Of these, we selected Rab4A, Rab5A and Rab7 since they resulted in the strongest phenotype. Interestingly, the Rab proteins identified in the screen are compatible with a hypothesis that aSyn is internalized via an active endocytic mechanism that then sorts the protein into vesicular

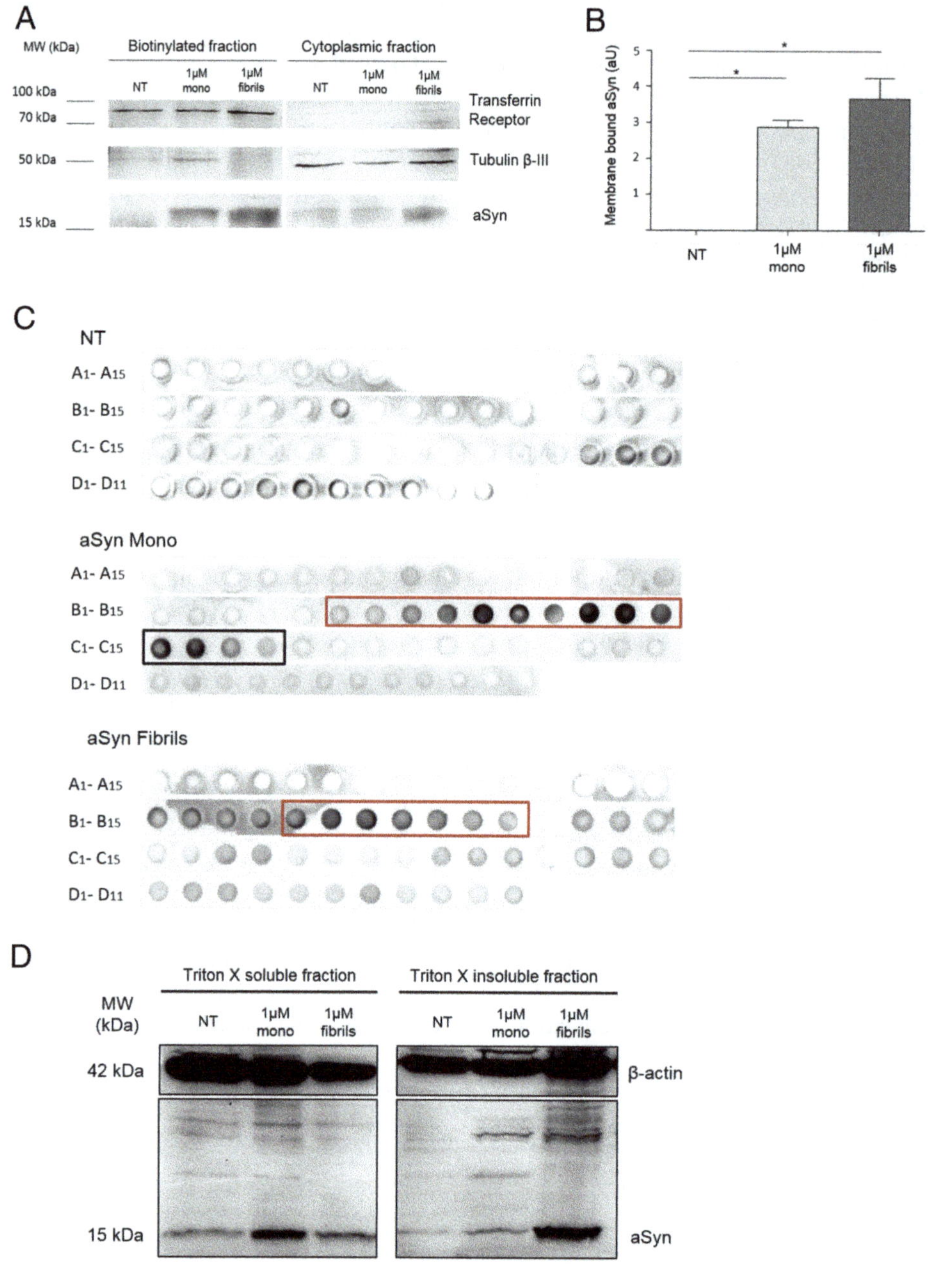

Fig. 2 aSyn associates with membranes and forms high molecular weight species during the internalization process in H4 cells. **a** Immunoblot of the biotinylation assay of cells treated with aSyn monomers or fibrils (tubulin is used as a loading control). **b** Quantification of the levels of aSyn present in the biotinylated fraction (membrane-associated aSyn). Statistical test was performed using one-way ANOVA followed by Tukey's post-hoc tests, *$p < 0.01$ **c** Dot blot of the size exclusion chromatography fractions of lysates of untreated cells, cells treated with 1 μM of aSyn monomers, and cells treated with 1 μM of aSyn fibrils. The black box highlights monomeric aSyn, while the red boxes highlight the presence of high molecular weight species of aSyn. **d** Triton X-100 fractionation, with the soluble (left panel) and the insoluble fractions (right panel), treated as described

compartments, such as endosomes and lysosomes [57]. Therefore, we next focused our study on the these Rab proteins.

First, we assessed the degree of colocalization of aSyn and Rab5A-GFP, or Rab7-GFP, in cells treated with aSyn monomers or fibrils (Fig. 3a). The colocalization was quantified using the Coloc2 plugin of ImageJ Software (Fig. 3b). In cells treated with aSyn monomers, we observed a strong colocalization between aSyn (in red) and Rab5A-GFP vesicles (in green) (Fig. 3a, left column, central panel), as well as a partial, although weaker, colocalization with Rab7-GFP (Fig. 3a, right column, central panel). Interestingly, the colocalization was not observed when cells were treated with

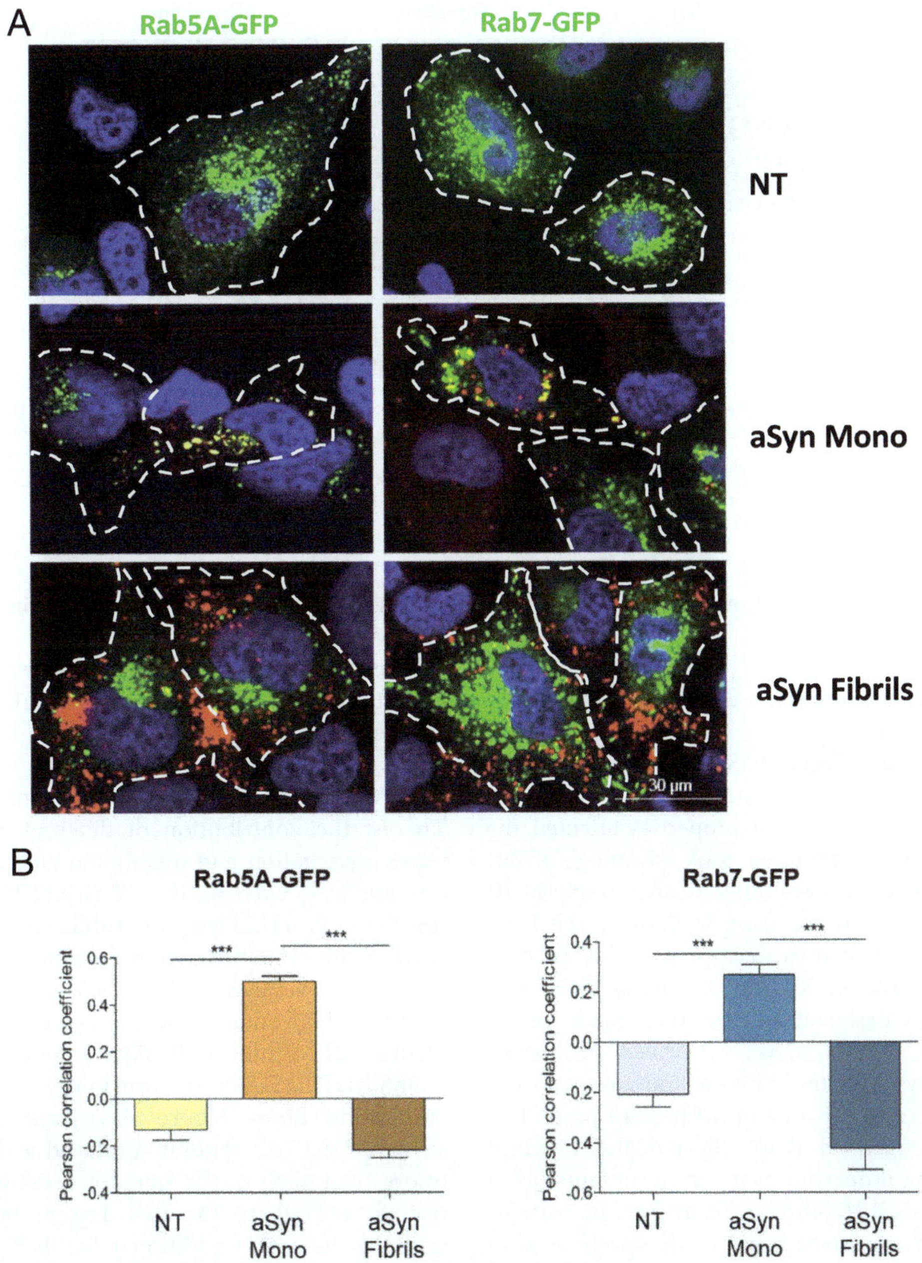

Fig. 3 aSyn partially colocalizes with Rab5A-GFP and Rab7-GFP in H4 cells. **a** ICC of cells transfected with Rab5A-GFP (right side of the panel) or with Rab7-GFP (left side) and then treated with 1 µM of aSyn monomers or fibrils. **b** Pearson correlation coefficient reveals colocalization of aSyn and Rab5A, and of aSyn and Rab7 in cells treated with aSyn monomers, but not with fibrils. Scale bar: 30 µm

aSyn fibrils. This supports the idea that the internalization and sorting of aSyn monomers and fibrils is different, as one might expect given their distinct biochemical properties.

aSyn form inclusions in Rab4A-positive compartments

Next, we examined the interplay between Rab4A and aSyn. We found no effect on the distribution of Rab4A-GFP in cells treated with aSyn fibrils. Likewise, we also found no colocalization between Rab4A-GFP and aSyn in those cells (Fig. 4, right panel). In contrast, when Rab4A-GFP-expressing cells were treated with aSyn monomers, we observed a prominent increase in the size of endosomes, as well as a massive internalization of aSyn that accumulated in compartments surrounded by large, abnormal rings of Rab4A (Fig. 4, central panel on the top and lower panels). This change in the size of early endosomes suggested that exposure to aSyn monomers altered the

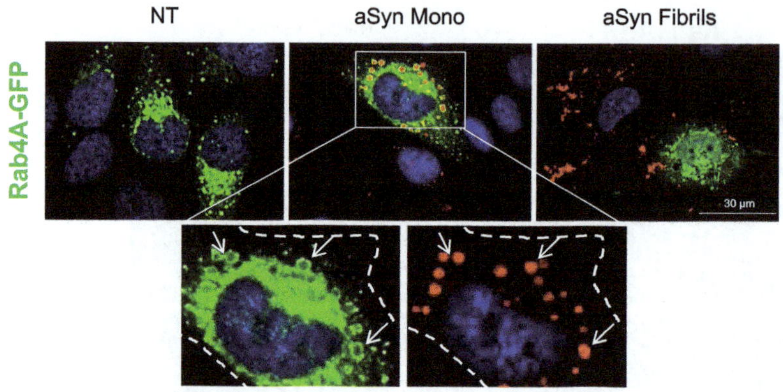

Fig. 4 aSyn accumulates in Rab 4A-positive vesicles. ICC on H4 cells transfected with Rab4A-GFP and treated with 1 µM of aSyn monomers or fibrils. Inset: zoom and separated channels of Rab4A-GFP aSyn. Arrows point to the large inclusions aSyn (in red, panel on the right) matching with the GFP-positive Rab4A vesicles (in green, panel on the left). Scale bar: 30 µm

normal biology of Rab4A and, therefore, the endosome-related trafficking processes.

Membrane binding properties are essential for the internalization of aSyn

Based on the stronger effects of aSyn monomers, we decided to focus on the effects of monomeric aSyn. To investigate whether intrinsic aSyn properties affected the internalization of the protein, we took advantage of different aSyn mutants that have different membrane binding abilities. Specifically, we used WT aSyn, the aSyn A30P familial mutant, known to display weaker binding to membranes [4, 12, 29, 30, 61], and the artificial mutant (A11P/V70P) designed to severely impair membrane binding [8, 10]. First, we performed membrane biotinylation assays with the different mutants, and detected a clear trend in the amount of protein present in the biotinylated fractions that reflected the different membrane binding properties of the aSyn mutants (aSyn A30P and aSyn A11P/V70P) (Fig. 5a and b). In particular, we detected a consistent trend in the levels of aSyn dimeric species in the biotinylated fraction, suggesting that membrane binding is important for dimerization and aggregation of aSyn.

Next, we tested the ability of the two aSyn mutants to enter cells and accumulate in intracellular inclusions, via immunocytochemistry (ICC) approach. We observed a significant reduction in the accumulation of aSyn in inclusions in cells treated with A30P or A11P/V70P aSyn mutants when compared with cells treated with WT aSyn (Fig. 5c), consistent with a difference in the internalization of the mutants. We further quantified the internalization of the different variants of aSyn using immunoblot analysis. We confirmed a strong difference in the internalization of the various mutants (Fig. 5d and e), although the amount of aSyn present in the medium was identical (Fig. 5d).

Taken together, these results suggest that membrane binding is essential for the internalization and, therefore, for the formation of intracellular aSyn inclusions.

aSyn A11P/V70P is unable to bind membranes

To test the contribution of different aSyn regions towards aggregation and membrane binding, several artificial mutants, such as the aSyn A11P/V70P have been designed [9, 11]. First, we validated the effect of the A11P/V70P mutation on the biding of aSyn to small unilamellar vesicles (SUVs) (5:3:12 mixture of DOPE:DOPS:DOPC) using NMR spectroscopy (Additional file 3: Figure S2). While WT aSyn interacted with DOPE:DOPS:DOPC SUVS, as previously described [22, 34] (Additional file 4: Figure S3A, panel on the left), the aSyn A11P/V70P mutant displayed a drastic reduction (more than 80%) of the signal broadening starting from residue 11 (where the first Pro mutation is located) until residue 140 (Additional file 4: Figure S3A, panel on the right), confirming the impairment in membrane binding.

Next, we tested the effect of aSyn A11P/V70P in the cellular model, in comparison to WT aSyn, and also in the presence or absence of Rab4A-GFP, as we had observed increased internalization of aSyn in H4 cells. Using immunoblot analysis and ICC, we found that the internalization A11P/V70P aSyn was negligible when compared to that of WT aSyn (Additional file 4: Figure S3B-D). Importantly, we did not detect internalization when we increased the concentration of A11P/V70P aSyn or when we overexpressed Rab4A-GFP, conditions which significantly increased the levels of intracellular WT aSyn (Additional file 4: Figure S3B-D).

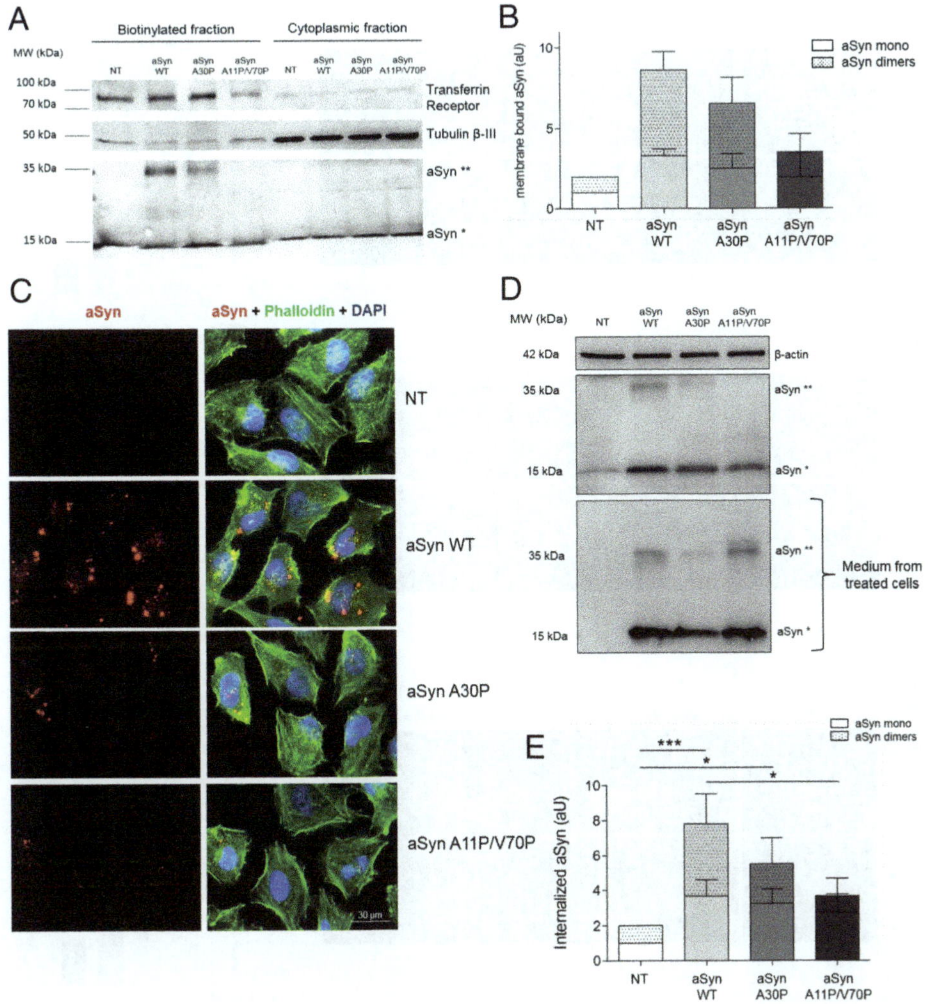

Fig. 5 Membrane binding is essential for the aSyn internalization and inclusion formation. **a** and **b** Immunoblot and quantification of the comparison of WT aSyn and two mutants (A30P and A11P/V70P) with different membrane-binding properties using a biotinylation assay. Dotted bars refer to the band corresponding to aSyn dimers (aSyn**), and clear bars refer to aSyn monomers (aSyn*). **c** ICC and **d** immunoblotting of non-treated (NT) cells, or cells treated with WT, A30P, or A11P/V70P aSyn for 24 h. Scale bar: 30 μm. **e** Quantification of the immunoblot in panel D. Dotted bars refer to the band corresponding to aSyn dimers (aSyn**), and clear bars refer to aSyn monomers (aSyn*). Statistical tests were performed using one-way ANOVA with repeated-measures for grouped analysis, followed by Tukey's post-hoc tests. Data were expressed as mean ± SEM and a 0.5% general significance level was defined, with significance levels as follows: *: $p < 0.05$; **: $p < 0.01$; ***: $p < 0.001$. Statistical significance is indicated with the symbol "#" for the monomers, "+" for the dimers, and "*" for the combination between monomers and dimers

The endocytic pathway is involved in the internalization of aSyn

In order to investigate whether mutants with different membrane binding properties altered the internalization and intracellular fate of internalized aSyn, we used cells overexpressing Rab4A-GFP, Rab5A-GFP, or a constitutively active (CA) mutant of Rab5A-GFP. As described above, WT aSyn was readily internalized and accumulated in Rab4A-GFP-positive vesicles. In contrast, the internalization of the artificial A11P/V70P aSyn mutant was strongly impaired. Curiously, the PD-associated mutant A30P displayed an intermediate phenotype (Fig. 6a-c).

Interestingly, the levels of internalized aSyn (monomers and dimers) were higher in cells expressing these Rab proteins than in naïve cells (shown above in Fig. 5d-e), suggesting that increased levels of Rab4A altered the dynamics of internalization and dimerization of aSyn. The same trend was observed in in cells overexpressing Rab5A-GFP indicating, once again, that stimulation of the early steps of endosome formation increased the internalization of aSyn, as long as the membrane binding properties of the protein are preserved (Fig. 6d-f). In cells overexpressing Rab5A, we also observed an increase in the levels of aSyn dimeric species.

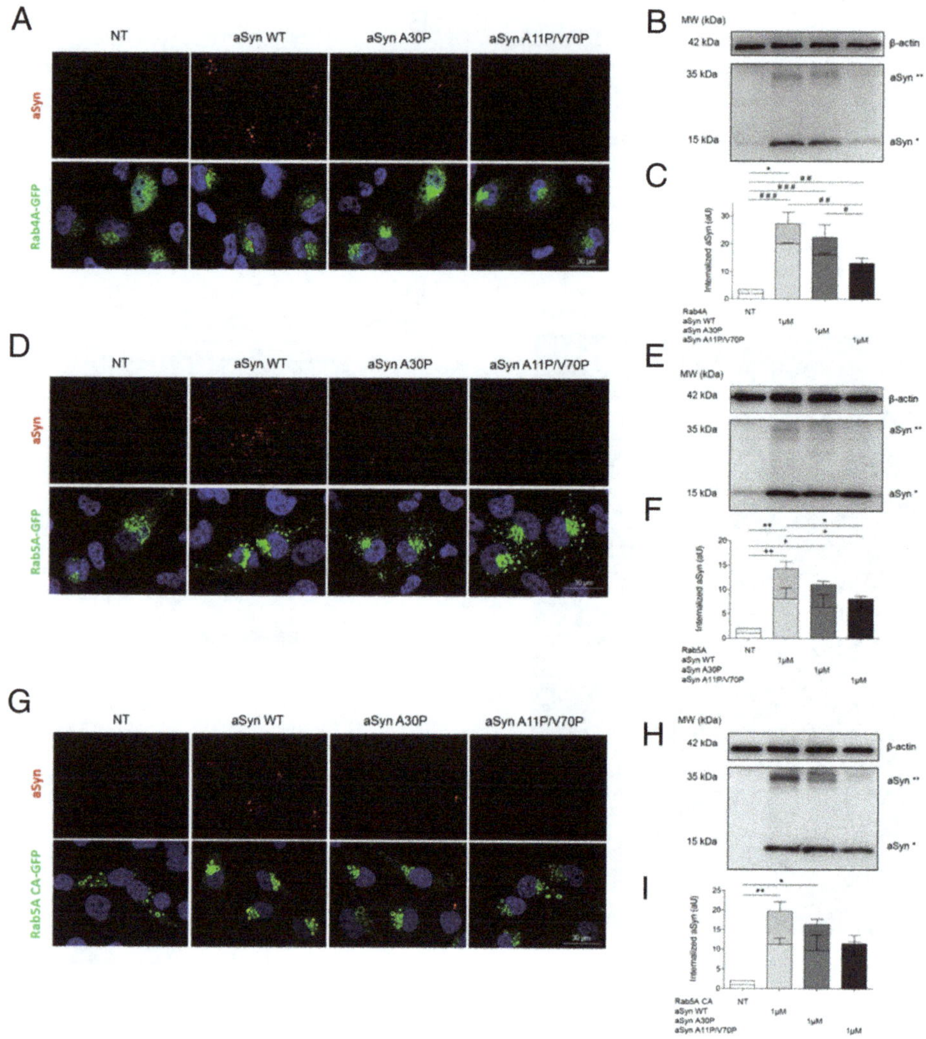

Fig. 6 The A30P and A11P/V70P aSyn mutants are less internalized than WT aSyn. **a** ICC and **b** Immunoblotting of cells transfected with Rab4A-GFP and treated as in experiments shown in Fig. 5. **d** and **e** ICC and Immunoblotting of cells transfected with Rab5A-GFP and treated as above. **g** and **h** ICC and Immunoblotting of cells transfected with Rab5ACA-GFP (constitutively active) and treated as above. **c**, **f** and **i** Quantifications of the immunoblots in panels **b**, **e** and **h**. Dotted bars refer to the band corresponding to aSyn dimers (aSyn**), and clear bars refer to aSyn monomers (aSyn*). Statistical tests were performed using one-way ANOVA with repeated-measures for grouped analysis, followed by Tukey's post-hoc tests. Data are expressed as mean ± SEM and a 0.5% general significance level was defined, with significance levels as follows: *: $p < 0.05$; **: $p < 0.01$; ***: $p < 0.001$. Statistical significance is indicated with the symbol "#" for the monomers, "+" for the dimers, and "*" for the combination between monomers and dimers. Scale bar: 30 μm

Finally, to confirm the functional involvement of Rab5A on the internalization of aSyn, we used a mutant in which the GTPase activity is deregulated, resulting in permanent activation - constitutively active mutant - Rab5ACA-GFP (Fig. 6g-i). In cells expressing this mutant Rab5A, we found overall higher levels of aSyn internalization, further confirming the role of the endocytic pathway in the internalization of aSyn.

Rab7 sorts aSyn for degradation and reduces its intracellular accumulation

Next, we investigated the intracellular fate of internalized aSyn along the endocytic pathway by using Rab7-GFP as a marker. Cells expressing Rab7-GFP were treated with WT, A30P, or A11P/V70P aSyn mutants, and analysed by ICC and immunoblotting, as described above. Surprisingly, we found that the internalization of aSyn, and the formation of dimers, was significantly reduced in cells overexpressing Rab7, and that there were no differences in internalization between WT aSyn or the two mutants. (Fig. 7a-c).

We hypothesized that this effect could be due to the sorting of aSyn for degradation in the lysosome, due to increased stimulation of the lysosomal pathway by overexpression of Rab7. To test this, we repeated the experiment in cells expressing a dominant negative (DN)

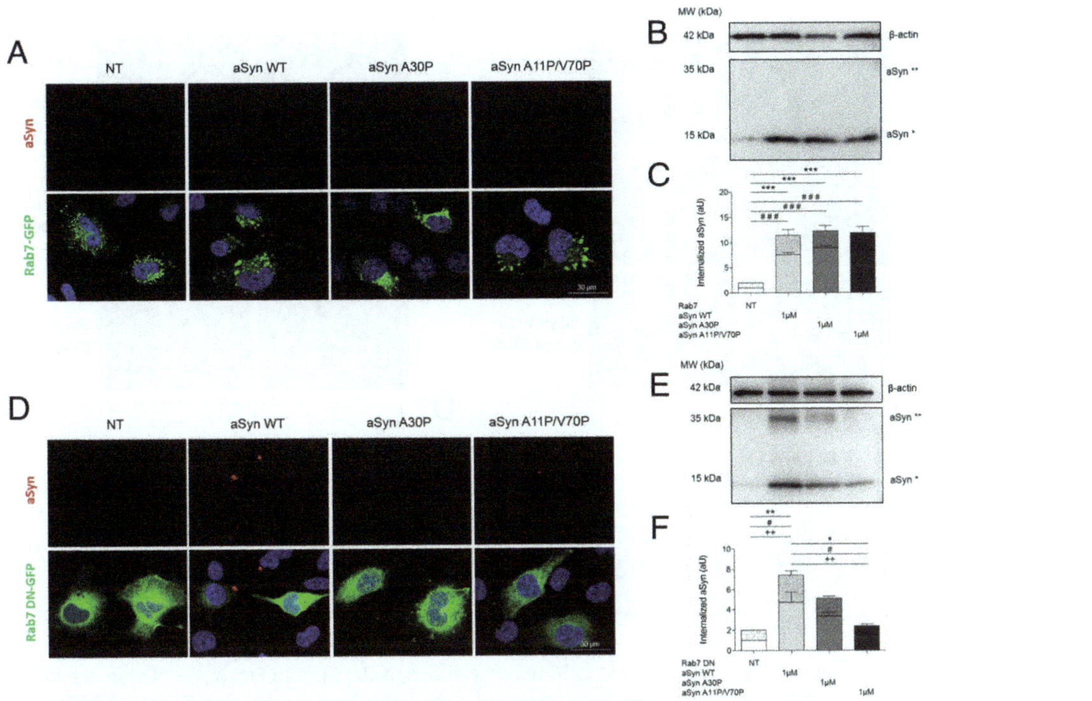

Fig. 7 Rab7 reduces the formation of dimers in cells treated with WT aSyn monomers. **a** ICC and **b** Immunoblotting of H4 cells transfected with Rab7-GFP and treated as described above. **d** and **e** The overexpression of the Rab7 dominant negative (DN) does not affect the degradation of the internalized aSyn. **c** and **f** Quantifications of the immunoblots in panels B and E. Dotted bars refer to the band corresponding to aSyn dimers (aSyn**), and clear bars refer to aSyn monomers (aSyn*). Statistical tests were performed using one-way ANOVA with repeated-measures for grouped analysis, followed by Tukey's post-hoc tests. Data were expressed as mean ± SEM and a 0.5% general significance level was defined, with significance levels as follows: *: $p < 0.05$; **: $p < 0.01$; ***: $p < 0.001$. The significance is shown with the symbol "#" for the monomers, with the symbol "+" for the dimers and with the symbol "*" for the sum between monomers and dimers. Scale bar: 30 μm

mutant of Rab7 (Rab7DN-GFP) that impairs its activity. Interestingly, we found that the internalization and dimerization of aSyn was restored to the initial levels, suggesting that the Rab7DN mutant blocked the sorting of aSyn to the lysosome (Fig. 7d-f).

The internalization of aSyn is mediated by dynamin

Next, we investigated the mechanism involved in the internalization of aSyn by using two well established chemical blockers of endocytosis: PitStop2 (PitStop) and Dyngo 4A (Dyngo).

Pitstop is a selective inhibitor of clathrin-mediated endocytosis (CME) [50, 53], while Dyngo blocks all dynamin-dependent endocytic mechanisms [40].

Naïve cells or cells overexpressing Rab4A-GFP were treated with each of the two compounds for 30 min prior to the treatment with aSyn monomers, and were then incubated together with aSyn for 24 h and processed for ICC or immunoblotting, as described above.

In naive cells, Dyngo efficiently blocked the internalization of aSyn (Additional file 5: Figure S4A). Yet, the opposite effect was observed using PitStop, which increased the accumulation of intracellular aSyn (Additional file 5: Figure S4B-D).

Similarly, in cells overexpressing Rab 4A, Dyngo prevented the internalization and accumulation of aSyn in Rab4A-surrounded vesicles. In contrast, PitStop failed to produce a significant effect (Fig. 8a-b).

Internalized aSyn is degraded by lysosomes

To investigate the fate of internalized aSyn, we tested whether blocking lysosomal and autophagic function would affect the levels of internalized aSyn. We treated cells expressing Rab7-GFP with bafilomycin A1 or chloroquine for 30 min, and then added monomeric aSyn.

Cells were then incubated for 24 h, and then processed for ICC analysis. We confirmed that blocking lysosomal acidification and consequently autophagy inhibited the degradation of internalized aSyn and led to its accumulation, possibly in late endosomes and lysosomes (Fig. 8c-d). Identical results were obtained in naïve cells (Additional file 5: Figure S4, A-D). Interestingly, treatment with chloroquine resulted in stronger accumulation of aSyn than that observed with Bafilomycin A1, consistent with chloroquine being more potent inhibitor then bafilomycin A1.

Taken together, our results suggest that aSyn is internalized via dynamin-mediated endocytosis and trafficked

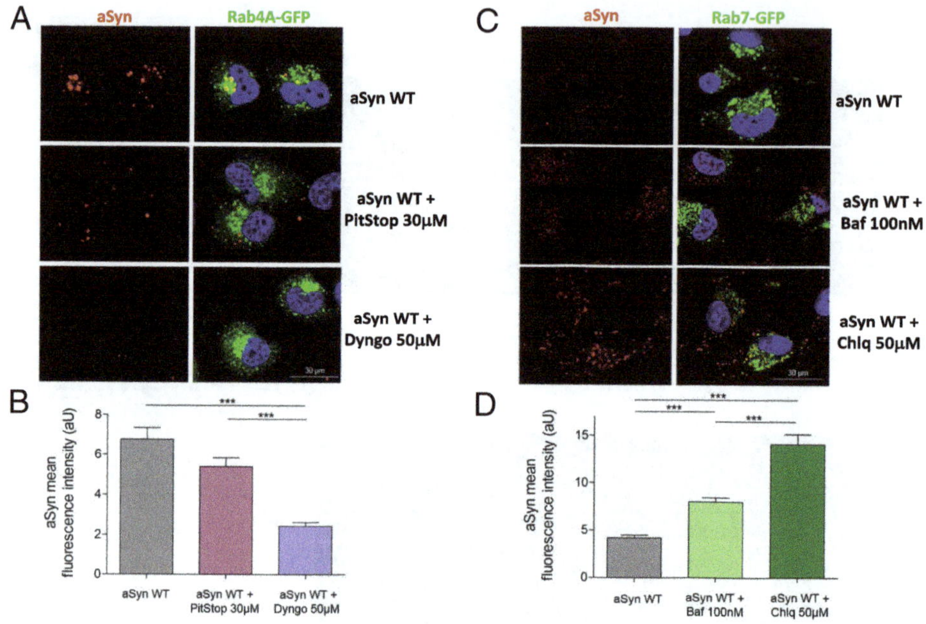

Fig. 8 Blocking endocytosis and autophagy modulates the internalization and degradation of aSyn. **a** ICC of Rab4A-GFP expressed in H4 cells treated with 1 µM aSyn monomers and with vehicle, PitStop 30 µM or with Dyngo 50 µM. Both PitStop and Dyngo are inhibitors of the endocytic processes. **b** Quantification of the aSyn mean fluorescence intensity in the three conditions. **c** ICC of Rab7-GFP-expressing cells treated with 1 µM aSyn monomers and with vehicle, Bafilomycin 100 nM (Baf 100 nM), or with Chloroquine 50 µM (Chlq 50 µM). Bafilomycin and Chloroquine are inhibitors of lysosomal acidification and, therefore, autophagy-lysosome pathway (ALP). (D) Quantification of the aSyn mean fluorescence intensity in the three conditions. Scale bar: 30 µm

via the endocytic pathway and then sorted and targeted for degradation via the autophagy-lysosomal system.

Discussion

A key premise for the prion-like spreading hypothesis of aSyn pathology in PD and other synucleinopathies is that aSyn assemblies are taken up by a recipient cell, are sorted, and then seed the aggregation of endogenous aSyn, thereby propagating pathology [6, 7]. Nevertheless, the precise molecular mechanisms involved are still elusive. Most studies have focused on the understanding of the mechanisms through which aSyn can be released from cells. However, present understanding of the uptake and fate of internalized aSyn are even more limited. Here, we focused on the initial steps of this complex process, and investigated how membrane binding properties might affect the internalization and processing of aSyn in the cell. Using a simple cell-based paradigm, we established that both monomeric and fibrillar aSyn proteins can interact with the plasma membrane and be internalized, accumulating in high molecular weight species.

We also investigated the role of aSyn membrane binding on internalization by using mutants that are known to affect the membrane binding properties of aSyn. The PD-associated mutant A30P displays reduced membrane interactions and reduced internalization, when compared to WT aSyn. Structurally, the substitution of the alanine by the proline disrupts the α-helical domains formed in the N-terminal and central regions of aSyn, thereby affecting the ability of the protein to interact and bind to membranes. Consistently, we found that the artificial mutant A11P/V70P, designed to disrupt membrane binding [9, 11], indeed disrupts the N-terminal region and compromises membrane binding, thereby impairing internalization of aSyn. Interestingly, it has been widely suggested that the aSyn A30P mutant may cause PD by affecting at least partially different cellular pathways that those affected by WT aSyn. Our study is consistent with this idea, and forms now the basis for in vivo investigations of the prion-like spreading ability of this mutant, which has not been documented, since the studies performed thus far focused on WT aSyn [57–60].

Importantly, our findings suggest that one cannot dismiss that, even monomeric aSyn might be sufficient to initiate the process of aggregation and therefore, the spreading of aSyn pathology. Only two studies have reported the internalization of aSyn monomers and the involvement of the endocytic pathway, although the purpose and the readouts used in those studies were different [1, 58]. Interestingly, a comparable study reported consistent results regarding tau, suggesting that extracellular monomeric tau is sufficient to initiate the spreading of tau pathology [41]. However, one caveat of our study is the concentration of aSyn used, which is higher than that used in several other studies [30, 61].

Nevertheless, our study establishes proof of concept that, perhaps a local and even just temporary increase in the concentration of aSyn, caused for example by dying cells in the brain, might lead to the release of monomeric aSyn that could be internalized by neighbouring cells, thereby initiating the spreading of pathology.

Using selective inhibitors of the endocytic pathways, such as Dyngo and PitStop, we demonstrate that aSyn is internalized in a dynamin-dependent process, but not through CME. It is possible that when clathrin-mediated mechanisms are blocked (PitStop-related effects), the clathrin-independent processes (blocked by Dyngo) which are presumably involved in the internalization of aSyn are exacerbated. In this way, a larger amount of aSyn can be internalized in naïve cells (Additional file 5: Figure S4). Additional studies are required to further dissect the molecular mechanisms involved in aSyn endocytosis.

In an unbiased screen of Rab proteins, we found that, once internalized, aSyn partially colocalizes with Rab5A and Rab7, suggesting that the endosomal pathway is involved in the sorting and processing of aSyn. Furthermore, we demonstrated that Rab4A modulates the internalization of aSyn, and surrounds the internalized protein. Both Rab4A and Rab5A are localized in the early endosomes, contributing to the recycling/transport of proteins to the plasma membrane, respectively. Rab7 is localized in the late endosomes, lysosomes and phagosomes, contributing to the fusion of late endosome with lysosome [27]. Recent studies suggest that, even when different Rabs localize in the same compartment, they can occupy distinct membrane microdomains with different functions, called Rab domains. These are dynamic structures but, interestingly, do not mix significantly over time [52]. One of the possible explanations for this segregation mechanism is that it could be, at least in part, mediated by effector proteins and their association with additional molecules in the organelle membranes [52]. Consistently, a recent study showed that Rab4A is essential for the recruitment of adaptor proteins to the tubular subdomain of the early endosomal membrane, orchestrating a signalling cascade that initiates the

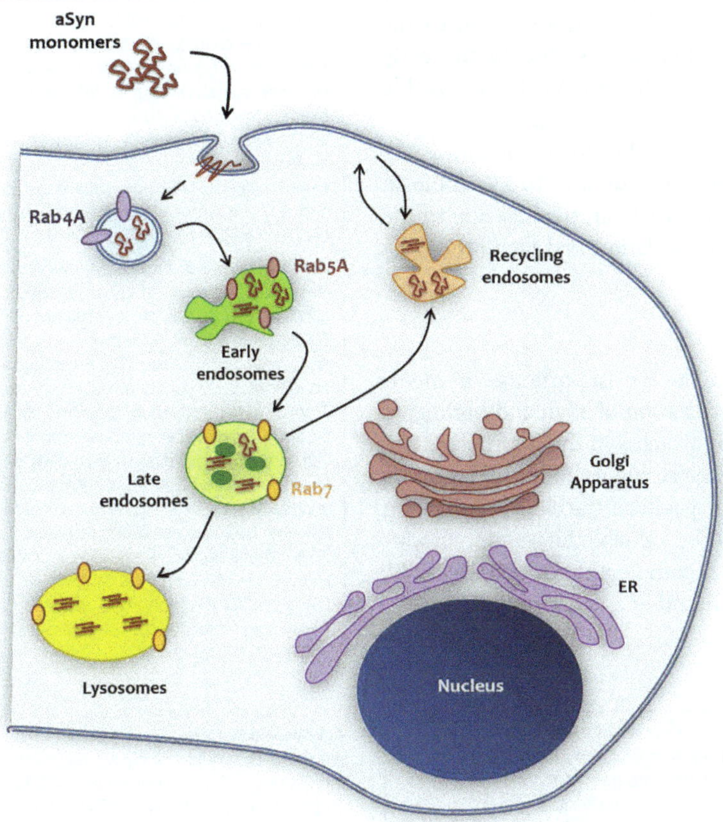

Fig. 9 Model for the proposed mechanism of aSyn internalization and interaction with the membrane and trafficking machinery. After interacting with the plasma membrane, aSyn monomers are internalized through the endocytic pathway. Rab4A plays an important role in the sorting of aSyn and on the transport from/to the plasma membrane. aSyn is then found in the early endosome and, it is possible that during the progress from early to late endosome, due to acidification of the environment, aSyn monomers start to aggregate and form high molecular weight species. Afterwards, late endosomes fuse with lysosomes, where aSyn can be degraded. It is possible that, during any of those steps, aSyn is released to the cytoplasm and to the extracellular milieu, leading to its potential accumulation and spreading of pathology

formation of a sorting platform from which multiple classes of vesicular carriers emerge [18]. Based on the perturbation of the morphology of Rab4A vesicles due to the interaction with aSyn, we hypothesize that this particular GTPase cascade nucleated by Rab4A is actively involved into the internalization of aSyn. Future studies should focus on the precise molecular link between Rab4A and aSyn, and with other potential effector proteins.

Our data are also consistent with the idea that aSyn aggregation may be favoured by the low pH - that is reported to vary between 5.0 and 6.0 - in endosomes [14], as well as by molecular crowding due to the presence of acidic endosomal proteins. It is also possible that aSyn aggregation occurs due to problems in the late endosome-lysosomal compartments, which we found to be involved in the degradation of aSyn. Thus, impaired autophagic degradation of aSyn could lead to the accumulation of aggregation-competent species [19].

Our data enable us to propose a model for the internalization of aSyn, based on the interaction with trafficking machinery components (Fig. 9). After interacting with the membrane, aSyn monomers are internalized through the endocytic pathway, where Rab4A plays an important role on the protein sorting and on the transport from/to the plasma membrane. aSyn is thereafter sorted to the early endosome (in colocalization with Rab5A). It is possible that during the progress from early to late endosome (where it colocalizes with Rab7), due to pH acidification, aSyn monomers start to oligomerize, and to assemble in high molecular weight species, that can then escape to the cytoplasm, leading to its accumulation and facilitating the spreading of aSyn pathology.

Conclusions

In total, our study emphasizes the importance of membrane binding for the internalization of aSyn and highlights the fundamental role of Rab proteins in the internalization, sorting, and processing of aSyn, suggesting that targeting specific Rab proteins and/or specific intracellular trafficking components might prove to be valuable targets for modulating the spreading of aSyn pathology and, consequently, disease progression in PD and other synucleinopathies.

Additional files

Additional file 1: Figure S1. Characterization of recombinant aSyn monomers. (A) Fractions collected upon protein separation on Superose 6 10/300 size exclusion column. (B) Chromatogram of recombinant aSyn monomers showing the fractions in which monomeric aSyn was recovered. (PDF 190 kb)

Additional file 2: Table S1. Results of the Rab protein screen. Rab-GTPase family members selected in a screen where we assessed alterations in the subcellular distribution of the Rab protein or the colocalization with aSyn in cells treated with aSyn monomers or fibrils. In the column "morphology", a "11% more Rab-vesicles" statement means that in the 11% of the cells analysed, the localization of Rabs is more vesicular (suggesting an increase of 11% in the active, GTP-bound Rab protein) compared to the localization pattern shown in naïve cells. In contrast, the statement "15% more cytosolic Rab" indicates that, in this case, 15% of the cells analysed showed an increase in the cytosolic, diffuse localization of Rab protein when compared to the naive cells (suggesting an increase of 15% in the inactive, GDP-bound Rab protein). (PDF 396 kb)

Additional file 3: Figure S2. Membrane binding and internalization of the A11P/V70P aSyn mutant. (A) Membrane binding properties of WT (left) and of A11P/V70P aSyn (right) in the presence of artificial small unilamellar vesicles membranes (SUVs) [1:100 protein:SUVs ratio]. (B) Immunoblotting of Rab 4A-GFP-expressing cells treated with 1 μM or 5 μM of WT or A11P/V70P aSyn. (C) Quantification of the immunoblots. Dotted bars refer to the band corresponding to aSyn dimers (aSyn**), and clear bars refer to aSyn monomers (aSyn*). Statistical tests were performed using one-way-analysis of variance (ANOVA) with repeated-measures for grouped analysis, followed by Tukey's post-hoc tests. Data were expressed as mean ± SEM and a 0.5% general significance level was defined, with significance levels as follows: *: $p < 0.05$; **: $p < 0.01$; ***: $p < 0.001$. Significance is shown with the symbol "#" for the monomers, with the symbol "+" for the dimers and with the symbol "*" for the sum between monomers and dimers. (D) ICC of H4 cells transfected with Rab 4A-GFP and treated with 1 μM or 5 μM of aSyn wild type and aSyn A11P/V70P. Scale bar: 30 μm. (PDF 3696 kb)

Additional file 4: Figure S3. Dyngo blocks whereas PitStop enhances the internalization of aSyn. (A) ICC of H4 cells treated with 1 μM aSyn monomers and with vehicle, PitStop 30 μM or Dyngo 50 μM. Both PitStop and Dyngo are inhibitors of the endocytic processes. (B) Quantification of the aSyn mean fluorescence intensity in the three conditions. Scale bar: 30 μm. (C) Immunoblotting of H4 cells treated with different concentrations of PitStop and Dyngo. (D) Quantification of the immunoblot. Dotted bars refer to the band corresponding to aSyn dimers (aSyn**), and clear bars refer to aSyn monomers (aSyn*). Statistical tests were performed using one-way-analysis of variance (ANOVA), with repeated-measures for grouped analysis, followed by Tukey's post-hoc tests. Data were expressed as mean ± SEM and a 0.5% general significance level was defined, with significance levels as follows: *: $p < 0.05$; **: $p < 0.01$; ***: $p < 0.001$. Significance is shown with the symbol "#" for the monomers, with the symbol "+" for the dimers and with the symbol "*" for the sum between monomers and dimers. Scale bar: 30 μm (PDF 1051 kb)

Additional file 5: Figure S4. Blocking of autophagy inhibits the degradation of aSyn. (A) ICC of H4 cells treated with 1 μM aSyn monomers and with vehicle, Bafilomycin 100 nM (Baf 100 nM), or with Chloroquine 50 μM (Chlq 50 μM). Bafilomycin and chloroquine are inhibitors of the ALP. (B) Quantification of the aSyn mean fluorescence intensity in the three conditions. Scale bar: 30 μm. (C) Immunoblotting of H4 cells treated with 1 μM aSyn WT and incubated with Bafilomycin 100 nM or Chloroquine 50 μM. (D) quantification of the immunoblot in panel C. Dotted bars refer to the band corresponding to aSyn dimers (aSyn**), and clear bars refer to aSyn monomers (aSyn*). Statistical tests were performed using one-way-analysis of variance (ANOVA), with repeated-measures for grouped analysis, followed by Tukey's post-hoc tests. Data is expressed as mean ± SEM and a 0.5% general significance level was defined, with significance levels as follows: *: $p < 0.05$; **: $p < 0.01$; ***: $p < 0.001$. Significance is shown with the symbol "#" for the monomers, with the symbol "+" for the dimers and with the symbol "*" for the sum between monomers and dimers. Scale bar: 30 μm. (PDF 1301 kb)

Acknowledgements

We thank Dr. Andrés Binolfi for technical assistance with NMR and fruitful discussions. TFO is supported by the DFG Center for Nanoscale Microscopy and Physiology of the Brain (CNMPB) and by SFB1286. TFO is also supported by a grant from La Marató de TV3, Spain. IM is supported by Schram-Stiftung T287/25457 and Deutsche Forschungsgemeinschaft (Emmy Noether Young Investigator Award MI-1702/1 and SFB1190/P2).

Funding

TFO is supported by the DFG Center for Nanoscale Microscopy and Physiology of the Brain (CNMPB) and by SFB1286. TFO is also supported by a grant from La Marató de TV3, Spain. IM is supported by Schram-Stiftung T287/25457 and Deutsche Forschungsgemeinschaft (Emmy Noether Young Investigator Award MI-1702/1 and SFB1190/P2).

Authors' contributions

CM conducted experiments, analysed data, and wrote the manuscript. MH, EG, TLF, AVP, MAS, CD and TB conducted experiments. CD, COF and IM designed experiments and revised the manuscript. TFO designed experiments, supervised the project, and wrote the manuscript. All authors read and approved the final manuscript.

Competing interests

The authors declare that they have no competing interests.

Author details

[1]Department of Experimental Neurodegeneration, Center for Biostructural Imaging of Neurodegeneration, Center for Nanoscale Microscopy and Molecular Physiology of the Brain, University Medical Center Goettingen, 37073 Göttingen, Germany. [2]Max Planck Laboratory for Structural Biology, Chemistry and Molecular Biophysics of Rosario (MPLbioR, UNR-MPIbpC) and Instituto de Investigaciones para el Descubrimiento de Fármacos de Rosario (IIDEFAR, UNR-CONICET), Universidad Nacional de Rosario, Ocampo y Esmeralda, S2002LRK Rosario, Argentina. [3]European Neuroscience Institute, University Medical Center Göttingen, Göttingen, Germany. [4]Trans-synaptic Signaling Group, European Neuroscience Institute, Grisebachstrasse 5, 37077 Goettingen, Germany. [5]Max Planck Institute for Experimental Medicine, Göttingen, Germany. [6]Institute of Neuroscience, The Medical School, Newcastle University, Framlington Place, Newcastle Upon Tyne NE2 4HH, UK.

References

1. Ahn KJ, Paik SR, Chung KC, Kim J (2006) Amino acid sequence motifs and mechanistic features of the membrane translocation of a-synuclein. J Neurochem 97:265–279
2. Appel-Cresswell S, Vilarino-Guell C, Encarnacion M, Sherman H, Yu I, Shah B et al (2013) Alpha-synuclein p.H50Q, a novel pathogenic mutation for Parkinson's disease. Mov. Disord 28:811–813
3. Bartels T, Choi JG, Selkoe DJ, Hospital W (2012) a-Synuclein occurs physiologically as a helically folded tetramer that resists aggregation. Nature. 2011;477(7362):107-110. https://doi.org/10.1038/nature10324
4. Bertoncini CW, Fernández CO, Griesinger C, Jovin TM, Zweckstetter M (2005) Familial mutants of α-synuclein with increased neurotoxicity have a destabilized conformation. J Biol Chem 280:30649–30652
5. Breda C, Nugent ML, Estranero JG, Kyriacou CP, Outeiro TF, Steinert JR et al (2015) Rab11 modulates α-synuclein-mediated defects in synaptic transmission and behaviour. Hum Mol Genet 24:1077–1091
6. Brundin P, Melki R (2017) Prying into the prion hypothesis for Parkinson's disease. J Neurosci 37:9808–9818
7. Brundin P, Melki R, Kopito R (2010) Prion-like transmission of protein aggregates in neurodegenerative disorders. Nat. Rev. Mol. Cell Biol. 11:301–307
8. Burré J, Sharma M, Südhof TC (2012) Systematic mutagenesis of aSyn reveals distinct sequence requirements for physiological and pathological activities. J Neurochem 15227:15242
9. Burré J, Sharma M, Südhof TC (2014) a-synuclein assembles into higher-order multimers upon membrane binding to promote SNARE [Internet]. Proc. Natl. Acad. Sci. U. S. A 111:E4274–E4283. Available from: http://www.pubmedcentral.nih.gov/articlerender.fcgi?artid=4210039&tool=pmcentrez&rendertype=abstract
10. Burré J, Sharma M, Südhof TC (2015) Definition of a molecular pathway mediating a-Synuclein. J Neurosci 35:5221–5232
11. Burré J, Sharma M, Tsetsenis T, Buchman V, Südhof TC (2010) α -Synuclein Promotes SNARE-Complex Assembly in vivo and in vitro. Science (80-) 329: 1663–1667
12. Bussell R, Eliezer D (2001) Residual structure and dynamics in Parkinson's disease-associated mutants of a-Synuclein. J Biol Chem 276:45996–46003
13. Bussell RJ, Eliezer D (2003) A structural and functional role for 11-mer repeats in a -Synuclein and other exchangeable lipid binding proteins. J Mol Biol 2836:763–778
14. Casey JR, Grinstein S, Orlowski J (2010) Sensors and regulators of intracellular pH. Nat. Rev. Mol. Cell Biol. 11:50–61
15. Chandra S, Chen X, Rizo J, Jahn R, Su TC (2003) A broken a-helix in folded a-Synuclein. J Biol Chem 278:15313–15318
16. Chartier-Harlin M-C, Kachergus J, Roumier C, Mouroux V, Douay X, Lincoln S et al (2004) α-Synuclein locus duplication as a cause of familial Parkinson's disease. Lancet 364:1169–1171
17. Chutna O, Gonçalves S, Villar-Piqué A, Guerreiro P, Marijanovic Z, Mendes T et al (2014) The small GTPase Rab11 co-localizes with α-synuclein in intracellular inclusions and modulates its aggregation, secretion and toxicity. Hum Mol Genet 23:6732–6745
18. D'Souza RS, Semus R, Billings EA, Meyer CB, Conger K, Casanova JE (2014) Rab4 orchestrates a small GTPase cascade for recruitment of adaptor proteins to early endosomes. Curr Biol 24:1187–1198
19. Danzer KM, Kranich LR, Ruf WP, Cagsal-Getkin O, Winslow AR, Zhu L et al (2012) Exosomal cell-to-cell transmission of alpha synuclein oligomers. Mol Neurodegener 7:42
20. Elia G (2008) Biotinylation reagents for the study of cell surface proteins. Proteomics 8:4012–4024
21. Ferreon ACM, Gambin Y, Lemke EA, Deniz AA (2009) Interplay of a-synuclein binding and conformational switching probed by single-molecule fluorescence. Proc Natl Acad Sci 106:5645–5650
22. Fonseca-Ornelas L, Eisbach SE, Paulat M, Giller K, Fernández CO, Outeiro TF et al (2014) Small molecule-mediated stabilization of vesicle-associated helical α-synuclein inhibits pathogenic misfolding and aggregation. [Internet]. Nat. Commun 5:5857. Available from: http://www.ncbi.nlm.nih.gov/pubmed/25524885
23. Giasson BI, Murray IVJ, Trojanowski JQ, Lee VM (2001) A hydrophobic stretch of 12 amino acid residues in the middle of a-Synuclein is essential for filament assembly. J Biol Chem 276:2380–2386
24. Golde TE, Borchelt DR, Giasson BI, Lewis J (2013) Thinking laterally about neurodegenerative proteinopathies. J Clin Invest 123:1847–1855
25. Gonçalves SA, Macedo D, Raquel H, Simões PD, Giorgini F, Ramalho JS et al (2016) shRNA-based screen identifies endocytic recycling pathway components that act as genetic modifiers of alpha-Synuclein aggregation, secretion and toxicity. PLoS Genet 12:1–26
26. Hoyer W, Cherny D, Subramaniam V, Jovin TM (2004) Impact of the Acidic C-Terminal Region Comprising Amino Acids 109–140 on a-Synuclein Aggregation in Vitro. Biochemistry:16233–16242
27. Hutagalung AA, Novick PJ (2011) Role of Rab GTPases in membrane traffic and cell physiology. Physiol Rev 91:119–149
28. Jao CC, Hegde BG, Chen J, Haworth IS, Langen R (2008) Structure of membrane-bound alpha-synuclein from site-directed spin labeling and computational refinement. Proc Natl Acad Sci U S A 105:19666–19671
29. Jensen PH, Nielsen MS, Jakes R, Dotti G, Goedert M (1998) Binding of alpha-synuclein to brain vesicles is abolished by familial parkinsons-disease mutation. J Biol Chem 273:26292–26294
30. Jo E, Fuller N, Rand RP, St George-Hyslop P, Fraser PE (2002) Defective membrane interactions of familial Parkinson's disease mutant A30P α-synuclein 1 1Edited by I. B Holland J Mol Biol 315:799–807
31. Kim TD, Paik SR, Yang C-H (2002) Structural and functional implications of C-terminal regions of alpha-synuclein. Biochemistry 41:13782–13790
32. Krüger R, Kuhn W, Müller T, Woitalla D, Graeber M, Epplen JT et al (1998) Ala30Pro mutation in the gene encoding a-synuclein in Parkinson's. Nat Genet 18:231–236
33. Lamberto GR, Binolfi A, Orcellet ML, Bertoncini CW, Zweckstetter M, Griesinger C et al (2009) Structural and mechanistic basis behind the inhibitory interaction of PcTS on alpha-synuclein amyloid fibril formation. Proc Natl Acad Sci U S A 106:21057–21062
34. Lee JH, Ying J, Bax A (2016) Nuclear magnetic resonance observation of a-Synuclein membrane interaction by monitoring the acetylation reactivity of its lysine side chains. Biochemistry 55:4949–4959
35. Lesage S, Anheim M, Letournel F, Bousset L, Honoré A, Rozas N et al (2013) G51D α-synuclein mutation causes a novel parkinsonian-pyramidal syndrome. Ann Neurol 73:459–471
36. Lesage S, Bras J, Cormier-Dequaire F, Condroyer C, Nicolas A, Darwent L et al (2015) Loss-of-function mutations in *RAB39B* are associated with typical early-onset Parkinson disease. Neurol Genet 1:e9
37. Luk KC, Kehm V, Carroll J, Zhang B, Brien PO, John Q et al (2013) Pathological a-synuclein transmission initiates parkinson-like Neurodegeneration in non-transgenic mice. Science (80-) 338:949–953

38. Luk KC, Song C, O'Brien P, Stieber A, Branch JR, Brunden KR et al (2009) Exogenous alpha-synuclein fibrils seed the formation of Lewy body-like intracellular inclusions in cultured cells. Proc Natl Acad Sci U S A 106:20051–20056
39. Maltsev AS, Ying J, Bax A (2012) Impact of N-terminal acetylation of a-synuclein on its random coil and lipid binding properties. Biochemistry 51:5004–5013
40. Mccluskey A, Daniel JA, Hadzic G, Chau N, Clayton EL, Mariana A et al (2013) Building a better dynasore: the dyngo compounds potently inhibit dynamin and endocytosis. Traffic 14:1272–1289
41. Michel CH, Kumar S, Pinotsi D, Tunnacliffe A, George-Hyslop PS, Mandelkow E et al (2014) Extracellular monomeric tau protein is sufficient to initiate the spread of tau protein pathology. J Biol Chem 289:956–967
42. Miotto MC, Valiente-Gabioud AA, Rossetti G, Zweckstetter M, Carloni P, Selenko P et al (2015) Copper binding to the N-terminally acetylated, naturally occurring form of alpha-Synuclein induces local helical folding. J Am Chem Soc 137:6444–6447
43. Park SM, Jung HY, Kim TD, Park JH, Yang C, Kim J. Distinct Roles of the N-terminal-binding Domain and the C-terminal-solubilizing Domain of A-Synuclein , a Molecular Chaperone. J Biol Chem 2002; 277: 28512–28520
44. Pasanen P, Myllykangas L, Siitonen M, Raunio A, Kaakkola S, Lyytinen J et al (2014) A novel α-synuclein mutation A53E associated with atypical multiple system atrophy and Parkinson's disease-type pathology. Neurobiol Aging 35:1–5
45. Polymeropoulos MH (1997) Mutation in the -synuclein gene identified in families with Parkinson's disease. Science (80-.) 276:2045–2047
46. Scheurer SB, Rybak JN, Roesli C, Brunisholz RA, Potthast F, Schlapbach R et al (2005) Identification and relative quantification of membrane proteins by surface biotinylation and two-dimensional peptide mapping. Proteomics 5: 2718–2728
47. Scott DA, Tabarean I, Tang Y, Cartier A, Masliah E (2010) A pathologic cascade leading to synaptic dysfunction in a-synuclein induced neurodegeneration. J Neurosci 30:8083–8095
48. Shi M, Shi C, Xu Y (2017) Rab GTPases: the key players in the molecular pathway of Parkinson's disease. Front Cell Neurosci 11:1–8
49. Singleton AB, Farrer M, Johnson J, Singleton A, Hague S, Kachergus J et al (2003) a-Synuclein locus triplication causes Parkinson ' s Disease. Science (80-.) 302:841
50. Smith CM, Haucke V, McCluskey A, Robinson PJ, Chircop M (2013) Inhibition of clathrin by pitstop 2 activates the spindle assembly checkpoint and induces cell death in dividing HeLa cancer cells. Mol Cancer 12:1–15
51. Snead D, Eliezer D (2014) Alpha-Synuclein function and dysfunction on cellular membranes. Exp Neurobiol 23:292
52. Sönnichsen B, De Renzis S, Nielsen E, Rietdorf J, Zerial M (2000) Distinct membrane domains on endosomes in the recycling pathway visualized by multicolor imaging of Rab4, Rab5, and Rab11. J Cell Biol 149:901–913
53. Soohoo AL, Puthenveedu Ma (2013) Divergent modes for cargo-mediated control of clathrin-coated pit dynamics. Mol Biol Cell 24:1725–1734 S1-12
54. Souza MY, Giasson BI, Lee VM, HI Y (2000) Chaperone-like activity of synucleins 474:116–119
55. Spillantini MG, Crowther RA, Jakes R, Hasegawa M, Goedert M (1998) alpha-Synuclein in filamentous inclusions of Lewy bodies from Parkinson's disease and dementia with lewy bodies. Proc Natl Acad Sci U S A 95:6469–6473
56. Spillantini MG, Schmidt ML, Lee VM-Y, Trojanowski JQ, Goedert M (1997) a-Synuclein in Lewy bodies. Nature. 1997;388(6645):839–840
57. Stenmark H (2009) Rab GTPases as coordinators of vesicle traffic. Nat Rev Mol Cell Biol 10:513–525
58. Sung JY, Kim J, Paik SR, Park JH, Ahn YS, Chung KC (2001) Induction of neuronal cell death by Rab5A-dependent endocytosis of a-Synuclein. J Biol Chem 276:27441–27448
59. Takeda A, Mallory M, Sundsmo M, Honer W, Hansen L, Masliah E (1998) Abnormal accumulation of NACP/α-synuclein in neurodegenerative disorders. Am J Pathol 152:367–372
60. Trojanowski JQ, Goedert M, Iwatsubo T, Lee VM (1998) Fatal attractions: abnormal protein aggregation and neuron death in Parkinson's disease and Lewy body dementia. Cell Death Differ 5:832–837
61. Tsigelny IF, Sharikov Y, Kouznetsova VL, Greenberg JP, Wrasidlo W, Overk C et al (2015) Molecular determinants of α-synuclein mutants' oligomerization and membrane interactions. ACS Chem Neurosci 6:403 416
62. Villar-Piqué A, Lopes da Fonseca T, Sant Anna R, Szegö ÉM, Fonseca-Ornelas L, Pinho R et al (2016) Environmental and genetic factors support the dissociation between α-synuclein aggregation and toxicity. Proc. Natl. Acad. Sci. U. S. A 113:E6506–E6515. http://www.ncbi.nlm.nih.gov/pubmed/27708160, http://www.pubmedcentral.nih.gov/articlerender.fcgi?artid=PMC5081623
63. Volpicelli-daley L a, Luk KC, Patel TP, Tanik SA, Dawn M, Stieber A et al (2011) Exogenous aSynuclein fibrils induce Lewy body pathology leading to synaptic dysfunction and neuron death. Neuron 72:57–71
64. Wakabayashi K, Yoshimoto M, Tsuji S, Takahashi H (1998) A-Synuclein immunoreactivity in glial cytoplasmic inclusions in multiple system atrophy. Neurosci. Lett 249:180–182
65. Wang W, Perovic I, Chittuluru J, Kaganovich A, Nguyen LTT, Liao J et al (2011) A soluble a-synuclein construct forms a dynamic tetramer. Proc Natl Acad Sci 108:17797–17802
66. Wilson GR, Sim JCH, McLean C, Giannandrea M, Galea CA, Riseley JR et al (2014) Mutations in RAB39B cause X-linked intellectual disability and early-onset parkinson disease with α-synuclein pathology. Am J Hum Genet 95: 729–735
67. Zarranz JJ, Alegre J, Gómez-Esteban JC, Lezcano E, Ros R, Ampuero I et al (2004) The new mutation, E46K, of alpha-synuclein causes Parkinson and Lewy body dementia. Ann Neurol 55:164–173

5

The complement system in glioblastoma multiforme

T. A. M. Bouwens van der Vlis[1,2,3], J. M. Kros[3*], D. A. M. Mustafa[3], R. T. A. van Wijck[4], L. Ackermans[2], P. M. van Hagen[4,5] and P. J. van der Spek[1,3]

Abstract

The human complement system is represents the main effector arm of innate immunity and its ambivalent function in cancer has been subject of ongoing dispute. Glioma stem-like cells (GSC) residing in specific niches within glioblastomas (GBM) are capable of self-renewal and tumor proliferation. Recent data are indicative of the influence of the complement system on the maintenance of these cells. It appears that the role of the complement system in glial tumorigenesis, particularly its influence on GSC niches and GSC maintenance, is significant and warrants further exploration for therapeutic interventions.

Introduction

Traditionally, the human complement system is regarded to be a main effector arm of the innate immunity and comprises around thirty soluble and membrane-associated proteins. Innate immunity forms the first line of defense against invading micro-organisms, and is ancient compared to the adaptive immune system. Analogues of components of the mammalian alternative complement pathway have been identified in deuterostomes/protostomes over 1000 million years ago, whereas the first molecules from the jawed-invertebrate specific adaptive immune system emerged at least 400 million years later [63]. After breaching the host's environmental barriers, invading microbes are detected by the pattern recognition molecules (PRM) of the classical (C1q) and lectin (MBL, ficolins) complement pathways (CP, LP) [75, 96]. (Fig. 1). Further, activation of the complement cascade through the AP is achieved through insufficient inhibition of spontaneous hydrolysis of C3 (C3-H20) by the microbe (AP). All three activation pathways converge at the level of C3 which, after formation of the C3 and C5 convertase complexes, continues with the formation of the terminal complement complex (TCC) either as the pore-like membrane attack complex (MAC) or as cell-activating sC5b-9 [51]. MAC assembly in the cell membrane causes prompt colloid osmotic lysis [56, 59].

Over a century after the initial discovery of the complement system by Buchner et al. it was realized that the actions of complement are not constrained to an effector mechanism of the innate immunity, but are also involved in directing the adaptive immune response, angiogenesis, tissue regeneration, fat metabolism and development of the central nervous system (CNS) [75]. The complement system acts as intricate immune surveillance system that is able to discriminate between healthy host tissue, cellular debris, apoptotic cells and foreign intruders (Fig. 1) and contributes to a large variety of inflammatory-, immune-, ischemic-, age-related pathologic processes of the CNS [75]. A wide repertoire of specific complement inhibitors has been developed against a variety of diseases of which eculizumab, an antibody against C5 and C1 (C1-INH) became FDA approved for the treatment of paroxysmal nocturnal hemoglobinuria, atypical hemolytic uremic syndrome and hereditary angiogoedema [1]. In recent years it became clear that many factors of the complement system are expressed in the brain [60, 93]. The production of the complement proteins in CNS is constrained to microglia, oligodendrocytes, astrocytes and, to a lesser extent, ependymal cells [93]. The complement system appears to be a key player in CNS homeostasis as complement effector mechanisms have been identified in neurogenesis and regulating synaptic pruning [54]. Various pathological conditions cause an imbalance between complement activation and inhibition.

* Correspondence: j.m.kros@erasmusmc.nl
[3]Department of Pathology, Erasmus University Medical Center, Wytemaweg 80, 3015, GE, Rotterdam, The Netherlands
Full list of author information is available at the end of the article

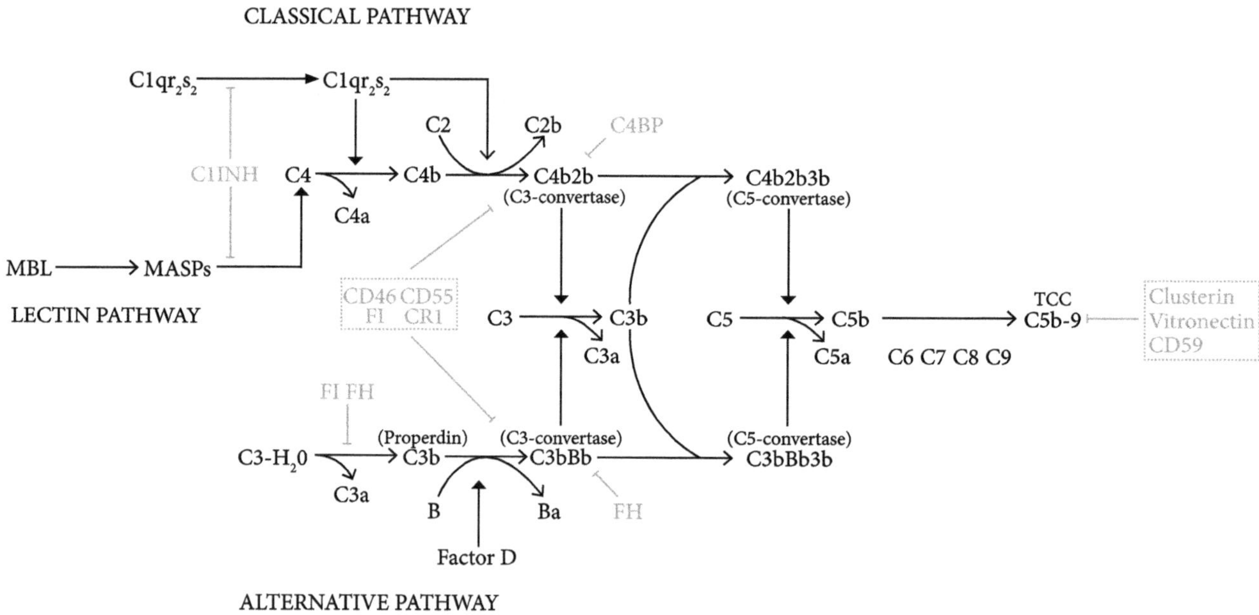

Fig. 1 Schematic and simplified representation of the complement system. Complement regulatory proteins, both fluid-phase and membrane-bound are coloured light grey. *C1INH: C1-inhibitor; CD46: Membrane Cofactor Protein; CD55: Complement decay-accelerating factor; FI: Complement factor I; CR1: Complement receptor type 1; FH: Complement factor H; C4BP: C4bbinding protein*

Activation of the complement system contributes to a wide variety of CNS diseases including Alzheimer disease, CNS inflammation, traumatic brain injury and tumors [6, 55]. The cross-link between inflammation and cancer is generally accepted as chronic and insidious inflammation is recognized to play decisive roles at different stages of tumor development, including initiation, promotion, malignant conversion, invasion, and metastasis [28, 52]. The actions of the immune system against tumors in progress can be referred to as the cancer immunoediting process, which is composed of three distinct phases: elimination, equilibrium and escape. During the elimination phase and the equilibrium phase the immunological response is able to prevent tumor progression. In contrast, during the escape phase acquired adaptations of malignant cells and the host immune system response allow for expansion of the tumor cell population [94]. By acting as an intrinsic effector mechanism and by forming a functional bridge between the innate and the adaptive immune system, the complement system is an integral component of the anti-tumor immune response [94]. Complement activation following recognition of damage-associated molecular patterns (DAMPs) expressed by tumor cells, or improper regulation, allow for a potent anti-tumor response [37]. The potent antitumor response by complement has been utilized for antibody-based cancer immunotherapies by eliciting complement-dependent cytotoxicity, exemplified by the use of rituximab and ofatumumab in the treatment of B cell lymphomas and chronic lympocytic leukaemia, respectively [89]. However, the complement system also shows another face. Recent pre-clinical cancer models showed that the activated complement system contributes to a tumor facilitating micro-environment [1]. This adverse capacity seems to be a consequence of imbalanced, rather than physiological, complement activation [74]. Various studies have reported significant reductions of orthotopic tumor growth following complement system inhibition within the cascade [74]. Further, diverse complement effectors are implicated in other cancer-related phenomena as sustained proliferative signaling, angiogenesis and invasion and metastasis [33, 74].

Contributory to treatment resistance of glial neoplasms is the presence of glioma stem-like cells (GSCs) [85]. GSCs reside in specific anatomical niches within the tumor and propagate glioma repopulation by converting into either a differentiated tumor cell, or a new cancer stem cell [46]. The maintenance of GSCs requires specific intrinsic factors within the cells and various paracrine cues from adjacent cells [46]. The complement system represents an as yet unidentified effector in GSC maintenance, and unraveling its interplay will reveal new targets for therapeutic intervention.

Complement and GSC maintenance: Intrinsic regulation

Factors that are involved in GSC maintenance comprise of metabolic, genetic and epigenetic regulatory mechanisms [90]. Although the mechanisms underlying GSC plasticity are largely unknown, several intrinsic regulatory mechanisms are known to be involved in reprogramming differentiated GBM cells into stem-like cells. Among these are Sex Determining Region Y -Box 2 (SOX-2) [88], signal transducer and activator of transcription 3 (STAT-3), octamer-binding transcription factor 4 (OCT-4) and mammalian target of rapamycin (mTOR) signaling [23, 82]. The GSCs maintain their multipotent state through autocrine stimulation of the C3a- and C5a-receptors on the plasma membrane by secretion of alternative pathway C3-convertase components (C3, factor D and factor B) and subsequent extracellular cleavage of C3, as observed in resting T-cells (Fig. 2) [87]. The C3 and C5 convertases (Fig. 1) are responsible for the release of their respective bioactive fragments C3a and C3b, and C5a and C5b. The anaphylotoxins C3a and C5a signal through the G protein coupled receptors C3aR and C5aR (CD88) respectively. Interaction of several downstream signal transduction pathways followed by C3aR and C5aR activation with recognized GSC regulatory mechanisms effectors may therefore aid to GSC maintenance. Figure 2 presents a schematic overview of the interaction of autocrine derived complement with GSC regulatory mechanisms. C3a-C3aR interaction activates STAT-3 and causes an increase of Wnt2b and SOX-2 expression in a serine protease MAPK dependent fashion, as was shown in a model of retinal regeneration [36]. C5aR1 activation contributes to the maintenance of pluripotency of OCT-4 positive human induced pluripotent stem cells (hPSC) through extracellular signal-regulated protein kinases 1 and 2 (ERK1/2) activation [35]. In vitro administration of recombinant C5a to C5aR expressing gastric cancer cells promotes the activation of PI3K/Akt and downregulates p21 activation [11]. Inhibition of p21 is a key mechanism of GSC self-renewal and prevention of differentiation [99].

Further, complement C3 activation is not limited to the extracellular space because intracellular C3 activation is an ubiquitous event within human cells [50]. Resting T-lymphocytes contain intracellular pools of C3 that are activated by cysteine protease cathepsin-L. This 'tonic' intracellular C3a generation engages C3aR present on lysosomes and sustains the basal mammalian target of rapamycin (mTOR) activation required for T-cell survival [50]. Activation of the PI3K/Akt/mTOR signaling cascade is involved in GSC phenotype maintenance [18]. In addition, GBM tissue shows the profound presence of cathepsin L, which has been shown to negatively affect apoptosis and promotes invasion [104]. Intracellular C3a activation by cathepsin-L may provide for an as yet unidentified additional intrinsic GSC effector mechanism. Noteworthy is the critical role of autocrine derived C5a-C5aR1 signalling in affecting the

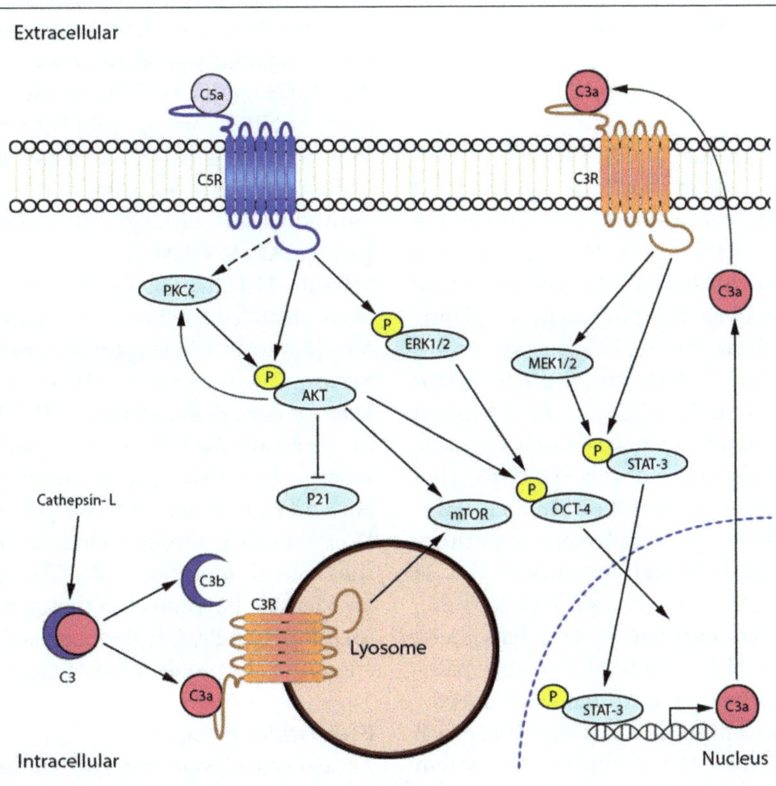

Fig. 2 Proposed interaction of complement C3a and C5a with GSC regulatory mechanisms. C5a-C5aR interaction activates PI3K/Akt/mTOR signaling and PKCζ but suppresses p21 with subsequent OCT-4 activation. Intracellular activation of C3a by cathepsin-L may occur, thereby sustaining basal mTOR activation. Either intracellular or extracellular derived C3a phosphorylates STAT-3 and causes an increase of SOX-2 expression

fate of Neural Progenitor Cells (NPCs). C5a was found actively secreted within neural rosettes [16]. Atypical protein kinase C zeta (PKCζ) dependent C5a-C5aR1 signaling regulates the apicobasal polarity of the NPC, thereby maintaining symmetrical self-renewing cell division [16]. PKCζ inhibition reverses C5a-C5aR induced p42/44 phosphorylation (ERK) and attenuates the mitotic activity of NPCs [16]. PKCζ is overexpressed in GBM cells and its downregulation inhibits in vitro migration and invasion of GBM cells. Another mechanism through which the GSC could generate C5 in order to self-activate C5aR is by expressing cell membrane-bound serine protease, as was found in different cancer cell types [62].

Complement and GSC maintenance

The neuropil constitutes the extracellular matrix of the brain in which endothelial and perivascular cells, microglia, tumor-associated macrophages and non-neoplastic astrocytes are present and networks of cytokines and growth factors are active [32]. Together with neoplastic cells the microenvironment coalesces into the tumor mass. GSC are enriched in areas of hypoxia, in perivascular niches and at the invasive edge of the tumor [46]. A graphical representation of the various cells involved is provided in Fig. 2. The actions of the complement factors within the various glioma niches are summarized in Table 1.

Hypoxic niche

The inadequate neo-angiogenesis in GBM results in hypoxic areas within the tumor. The neoplastic cells respond to low oxygen levels by the expression of members of the hypoxia inducible factor (HIF) family of transcriptional factors [49]. HIFs are upregulated in GSC and its forced expression induces a stem-cell like phenotype in glioma cells [49]. Transcriptional targets of HIFs include angiogenic genes like Vascular Endothelial Growth Factor (VEGF) as well as stem cell markers [49]. Areas of hypoxia optimally accommodate complement activation as provide for damage-associated molecular patterns (DAMPs) that are recognized by C1q. Hypoxic conditions induce (HIF-dependent) down-regulation of complement regulatory genes CD55, CD46 and factor H and upregulate C3, C3a and C3aR and enhance C3a-C3aR engagement [27, 66]. The constituents of the complement system have been identified to interact with HIF associated signaling pathways and may therefore act as an additional effector mechanism in HIF dependent GSC survival, self-renewal and tumor growth. Firstly, the complement system contributes to facilitate HIF transcription through STAT-3 activation that is critical for the transcription of HIF-1α in GSCs and tumor-associated myeloid cells [69]. The production of reactive oxygen species, as a result of overexpression of nicotinamide adenine dinucleotide phosphate oxidase 4 (NOX-4), was identified as the molecular mechanism underlying hypoxia-induced STAT-3 activation in GBM cells [103]. In a model of renal ischemia/reperfusion injury, oxidative stress induces an increased expression of NOX-4 in tubular cells and NOX-2 infiltrating monocytes and myeloid dendritic cells [84]. This effect is dramatically reduced after the administration of the complement 1-inhibitor (C1-INH). In vitro administration of C3a to cultured proximal tubular cells induces NOX-4 expression regardless of hypoxic conditions [84].

Secondly, through C3aR and C5aR interaction on the GSC, complement may provide for additional signal transduction pathways for PI3K- or mitogen-activated protein kinase (MAPK)/ERK1/2-dependent HIF-1α protein translation [68, 69]. HIF-1α and the components of the complement cascade converge at the level of the Notch signaling pathway. Notch activation restricts glioma cell differentiation and stimulates astrocytes into a neural stem-like cell state [69]. HIF-1/2α driven GSC maintenance requires Notch signaling [69]. In resting T-cells, CD46 sequesters the Notch ligand Jagged-1, thereby preventing the interaction between Jagged-1 and Notch that activates T-cells [48]. Hypoxia-mediated downregulation of the expression of CD46 or CD46-C3b interaction following complement activation may allow for Notch-Jagged-1 interaction. A direct contribution of CD46 downregulation in maintaining the undifferentiated state of the GSC remains to be elucidated. C3a inhibits SDF-1α induced neuronal differentiation of NPCs through ERK1/2 phosphorylation regulation [83]. SDF-1α is a HIF-1α target gene in GBM cells [22]. Importantly, SDF-1α induces recruitment of bone-marrow derived CD45+ myeloid cells, endothelial and pericyte progenitor cells to GBM [22].

Lastly, HIF-1α modulated, Wnt/ β-catenin activation has been identified to stimulate GSC differentiation and therefore promotes a less-aggressive, neuronal tumor phenotype. Subsequent β-catenin mediated Notch inhibition further allows for GSC differentiation [71]. The role of Wnt activation in regulating the GSC state remains controversial as many reports claim that Wnt activation promotes GSC maintenance and expansion [42]. C1q is an activator of canonical Wnt signalling through binding with the Fz-receptor and subsequent induction of C1s dependent cleavage of low-density lipoprotein receptor-related protein 6 (LRP6) [61]. Intriguingly, C1q-induced activation of Wnt signalling attenuates the proliferation of muscle stem cells [61].

Perivascular niche

As a result of vigorous and abnormal angiogenesis, local regions of hypoxia develop with subsequent complement activation, facilitating complement mediated GSC regulation. The actions of the complement system extend into the perivascular niche, where GSC survival and

Table 1 Complement protein in the GBM tumor

Complement proteins	Niche	Mechanism	Proposed effect
C1q	GSC	Wnt activation	Stimulate GSC differentiation
	Perivascular	Priming SDF-1 gradient	GSC migration
	Perivascular	gC1qR interaction	Potentiate tumor cell invasiveness, chemoattractant
	Perivascular	cC1qR interaction (IL-8, MCP-1 and IL-6 secretion)	GSC migration
	Microenvironment	GAM-M2 induction	Immunosuppression
	Invasive	gC1qR (Bradykinin induction)	Tumor cell invasiveness
MBL	Microenvironment	GAM-M2 induction	Immunosuppression
C3	Microenvironment	MDSC recruitment	Immunosuppression
C3b	Microenvironment	Treg induction (ligand for CD46)	Immunosuppression
	Microenvironment	GAM-M2 induction	Immunosuppression
C3a/C3aR	Hypoxic	STAT-3 activation	GSC maintenance
	GSC	mTOR activation	GSC maintenance
	Hypoxic	NOX-4 activation, (STAT-3, HIF-α)	GSC maintenance
	GSC	SOX-2 activation	GSC maintenance
	Microenvironment	Chemotaxis immune cells	
CD46	Hypoxic	Jagged-1-Notch disruption	GSC maintenance
	Perivascular		
C5a/C5aR	Invasive nice	PKCζ activation	Potentiate tumor cell invasiveness
	GSC	PI3K/Akt/mTOR	GSC maintenance
	GSC	P21 inhibition	GSC maintenance
	GSC	OCT-4 expression	GSC maintenance
	Perivascular	MMP-9, MT1-MMP activation	Increase tumor cell invasiveness
	Perivascular	MCP-1 secretion	Increase tumor cell invasiveness
	Perivascular	TGF-β	GSC differentation into vascular pericytes
	Perivascular	NO secretion (iNOS/eNOS induction)	GSC maintenance
	Perivascular	VEGF expression	Vascular tube formation
	Microenvironment	Chemotaxis immune cells	
	Microenvironment	Treg induction (High concentration)	Immunosuppression
	Microenvironment	GAM-M1 activation (balanced C activation)	Anti-tumor response
	Microenvironment	MDSC recruitment (ROS)	Immunosuppression
C5b-C9	Perivascular	bFGF release	GSC dedifferentiation

angiogenesis are mediated. In the adult mammalian brain, neuronal stem cells reside in two germinal niches of which the subventricular zone is highly vascularized. Adult stem cells lie in close proximity to the vasculature, where communication occurs through direct cell-cell interactions and soluble secreted cues [81]. In contrast to germinal niches, where the rate of stem cell proliferation is low, the perivascular niche within GBM contributes to the generation of the GSCs and tumor growth [9]. Besides endothelial cells (EC) and GSCs, major cell types that have been recognized to reside in the perivascular niche include pericytes, immune cells, fibroblasts and astrocytes, all of which provide an additional biological source of complement proteins [9, 93]. The complement system establishes an active interplay with the perivascular niche constituents and GSCs and fulfills an active role in the migration of GSC and promotes angiogenesis.

Several components of the (activated) complement system are powerful chemoattractants attracking blood borne cells and GSCs to the perivascular niche [75]. The complement system potentiates the migration of GSC towards the perivascular niche just like it does to mesenchymal stem cells (MSCs), NPCs and hematopoietic stem/progenitor cells (hPSC). In GBM, GSCs move towards the tumor vasculature and engage in cell-cell contact [9]. The interaction of Stromal Cell-Derived Factor 1 (SDF1) with its receptor C-X-C Motif Chemokine Receptor 4 (CXCR4) is operative to guide the tumor cells to the peri-endothelial space [72]. C1q primes chemotactic SDF-1-dependent migration of human umbilical cord blood derived- Mesenchymal Stem Cells (MSC), in part by upregulating the expression of CXCR4 and through interaction with its globular heads binding receptor (gC1qR), which is ubiquitously expressed [70]. The activated complement system potentiates the SDF1-CXCR4 chemotaxis, independent of C3aR, as has been observed in hematopoietic stem/progenitor cells (HSPCs) [38]. C3a modulates concentration-dependent SDF1-CXCR4 induced migration of NPCs [83]. C5a attracts MSCs in a C5aR dependent fashion and its degradation product $C5a_{desArg}$ causes an increased secretion of MMP-9 and MT1-MMP, creating a highly proteolytic microenvironment in favor of cell-migration [79]. In addition, IL-8 secreted by ECs stimulates GSC migration and maintains its stemness properties, in part by upregulating the expression of its cognate receptors CXCR1 and CXCR2 [39]. The interaction of C1q with its receptor present on the EC (cC1qR) initiates the release of IL-8, macrophage chemoattractant protein-1 (MCP-1) and IL-6, that contribute to the homing of GSCs and MSCs [92, 97]. IL-8 mediated GSC homing and maintenance is amplified by C5a as observed in whole blood cells and in human dermal microvascular endothelial cells (HVEC-d) [97]. Lastly, C5a contributes to GSC migration by increasing the expression of MCP-1/CCL2, as observed in HVEC-d in a dose dependent matter [102].

Complement modulated Perivascular niche-GSC interaction

GSCs form intimate contacts along the length of the endothelial tube, essential for their survival and inducing secretion of soluble factors by ECs that keep the GSCs in an undifferentiated state [80]. Several pathways constitute EC orchestrated GSC regulation, including the transforming growth factor-β (TGF-β) pathway [80]. EC-derived TGF-β induces the differentiation of GSC into pericytes that contribute to vascular formation in GBM [12]. Crosstalk of TGF-β signaling and complement activation is observed in various cells. C5a has been shown to upregulate TGF-β transcript expression, and vice-versa TGF-β upregulates the expression of C5aR [29]. Further, TGF-β and C5a signaling converge downstream at the level of SMAD independent pathways including, PI3K/AKT/mTOR and MAPK/ERK1/2 signaling pathways [77, 105]. A second pathway through which the GSC phenotype is maintained is the nitric oxide (NO) signaling [10]. The biological source of NO is EC-derived by the expression of endothelial NO synthase (eNOS) or alternatively, through the expression of inducible NOS (iNOS) by the GSC [10]. Activation of eNOS requires phosphorylation of AKT, suggestive of a contributory role for the activated complement system [21]. The complement system mediates the expression of iNOS and NO levels as shown in models of gastrointestinal ischemia by inhibition of C3 and C5 [57]. Lastly, EC derived basic Fibroblast Growth Factor (bFGF) induces the reversion of differentiated GBM cells [25]. However, the underlying mechanism resulting in functional expression of bFGF remains poorly defined, given that hypoxia does not induce bFGF expression in human vascular smooth muscles cells [7]. Interestingly, small amounts of C5b-9 (membrane attack complex) releases bFGF from human umbilical vein endothelial cells (HVEC) [4]. Active and inactive forms of C1s are found to form aggregates with bFGF, thereby reducing its activity [76].

The complement system influences GSC mediated angiogenesis

Indications for the close interaction of GSC and endothelial cells emerged from the finding that xenotransplanted tumors derived from GSCs were characterized by widespread angiogenesis that was not encountered in their non-GSC counterparts [3]. GSCs secrete VEGF and treatment with bevacizumab blocks the pro-angiogenic effects of VEGF by hampering microvascular endothelial cell migration and vascular tube formation and inhibiting the growth and vascularity of GSC derived xenotransplants [3]. Bioactive complement products have been identified as important effectors in pathological neovascularization in age-related macular degeneration (ARMD), diabetic retinopathy, and retinopathy of prematurity [100]. Interaction of C5a with its receptor C5aR1 induces VEGF expression in a dose-dependent matter in retinal pigmented epithelium (RPE) in-vivo and in-vitro [15]. The induction of oxidative stress in RPE cells reduces the surface expression of DAF, CD55 and CD59 and impairs complement regulation at the level of factor H, resulting in complement activation and complement-dependent VEGF expression [91]. Consequently, inhibiting the AP using a recombinant factor H reduces the expression of VEGF and subsequent angiogenesis in a mouse model of choroidal neovascularization [91]. Conversely, the inhibition of VEGF causes a decrease of the complement inhibitory proteins (CIPs) factor H, CD46 and CD59 in human RPE-cells and glomerular endothelial cells through VEGFR2/PKC-α/CREB signaling [44]. These observations imply that VEGF protects neo-angiogenesis by local inhibition of the complement system. It remains undetermined whether complement activation directly contributes to VEGF expression or VEGF suppresses complement activation through CIP induction. In a mouse model of ovarian cancer, C3 and C5aR were shown to be closely involved in neo-angiogenesis [63]. Tumors derived from partial C3, C5aR and complete C5aR knock out mice displayed decreased microvascular density compared to their WT-littermates [63]. Additional in vivo assays showed significant impairment of angiogenesis for complete C3 and C5aR knock-out mice. Interestingly, direct functional effect of C5a comparable to VEGF-A on tube formation of endothelial cells was also observed. This effect was found to be reversible using the C5aR inhibitor PMX-53. PMX-53 also significantly impaired $VEGF_{165}$ mediated HMEC tube formation [63]. In addition to C3 and C5aR, microvascular density was significant decreased in tumors in C1q deficient mice bearing a syngeneic B16 melanoma compared to their WT-littermates [8].

Complement and immune cell crosstalk in the perivascular niche

Activation of the complement system by means of C3a and C5a plays an important role in the inflammatory process by recruiting immune cells such as mast cells, monocytes, macrophages, neutrophils, MDSCs and adaptive immune cells including T cells [75]. The BBB consists of highly specialized endothelial cells that communicate with pericytes and astrocytes to protect the CNS from the chemical variations in the bloodstream, and establishes a strictly controlled interface for immune cell trafficking. In GBM the BBB's integrity is disrupted due to the abnormal tumor microvasculature, resulting in an increased vascular permeability and consequently, an increase in immune cell infiltration including monocyte-derived cells, microglia and T-lymphocytes [19, 24]. C5a/C5aR neutralization alleviates the BBB breakdown in models of traumatic brain injury and systemic lupus erythematodus and it is likely that the activated complement system also affects the BBB in GBM, with possible consequences for the passage of immune cells [40].

Lymphocytic infiltration and PD-1

In glioma, tumor infiltrating lymphocytes (TILs) consisting of CD4+ and CD8+ cells are present [65]. Glioma TILs show a predominant regulatory T-cell population (CD4 + CD25 + Foxp3+) [65]. Regulatory T cells (Tregs) are believed to be the primary regulators of immunosuppression in the glioma microenvironment [65]. The proportions of CD3+ and CD8+ over Foxp3+ cells reportedly correlate with the clinical course of GBM patients [78]. The activated complement system by means of CD46 may account for an increased proportion of Tregs. The C3 cleavage fragment, C3b, is a natural ligand for CD46 on T cells. Stimulation of naïve CD4+ T cells with anti-CD46 monocolonal antibodies (mAb) or C3b dimers in the presence of IL-2 induces a differentiation towards a IL-10 producing type 1 regulatory T cell (Tr1) [45]. However, CD4+ Foxp3+ regulatory T cells present in GBM are predominantly thymus derived (tTregs) rather than peripheral induced IL-10 producing regulatory T-cells [95]. In the presence of CD46 stimulation, cell contact-mediated tTreg function is impaired [47]. Instead, tTregs differentiate to IL-10 secreting Tr1 cells [47]. In several human cancers a potent immunosuppressive subpopulation of IL-10 producing Tregs has been identified and these Tregs suppress CD8+ T-cell effector functions which is associated with poor survival [64].

In models of melanoma and non-small cell lung cancer combined with genetic ablation or mAb blocking of programmed death 1/programmed death ligand 1 (PD-1/PD-L1) and C3aR appears to be more effective in restraining tumor growth than only blocking PD-1 therapy alone [2]. In glioma, the expression of PD-L1 is correlated with glioma grade and has been identified as a negative prognostic factor. Recently, therapeutic blockade of PD-1 in the GL-261 murine glioma model induced an impressive prolonged survival, with TILs showing a shift towards CD8+ T cells [20]. The dual role of complement activation in the tumor micro-environment was illustrated by tumor progression in tumor-bearing mice with either high- or low C5a-producing syngeneic lymphoma cells [30]. High C5a producing tumors showed a significant increased tumor progression associated with an overall decrease CD4+ and CD8+ T cells in the tumor [30]. Further, it was shown that in vitro polarization of CD4+ cells is observed to be C5a concentration dependent. A low C5a concentration promotes Th1 cell differentiation while high concentrations (> 500 ng/ml) promotes Treg induction [30]. Taken together, imbalanced complement activation may be associated with an immunosuppressive micro-environment and is therefore contributory to tumor progression.

Glioma associated microglia and macrophages (GAMs)

Glioma associated microglia and macrophages (GAMs) are considered to be the most prominent glioma-infiltrating immune cells, constituting up to 30% of all immune cells within the tumor microenvironment. GAMs are recruited into the tumor microenvironment through various glioma derived factors that contribute to polarization from a tumor-suppressive to a tumor-promoting phenotype. Activated complement factors may contribute to the constant recruitment of GAMs to the tumor micro-environment, either directly, or through interactions with glioma factors, and influence their polarization. Several factors have been recognized to attract GAMs to the tumor environment, among which are MCP-1 and SDF-1 [31]. C1q and C5a have been shown to potentiate the actions of these chemokines [67, 70]. Upon lipopolysaccharide (LPS), IFNγ, or C3a and C5a induced activation (M1) GAMs produce inflammatory mediators, phagocytose tumor cells, present antigens to immune cells and induce a T-cell response [34]. In contrast, GAMs activated through factors such as CSF-1,

IL-10, TGF-β and the complement opsonins C1q and C3b acquire a pro-tumorigenic phenotype (M2) which include promoting and facilitating immunosuppression, glioma proliferation and invasiveness [98]. Unbalanced complement activation induces Il-10 and TGF-β expression, thereby promoting a pro-tumorigenic phenotype [29, 47]. Within the scope of autoimmunity the complement opsonins C1q and MBL regulate macrophage polarization towards a M2 macrophage phenotype [5, 26].

Myeloid-derived suppressor cells (MDSC)

Myeloid-derived suppressor cells (MDSCs) are a heterogeneous population of cells containing myeloid cells in various differentiation stages. MDSCs are found to be significantly increased in the peripheral blood of glioblastoma patients [73]. MDSCs are believed to elicit distinct immunosuppressive actions within the glioma micro-environment including generation of oxidative stress through the production of ROS and thereby inducing T-cell inhibition [24]. Intriguingly, in a TC-1 syngeneic model of cervical cancer in mice, pharmacological inhibition of C5aR resulted in an increase in CD8+ T-cell infiltration along with deceleration of tumor growth comparable to the effects of paclitaxel [53]. C5a mediated suppression of the antitumor CD8+ T-cell response is associated with an increase of MDSC in the tumor microenvironment and a subsequent increase in ROS production [53]. In a syngeneic mouse model for lung cancer, MDSCs appeared to be reduced in a subpopulation analysis of splenocytes after C5aR blockade [14]. Concordantly, in a murine model of ovarian cancer C3 silencing increases the number infiltrating CD8+ T-cells infiltrating the tumor by 10-fold and reduced the number of MDSCs by 80% [13]. However, the observed reduction in tumor growth was found to be independent on the number of CD8+ T-cells [13].

Invasive niche

High-grade gliomas show aggressive invasiveness in two compartments: the perivascular space and the brain parenchyma [17]. The bradykinin and the SDF-1/CXCR4 axis act as a chemoattractant that guide glioma cells towards blood vessels [58]. The bradykinin-forming cascade and the

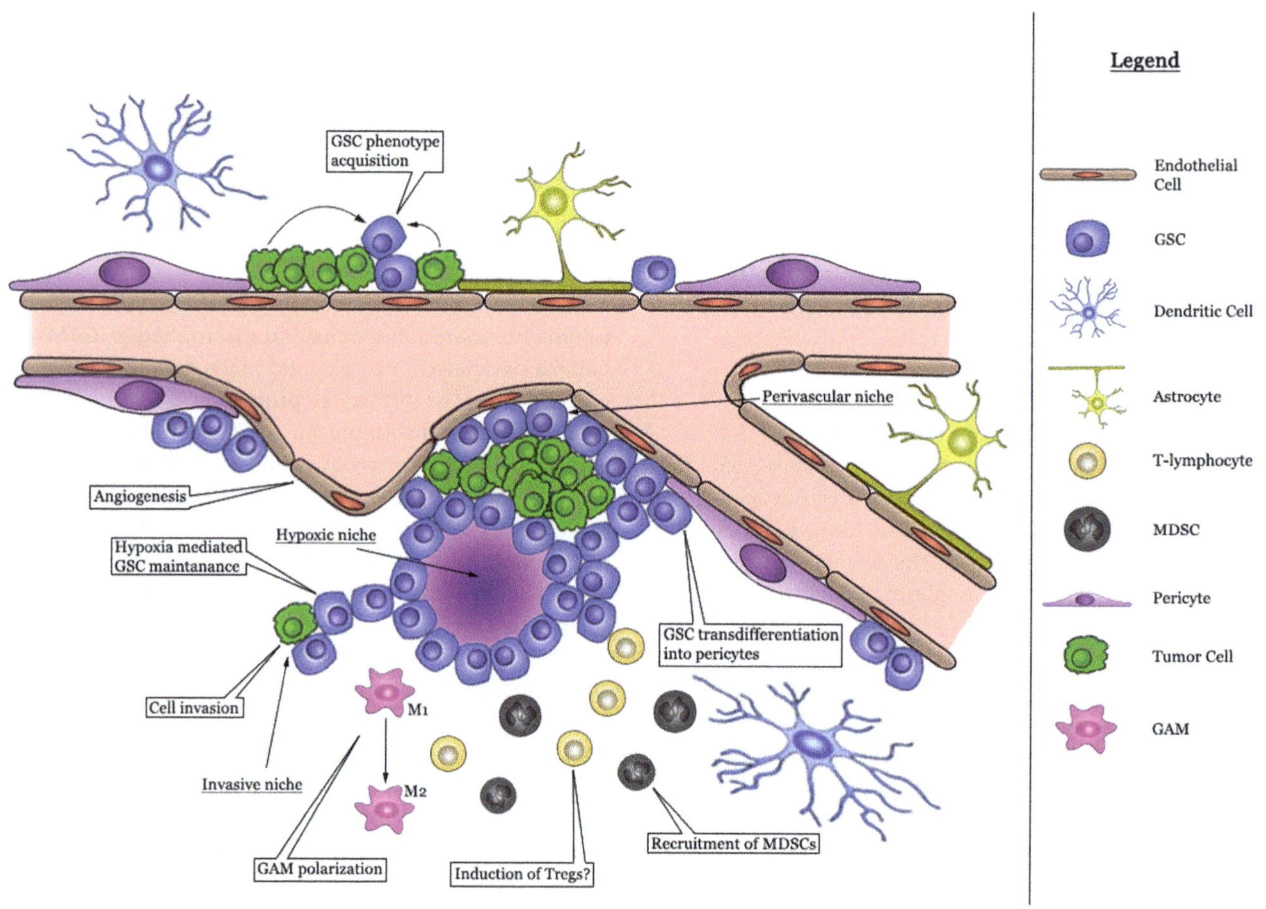

Fig. 3 Graphical summary of the potential actions of the complement system in the glial tumor microenvironment

classical complement pathway share many elements, including cross-activation, shared binding proteins and control mechanisms [43]. C1 inhibitor (C1 INH) inhibits the bradykinin-forming cascade at several levels: its local absence at the glioma invasive edge may further initiate the classic complement cascade by activating C1r [43]. Vice-versa, the presence of the ubiquitously expressed receptor for the globular head of C1q (gC1qR) allows for bradykinin production on the endothelial cell surface [43]. The activated complement system further stimulates glioma cell migration through facilitating local degradation of extracellular matrix proteins. In C5aR expressing colon and bile duc cancer cells C5a enhances cell invasiveness by increasing the expression of several matrix metalloproteinases (MMPs), including MMP-1 and MMP-9 in C5aR expressing colon and bile duct cancer cells [62]. Soluble C5b-9, which is as a pro-inflammatory mediator, induces MMP-2 expression by microglia upon activation [101]. The molecular regulation of MMP-2 and MMP-9 is poorly characterized. The activated complement system signaling may provide an alternative mechanism for the NF-κB mediated overexpression of MMP-2 and MMP-9 in glioma cells by means of C5a-C5aR [86]. C1q may activate the canonical Wnt/β-catenin signaling pathway, which increases MMP-2 and MMP-9 expression by glioma cells [41].

Conclusion remarks and future directions

A wide variety of molecular pathways that are characteristic of the aggressive nature of GBM are known to be induced or modulated by the activated complement system (Table 1 and Fig. 3). The unbalanced activation in the tumor microenvironment may suppress the antitumor inflammatory response. There is increasing data that the complement system could serve as an unidentified intrinsic and extrinsic regulatory mechanism for GSC maintenance, thereby supporting treatment resistance. The activated complement system may contribute to evade the immune response, activate invasion and induce angiogenesis. Because several tumor promoting pathways are activated by complement the elucidation of the role of complement activation is necessary to discover new targets for therapy in malignant glioma.

Abbreviations

AP: Alternative Pathway; BBB: Blood Brain Barrier; bFGF: basic Fibroblast Growth Factor; C1-INH: C1-inhibitor; C3aR: C3a Receptor; C5aR1: C5a Receptor-1; C5b-9: Membrane Attack Complex; cC1qR: collagen C1q Receptor (Calreticulin); CNS: Central Nervous System; CREB: Cellular Transcription Factor; CSF 1: Colony Stimulating Factor 1; CXCR1: C-X-C Motif Chemokine Receptor 1; CXCR2: C-X-C Motif Chemokine Receptor 2; CXCR4: C-X-C Motif Chemokine Receptor 4; DAMP: Damage-Associated Molecular Pattern; EC: Endothelial Cells; eNOS: endothelial Nitric Oxide synthase; ERK1/2: Extracellular-signal regulated kinases; Fz: Frizzled; GBM: Glioblastoma Multiforme; gC1qR: Globular C1q Receptor; GSC: Glioma Stem-like Cell; HIF: Hypoxia Inducible Factor; HMEC: Human Mammary Epithelial Cells; hPSC: human induced Pluripotent Stem Cells; HVEC: Human Umbilical Vein Endothelial Cells; HVEC-d: Human Dermal Microvascular Endothelial Cells; IFNγ: Interferon-γ; LPS: lipopolysaccharide; LRP6: Low-density lipoprotein receptor-related protein 6; MAPK: Mitogen-Activated Protein Kinase; MBL: Mannose Binding Lectin; MCP-1: Macrophage Chemoattractant Protein-1; MMP-9: Matrix metallopeptidase 9; MSC: mesenchymal stem cells; MSDC: Myeloid-Derived Suppressor Cell; MT1-MMP: Membrane-Type 1 Matrix MetalloProteinase; mTOR: mammalian target of rapamycin; NO: Nitric Oxide; NOX-2: Nicotinamide adenine dinucleotide phosphate OXidase 2; NOX-4: Nicotinamide adenine dinucleotide phosphate OXidase 4; NPC: Neural Progenitor Cell; OCT-4: Octamer-Binding Transcription Factor-4; PD-1: Programmed Cell Death protein 1; PD-L1: Programmed Cell Death ligand 1; PI3K: Phosphatidylinositol 3-kinase; PKC-α: Protein kinase C-α; PKCζ: Protein Kinase C zeta; ROS: Reactive Oxygen Species; SDF-1α: Stromal Cell-Derived Factor 1α; SOX-2: Sex Determining Region Y -Box 2; STAT-3: Signal Transducer and Activator of Transcription-3; TGF-β: Transforming Growth Factor-β; TIL: Tumor Infiltrating Lymphocytes; VEGF: Vascular Endothelial Growth Factor; VEGFR2: Vascular Endothelial Growth Factor Receptor 2; Wnt: Wingless-related integration site

Authors' contributions

TB conceived of the presented idea. TB wrote the manuscript with support from JK, LA, RW, DM and PH. PS and JK supervised the project. All authors read and approved the final manuscript.

Competing interest

The authors whose names are listed above certify that they have no affiliations with or involvement in any organization or entity with any fi nancial interest (such as honoraria; educational grants; participation in speakers' bureaus; membership, employment, consultancies, stock ownership, or other equity interest; and expert testimony or patent-licensing arrangements), or non-financial interest (such as personal or professional relationships, affiliations, knowledge or beliefs) in the subject matter or materials discussed in this manuscript.

Author details

[1]Department of Bioinformatics, Erasmus University Medical Center, Wytemaweg 80, 3015, GE, Rotterdam, The Netherlands. [2]Department of Neurosurgery, Maastricht University Medical Center, P. Debyelaan 25, 6229, HX, Maastricht, The Netherlands. [3]Department of Pathology, Erasmus University Medical Center, Wytemaweg 80, 3015, GE, Rotterdam, The Netherlands. [4]Department of Internal Medicine, Erasmus University Medical Center, Wytemaweg 80, 3015, GE, Rotterdam, The Netherlands. [5]Department of Immunology, Erasmus University Medical Center, Wytemaweg 80, 3015, GE, Rotterdam, The Netherlands.

References

1. Afshar-Kharghan V (2017) The role of the complement system in cancer. J Clin Invest 127:780–789. https://doi.org/10.1172/jci90962
2. Ajona D, Ortiz-Espinosa S, Moreno H, Lozano T, Pajares MJ, Agorreta J, Bertolo C, Lasarte JJ, Vicent S, Hoehlig K et al (2017) A combined PD-1/C5a blockade synergistically protects against lung Cancer growth and metastasis. Cancer discovery 7:694–703. https://doi.org/10.1158/2159-8290.cd-16-1184
3. Bao S, Wu Q, Sathornsumetee S, Hao Y, Li Z, Hjelmeland AB, Shi Q, McLendon RE, Bigner DD, Rich JN (2006) Stem cell-like glioma cells promote tumor angiogenesis through vascular endothelial growth factor. Cancer Res 66:7843–7848. https://doi.org/10.1158/0008-5472.can-06-1010
4. Benzaquen LR, Nicholson-Weller A, Halperin JA (1994) Terminal complement proteins C5b-9 release basic fibroblast growth factor and platelet-derived growth factor from endothelial cells. J Exp Med 179:985–992
5. Bohlson SS, O'Conner SD, Hulsebus HJ, Ho MM, Fraser DA (2014) Complement, c1q, and c1q-related molecules regulate macrophage polarization. Front Immunol 5:402. https://doi.org/10.3389/fimmu.2014.00402
6. Bouwens TA, Trouw LA, Veerhuis R, Dirven CM, Lamfers ML, Al-Khawaja H (2015) Complement activation in glioblastoma multiforme pathophysiology: evidence from serum levels and presence of complement activation products in tumor tissue. J Neuroimmunol 278:271–276. https://doi.org/10.1016/j.jneuroim.2014.11.016

7. Brogi E, Wu T, Namiki A, Isner JM (1994) Indirect angiogenic cytokines upregulate VEGF and bFGF gene expression in vascular smooth muscle cells, whereas hypoxia upregulates VEGF expression only. Circulation 90:649–652
8. Bulla R, Tripodo C, Rami D, Ling GS, Agostinis C, Guarnotta C, Zorzet S, Durigutto P, Botto M, Tedesco F (2016) C1q acts in the tumour microenvironment as a cancer-promoting factor independently of complement activation. Nat Commun 7:10346. https://doi.org/10.1038/ncomms10346
9. Calabrese C, Poppleton H, Kocak M, Hogg TL, Fuller C, Hamner B, Oh EY, Gaber MW, Finklestein D, Allen M et al (2007) A perivascular niche for brain tumor stem cells. Cancer Cell 11:69–82
10. Charles N, Ozawa T, Squatrito M, Bleau AM, Brennan CW, Hambardzumyan D, Holland EC (2010) Perivascular nitric oxide activates notch signaling and promotes stem-like character in PDGF-induced glioma cells. Cell Stem Cell 6:141–152. https://doi.org/10.1016/j.stem.2010.01.001
11. Chen J, Li GQ, Zhang L, Tang M, Cao X, Xu GL, Wu YZ (2018) Complement C5a/C5aR pathway potentiates the pathogenesis of Gastric Cancer by down-regulating p21 expression. Cancer Lett 412:30–36. https://doi.org/10.1016/j.canlet.2017.10.003
12. Cheng L, Huang Z, Zhou W, Wu Q, Donnola S, Liu JK, Fang X, Sloan AE, Mao Y, Lathia JD et al (2013) Glioblastoma Stem Cells generate vascular pericytes to support vessel function and tumor growth. Cell 153:139–152. https://doi.org/10.1016/j.cell.2013.02.021
13. Cho MS, Vasquez HG, Rupaimoole R, Pradeep S, Wu S, Zand B, Han HD, Rodriguez-Aguayo C, Bottsford-Miller J, Huang J et al (2014) Autocrine effects of tumor-derived complement. Cell Rep 6:1085–1095
14. Corrales L, Ajona D, Rafail S, Lasarte JJ, Riezu-Boj JI, Lambris JD, Rouzaut A, Pajares MJ, Montuenga LM, Pio R (2012) Anaphylatoxin C5a creates a favorable microenvironment for lung cancer progression. J Immunol 189:4674–4683
15. Cortright DN, Meade R, Waters SM, Chenard BL, Krause JE (2009) C5a, but not C3a, increases VEGF secretion in ARPE-19 human retinal pigment epithelial cells. Curr Eye Res 34:57–61. https://doi.org/10.1080/02713680802546658
16. Coulthard LG, Hawksworth OA, Li R, Balachandran A, Lee JD, Sepehrband F, Kurniawan N, Jeanes A, Simmons DG, Wolvetang E et al (2017) Complement C5aR1 Signaling Promotes Polarization and Proliferation of Embryonic Neural Progenitor Cells through PKCzeta. J Neurosci 37:5395–5407. https://doi.org/10.1523/jneurosci.0525-17.2017
17. Cuddapah VA, Robel S, Watkins S, Sontheimer H (2014) A neurocentric perspective on glioma invasion. Nat Rev Neurosci 15:455–465. https://doi.org/10.1038/nrn3765
18. Daniele S, Costa B, Zappelli E, Da Pozzo E, Sestito S, Nesi G, Campiglia P, Marinelli L, Novellino E, Rapposelli S et al (2015) Combined inhibition of AKT/mTOR and MDM2 enhances glioblastoma Multiforme cell apoptosis and differentiation of cancer stem cells. Scientific reports 5:9956. https://doi.org/10.1038/srep09956
19. Davies DC (2002) Blood-brain barrier breakdown in septic encephalopathy and brain tumours. J Anat 200:639–646
20. Dejaegher J, Verschuere T, Vercalsteren E, Boon L, Cremer J, Sciot R, Van Gool SW, De Vleeschouwer S (2017) Characterization of PD-1 upregulation on tumor-infiltrating lymphocytes in human and murine gliomas and preclinical therapeutic blockade. Int J Cancer 141: 1891–1900 Doi https://doi.org/10.1002/ijc.30877
21. Dimmeler S, Fleming I, Fisslthaler B, Hermann C, Busse R, Zeiher AM (1999) Activation of nitric oxide synthase in endothelial cells by Akt-dependent phosphorylation. Nature 399:601–605. https://doi.org/10.1038/21224
22. Du R, Lu KV, Petritsch C, Liu P, Ganss R, Passegue E, Song H, Vandenberg S, Johnson RS, Werb Z et al (2008) HIF1alpha induces the recruitment of bone marrow-derived vascular modulatory cells to regulate tumor angiogenesis and invasion. Cancer Cell 13:206–220. https://doi.org/10.1016/j.ccr.2008.01.034
23. Du Z, Jia D, Liu S, Wang F, Li G, Zhang Y, Cao X, Ling EA, Hao A (2009) Oct4 is expressed in human gliomas and promotes colony formation in glioma cells. Glia 57:724–733. https://doi.org/10.1002/glia.20800
24. Eder K, Kalman B (2015) The dynamics of interactions among immune and glioblastoma cells. NeuroMolecular Med 17:335–352. https://doi.org/10.1007/s12017-015-8362-x
25. Fessler E, Borovski T, Medema JP (2015) Endothelial cells induce cancer stem cell features in differentiated glioblastoma cells via bFGF. Mol Cancer 14:157
26. Fraser DA, Bohlson SS, Jasinskiene N, Rawal N, Palmarini G, Ruiz S, Rochford R, Tenner AJ (2006) C1q and MBL, components of the innate immune system, influence monocyte cytokine expression. J leukoc biol 80:107–116. https://doi.org/10.1189/jlb.1105683
27. Greijer AE, van der Groep P, Kemming D, Shvarts A, Semenza GL, Meijer GA, van de Wiel MA, Belien JA, van Diest PJ, van der Wall E (2005) Up-regulation of gene expression by hypoxia is mediated predominantly by hypoxia-inducible factor 1 (HIF-1). J Pathol 206:291–304. https://doi.org/10.1002/path.1778
28. Grivennikov SI, Greten FR, Karin M (2010) Immunity, inflammation, and cancer. Cell 140:883–899. https://doi.org/10.1016/j.cell.2010.01.025
29. Gu H, Mickler EA, Cummings OW, Sandusky GE, Weber DJ, Gracon A, Woodruff T, Wilkes DS, Vittal R (2014) Crosstalk between TGF-beta1 and complement activation augments epithelial injury in pulmonary fibrosis. FASEB J 28:4223–4234. https://doi.org/10.1096/fj.13-247650
30. Gunn L, Ding C, Liu M, Ma Y, Qi C, Cai Y, Hu X, Aggarwal D, Zhang HG, Yan J (2012) Opposing roles for complement component C5a in tumor progression and the tumor microenvironment. J Immunol 189:2985–2994. https://doi.org/10.4049/jimmunol.1200846
31. Hambardzumyan D, Gutmann DH, Kettenmann H (2016) The role of microglia and macrophages in glioma maintenance and progression. Nat Neurosci 19:20–27. https://doi.org/10.1038/nn.4185
32. Hanahan D, Coussens LM (2012) Accessories to the crime: functions of cells recruited to the tumor microenvironment. Cancer Cell 21:309–322. https://doi.org/10.1016/j.ccr.2012.02.022
33. Hanahan D, Weinberg RA (2011) Hallmarks of cancer: the next generation. Cell 144:646–674. https://doi.org/10.1016/j.cell.2011.02.013
34. Hanisch UK, Kettenmann H (2007) Microglia: active sensor and versatile effector cells in the normal and pathologic brain. Nat Neurosci 10:1387–1394. https://doi.org/10.1038/nn1997
35. Hawksworth OA, Coulthard LG, Taylor SM, Wolvetang EJ, Woodruff TM (2014) Brief report: complement C5a promotes human embryonic stem cell pluripotency in the absence of FGF2. Stem cells 32:3278–3284. https://doi.org/10.1002/stem.1801
36. Haynes T, Luz-Madrigal A, Reis ES, Echeverri Ruiz NP, Grajales-Esquivel E, Tzekou A, Tsonis PA, Lambris JD, Del Rio-Tsonis K (2013) Complement anaphylatoxin C3a is a potent inducer of embryonic chick retina regeneration. Nat Commun 4: 2312 Doi https://doi.org/10.1038/ncomms3312
37. Hernandez C, Huebener P, Schwabe RF (2016) Damage-associated molecular patterns in cancer: a double-edged sword. Oncogene 35:5931–5941. https://doi.org/10.1038/onc.2016.104
38. Honczarenko M, Ratajczak MZ, Nicholson-Weller A, Silberstein LE (2005) Complement C3a enhances CXCL12 (SDF-1)-mediated chemotaxis of bone marrow hematopoietic cells independently of C3a receptor. J Immunol 175:3698–3706
39. Infanger DW, Cho Y, Lopez BS, Mohanan S, Liu SC, Gursel D, Boockvar JA, Fischbach C (2013) Glioblastoma stem cells are regulated by interleukin-8 signaling in a tumoral perivascular niche. Cancer Res 73:7079–7089
40. Jacob A, Alexander JJ (2014) Complement and blood-brain barrier integrity. Mol Immunol 61:149–152. https://doi.org/10.1016/j.molimm.2014.06.039
41. Kahlert UD, Maciaczyk D, Doostkam S, Orr BA, Simons B, Bogiel T, Reithmeier T, Prinz M, Schubert J, Niedermann G et al (2012) Activation of canonical WNT/beta-catenin signaling enhances in vitro motility of glioblastoma cells by activation of ZEB1 and other activators of epithelial-to-mesenchymal transition. Cancer Lett 325:42–53. https://doi.org/10.1016/j.canlet.2012.05.024
42. Kalani MY, Cheshier SH, Cord BJ, Bababeygy SR, Vogel H, Weissman IL, Palmer TD, Nusse R (2008) Wnt-mediated self-renewal of neural stem/progenitor cells. Proc Natl Acad Sci U S A 105:16970–16975. https://doi.org/10.1073/pnas.0808616105
43. Kaplan AP, Ghebrehiwet B (2010) The plasma bradykinin-forming pathways and its interrelationships with complement. Mol Immunol 47:2161–2169. https://doi.org/10.1016/j.molimm.2010.05.010
44. Keir LS, Firth R, Aponik L, Feitelberg D, Sakimoto S, Aguilar E, Welsh GI, Richards A, Usui Y, Satchell SC et al (2017) VEGF regulates local inhibitory complement proteins in the eye and kidney. J Clin Invest 127:199–214. https://doi.org/10.1172/jci86418
45. Kemper C, Chan AC, Green JM, Brett KA, Murphy KM, Atkinson JP (2003) Activation of human CD4+ cells with CD3 and CD46 induces a T-regulatory cell 1 phenotype. Nature 421:388–392. https://doi.org/10.1038/nature01315
46. Lathia JD, Heddleston JM, Venere M, Rich JN (2011) Deadly teamwork: neural cancer stem cells and the tumor microenvironment. Cell Stem Cell 8: 482–485. https://doi.org/10.1016/j.stem.2011.04.013
47. Le Buanec H, Gougeon ML, Mathian A, Lebon P, Dupont JM, Peltre G, Hemon P, Schmid M, Bizzini B, Kunding T et al (2011) IFN-alpha and CD46 stimulation are associated with active lupus and skew natural T regulatory

cell differentiation to type 1 regulatory T (Tr1) Cell. Proc Natl Acad Sci U S A 108:18995–19000. https://doi.org/10.1073/pnas.1113301108
48. Le Friec G, Sheppard D, Whiteman P, Karsten CM, Shamoun SA, Laing A, Bugeon L, Dallman MJ, Melchionna T, Chillakuri C et al (2012) The CD46-Jagged1 interaction is critical for human TH1 immunity. Nat Immunol 13:1213–1221
49. Li Z, Bao S, Wu Q, Wang H, Eyler C, Sathornsumetee S, Shi Q, Cao Y, Lathia J, RE ML et al (2009) Hypoxia-inducible factors regulate tumorigenic capacity of glioma stem cells. Cancer Cell 15:501–513. https://doi.org/10.1016/j.ccr.2009.03.018
50. Liszewski MK, Kolev M, Le Friec G, Leung M, Bertram PG, Fara AF, Subias M, Pickering MC, Drouet C, Meri S et al (2013) Intracellular complement activation sustains T cell homeostasis and mediates effector Differentiation. Immunity 39:1143–1157. https://doi.org/10.1016/j.immuni.2013.10.018
51. Mamidi S, Hone S, Kirschfink M (2017) The complement system in cancer: ambivalence between tumour destruction and promotion. Immunobiology 222:45–54. https://doi.org/10.1016/j.imbio.2015.11.008
52. Mantovani A, Allavena P, Sica A, Balkwill F (2008) Cancer-related inflammation. Nature 454:436–444. https://doi.org/10.1038/nature07205
53. Markiewski MM, DeAngelis RA, Benencia F, Ricklin-Lichtsteiner SK, Koutoulaki A, Gerard C, Coukos G, Lambris JD (2008) Modulation of the antitumor immune response by complement. Nat Immunol 9:1225–1235. https://doi.org/10.1038/ni.1655
54. Mastellos DC (2014) Complement emerges as a masterful regulator of CNS homeostasis, neural synaptic plasticity and cognitive function. Exp Neurol 261:469–474. https://doi.org/10.1016/j.expneurol.2014.06.019
55. McGeer PL, Lee M, McGeer EG (2017) A review of human diseases caused or exacerbated by aberrant complement activation. Neurobiol Aging 52:12–22. https://doi.org/10.1016/j.neurobiolaging.2016.12.017
56. Merle NS, Church SE, Fremeaux-Bacchi V, Roumenina LT (2015) Complement system part I - molecular mechanisms of activation and regulation. Front Immunol 6:262. https://doi.org/10.3389/fimmu.2015.00262
57. Montalto MC, Hart ML, Jordan JE, Wada K, Stahl GL (2003) Role for complement in mediating intestinal nitric oxide synthase-2 and superoxide dismutase expression. Am J Physiol Gastrointest Liver Physiol 285:G197–G206. https://doi.org/10.1152/ajpgi.00029.2003
58. Montana V, Sontheimer H (2011) Bradykinin promotes the chemotactic invasion of primary brain tumors. J Neurosci 31:4858–4867. https://doi.org/10.1523/jneurosci.3825-10.2011
59. Morgan BP, Boyd C, Bubeck D (2017) Molecular cell biology of complement membrane attack. Seminars in cell & developmental biology: Doi. https://doi.org/10.1016/j.semcdb.2017.06.009
60. Morgan BP, Gasque P (1996) Expression of complement in the brain: role in health and disease. Immunol Today 17:461–466
61. Naito AT, Sumida T, Nomura S, Liu ML, Higo T, Nakagawa A, Okada K, Sakai T, Hashimoto A, Hara Y et al (2012) Complement C1q activates canonical Wnt signaling and promotes aging-related phenotypes. Cell 149:1298–1313
62. Nitta H, Murakami Y, Wada Y, Eto M, Baba H, Imamura T (2014) Cancer cells release anaphylatoxin C5a from C5 by serine protease to enhance invasiveness. Oncol Rep 32:1715–1719. https://doi.org/10.3892/or.2014.3341
63. Nunez-Cruz S, Gimotty PA, Guerra MW, Connolly DC, Wu YQ, DeAngelis RA, Lambris JD, Coukos G, Scholler N (2012) Genetic and pharmacologic inhibition of complement impairs endothelial cell function and ablates ovarian cancer neovascularization. Neoplasia 14:994–1004
64. O'Garra A, Barrat FJ, Castro AG, Vicari A, Hawrylowicz C (2008) Strategies for use of IL-10 or its antagonists in human disease. Immunol Rev 223:114–131. https://doi.org/10.1111/j.1600-065X.2008.00635.x
65. Ooi YC, Tran P, Ung N, Thill K, Trang A, Fong BM, Nagasawa DT, Lim M, Yang I (2014) The role of regulatory T-cells in glioma immunology. Clin Neurol Neurosurg 119:125–132. https://doi.org/10.1016/j.clineuro.2013.12.004
66. Pedersen ED, Froyland E, Kvissel AK, Pharo AM, Skalhegg BS, Rootwelt T, Mollnes TE (2007) Expression of complement regulators and receptors on human NT2-N neurons--effect of hypoxia and reoxygenation. Mol Immunol 44:2459–2468. https://doi.org/10.1016/j.molimm.2006.10.022
67. Peerschke EI, Ghebrehiwet B (2014) cC1qR/CR and gC1qR/p33: observations in cancer. Mol Immunol 61:100–109
68. Perianayagam MC, Balakrishnan VS, King AJ, Pereira BJ, Jaber BL (2002) C5a delays apoptosis of human neutrophils by a phosphatidylinositol 3-kinase-signaling pathway. Kidney Int 61:456–463. https://doi.org/10.1046/j.1523-1755.2002.00139.x
69. Qiang L, Wu T, Zhang HW, Lu N, Hu R, Wang YJ, Zhao L, Chen FH, Wang XT, You QD et al (2012) HIF-1alpha is critical for hypoxia-mediated maintenance of glioblastoma Stem Cells by activating Notch signaling pathway. Cell Death Differ 19:284–294. https://doi.org/10.1038/cdd.2011.95
70. Qiu Y, Marquez-Curtis LA, Janowska-Wieczorek A (2012) Mesenchymal stromal cells derived from umbilical cord blood migrate in response to complement C1q. Cytotherapy 14:285–295
71. Rampazzo E, Persano L, Pistollato F, Moro E, Frasson C, Porazzi P, Della Puppa A, Bresolin S, Battilana G, Indraccolo S et al (2013) Wnt activation promotes neuronal differentiation of glioblastoma. Cell Death Dis 4:e500. https://doi.org/10.1038/cddis.2013.32
72. Rao S, Sengupta R, Choe EJ, Woerner BM, Jackson E, Sun T, Leonard J, Piwnica-Worms D, Rubin JB (2012) CXCL12 mediates trophic interactions between endothelial and tumor cells in glioblastoma. PLoS One 7:e33005
73. Raychaudhuri B, Rayman P, Ireland J, Ko J, Rini B, Borden EC, Garcia J, Vogelbaum MA, Finke J (2011) Myeloid-derived suppressor cell accumulation and function in patients with newly diagnosed glioblastoma. Neuro-Oncology 13:591–599
74. Reis ES, Mastellos DC, Ricklin D, Mantovani A, Lambris JD (2017) Complement in cancer: untangling an intricate relationship. Nature reviews Immunology: Doi. https://doi.org/10.1038/nri.2017.97
75. Ricklin D, Hajishengallis G, Yang K, Lambris JD (2010) Complement: a key system for immune surveillance and homeostasis. Nat Immunol 11:785–797. https://doi.org/10.1038/ni.1923
76. Sakiyama H, Kaji K, Nakagawa K, Nagino K (1998) Inhibition of bFGF activity by complement C1s: covalent binding of C1s with bFGF. Cell Biochem Funct 16:159–163
77. Sarma JV, Ward PA (2012) New developments in C5a receptor signaling. Cell health and cytoskeleton 4:73–82. https://doi.org/10.2147/chc.s27233
78. Sayour EJ, McLendon P, McLendon R, De Leon G, Reynolds R, Kresak J, Sampson JH, Mitchell DA (2015) Increased proportion of FoxP3+ regulatory T cells in tumor infiltrating lymphocytes is associated with tumor recurrence and reduced survival in patients with glioblastoma. Cancer immunology, immunotherapy : CII 64:419–427. https://doi.org/10.1007/s00262-014-1651-7
79. Schraufstatter IU, Discipio RG, Zhao M, Khaldoyanidi SK (2009) C3a and C5a are chemotactic factors for human mesenchymal stem cells, which cause prolonged ERK1/2 phosphorylation. Journal of immunology (Baltimore, Md : 1950) 182: 3827–3836
80. Sharma A, Shiras A (2016) Cancer stem cell-vascular endothelial cell interactions in glioblastoma. Biochem Biophys Res Commun 473:688–692. https://doi.org/10.1016/j.bbrc.2015.12.022
81. Shen Q, Goderie SK, Jin L, Karanth N, Sun Y, Abramova N, Vincent P, Pumiglia K, Temple S (2004) Endothelial cells stimulate self-renewal and expand neurogenesis of neural stem cells. Science 304:1338–1340
82. Sherry MM, Reeves A, Wu JK, Cochran BH (2009) STAT3 is required for proliferation and maintenance of multipotency in glioblastoma Stem Cells. Stem cells (Dayton, Ohio) 27:2383–2392. https://doi.org/10.1002/stem.185
83. Shinjyo N, Stahlberg A, Dragunow M, Pekny M, Pekna M (2009) Complement-derived anaphylatoxin C3a regulates in vitro differentiation and migration of neural progenitor cells. Stem cells (Dayton, Ohio) 27:2824–2832
84. Simone S, Rascio F, Castellano G, Divella C, Chieti A, Ditonno P, Battaglia M, Crovace A, Staffieri F, Oortwijn B et al (2014) Complement-dependent NADPH oxidase enzyme activation in renal ischemia/reperfusion injury. Free Radic Biol Med 74:263–273
85. Singh SK, Hawkins C, Clarke ID, Squire JA, Bayani J, Hide T, Henkelman RM, Cusimano MD, Dirks PB (2004) Identification of human brain tumour initiating cells. Nature 432:396–401. https://doi.org/10.1038/nature03128
86. Speidl WS, Kastl SP, Hutter R, Katsaros KM, Kaun C, Bauriedel G, Maurer G, Huber K, Badimon JJ, Wojta J (2011) The complement component C5a is present in human coronary lesions in vivo and induces the expression of MMP-1 and MMP-9 in human macrophages in vitro. FASEB j 25:35–44
87. Strainic MG, Liu J, Huang D, An F, Lalli PN, Muqim N, Shapiro VS, Dubyak GR, Heeger PS, Medof ME (2008) Locally produced complement fragments C5a and C3a provide both costimulatory and survival signals to naive CD4+ T cells. Immunity 28:425–435. https://doi.org/10.1016/j.immuni.2008.02.001
88. Suva ML, Rheinbay E, Gillespie SM, Patel AP, Wakimoto H, Rabkin SD, Riggi N, Chi AS, Cahill DP, Nahed BV et al (2014) Reconstructing and reprogramming the tumor-propagating potential of glioblastoma stem-like Cell. Cell 157:580–594. https://doi.org/10.1016/j.cell.2014.02.030
89. Taylor RP, Lindorfer MA (2016) Cytotoxic mechanisms of immunotherapy: harnessing complement in the action of anti-tumor monoclonal antibodies. Semin Immunol 28:309–316. https://doi.org/10.1016/j.smim.2016.03.003

90. Thomas TM, Yu JS (2017) Metabolic regulation of glioma stem-like cells in the tumor micro-environment. Cancer Lett 408:174–181. https://doi.org/10.1016/j.canlet.2017.07.014
91. Thurman JM, Renner B, Kunchithapautham K, Ferreira VP, Pangburn MK, Ablonczy Z, Tomlinson S, Holers VM, Rohrer B (2009) Oxidative stress renders retinal pigment epithelial cells susceptible to complement-mediated injury. J Biol Chem 284:16939–16947. https://doi.org/10.1074/jbc.M808166200
92. van den Berg RH, Faber-Krol MC, Sim RB, Daha MR (1998) The first subcomponent of complement, C1q, triggers the production of IL-8, IL-6, and monocyte chemoattractant peptide-1 by human umbilical vein endothelial Cell. J Immunol 161:6924–6930
93. Veerhuis R, Nielsen HM, Tenner AJ (2011) Complement in the brain. Mol Immunol 48:1592–1603. https://doi.org/10.1016/j.molimm.2011.04.003
94. Vesely MD, Kershaw MH, Schreiber RD, Smyth MJ (2011) Natural innate and adaptive immunity to cancer. Annu Rev Immunol 29:235–271. https://doi.org/10.1146/annurev-immunol-031210-101324
95. Wainwright DA, Sengupta S, Han Y, Lesniak MS (2011) Thymus-derived rather than tumor-induced regulatory T cells predominate in brain tumors. Neuro-Oncology 13:1308–1323. https://doi.org/10.1093/neuonc/nor134
96. Walport MJ (2001) Complement. First of two parts. N Engl J Med 344:1058–1066. https://doi.org/10.1056/nejm200104053441406
97. Wang L, Han G, Wang R, Chen G, Xu R, Xiao H, Li X, Geng S, Li Y, Li X et al (2010) Regulation of IL-8 production by complement-activated product, C5a, in vitro and in vivo during sepsis. Clin Immunol 137:157–165
98. Wurdinger T, Deumelandt K, van der Vliet HJ, Wesseling P, de Gruijl TD (2014) Mechanisms of intimate and long-distance cross-talk between glioma and myeloid cells: how to break a vicious cycle. Biochim Biophys Acta 1846: 560–575 Doi https://doi.org/10.1016/j.bbcan.2014.10.003
99. Yan K, Wu Q, Yan DH, Lee CH, Rahim N, Tritschler I, DeVecchio J, Kalady MF, Hjelmeland AB, Rich JN (2014) Glioma cancer stem cells secrete Gremlin1 to promote their maintenance within the tumor hierarchy. Genes Dev 28: 1085–1100. https://doi.org/10.1101/gad.235515.113
100. Yanai R, Thanos A, Connor KM (2012) Complement involvement in neovascular ocular diseases. Adv Exp Med Biol 946:161–183. https://doi.org/10.1007/978-1-4614-0106-3_10
101. Yang C, Yang L, Liu Y (2014) Soluble complement complex C5b-9 promotes microglia activation. J Neuroimmunol 267:16–19. https://doi.org/10.1016/j.jneuroim.2013.11.007
102. Yang YH, Tsai IJ, Chang CJ, Chuang YH, Hsu HY, Chiang BL (2015) The interaction between circulating complement proteins and cutaneous microvascular endothelial cells in the development of childhood Henoch-Schonlein purpura. PLoS One 10:e0120411
103. Yu MO, Park KJ, Park DH, Chung YG, Chi SG, Kang SH (2015) Reactive oxygen species production has a critical role in hypoxia-induced Stat3 activation and angiogenesis in human glioblastoma. J Neuro-Oncol
104. Zajc I, Hreljac I, Lah T (2006) Cathepsin L affects apoptosis of glioblastoma cells: a potential implication in the design of cancer therapeutics. Anticancer Res 26:3357–3364
105. Zhang YE (2009) Non-Smad pathways in TGF-beta signaling. Cell Res 19: 128–139. https://doi.org/10.1038/cr.2008.328

Somatostatin receptor 2A protein expression characterizes anaplastic oligodendrogliomas with favorable outcome

Romain Appay[1,3], Emeline Tabouret[2,3], Mehdi Touat[4,5], Catherine Carpentier[5], Carole Colin[3], François Ducray[6,7], Ahmed Idbaih[4,5], Karima Mokhtari[5,8], Emmanuelle Uro-Coste[9,10], Caroline Dehais[4], Dominique Figarella-Branger[1,3*] and the POLA network

Abstract

Diffuse gliomas are classified according to the 2016 WHO Classification of Tumors of the Central Nervous System, which now defines entities by both histology and molecular features. Somatostatin receptor subtype 2A (SSTR2A) expression has been reported in various solid tumors as associated with favorable outcomes. Its expression has been reported in gliomas with uncertain results regarding its prognostic value. The objective of this study was to assess the prognostic impact of SSTR2A protein expression in a large cohort of grade III and IV gliomas classified according to the updated 2016 WHO classification. We further validated our result with an independent cohort of low grade glioma using dataset generated by The Cancer Genome Atlas (TCGA) Research Network.
We analyzed clinical and molecular data from 575 patients. SSTR2A protein expression was evaluated using immunohistochemistry on tissue microarrays. High expression of SSTR2A protein associated with the anaplastic oligodendroglioma *IDH*-mutant and 1p/19q-codeleted subgroup ($p < 0.001$). Among these tumors, SSTR2A protein expression was significantly associated with a lower proliferative index, the absence of microvascular proliferation and the absence of necrosis ($p < 0.001$). Furthermore SSTR2A protein expression associated with better overall survival ($p = 0.007$) and progression-free survival ($p = 0.01$) in both univariate and multivariate analysis when adjusted by the age, the presence of necrosis and the mitotic index. Similar results were obtained regarding SSTR2 mRNA expression in the TCGA low grade glioma, subtype *IDH*-mutant and 1p/19q-codeleted, dataset. SSTR2A might represent an attractive biomarker and therapeutic target in anaplastic oligodendroglioma *IDH*-mutant and 1p/19q-codeleted specific subgroup. Understanding the implicated molecular pathways may represent a step forward to improve therapeutic approaches.

Keywords: Somatostatin receptor subtype 2A (SSTR2A), Glioma, Biomarker, Therapeutic target

Introduction

Diffuse gliomas are the most common primary brain cancers. They are classified according to the 2016 WHO (World Health Organization) Classification of Tumors of the Central Nervous System (CNS), which combines for the first time histological and molecular features in an "integrated diagnosis" [20]. This novel classification system incorporates mutations in the isocitrate deshydrogenase 1 and 2 genes (*IDH1* and *IDH2*) and the whole-arm chromosomal loss of 1p and 19q (1p/19q-codeletion), which are both required to be present for confirming the diagnosis of oligodendroglioma. *IDH* mutations are the key genetic alterations characterizing grade II and III gliomas and glioblastomas with favorable outcome [37]. Diagnostic strategy and therapeutic management depend on each subtype and the identification of distinct prognostic

* Correspondence: dominique.figarella-branger@univ-amu.fr
[1]APHM, Hôpital de la Timone, Service d'Anatomie Pathologique et de Neuropathologie, Marseille, France
[3]Aix-Marseille Univ, CNRS, INP, Inst Neurophysiopathol, Marseille, France
Full list of author information is available at the end of the article

subgroups among gliomas belonging to the same histo-molecular category is crucial to open perspectives of therapeutic development.

Somatostatin (SST), also known as growth hormone-inhibiting hormone (GHIH), was first described in 1968 as a hormone secretion [18]. The effects of SST are mediated through its interaction with somatostatin receptors (SSTR), a family of G protein-coupled receptors consisting of 6 different subtypes (SSTR1, 2A, 2B, 3, 4 and 5) [26, 32].

SSTR2A is the predominant subtype. Its expression has been reported in various solid tumors as associated with favorable outcomes [1, 19, 23, 25, 28]. SSTRs are commonly expressed on neuroendocrine tumors (NETs). In NETs, the expression of SSTR2A by tumor cells is of interest for both diagnostic and therapeutic strategy. Indeed, SSTR2A is a target for radiolabeled imaging (OCTREOSCAN, PET 68Ga-DOTATOC) as well as therapy using SST analogs labelled with β-emitting isotopes (90Y-DOTATOC and 177Lu-DOTATATE) [2, 5, 29]. In addition, SST analogs (Octreotide and Lanreotide) are used to inhibit the release of hormones and control secretory symptoms [1, 13, 14, 16, 26]. Interestingly, recent studies demonstrated that SST analogs can also inhibit growth of SSTRs-dependent tumors by regulating intracellular signaling pathways, including dephosphorylation of actors implicated in the mitogen-activated protein kinase (MAPK) signaling and induction of apoptosis [13, 26, 32].

Few studies have previously reported the expression of SSTR2A in gliomas with discrepant results regarding their association with grade [11, 17, 21, 26]. In a recent study, Kiviniemi et al. [17] reported high expression of SSTR2A protein predominant in oligodendrogliomas in a cohort of 184 gliomas classified according to the specific molecular signatures of the updated WHO classification. Furthermore, they reported a survival benefit in gliomas with high expression of SSTR2A protein. However, this difference might be related to the association between SSTR2A and the oligodendroglioma subtype and it is not clear whether the level of SSTR2A expression has prognostic significance among the oligodendroglioma subgroup.

In France, since 2008, the POLA network provides a centralized review and molecular analysis of de novo adult high-grade glioma with an oligodendroglial component. Using the tissue samples and dataset provided by this network, our objective was to assess the prognostic impact of the SSTR2A protein expression in a large cohort of grade III and IV gliomas. We further validated our result with an independent cohort using dataset generated by the TCGA Research Network [8].

Materials and methods

Study population

A total number of 575 patients from the French nation-wide POLA cohort were included in this study. Inclusion criteria were the written consent of the patient for clinical data collection and genetic analysis according to national and POLA network policies, sufficient tissue material for molecular studies allowing classification according to the WHO 2016 (i.e. evaluation of the *IDH* mutation and 1p/19q-codeletion status) and an established diagnosis of high grade glioma (WHO grade III or IV).

IDH mutation status was evaluated using automated immunohistochemistry (IHC) and direct sequencing using the Sanger method as previously described [30]. The genomic profile and assessment of the 1p/19q-codeletion status was determined based on single nucleotide polymorphism (SNP) arrays, comparative genomic hybridization (CGH) arrays, or microsatellite marker analysis as previously described [30].

Anaplastic oligodendroglioma, *IDH*-mutant, 1p19q-codeleted were classified into 3 pathological subgroups based on mitotic index, microvascular proliferation (MVP), and necrosis [12]. Group 1, involved cases with more than 5 mitoses per 10-high power field (HPF), no MVP, and no necrosis, group 2 displayed MVP but no necrosis, and group 3 showed MVP and necrosis.

Proliferative index was evaluated using Ki67 antibody (clone Mib1; 1:100; Dako) and scored as percentage by counting the immunostained nuclei of 400 cells in the most positive area.

Clinical data collected included gender, age at surgery, extend of surgical resection, preoperative Karnofsky performance scale (KPS) and post-operative treatment. Clinical information regarding the outcome was recorded as follow: progression-free survival (PFS) was defined as the time from the date of surgery to recurrence or death from any cause, censored at the date of the last documented disease evaluation. Overall survival (OS) was defined as the time from the date of surgery to death from any cause, censored at the date of last contact.

Tissue microarray (TMA) design

Tissue microarrays (TMA) were constructed from routinely processed formalin-fixed paraffin-embedded tumor material. Areas of viable and representative tumor, away from foci of necrosis and from grade II areas when present, following review of all blocks were marked by a pathologist (DFB) prior to inclusion into the TMA. In order to accurately study tumor heterogeneity, for each case three replicate 0.6 mm cores were sampled from different tumor areas.

SSTR2A immunohistochemistry and evaluation on TMA

Immunohistochemical detection of SSTR2A protein expression was performed on 5 μm thick TMA sections with Ventana Benchmark XT. The monoclonal SSTR2A antibody (clone UMB1) was purchased from Abcam and used at 1/4000 dilution. A Benchmark Ventana autostainer (Ventana Medical Systems SA, Illkirch, France) was used for detection and TMA slides were simultaneously immunostained to avoid inter-manipulation variability.

Immunostaining was scored for 185 triplicates by two independent pathologists (RA & DFB). In almost all cases the staining intensity was roughly similar among the different cores of the triplicate, thus the core demonstrating the stronger immunoreaction was analyzed to establish an immunoreactive score (IRS) as previously described by Casar-Borota et al. [9]. Staining intensity was scored as 0 (no immunostaining), 1 (weak), 2 (moderate), or 3 (strong). The percentage of immunoreactive cells was scored as 0 (none), 1 (< 10%), 2 (10–50%), 3 (51–80%), or 4 (> 80%). Multiplication of the staining intensity score and the percent immunoreactivity score resulted in an IRS score, which ranged from 0 to 12. Because the results were highly concordant among the two pathologists, the 390 remaining triplicates were scored by a single pathologist (RA).

The median IRS score based on analysis of all cases in the cohort was equal to 1 and corresponded to a cutoff between positive (IRS ≥ 1) and negative expression (IRS = 0). The third quartile (IRS = 4) was used as the cutoff to define positive cases with high expression (IRS ≥ 4) versus positive cases with low expression (1 ≤ IRS < 4). Representative images of different IRS are shown in Fig. 1.

Public data acquisition

Informed consent and ethical approval was obtained when The National Cancer Institute (NCI) and National Human Genome Research Institute (NHGRI) had collected tissue for TCGA [8]. All sequencing or clinical datasets used in the present study are publicly available to researchers without restrictions or limitations according to TCGA policy. Therefore, the results shown here are in whole based upon data generated by the TCGA Research Network: http://cancergenome.nih.gov/. We used GlioVis data portal for visualization and analysis of brain tumor expression datasets [6] (http://gliovis.bioinfo.cnio.es/), we limited our analysis to data from adult patients with LGG.

Statistical analysis

For statistical analysis, the Chi-square test was used to compare qualitative variables when they were scored as positive or negative. Continuous variables were compared using the Mann–Whitney U test. The Kaplan–Meier method was used to estimate survival distributions. Log-rank tests were used for univariate comparisons. Age at diagnosis, KPS, presence of necrosis and proliferative index were used to build the multivariate Cox proportional hazard backward models. All statistical tests were two-sided, and the threshold for statistical significance was $p = 0.05$. Statistical analysis was done using IBM SPSS statistics software version 23 (IBM SPSS Inc., Chicago, IL, USA).

Results

Clinicopathologic characteristics

We evaluated 227 anaplastic oligodendroglioma *IDH*-mutant and 1p/19q-codeleted. In addition, we analyzed 86 anaplastic astrocytoma (80 *IDH*-mutant and 6 *IDH*-wildtype) and 262 glioblastomas (124 *IDH*-mutant and 138 *IDH*-wildtype). Overall 575 patients were included in this study. Patient characteristics are presented in Table 1.

Scoring of SSTR2A immunohistochemistry and its association with tumor entity

Expression (any level; IRS ≥ 1) of SSTR2A was detected in 59% (337/575) of gliomas. The distribution of

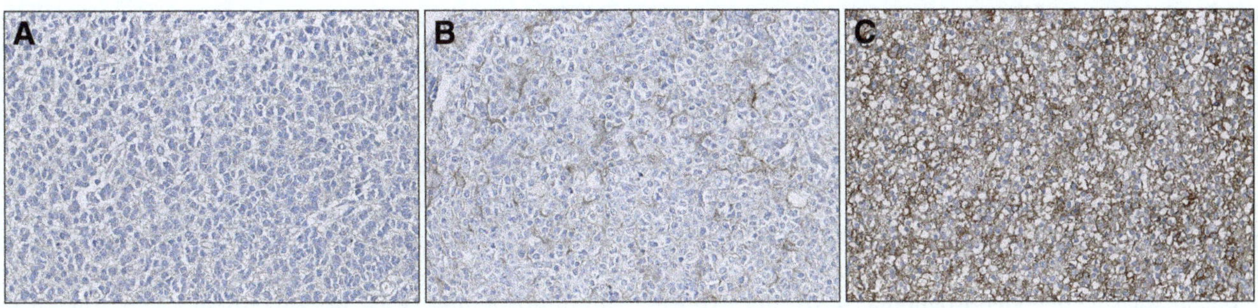

Fig. 1 Immunohistochemical staining of SSTR2A protein in anaplastic oligodendroglioma *IDH*-mutant and 1p/19q-codeleted **(a)** Negative staining (IRS = 0) **(b)** Weak staining intensity (score = 1) in 20% of the cells (score = 2) corresponding to positive cases with low expression (IRS = 3) **(c)** High staining intensity (score = 3) in 100% of the cells (score = 4) corresponding to positive cases with high expression (IRS = 12)

Table 1 Characteristics of the patient population ($N = 583$)

	Anaplastic astrocytoma IDH-wildtype *N* (% of Total)	Glioblastoma IDH-wildtype *N* (% of Total)	Anaplastic astrocytoma IDH-mutant *N* (% of Total)	Glioblastoma IDH-mutant *N* (% of Total)	Anaplastic oligodendroglioma IDH-mutant and 1p/19q-codeleted *N* (% of Total)
Total	6	138	80	124	227
Median age, years (range)	52 (26–60)	59 (16–83)	37 (18–75)	38 (19–78)	48 (19–80)
Gender					
Female	2 (33)	60 (44)	35 (44)	53 (43)	101 (44)
Male	4 (67)	78 (56)	45 (56)	71 (57)	126 (56)
Median preoperative KPS% (range)	90 (90–100)	80 (40–100)	90 (60–100)	90 (50–100)	90 (50–100)
Resection					
Gross total resection	1 (17)	57 (41)	18 (23)	34 (27)	81 (36)
Subtotal resection	2 (33)	37 (27)	38 (48)	43 (35)	58 (26)
Biopsy	2 (33)	25 (18)	9 (11)	35 (28)	66 (29)
Unknown	1 (17)	19 (14)	15 (19)	12 (10)	22 (10)
Adjuvant treatment					
None	0	3 (2)	3 (4)	1 (1)	9 (4)
RT alone	1 (17)	3 (2)	5 (6)	3 (2)	42 (19)
RT + PCV	0	2 (1)	15 (19)	16 (13)	72 (32)
PCV alone	0	2 (1)	0	0	6 (3)
RT + TMZ	5 (83)	104 (75)	38 (48)	73 (59)	70 (31)
TMZ alone	0	5 (4)	3 (4)	4 (3)	5 (2)
No data	0	19 (14)	16 (20)	27	23 (10)
IRS Score					
IRS = 0	5 (83)	86 (62)	31 (39)	47 (38)	69 (30)
1 ≥ IRS < 4	0	33 (24)	27 (34)	46 (37)	51 (22)
IRS ≥ 4	1 (17)	19 (14)	22 (27)	31 (25)	107 (47)

Abbreviations: *IRS* Immunoreactive score; *KPS* Karnofsky Performance Status Scale; *PCV* Procarbazine + Lomustine + Vincristine; *RT* Radiotherapy; *TMZ* Temozolomide

SSTR2A protein expression according to gliomas subtype is shown in Fig. 2. SSTR2A protein expression was significantly associated with *IDH* mutation (66% of *IDH*-mutant tumors were positive for SSTR2A expression versus 39% of *IDH*-wild type, $p < 0.001$). High expression of SSTR2A (IRS score ≥ 4) was detected in 31% (180/575) of gliomas. High expression of SSTR2A was significantly associated with anaplastic oligodendroglioma, *IDH*-mutant and 1p/19q-codeleted and was found in approximatively half of the studied samples whereas it was uncommon in astrocytoma and glioblastoma independently of the presence of *IDH*-mutation ($p < 0.001$).

Association between expression of SSTR2A protein and the anaplastic oligodendroglioma *IDH*-mutant and 1p/19q-codeleted pathological subgroups

In the anaplastic oligodendroglioma subgroup, expression levels of SSTR2A protein varied depending on the pathological subgroup as shown in Fig. 3a ($p < 0.001$). High SSTR2A expression was detected in the majority (70%) of anaplastic oligodendroglioma belonging to the pathological group 1. It was less frequent in group 2 (50%) and only one-third of the anaplastic oligodendroglioma of group 3 presented with high SSTR2A expression. In accordance with the previous result, the absence of SSTR2A expression was uncommon among the patients of the group 1 (5%) whereas approximately half of the group 3 presented with no SSTR2A expression.

We selected six representative cases (i.e. two negative for SSTR2A expression, IRS = 0; two positive with low expression, IRS = 3; and two positive with high expression, IRS = 9) from anaplastic oligodendroglioma, *IDH*-mutant and 1p/19q-codeleted belonging to the group 3 and evaluated intratumoral heterogeneity of SSTR2A expression on whole tumor sample. For each selected case, the percentage and intensity of the immunostaining was in agreement with the result obtained on TMA. One of the two cases evaluated with an IRS score equal to 0 showed focal positivity away from the area selected for the TMA (this area

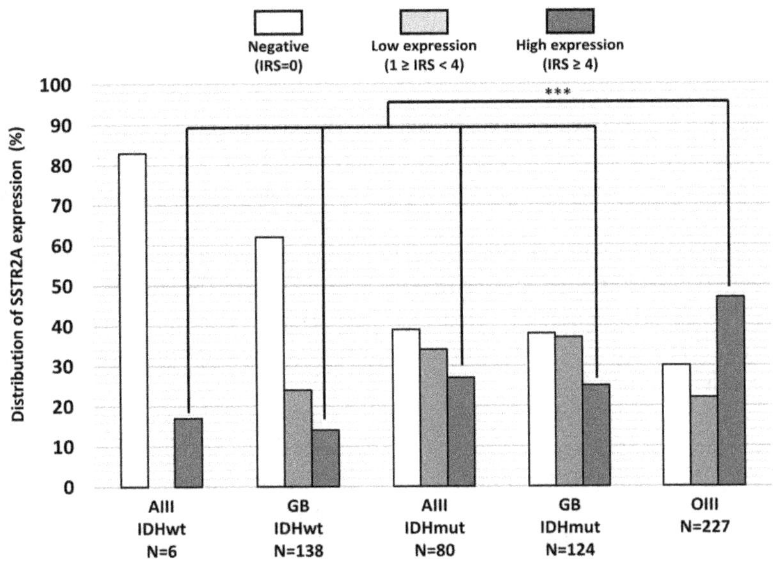

Fig. 2 Distribution of SSTR2A protein expression according to tumor subtypeAbbreviations: AIII IDHwt, Anaplastic astrocytoma *IDH*-wildtype; GB IDHwt, Glioblastoma *IDH*-wildtype; AIII IDHmut, Anaplastic astrocytoma *IDH*-mutant; GB IDHmut, Glioblastoma *IDH*-mutant; OIII, Anaplastic oligodendroglioma, *IDH*-mutant and 1p/19q-codeleted. *** = $p < 0.001$

corresponded to a much less aggressive part of the tumor) but the vast majority of the sample was negative, the second one was completely negative. In the other four cases, the SSTR2A protein expression was in keeping with results evaluated on TMA and always absent in tumor cells located in the vicinity of foci of palisading necrosis (data not shown). Therefore, the low SSTR2A expression in group 3 is not related to necrosis per se but likely represent the acquisition of a more aggressive phenotype. It should be noted that SSTR2A protein expression was also observed in neurons within the peripheral cortex.

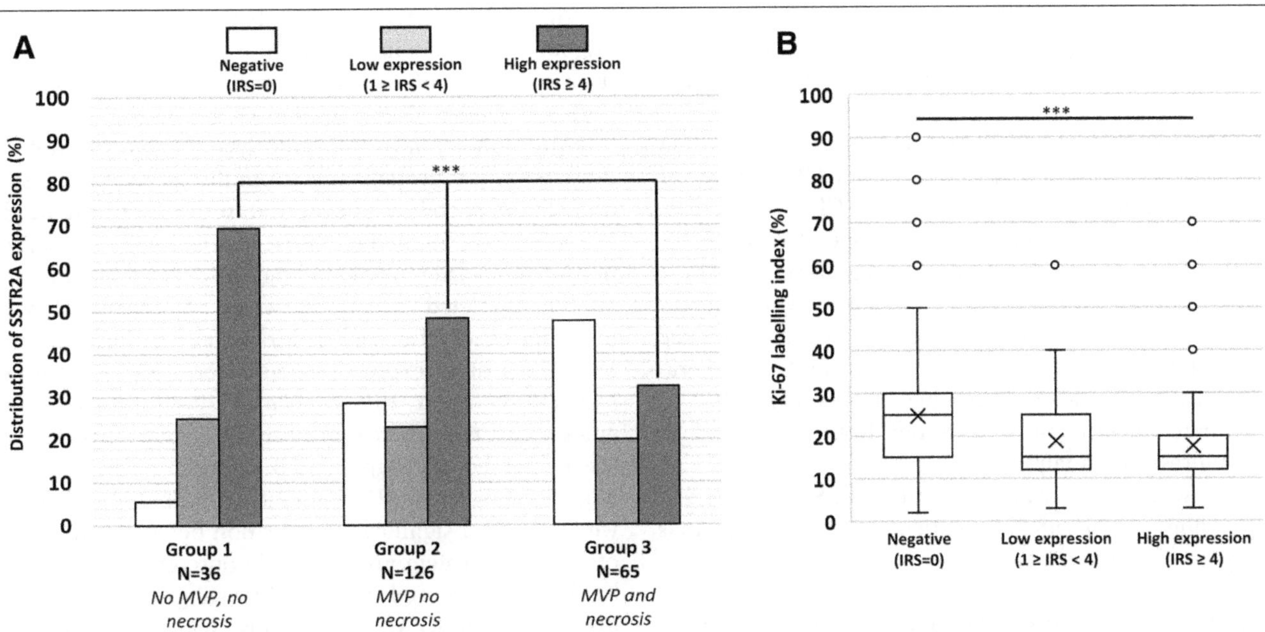

Fig. 3 a Distribution of SSTR2A protein expression according to pathological groups in anaplastic oligodendrogliomas, *IDH*-mutant and 1p/19q-codeleted. Abbreviation: MVP, microvascular proliferation **b** Ki-67 labelling index according to SSTR2A protein expression among the anaplastic oligodendroglioma, *IDH*-mutant and 1p/19q-codeleted. *** = $p < 0.001$

Association between expression of SSTR2A protein and proliferative index in anaplastic oligodendroglioma, IDH-mutant and 1p/19q-codeleted

To examine whether SSTR2A protein expression was associated with differences in the proliferative index we compared quantitative evaluation of Ki-67 (MIB-1) labelling index using immunohistochemistry. As presented in Fig. 3b, anaplastic oligodendroglioma, *IDH*-mutant and 1p/19q-codeleted with high or low SSTR2A protein expression presented with a lower Ki-67 labelling index when compared to SSTR2A negative gliomas (median Ki-67 expression = 15% in SSTR2A positive gliomas versus 26% in SSTR2A negative, $p < 0.001$).

Association between SSTR2A protein expression and survival in anaplastic oligodendroglioma, *IDH*-mutant and 1p/19q-codeleted

We further analyzed the prognostic significance of SSTR2A expression in gliomas. Among the anaplastic oligodendroglioma, *IDH*-mutant and 1p/19q-codeleted, SSTR2A protein expression is prognostic for PFS and OS (Fig. 4a). Both low and high SSTR2A expressive anaplastic oligodendroglioma presented with better OS ($p = 0.022$) and PFS ($p = 0.017$) when compared to negative gliomas. No significant prognostic difference was observed between low expression and high expression in terms of PFS ($p = 0.293$) and OS ($p = 0.280$).

Accordingly, expression of SSTR2A protein (any level, IRS ≥ 1) was significantly associated with longer PFS ($p = 0.010$) and OS ($p = 0.007$) among the subgroup of anaplastic oligodendroglioma, IDH-mutant and 1p/19q-codeleted (Fig. 4b).

In multivariate analysis, expression of SSTR2A (any level, IRS ≥ 1) was also significantly associated with better OS when adjusted by the age (HR = 0.414; 95% CI, 0.185–0.929; $p = 0.033$), by the presence of necrosis (HR = 0.391; 95% CI, 0.174–0,877; $p = 0.023$) or by the proliferative index (HR = 0.411; 95% CI, 0.176–0.959; $p = 0.04$). When adjusted by the preoperative KPS, the result did not reach statistical significance (HR = 0.413; 95% CI, 0.165–1.034; $p = 0.059$), which might be attributed to an insufficient number of collected data. The extend of surgical resection and postoperative treatment did not reached a p-value > 0.2 in univariate analysis thus have not been integrated in the multivariate analysis.

No association between survival and SSTR2A protein expression was observed in patients with other gliomas subtypes (data not shown).

TCGA low-grade glioma RNA-seq data as confirmation dataset

As presented in Fig. 5a, among the independent TCGA low-grade glioma dataset we observed a higher SSTR2 mRNA expression in *IDH*-mutant glioma when compared to *IDH*-wild type ($p < 0.001$). Furthermore, *IDH*-mutant and 1p/19q-codeleted gliomas presented with the highest level of SSTR2 mRNA expression ($p < 0.001$).

When categorized into two groups according to the median score, we observed a better overall survival in glioma with high SSTR2 mRNA expression ($p = 0.056$) among the low grade glioma *IDH*-mutant and 1p/19q-codeleted subgroup (Fig. 5b).

No association between survival and SSTR2 mRNA expression was observed in patient with *IDH*-mutant without 1p/19q-codeletion ($p = 0.478$) and *IDH*-wildtype gliomas ($p = 0.301$) (data not shown).

Discussion

Gliomas are the most common primary CNS tumors. Updated WHO classification of CNS tumors combines for the first time histological and molecular features for an integrated diagnosis. *IDH*-mutant gliomas display a more favorable outcome than the *IDH*-wildtype counterpart. However, despite aggressive treatment, *IDH*-mutant gliomas are characterized by a malignant transformation over time with a median survival of approximately 10 years. Therefore, the identification of distinct prognostic groups among *IDH*-mutant gliomas might be of interest to better stratified the patients and improve therapeutic approaches.

In the present study, we have evaluated the protein expression of SSTR2A by immunohistochemistry in a large cohort of gliomas classified according to the WHO 2016 classification. Because of the inclusion criteria defined by the POLA network (i.e. high-grade glioma with oligodendroglial component) it is worth noticing that the percentage of each category in our study does not reflect the normal distribution of gliomas.

In our cohort, most *IDH*-wild type gliomas did not express SSTR2A protein and a significant overexpression of SSTR2A protein was observed in the *IDH*-mutant gliomas. Among these, the highest expression was recorded in the anaplastic oligodendroglioma, *IDH*-mutant and 1p/19q-codeleted subgroup, which is consistent with previous observations [17]. Moreover, in these tumors, SSTR2A protein expression was associated with a lower proliferative index, the absence of microvascular proliferation and the absence of necrosis (group 1) while it is less expressed in group 2 and 3. Of interest, in anaplastic oligodendroglioma *IDH*-mutant and 1p/19q-codeleted, we observed a significant association between expression of SSTR2A protein and favorable outcome (as indicated by longer PFS and OS in this subgroup of tumors expressing SSTR2A). Importantly, association between SSTR2A expression and outcome remained significant in multivariate analysis adjusting for known prognostic factors in this subtype. Furthermore, similar results were

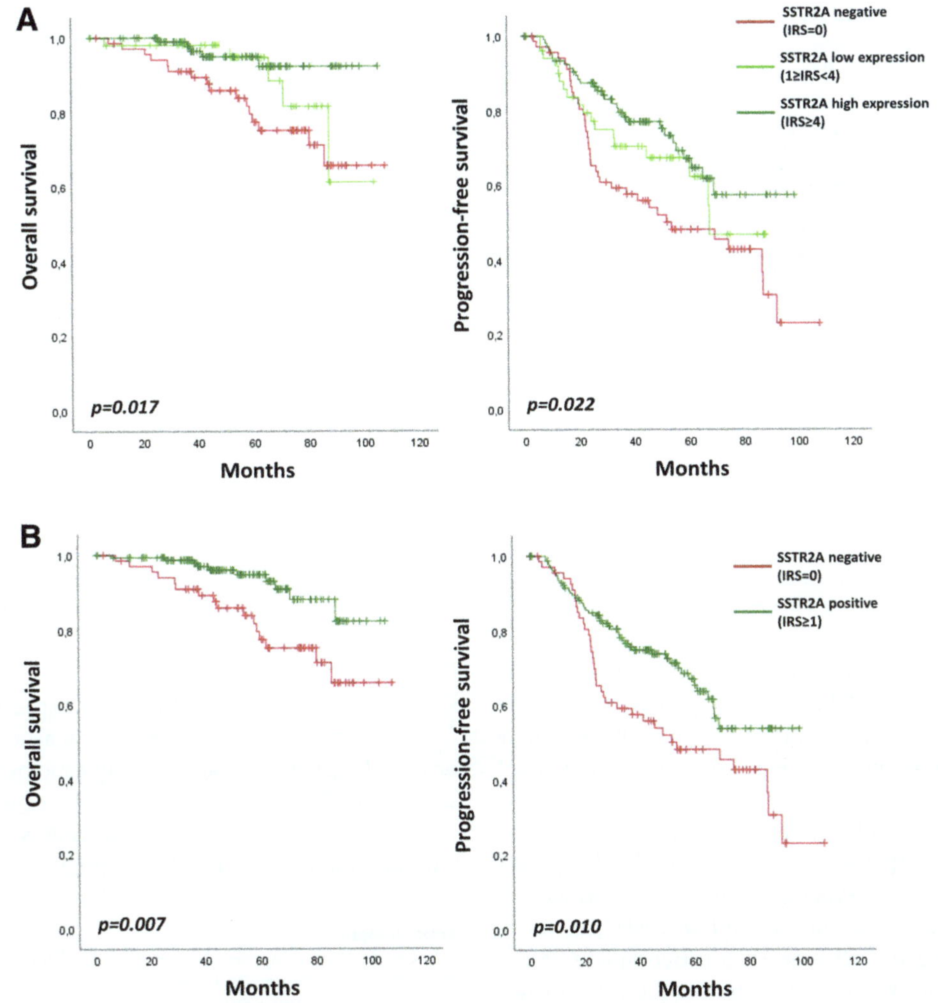

Fig. 4 Overall survival and Progression-free survival according to SSTR2A protein expression in anaplastic oligodendroglioma, *IDH*-mutant and 1p/19q-codeleted. **a** No SSTR2A expression (IRS = 0) versus low SSTR2A expression (1 ≤ IRS < 4) versus high SSTR2A expression (IRS ≥ 4). **b** Negative (IRS = 0) versus positive (IRS ≥ 1) SSTR2A expression

obtained regarding SSTR2 mRNA expression in an independent cohort using the low grade gliomas TCGA dataset.

Thus, our results indicate that the immunohistochemical expression of SSTR2A protein could serve as a prognostic biomarker among anaplastic oligodendroglioma, *IDH*-mutant and 1p/19q-codeleted.

SSTR2A is strongly expressed in neuroendocrine tumors and also in normal neurons according to the brain transcriptome database [7]. The high expression of SSTR2A in *IDH*-mutant adult high grade gliomas in comparison to *IDH*-wildtype gliomas is in accordance with the proneural subtype of *IDH*-mutant gliomas reported by the Cancer genome Atlas [33]. Importantly, among this group, the highest SSTR2A expression is recorded in anaplastic oligodendrogliomas *IDH*-mutant and 1p/19q-codeleted. This is in keeping with the neuronal differentiation of these tumors highlighted by ultrastructural and transcriptional studies [4, 10, 35]. Indeed, Ducray et al. [10] demonstrated that there is a strong correlation between 1p/19q-codeletion and the expression of proneural genes in malignant gliomas. Interestingly, in their cohort, SSTR2 was significantly overexpressed ($p = 0.0001$) in the 1p/19q-codeleted group when compared to the EGFR amplified high grade gliomas. Furthermore, Bielle et al. [4] reported the occurrence of neuronal intermediate progenitors (NIP) markers in a subset of anaplastic oligodendroglioma *IDH*-mutant and 1p/19q-codeleted especially in cases associated with necrosis ($p = 0.0034$). It is possible that NIP-high subgroup could result from tumor dedifferentiation. In the same line, we can postulate that loss of SSTR2A expression among anaplastic oligodendrogliomas *IDH*-mutant and 1p/19q-codeleted is correlated to the

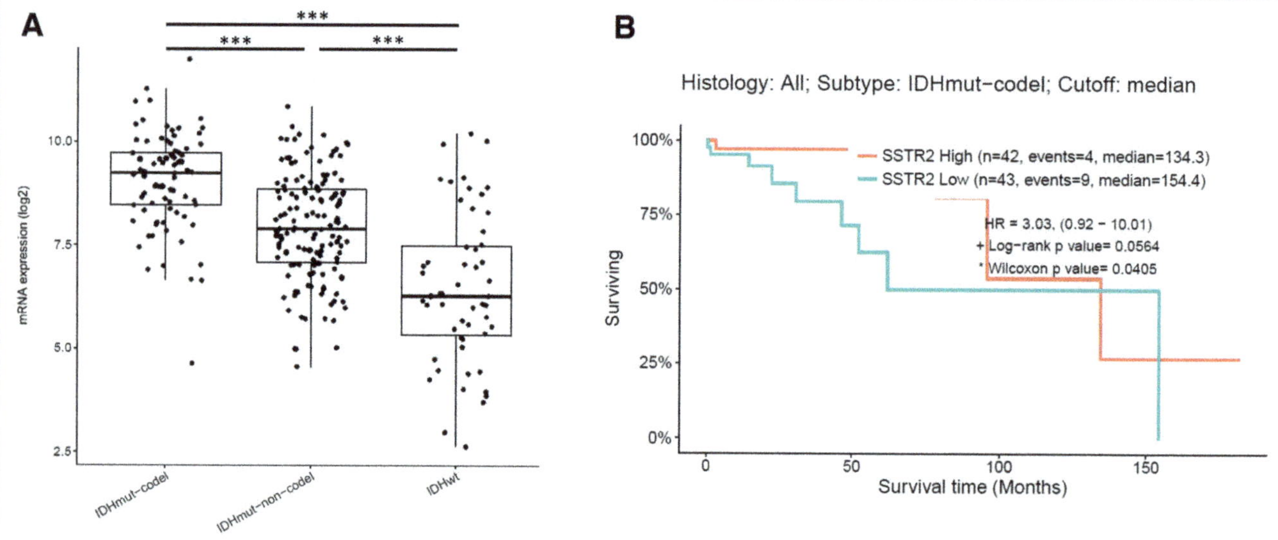

Fig. 5 **a** Relation between SSTR2 mRNA expression and subtype in low grade glioma patients from TCGA cohort. **b** Overall survival according to SSTR2 mRNA expression in patients with *IDH*-mutant and 1p/19q-codeleted low grade gliomas from the TCGA cohort. *** = $p < 0.001$

acquisition of a more immature neuronal phenotype. We could hypothesize that SSTR2A, through its inhibition of the MAPK-pathway and induction of apoptosis, may constitute a brake against in the acquisition of markers of aggressiveness.

Surgery, when possible, is the first line treatment for *IDH*-mutant and 1p/19q-codeleted oligodendroglioma and "watch-and-wait" strategies may be justified in patients with macroscopically complete resection or in young patients (aged < 40 years) with incomplete resection if the tumor is not symptomatic [36]. If further treatment beyond surgery is considered necessary the standard treatment is radiotherapy followed by procarbazine, lomustine and vincristine chemotherapy (PCV). "Watch-and-wait" strategies might be considered even with WHO grade III tumors [36]. According to our results, SSTR2A protein expression is an independent biomarker of good prognosis in both uni- and multivariate analysis that might guide the decision of a postsurgical treatment among the subgroup of anaplastic oligodendrogliomas, *IDH*-mutant and 1p/19q-codeleted.

Moreover, SSTR2A is an attractive therapeutic target. An inhibitory effect of SST analogs on cell migration and invasion has been demonstrated in various tumors [3, 24, 31, 34]. Moreover, therapy using radiolabeled SST analogs is commonly used as a therapeutic option for the treatment of metastatic or inoperable neuroendocrine tumors [5, 29]. Interestingly, encouraging results have been reported in some studies evaluating locoregional delivery of targeted radionuclide therapy in gliomas [15, 22, 27]. Our results suggest the identification of the specific oligodendroglioma subgroup as a possible target of therapies using SST analogs.

Conclusion

Our results demonstrate that high expression of SSTR2A is commonly associated with anaplastic oligodendrogliomas. Moreover, among the anaplastic oligodendroglioma subgroup SSTR2A expression is associated with a better OS. Therefore, SSTR2A might represent an attractive biomarker and target for these tumors.

Abbreviations

CNS: Central Nervous System;; HPF: High power field (HPF); IDH 1 & 2: Isocitrate deshydrogenase 1 and 2 genes; IRS: Immunoreactive score; KPS: Karnofsky performance status scale; MVP: Microvascular proliferation; NIP: Neuronal Intermediate Progenitor; OS: Overall survival; PCV: Procarbazine + Lomustine + Vincristine chemotherapy; PFS: Progression-free survival; RT: Radiotherapy; SST: Somatostatin; SSTR2A: Somatostatin receptor subtype 2A; TCGA: The Cancer Genome Atlas; TMA: Tissue microarray; TMZ: Temozolomide

Acknowledgements

We thank the ARTC-Sud patients's association (Association pour le Recherche sur les Tumeurs Cérébrales) and the Cancéropôle PACA.

POLA Network :

Amiens (C. Desenclos, H. Sevestre), Angers (P. Menei, A. Rousseau), Annecy (T. Cruel, S. Lopez), Besançon (M-l Mihai, A. Petit), Bicêtre (C. Adam, F. Parker), Brest (P. Dam-Hieu, I. Quintin-Roué), Bordeaux (S. Eimer, H. Loiseau), Caen (L. Bekaert, F. Chapon), Clamart (D. Ricard) Clermont-Ferrand (C. Godfraind, T. Khallil), Clichy (D. Cazals-Hatem, T. Faillot), Colmar (C. Gaultier, MC. Tortel), Cornebarrieu (I. Carpiuc, P. Richard), Créteil (W. Lahiani), Dijon (H. Aubriot-Lorton, F. Ghiringhelli), Lille (CA. Maurage, C. Ramirez), Limoges (EM. Gueye, F. Labrousse), Marseille (O Chinot), Montpellier (L. Bauchet, V. Rigau), Nancy (P. Beauchesne, G. Gauchotte), Nantes (M. Campone, D. Loussouarn), Nice (D. Fontaine, F. Vandenbos-Burel) , Nimes (A. Le Floch, P. Roger) Orléans (C. Blechet, M. Fesneau), Paris (A.Carpentier, JY Delattre (POLA Network National coordinator), S. Elouadhani-Hamdi, M. Polivka), Poitiers (D. Larrieu-Ciron,S. Milin), Reims (P. Colin, MD. Diebold), Rennes (D. Chiforeanu, E. Vauleon), Rouen (O. Langlois, A. Laquerriere), Saint-Etienne (F. Forest, MJ. Motso-Fotso), Saint-Pierre de la Réunion (M. Andraud, G. Runavot), Strasbourg (B. Lhermitte, G. Noel), Suresnes (S. Gaillard, C. Villa), Toulon (N. Desse), Tours (C. Rousselot-Denis, I. Zemmoura), Toulouse (E. Cohen-Moyal, E. Uro-Coste), Villejuif (F. Dhermain)

Funding
This work was funded by the French Institut National du Cancer (INCa, POLA network) and is part of the national program Cartes d'Identité des Tumeurs (CIT) (http://cit.ligue-cancer.net/), which is funded and developed by the Ligue nationale contre le cancer.

Authors' contributions
RA wrote the manuscript, performed the SSTR2A immunohistochemical examination & all the statistical analysis. ET, MT, AI, CC, AI & DFB were major contributors in writing the manuscript. All authors (RA, ET, MT, CC, CC, FD, AI, KM, EUC, CD, DFB) contributed to the data collection. All authors read and approved the final manuscript.

Competing interests
The authors declare that they have no competing interests.

Author details
[1]APHM, Hôpital de la Timone, Service d'Anatomie Pathologique et de Neuropathologie, Marseille, France. [2]APHM, Hôpital de la Timone, Service de Neurooncologie, Marseille, France. [3]Aix-Marseille Univ, CNRS, INP, Inst Neurophysiopathol, Marseille, France. [4]AP-HP, Hôpitaux Universitaires La Pitié Salpêtrière-Charles Foix, Service de Neurologie 2–Mazarin, Paris, France. [5]Sorbonne Université, Inserm, CNRS, UMR S 1127, Institut du Cerveau et de la Moelle épinière, ICM, AP-HP, Hôpitaux Universitaires La Pitié Salpêtrière - Charles Foix, Service de Neurologie 2-Mazarin, F-75013 Paris, France. [6]Hospices Civils de Lyon, Hôpital Pierre Wertheimer, Service de Neuro-oncologie, Bron, France. [7]Department of Cancer Cell Plasticity, Cancer Research Centre of Lyon, Inserm U1052, CNRS UMR5286, Lyon, France. [8]AP-HP, Hôpitaux Universitaires La Pitié Salpêtrière-Charles Foix, Service de Neuropathologie Raymond Escourolle, Paris, France. [9]CHU Toulouse, Hôpital Rangueil, Service d'Anatomie Pathologique et Histologie-Cytologie, Toulouse, France. [10]Inserm U1037, Centre de Recherche en Cancérologie de Toulouse, Université de Toulouse, Toulouse, France.

References
1. Abdel-Rahman O, Lamarca A, Valle JW, Hubner RA (2014) Somatostatin receptor expression in hepatocellular carcinoma: prognostic and therapeutic considerations. Endocr Relat Cancer 21:R485–R493
2. Ambrosini V, Fani M, Fanti S, Forrer F, Maecke HR (2011) Radiopeptide imaging and therapy in Europe. J Nucl Med 52(Suppl 2):42S–55S
3. Annunziata M, Luque RM, Durán-Prado M, Baragli A, Grande C, Volante M et al (2012) Somatostatin and somatostatin analogues reduce PDGF-induced endometrial cell proliferation and motility. Hum Reprod 27:2117–2129
4. Bielle F, Ducray F, Mokhtari K, Dehais C, Adle-Biassette H, Carpentier C et al (2017) Tumor cells with neuronal intermediate progenitor features define a subgroup of 1p/19q co-deleted anaplastic gliomas. Brain Pathol 27:567–579
5. Bodei L, Mueller-Brand J, Baum RP, Pavel ME, Hörsch D, O'Dorisio MS et al (2013) The joint IAEA, EANM, and SNMMI practical guidance on peptide receptor radionuclide therapy (PRRNT) in neuroendocrine tumours. Eur J Nucl Med Mol Imaging 40:800–816
6. Bowman RL, Wang Q, Carro A, Verhaak RGW, Squatrito M (2017) GlioVis data portal for visualization and analysis of brain tumor expression datasets. Neuro-Oncology 19:139–141
7. Cahoy JD, Emery B, Kaushal A, Foo LC, Zamanian JL, Christopherson KS et al (2008) A transcriptome database for astrocytes, neurons, and oligodendrocytes: a new resource for understanding brain development and function. J Neurosci 28:264–278
8. Cancer Genome Atlas Research Network, Brat DJ, RGW V, Aldape KD, WKA Y, Salama SR et al (2015) Comprehensive, Integrative Genomic Analysis of Diffuse Lower-Grade Gliomas. N Engl J Med 372:2481–2498
9. Casar-Borota O, Heck A, Schulz S, Nesland JM, Ramm-Pettersen J, Lekva T et al (2013) Expression of SSTR2a, but not of SSTRs 1, 3, or 5 in somatotroph adenomas assessed by monoclonal antibodies was reduced by octreotide and correlated with the acute and long-term effects of octreotide. J Clin Endocrinol Metab 98:E1730–E1739
10. Ducray F, Idbaih A, de Reyniès A, Bièche I, Thillet J, Mokhtari K et al (2008) Anaplastic oligodendrogliomas with 1p19q codeletion have a proneural gene expression profile. Mol Cancer 7:41
11. Dutour A, Kumar U, Panetta R, Ouafik L, Fina F, Sasi R et al (1998) Expression of somatostatin receptor subtypes in human brain tumors. Int J Cancer 76:620–627
12. Figarella-Branger D, Mokhtari K, Dehais C, Jouvet A, Uro-Coste E, Colin C et al (2014) Mitotic index, microvascular proliferation, and necrosis define 3 groups of 1p/19q codeleted anaplastic oligodendrogliomas associated with different genomic alterations. Neuro-Oncology 16:1244–1254
13. Florio T (2008) Molecular mechanisms of the antiproliferative activity of somatostatin receptors (SSTRs) in neuroendocrine tumors. Front Biosci 13:822–840
14. Gatto F, Hofland LJ (2011) The role of somatostatin and dopamine D2 receptors in endocrine tumors. Endocr Relat Cancer 18:R233–R251
15. Heute D, Kostron H, von Guggenberg E, Ingorokva S, Gabriel M, Dobrozemsky G et al (2010) Response of recurrent high-grade glioma to treatment with 90Y-DOTATOC. J Nucl Med 51:397–400
16. Kim HS, Lee HS, Kim WH (2011) Clinical significance of protein expression of cyclooxygenase-2 and somatostatin receptors in gastroenteropancreatic neuroendocrine tumors. Cancer Res Treat 43:181–188
17. Kiviniemi A, Gardberg M, Kivinen K, Posti JP, Vuorinen V, Sipilä J et al (2017) Somatostatin receptor 2A in gliomas: association with oligodendrogliomas and favourable outcome. Oncotarget 8:49123
18. Krulich L, Dhariwal AP, McCann SM (1968) Stimulatory and inhibitory effects of purified hypothalamic extracts on growth hormone release from rat pituitary in vitro. Endocrinology 83:783–790
19. Liu Y, Jiang L, Mu Y (2013) Somatostatin receptor subtypes 2 and 5 are associated with better survival in operable hepatitis B-related hepatocellular carcinoma following octreotide long-acting release treatment. Oncol Lett 6:821–828
20. Louis DN, Perry A, Reifenberger G, von Deimling A, Figarella-Branger D, Cavenee WK et al (2016) The 2016 World Health Organization classification of tumors of the central nervous system: a summary. Acta Neuropathol 131:803–820
21. Mawrin C, Schulz S, Pauli SU, Treuheit T, Diete S, Dietzmann K et al (2004) Differential expression of sst1, sst2A, and sst3 somatostatin receptor proteins in low-grade and high-grade astrocytomas. J Neuropathol Exp Neurol 63:13–19
22. Merlo A, Hausmann O, Wasner M, Steiner P, Otte A, Jermann E et al (1999) Locoregional regulatory peptide receptor targeting with the diffusible somatostatin analogue 90Y-labeled DOTA0-D-Phe1-Tyr3-octreotide (DOTATOC): a pilot study in human gliomas. Clin Cancer Res 5:1025–1033
23. Okuwaki K, Kida M, Mikami T, Yamauchi H, Imaizumi H, Miyazawa S et al (2013) Clinicopathologic characteristics of pancreatic neuroendocrine tumors and relation of somatostatin receptor type 2A to outcomes. Cancer 119:4094–4102
24. Pola S, Cattaneo MG, Vicentini LM (2003) Anti-migratory and anti-invasive effect of somatostatin in human neuroblastoma cells: involvement of Rac and MAP kinase activity. J Biol Chem 278:40601–40606
25. Raggi CC, Maggi M, Renzi D, Calabrò A, Bagnoni ML, Scaruffi P et al (2000) Quantitative determination of sst2 gene expression in neuroblastoma tumor predicts patient outcome. J Clin Endocrinol Metab 85:3866–3873
26. Reubi JC, Schonbrunn A (2013) Illuminating somatostatin analog action at neuroendocrine tumor receptors. Trends Pharmacol Sci 34:676–688
27. Schumacher T, Hofer S, Eichhorn K, Wasner M, Zimmerer S, Freitag P et al (2002) Local injection of the 90Y-labelled peptidic vector DOTATOC to control gliomas of WHO grades II and III: an extended pilot study. Eur J Nucl Med Mol Imaging 29:486–493
28. Song KB, Kim SC, Kim JH, Seo D-W, Hong S-M, Park K-M et al (2016) Prognostic value of somatostatin receptor subtypes in pancreatic neuroendocrine tumors. Pancreas 45:187–192
29. Strosberg J, El-Haddad G, Wolin E, Hendifar A, Yao J, Chasen B et al (2017) Phase 3 trial of 177Lu-Dotatate for midgut neuroendocrine tumors. N Engl J Med 376:125–135
30. Tabouret E, Nguyen AT, Dehais C, Carpentier C, Ducray F, Idbaih A et al (2016) Prognostic impact of the 2016 WHO classification of diffuse gliomas in the French POLA cohort. Acta Neuropathol 132:625–634

31. Tang X, Takekoshi S, Itoh J, Umemura S, Shoji S, Terachi T et al (2010) Somatostatin analogue inhibits the mobility of prostate carcinoma cells: a new therapeutic method for advanced prostate carcinoma. Int J Oncol 37:1077–1083
32. Theodoropoulou M, Stalla GK (2013) Somatostatin receptors: from signaling to clinical practice. Front Neuroendocrinol 34:228–252
33. Verhaak RGW, Hoadley KA, Purdom E, Wang V, Qi Y, Wilkerson MD et al (2010) Integrated genomic analysis identifies clinically relevant subtypes of glioblastoma characterized by abnormalities in PDGFRA, IDH1, EGFR, and NF1. Cancer Cell 17:98–110
34. Vitali E, Cambiaghi V, Zerbi A, Carnaghi C, Colombo P, Peverelli E et al (2016) Filamin-a is required to mediate SST2 effects in pancreatic neuroendocrine tumours. Endocr Relat Cancer 23:181–190
35. Vyberg M, Ulhøi BP, Teglbjaerg PS (2007) Neuronal features of oligodendrogliomas--an ultrastructural and immunohistochemical study. Histopathology 50:887–896
36. Weller M, van den Bent M, Tonn JC, Stupp R, Preusser M, Cohen-Jonathan-Moyal E et al (2017) European Association for Neuro-Oncology (EANO) guideline on the diagnosis and treatment of adult astrocytic and oligodendroglial gliomas. Lancet Oncol 18:e315–e329
37. Yan H, Parsons DW, Jin G, McLendon R, Rasheed BA, Yuan W et al (2009) IDH1 and IDH2 mutations in gliomas. N Engl J Med 360:765–773

7

The relevance of cerebrospinal fluid α-synuclein levels to sporadic and familial Alzheimer's disease

Daniel Twohig[1], Elena Rodriguez-Vieitez[2], Sigrid B. Sando[3,4], Guro Berge[4], Camilla Lauridsen[4], Ina Møller[3], Gøril R. Grøntvedt[3,4], Geir Bråthen[3,4], Kalicharan Patra[1], Guojun Bu[5], Tammie L. S. Benzinger[6], Celeste M. Karch[7], Anne Fagan[8], John C. Morris[8], Randall J. Bateman[8], Agneta Nordberg[2,9], Linda R. White[3,4], Henrietta M. Nielsen[1*] and for the Dominantly Inherited Alzheimer Network (DIAN)

Abstract

Accumulating evidence demonstrating higher cerebrospinal fluid (CSF) α-synuclein (αSyn) levels and αSyn pathology in the brains of Alzheimer's disease (AD) patients suggests that αSyn is involved in the pathophysiology of AD. To investigate whether αSyn could be related to specific aspects of the pathophysiology present in both sporadic and familial disease, we quantified CSF levels of αSyn and assessed links to various disease parameters in a longitudinally followed cohort ($n = 136$) including patients with sporadic mild cognitive impairment (MCI) and AD, and in a cross-sectional sample from the Dominantly Inherited Alzheimer's Network ($n = 142$) including participants carrying autosomal dominant AD (ADAD) gene mutations and their non-mutation carrying family members.

Our results show that sporadic MCI patients that developed AD over a period of two years exhibited higher baseline αSyn levels ($p = 0.03$), which inversely correlated to their Mini-Mental State Examination scores, compared to cognitively normal controls ($p = 0.02$). In the same patients, there was a dose-dependent positive association between CSF αSyn and the *APOEε4* allele. Further, CSF αSyn levels were higher in symptomatic ADAD mutation carriers versus non-mutation carriers ($p = 0.03$), and positively correlated to the estimated years from symptom onset ($p = 0.05$) across all mutation carriers. In asymptomatic (Clinical Dementia Rating < 0.5) PET amyloid-positive ADAD mutation carriers CSF αSyn was positively correlated to ^{11}C-Pittsburgh Compound-B (PiB) retention in several brain regions including the posterior cingulate, superior temporal and frontal cortical areas. Importantly, *APOEε4*-positive ADAD mutation carriers exhibited an association between CSF αSyn levels and mean cortical PiB retention ($p = 0.032$). In both the sporadic AD and ADAD cohorts we found several associations predominantly between CSF levels of αSyn, tau and amyloid-β_{1-40}.

Our results suggest that higher CSF αSyn levels are linked to AD pathophysiology at the early stages of disease development and to the onset of cognitive symptoms in both sporadic and autosomal dominant AD. We conclude that *APOEε4* may promote the processes driven by αSyn, which in turn may reflect on molecular mechanisms linked to the asymptomatic build-up of amyloid plaque burden in brain regions involved in the early stages of AD development.

Keywords: Alzheimer's disease, Mild cognitive impairment, alpha-synuclein, Biomarkers, APOEε4

* Correspondence: henrietta.nielsen@dbb.su.se
[1]Department of Biochemistry and Biophysics, Stockholm University, Svante Arrhenius väg 16B, 106 91 Stockholm, Sweden
Full list of author information is available at the end of the article

Introduction

Intracellular neurofibrillary tau tangles and extracellular amyloid-β (Aβ) plaques are the main neuropathological hallmarks of Alzheimer's disease (AD), however, at autopsy more than 50% of these patients exhibit concurrent α-synuclein (αSyn) pathology [1, 16, 31]. Conversely, it has been shown that Aβ plaques promote the development of cortical αSyn lesions in individuals with Parkinson's disease (PD), and up to 50% of PD with dementia patients develop Aβ and tau pathology sufficient for a secondary pathological diagnosis of AD [19, 46]. α-synuclein, a 140 amino-acid protein encoded by the *SNCA* gene, is abundantly expressed neuronal presynaptic terminals [20]. The peptide corresponding to residues 61–95 of αSyn was originally discovered 25 years ago in Aβ plaques and named the non-Aβ component (NAC) [58]. Follow-up studies from the same group further demonstrated NAC immune-reactivity in diffuse Aβ plaques which may result in the formation of mature compact Aβ plaques [33]. Results from these studies suggest that αSyn might be involved in the development of AD from the very early stages of Aβ pathology formation. In support, a recent study demonstrated a positive correlation between cerebrospinal fluid levels of αSyn and Aβ plaque deposition in cognitively normal individuals with subjective memory complaints. The authors speculated that CSF αSyn levels may be related to pathophysiological mechanisms occurring early in the preclinical phase of AD [59].

The relevance of correlating αSyn levels in the CSF to brain parenchymal concentration of αSyn in AD patients is nevertheless controversial due to the lack of autopsy-validated studies assessing αSyn levels in paired CSF and brain tissue samples. Ideally, advancements in radiotracer chemistry could allow for ante-mortem imaging studies in which CSF αSyn concentrations could be correlated to αSyn load in the antemortem brain, similar to PiB-PET correlations between CSF Aβ levels and brain amyloid plaque load [13]. Hence the exact relationship between the low CSF αSyn and brain parenchymal Lewy bodies as occurring in patients with synucleinopathies like PD [68] and dementia with Lewy bodies (DLB) [37] remains unknown. Similarly, the relevance of slightly increased CSF αSyn levels in AD patients [26] to αSyn pathology in the brain is also unclear.

In a study of brains from AD patients devoid of Lewy bodies, intracellular levels of soluble monomeric αSyn were two-fold higher than in controls and significantly correlated to cognitive decline [28]. Potential alterations of CSF αSyn levels in the same subjects were not investigated. Although αSyn can interact with a multitude of cellular components, of which many have been shown to potentially contribute to neurodegeneration [66], the molecular underpinnings of these findings in relation to the development of AD remain poorly understood. Results from studies of the AD mouse model Tg2576 crossed onto an *SNCA* knockout background proposed that loss of αSyn increased the Aβ plaque load in all forebrain areas by the age of 18 months [23]. In support, it was found that αSyn inhibited amyloid plaque formation in *APPPS1* mice co-expressing the *SNCA* gene. The authors of the same study further demonstrated that the seeding activity of injected Aβ containing brain homogenates was reduced, and that Aβ deposition was suppressed in grafted tissue from the [A30P]aSYN transgenic mouse model [2]. The mechanisms underlying the inverse association between Aβ plaque formation and αSyn have yet to be determined, and may be complicated by the existence of various αSyn species (monomers, oligomers, fibrils) which individually may have differing effects on AD pathophysiology. For example, Larson et al. recently reported that increasing oligomeric αSyn selectively decreased presynaptic proteins and cognitive performance in the bigenic J20xTgI2.2 AD mouse model where both monomeric and oligomeric αSyn increase with age, [27].

In addition to the overlapping neuropathological features between AD and synucleinopathies like PD and DLB, but not multiple system atrophy [42], these disorders also share the *APOEε4* allele as a strong genetic risk factor. The *APOEε4* allele increases the risk of AD by up to 15-fold and the risk of DLB by up to 6-fold [7, 10]. Furthermore, the *APOEε4* allele also increases the risk of PD and decreases the age of disease onset [30]. The exact molecular pathways underlying the increased risk of several neurodegenerative disorders in *APOEε4*-carriers have yet to be identified but the urgency to do so is clearly illustrated by the rather high frequency (~ 14%) of this allele in the normal population [32].

In the current study we aimed to assess the pathophysiological relevance of αSyn levels in CSF to the development of AD, including surrogate markers of AD neuropathology, in a clinical setting. We further aimed to investigate potential effects of the *APOEε*4 risk allele on identified relationships. For our purposes we used two cohorts, a longitudinally followed cohort of patients with sporadic MCI and AD to investigate potential relationships between αSyn and disease progression. The second cohort was comprised of a cross-sectional sample from the Dominantly Inherited Alzheimer's Disease Network (DIAN) including participants with autosomal dominant mutations in the *APP, PSEN1* and *PSEN2* genes causing early-onset autosomal dominant Alzheimer's disease (ADAD), and their non-mutation carrying relative control participants [3]. The latter cohort was included to enable specific analyses of relationships between αSyn and the development of symptoms in subjects that will de facto develop Alzheimer's disease. Furthermore, inclusion of subjects with ADAD mutations also enabled us to assess potential associations between CSF αSyn levels and specific mutations in the genes encoding key components in the Aβ peptide production pathway i.e. the amyloid precursor protein (APP) and the presenilins 1 and 2 (PS1 and PS2).

Materials and methods

Study participants

Subjects from two different cohorts were included in the present study. The longitudinally followed cohort totaling $n = 136$ subjects, including $n = 52$ subjects that were cognitively healthy controls and $n = 84$ patients who at baseline were diagnosed with sporadic MCI ($n = 57$) or AD ($n = 27$). Clinical diagnoses of MCI or AD were based on the International Working Group on Mild Cognitive Impairment Criteria [63] and the National Institute of Neurological and Communicative Disorders and Stroke and the Alzheimer's Disease and Related Disorders Association criteria [34, 35]. The clinical workup and results from parts of this cohort were previously reported [5, 6, 45]. Briefly, consenting participants were recruited by the Department of Neurology at Trondheim University Hospital in Trondheim, Norway on the conditions that individuals had adequate vision and hearing, did not exhibit high alcohol consumption or use of anti-coagulation medication, and had no psychiatric or malignant disease. Cognitively healthy control subjects were recruited from elderly caregivers who were not genetically related to the patients or from societies for retired people in central Norway. Cerebrospinal fluid was drawn at baseline and at intervals of 12-months follow-up examinations over two years. Control individuals had their CSF sampled at baseline only, due to restrictions pertaining to ethical permits; however, they were clinically assessed both at baseline and at the study-end. Cerebrospinal fluid sampling and CSF AD biomarker analysis (for research purposes only) including quantification of $A\beta_{1-40}$, $A\beta_{1-42}$, total tau (t-tau) and tau phosphorylated at Thr181 (p-tau) levels were previously described [6]. Both patient and control groups were *APOE* genotyped from blood samples and underwent cerebral volumetric 3-Tesla MRI (3T-MRI) brain imaging at baseline and after two years. Upon completion of the study, MCI patients were re-classified into two groups based on their baseline and two-year follow-up diagnoses. Patients diagnosed with MCI at baseline and who remained diagnosed with MCI after two years were classified as MCI-MCI ($n = 30$), while patients who were MCI at baseline but that fulfilled the clinical criteria for AD after two years were classified as MCI-AD ($n = 27$). The studies performed on this cohort were approved by the regional ethics committee in Trondheim, Norway (2010/226) and Stockholm, Sweden (2016/771–31/4) and carried out in agreement with the Helsinki Declaration.

Participants in the second cohort were a cross-sectional sample from the multi-site Dominantly Inherited Alzheimer's Network (DIAN) (https://dian.wustl.edu/our-research/observational-study/). The DIAN registry includes mutation carrying and non-mutation carrying adult biological children from families in which one parent carries an ADAD causative mutation in the *APP, PSEN1* or *PSEN2* genes. At inclusion, and during subsequent visits, DIAN participants were assessed neuropsychologically, biochemically (CSF and plasma), and underwent structural neuroimaging using 3T-MRI, and functional neuroimaging using PET combined with either fluorodeoxyglucose (FDG-PET) or PiB to quantify brain glucose metabolism and Aβ deposition respectively [4]. Updated datasets are generated biannually in what are termed dataset "freezes". For the current study we examined subjects from the DIAN dataset freeze-10 including a total of $n = 92$ ADAD mutation carriers: $n = 24$ *APP,* $n = 50$ *PSEN1,* and $n = 18$ *PSEN2* mutation carriers, as well as $n = 50$ non-mutation carriers who were genetically related to the mutation carriers but did not harbor ADAD mutations. Included subjects were considered as symptomatic if fulfilling the criteria for a Clinical Dementia Rating (CDR) score ≥ 0.5. The purposes of the current study involving DIAN participants were approved by the local ethics committee in Stockholm, Sweden (2016/2114–31/4) and the study was carried out in agreement with the Helsinki Declaration.

αSyn quantification

Quantification of CSF αSyn in samples from both cohorts was performed using a commercially available sandwich enzyme-linked immunosorbent assay (ELISA) (AnaSpec; California, USA) according to the manufacturer's instructions and using overnight incubation at 4 °C for the sample incubation step. Samples were assayed alongside freshly prepared standard curves for each individual assay. Standards, study subject and internal control samples were run in duplicates and averaged. Spike recovery experiments were performed to assess matrix effects by adding known concentrations of the supplier-provided αSyn standard to diluted CSF resulting in a spike recovery range of 99–122%. Intra- and inter-assay coefficients of variation were determined by repeated analysis of pooled CSF samples as internal controls resulting in coefficients of variation of < 15% respectively for the two lots of assays used to quantify αSyn in the two cohorts.

DIAN participant MRI image acquisition

Structural MRI acquisition was performed using the Alzheimer Disease Neuroimaging Initiative (ADNI) protocol [21, 22]. Participating sites were required to pass initial and regular follow-up quality control assessments to insure acquisition conformity. Each participant received an accelerated 3D sagittal T1-weighted MPRAGE on a 3T scanner. A high quality, whole-brain image with

1.1 × 1.1 × 1.2 mm voxels was acquired in approximately 5–6 min. Before further image processing, images were screened for artifacts and protocol compliance by the ADNI imaging core.

DIAN participant PET image acquisition and processing

A subset of $n = 132$ ADAD participants (mutation carriers $n = 85$, non-mutation carriers $n = 47$) underwent PiB-PET imaging. Each site underwent an initial evaluation to ensure compliance with common PiB-PET ADNI protocols. Amyloid imaging was performed with a bolus injection of ~15 mCi of PiB. Dynamic acquisition consisted of either a 70-min scan starting at injection or a 30-min scan beginning 40 min post injection. For analysis, the PiB-PET data within the common time frame between 40 and 70 min was used. The ADNI PET Core verified that all PET images were acquired using the established protocol and substantially free of artifacts.

Each subject's PET data were motion-corrected and registered to their MRI using methods described in detail elsewhere [11, 49]. For each participant, the T1-weighted MRI image was segmented into grey and white matter tissue maps. An inclusive binary gray matter mask was subsequently applied to the resulting atlas to obtain individual gray matter atlases. An automated quantitative image analysis approach was applied using regions of interest generated with FreeSurfer [14] using an in-house software previously described [36, 40, 54]. A regional spread function based approach for partial volume correction of PET data was subsequently implemented using FreeSurfer regions [48] to generate partial volume corrected regional PET data. Partial volume corrected PiB data were used to quantify PiB retention, as it was demonstrated to produce accurate results [53].

Regional partial volume corrected PiB retention data was evaluated in all FreeSurfer regions of interest in Standardized Uptake Value Ratio (SUVR) units, using the brainstem as reference, as this was reported as optimum quantification procedure for PiB-PET in DIAN studies [53]. Overall brain PiB retention was given by the mean cortical PiB retention, calculated as the average PiB SUVR across four cortical regions (prefrontal, gyrus rectus, lateral temporal, and precuneus) as previously used [44]. In addition, exploratory analyses were performed across forty-two bilateral regions of interest from the FreeSurfer atlas. PiB positivity was defined by a mean cortical SUVR uptake using partial volume corrected data and the brainstem as reference region ≥0.72 cutoff value, equivalent to an SUVR of ≥1.42 if using the cerebellar grey matter, as previously described [56].

Statistics

Normal distribution of the data was assessed by use of the Shapiro-Wilk test. Accordingly, statistical tests applied for cross-sectional group comparisons were either non-parametric (Kruskal-Wallis tests followed by pairwise comparisons with the Mann-Whitney *U* test) or parametric (analysis of variance [ANOVA] or analysis of covariance [ANCOVA] with age entered as covariate when relevant, followed by pairwise comparisons when appropriate, performed with the Student's *t*-test). The Bonferroni correction was applied to account for multiple comparisons.

In the longitudinal cohort, CSF αSyn levels did not follow a normal distribution and were therefore log-transformed to achieve normal distribution of the data for which parametric tests were used. Longitudinal changes were assessed using the repeated measures multivariate analysis of variance (MANOVA) approach, and correlation analyses were performed by use of the Spearman's rank correlation test on non-log-transformed data. Data is presented as mean ± standard deviation or as median with range (min-max). A *p*-value of ≤0.05 was considered statistically significant.

In order to compare CSF αSyn levels between subjects with AD pathological versus non-pathological CSF AD biomarker levels receiver operating characteristic (ROC) curve analyses were performed for the longitudinal cohort using the clinical status (cognitively healthy controls versus a baseline AD diagnosis) as the dichotomous variable to establish biochemical cutoff values. Cutoff values were determined based on the highest Youden index defined as *sensitivity + specificity – 1.*

In the DIAN cohort, CSF αSyn levels did not follow a normal distribution hence we log-transformed the data so that the association between αSyn with estimated years from symptom onset (EYO) was evaluated using the Pearson's correlation test. For brain PiB-PET imaging-related CSF αSyn analyses, linear regression was applied to predict regional PiB-retention as a function of CSF αSyn and EYO as independent variables, separately in ADAD mutation carriers and non-mutation carriers. The same linear regression model was applied after mutation carriers were stratified into asymptomatic (CDR < 0.5) and symptomatic (CDR ≥ 0.5) mutation carriers, as well as for the subset of PiB-positive asymptomatic mutation carriers. Additionally, in mutation carriers the interaction of αSyn and *APOEε4* status (positive/negative) was investigated to elucidate whether the relationship of regional PiB retention and αSyn is influenced by *APOEε4* status. Finally, regional PiB retention was modeled as a function of αSyn and EYO separately in *APOEε4*-positive and *APOEε4*-negative mutation carriers.

All statistical analyses concerning CSF αSyn levels and their association to parameters unrelated to brain imaging were performed in JMP software v.12.1.0 (SAS Institute Inc.). Statistical analyses regarding associations between CSF αSyn and amyloid PET data in the DIAN cohort were performed using the R v.3.1.0 software (The R Foundation for

Statistical Computing, http://www.r-project.org/). Graphical representations of significant linear associations were obtained with the ggplot2 package v.1.0.1, as implemented in R. The significance level for all statistical tests and models was set at $p \leq 0.05$. On the brain volumes, display of regression standardized β coefficients of the significant associations between CSF αSyn and regional PiB-PET were performed using BrainNet viewer [67] (http://www.nitrc.org/projects/bnv/) implemented in MATLAB.

Results

Descriptive statistics of the longitudinally followed cohort of sporadic patients

The demographics and clinical characteristics of the longitudinal cohort were in part previously published [45] and are shown in Table 1. The control group was significantly older than the MCI and AD patient groups, thus age was added as a covariate when including control subjects in the statistical analyses. At the two-year follow-up 48% of the subjects diagnosed with MCI at baseline had progressed to an AD diagnosis (MCI-AD), whereas 52% remained stable (MCI-MCI). The patient dropout rate at 24-months was 17% of the MCI-MCI patients, 7% of the MCI-AD patients, and 15% of the AD patients. As expected, the Mini-Mental State Examination (MMSE) scores differed significantly amongst the investigated groups with AD patients displaying the lowest scores ($p < 0.001$). In comparison to controls all individual patient groups exhibited altered levels of CSF AD biomarkers including $A\beta_{1-42}$, $A\beta_{1-40}$, $A\beta_{42/40}$, t-tau and p-tau. A much lower frequency of *APOEε4* carriers was observed in the control group (38%) compared to the MCI (66%) and AD patient groups (82%) ($p < 0.001$).

Higher baseline CSF αSyn levels in MCI patients converting to sporadic AD

We first sought to compare CSF αSyn levels between the diagnostic groups. At baseline, MCI patients that after 24 months fulfilled the criteria for an AD diagnosis

Table 1 Longitudinal cohort baseline characteristics

	Controls	MCI			Alzheimer's disease	p-value[c]
		All	MCI-MCI	MCI-AD		
N	57	54	27	27	27	
Sex (% f/m)	65/35	53/47	44/56	59/41	52/48	- [d]
Age at examination (yrs)[a]	68 ± 5	65 ± 6	66 ± 6	64 ± 4	64 ± 6	** [e]
MMSE[b]	30 (28–30)	28 (23–30)	28 (25–30)	27 (23–29)	23 (16–27)	*** [f]
APOEε4 carrier (%)	38 (*n* = 55)	66	63	74	81	*** [d]
CSF $A\beta_{1-42}$ (pg/mL)[b]	969.7 (500–1674) (*n* = 45)	546.5 (173–1269)	574.3 (173–1269)	539.0 (283–1060)	476.8 (212–1092)	*** [f]
CSF $A\beta_{1-40}$ (ng/mL) [b]	17.1 (11–41) (*n* = 41)	13.5 (4–31)	12.9 (4–31) (*n* = 26)	14.2 (8–23) (*n* = 23)	14.9 (7–29) (*n* = 26)	** [f]
CSF $A\beta_{42/40}$ [b]	0.059 (0.013–0.096) (*n* = 41)	0.042 (0.008–0.118)	0.049 (0.008–0.118) (*n* = 26)	0.039 (0.012–0.088) (*n* = 23)	0.034 (0.007–0.080) (*n* = 26)	*** [f]
CSF t-tau (pg/mL)[b]	269.0 (138–1314) (*n* = 46)	447.9 (99–2325)	315.2 (99–1057)	550.8 (163–2325)	624.2 (177–1540)	*** [f]
CSF p-tau (pg/mL)[b]	53.5 (33–135) (*n* = 46)	69.9 (16–169)	53.0 (16–131)	85.6 (37–169)	90.8 (28–157)	*** [f]

MCI = mild cognitive impairment MCI-MCI = MCI patients who remained MCI at the 24-month follow up
MCI-AD = MCI patients who converted to Alzheimer's disease at the 24-month follow up
MMSE = Mini-Mental State Examination score
a = mean ± standard deviation b = median (minimum-maximum)
c = *P*-value reflecting potential differences between the groups Controls, mild cognitive impairment and Alzheimer's disease d = χ^2 test e = one-way ANOVA
f = Kruskal-Wallis test - = non-significant
** = $p \leq 0.01$ *** = $p \leq 0.001$

(MCI-AD) exhibited the highest CSF αSyn levels compared to controls (Fig. 1a), while CSF αSyn levels did not significantly differ between controls and the non-converting MCI patients (MCI-MCI), and the AD group (Fig. 1a). No differences in CSF αSyn levels were noted between the patient groups at either the 12-month (Fig. 1b) or 24-month (Fig. 1c) follow-ups.

Association between *APOEε4* and higher CSF αSyn levels in sporadic AD patients

As *APOEε4* is a common genetic risk factor for both synucleinopathies and AD [7, 10, 30], we assessed the effect of *APOEε4* on CSF αSyn levels. Within the MCI patient group *APOEε4* carriers exhibited higher αSyn levels compared to non-carriers at the 24-month follow up (Fig. 2c). Additionally, MCI-AD *APOEε4* carriers exhibited higher CSF αSyn than non-carriers at the 12-month (Fig. 2e) and 24-month (Fig. 2f) follow-ups. The MCI-AD *APOEε4* carrying patient group also exhibited significantly higher CSF αSyn levels at baseline compared to *APOEε4* carrying MCI-MCI patients and controls (Fig. 3a), while no differences in CSF αSyn were found at the 12-month follow up (Fig. 3b). However, at 24-months MCI-AD *APOEε4* carriers again had higher CSF αSyn than MCI-MCI *APOEε4* carriers (Fig. 3c). This difference in CSF αSyn levels was absent when comparing *APOEε4* non-carrier patient groups at any time point (Fig. 3d-f). Intriguingly, the elevated CSF αSyn levels found in MCI-AD patients exhibited a dose-dependent relationship with the *APOEε4* allele at all three time points (Fig. 4a-c).

To assess any longitudinal changes in CSF αSyn levels occurring within- or between the patient groups we performed repeated measures MANOVAs spanning the intervals of: baseline to 12-months, baseline to 24-months, and 12-months to 24-months (Fig. 5). When all patients (diagnosed with MCI or AD at baseline) were grouped as either *APOEε4* carriers or non-carriers, *APOEε4* carriers showed increasing CSF αSyn levels (+ 17%) from baseline to 12-months, followed by a decrease (− 10%) from 12- to 24-months (Fig. 5a). Patients that did not carry the *APOEε4* variant consistently exhibited a longitudinal decline in CSF αSyn levels which reached statistical significance between the 12- and 24-months examinations (Fig. 5a), thus proposing potentially different pathological processes for *APOEε4* carriers versus non-carriers. These findings prompted us to also subdivide the MCI-AD patient group into *APOEε4* carriers or non-carriers. We found a significant increase in CSF αSyn (+ 5%) in MCI-AD *APOEε4* carriers from baseline to 12-months, while also observing elevated CSF αSyn in MCI-AD *APOEε4* carriers versus non-carriers from 12- to 24-months (Fig. 5b). Moreover, when the MCI-AD patient group was further subdivided by *APOEε4* zygosity we found significant differences in CSF αSyn levels from baseline to 12-months, and 12- to 24-months (Fig. 5c). A significant difference between *APOEε4* homozygotes, heterozygotes, and non-carriers was also found in the AD group between 12-months and 24-months (data not shown).

Links between CSF αSyn, AD biomarkers and cognition in sporadic patients

In the absence of established brain imaging tools to assess ante-mortem αSyn burden in the brain, we assessed potential correlations between CSF αSyn levels and established fluid surrogate markers of AD pathology and global cognitive test scores. At baseline, 12-months, and 24-months CSF αSyn showed a positive correlation to CSF $A\beta_{1-40}$, t-tau, and p-tau in all patient groups and controls (Table 2). All MCI patients showed a marked

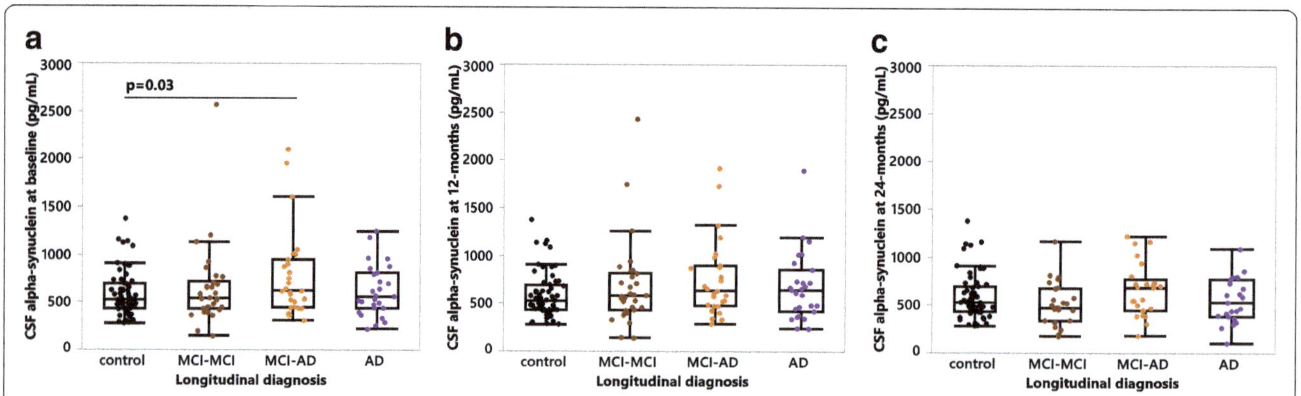

Fig. 1 CSF αSyn levels in sporadic MCI and Alzheimer's disease cohort. Quantification of CSF αSyn levels at (**a**) baseline, (**b**) 12-months and (**c**) 24-months in the four longitudinally diagnosed patient groups: control = cognitively healthy controls, MCI-MCI = MCI patients who remained MCI at the 24-month follow up, MCI-AD = MCI patients who converted to Alzheimer's disease at the 24-month follow up, AD = patients diagnosed with AD at baseline. *P*-values were calculated using a one-way ANCOVA of log-transformed data with age entered as a covariate with post-hoc testing by use of the student's *t*-test. Bonferroni correction was used to account for multiple comparisons (a: $n = 6$ comparisons, b-c: $n = 3$ comparisons). Results are displayed without log-transformation or age-correction (raw data)

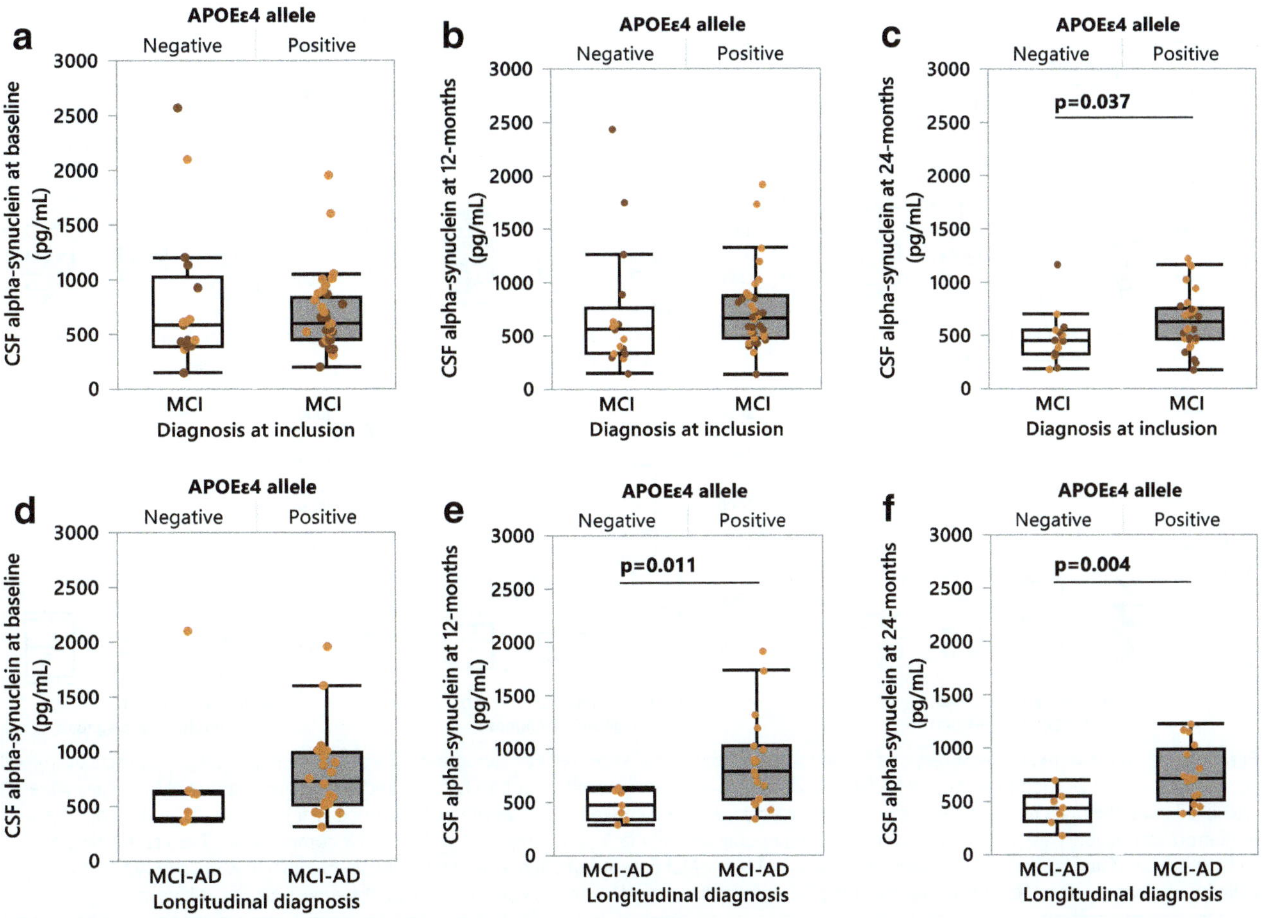

Fig. 2 CSF αSyn measured in *APOEε4*-positive versus *APOEε4*-negative sporadic MCI patients. **a-c** CSF αSyn quantified in MCI patients at baseline, 12- and 24-months respectively. *APOEε4*-positive patients are shown as grey shaded boxes, *APOEε4*-negative patients are white boxes. Orange data points represent patients longitudinally diagnosed as MCI-AD, brown points represent MCI-MCI patients. **d-f** CSF αSyn quantified in MCI-AD diagnosed patients at baseline, 12- and 24-months respectively. *P*-values were calculated using the one-way ANCOVA of log-transformed data, with age entered as a covariate with post-hoc testing by use of the student's *t*-test. Results are displayed without log-transformation or age-correction (raw data). MCI-AD = MCI patients who converted to Alzheimer's disease at the 24-month follow up

strengthening in the correlations of CSF αSyn to CSF $A\beta_{1-40}$, t-tau, and p-tau between baseline and 12-months, and were the strongest in the MCI-AD group compared to all other groups at 12-months (Table 2). In MCI-MCI patients the correlations of CSF αSyn to CSF $A\beta_{1-40}$ and p-tau strengthened from baseline to 24-months, while the MCI-AD group exhibited the strongest correlation between CSF αSyn and t-tau (all above Spearman's $\rho = 0.9$ respectively) (Table 2). The AD group exhibited a modest increase in the correlations between CSF αSyn and CSF t-tau and p-tau between baseline and 12-months, but showed weakening of the correlation between CSF αSyn and $A\beta_{1-40}$ over the same time interval. Additionally, at 24-months the AD group was the only group of patients at any time point to exhibit a significant correlation between CSF αSyn and $A\beta_{1-42}$. At 24-months however, the AD group exhibited the weakest correlations between CSF αSyn and the CSF AD biomarker panel. Intriguingly, our analyses revealed a significant association between CSF αSyn levels and cognition, in which higher CSF αSyn levels were significantly correlated to lower MMSE scores in the MCI-AD group at baseline (Table 2); moreover, the correlation was solely attributable to *APOEε4* carrying MCI-AD patients (ρ=-0.607, p=0.005) (non-carrier ρ=-0.618, p=0.140).

αSyn levels in sporadic patients with or without pathological CSF AD biomarkers

The MCI and AD patients enrolled in the longitudinal cohort were diagnosed according to clinical guidelines [34, 63] without workup of CSF AD biomarkers to support the clinical diagnosis, however, for research purposes the CSF biomarkers were assessed. With the rationale to compare CSF αSyn levels between patients with pathological versus without pathological CSF AD biomarker levels, we employed ROC analyses to produce cutoffs for the core CSF AD biomarkers $A\beta_{1-42}$, t-tau, p-tau and the

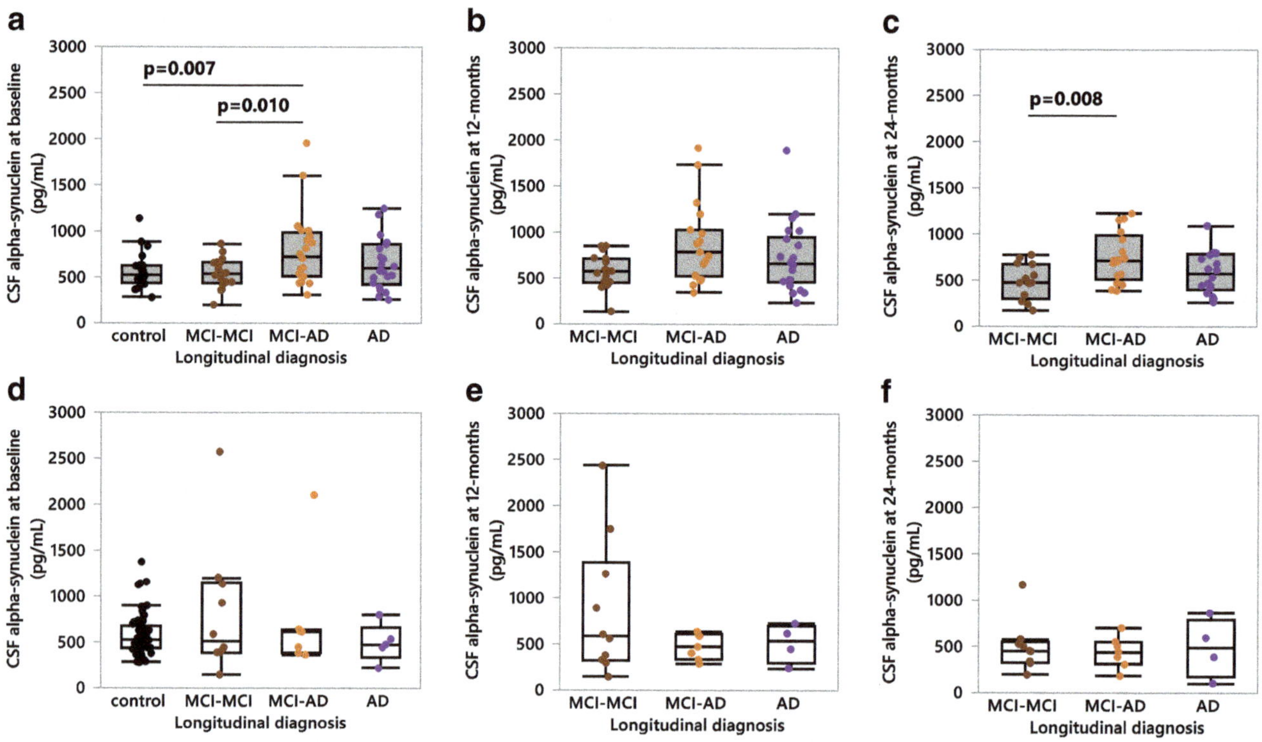

Fig. 3 CSF αSyn measured in the subjects of the longitudinal cohort based on *APOEε4* status. **a-c** *APOEε4*-positive (grey boxes) subjects examined at baseline, 12- and 24-months respectively. **d-f** *APOEε4*-negative patients examined at baseline, 12- and 24-months respectively. *P*-values were calculated using the one-way ANCOVA of log-transformed data, with age entered as a covariate with post-hoc testing by use of the student's *t*-test. Bonferroni correction was used to account for multiple comparisons (a: $n = 6$ comparisons, b-f: $n = 3$ comparisons). Results are displayed without log-transformation or age-correction (raw data). MCI-MCI = MCI patients who remained MCI at the 24-month follow up, MCI-AD = MCI patients who converted to Alzheimer's disease at the 24-month follow up, AD = patients diagnosed with Alzheimer's disease at baseline

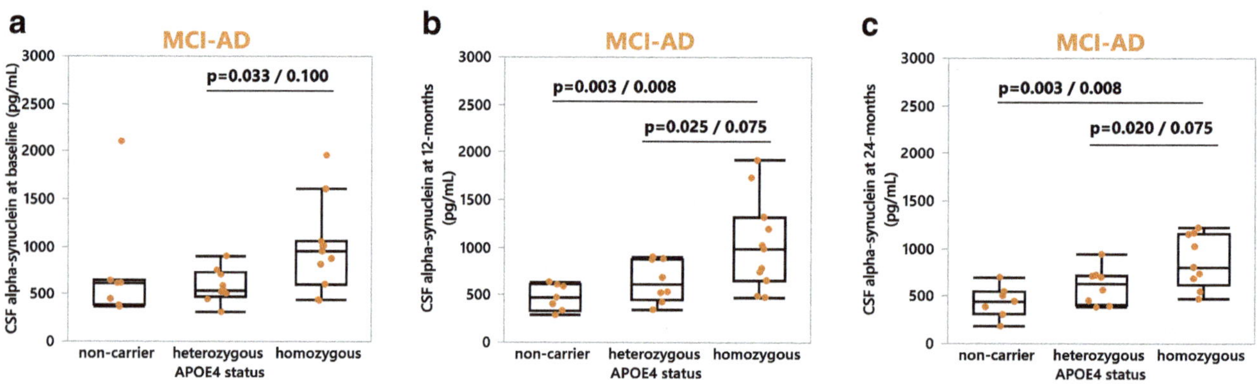

Fig. 4 Cerebrospinal fluid αSyn levels in MCI-AD patients at: **a** baseline (non-carrier $n = 7$, heterozygous $n = 9$, homozygous $n = 11$), **b** 12-months (non-carrier $n = 7$, heterozygous $n = 8$, homozygous $n = 11$), and **c** 24-months (non-carrier $n = 7$, heterozygous $n = 8$, homozygous $n = 9$). *P*-values are presented first as uncorrected followed by Bonferroni corrected (a-c: $n = 3$ comparisons). In **a** *P*-values were calculated using the non-parametric Kruskal-Wallis with post-hoc testing by use of the Mann-Whitney *U* test. Bonferroni correction was used to account for multiple comparisons. In (**b-c**) *P*-values were calculated using the one-way ANCOVA with age entered as a covariate with post-hoc testing by use of the student's *t*-test. Results are displayed without age-correction (raw data). MCI-AD = MCI patients who converted to Alzheimer's disease at the 24-month follow up

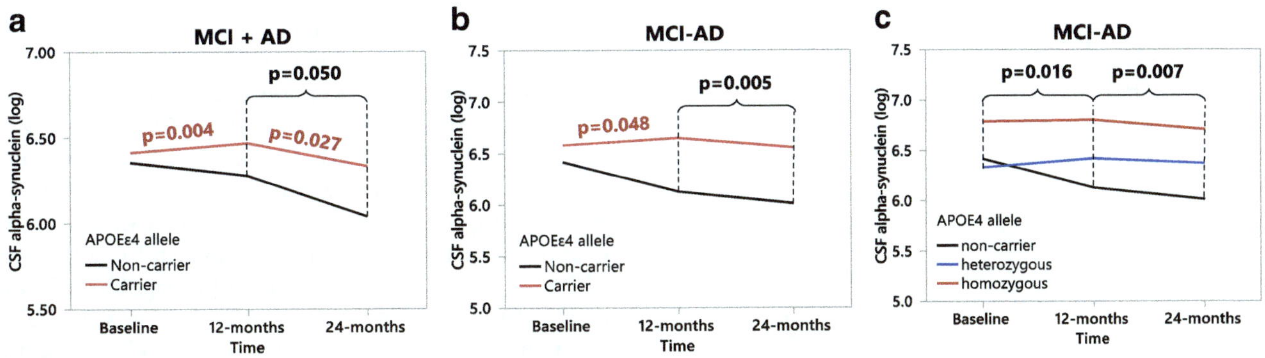

Fig. 5 Longitudinal assessment of mean CSF αSyn values at baseline, 12-months and 24-months in (**a**) a pooled MCI and Alzheimer's disease (AD) patient group classified as either *APOEε4* carriers (baseline to 12-months $n = 58$, 12-months to 24-months $n = 47$) or non-carriers (baseline to 12-months $n = 21$, 12-months to 24-months $n = 19$). **b** the MCI-AD patient group classified as either *APOEε4* carriers (baseline to 12-months $n = 19$, 12-months to 24-months $n = 16$) or non-carriers (baseline to 12-months $n = 7$, 12-months to 24-months $n = 7$), **c** the MCI-AD patient group classified as either *APOEε4* non-carriers (baseline to 12-months $n = 7$, 12-months to 24-months $n = 7$), heterozygotes (baseline to 12-months $n = 8$, 12-months to 24-months $n = 7$) or homozygotes (baseline to 12-months $n = 11$, 12-months to 24-months $n = 9$). *P*-values within patient groups over 1-year intervals are shown in red along the trend lines, and between-group *p*-values are indicated in black. *P*-values were calculated using repeated measures MANOVA analysis of log-transformed data. MCI-AD = MCI patients who converted to Alzheimer's disease at the 24-month follow up

conventional ratios thereof [47]. We then stratified the patients into groups with and without pathological CSF AD biomarker levels (Fig. 6). Cutoffs were determined by comparing AD biomarker levels between cognitively healthy controls and AD patients at baseline. The resulting ROC axes were labeled as sensitivity and 1-specificity; the accompanying results table provided the individual cutoff values, sensitivity, specificity, and the area under the curve (AUC) with its respective 95% confidence interval (95% CI) (Fig. 6). The cutoff values (Fig. 6) for each CSF biomarker were then used to compare CSF αSyn levels between CSF AD biomarker negative (CSF(−))

Table 2 Longitudinal cohort CSF αSyn correlation analyses[‡]

Time	Longitudinal diagnosis	MMSE	CSF $A\beta_{1-42}$	CSF $A\beta_{1-40}$	CSF $A\beta_{42/40}$	CSF t-tau	CSF p-tau
Baseline	Control (*n*=57)	-	- (*n*=41)	0.616*** (*n*=37)	-0.542*** (*n*=37)	0.871*** (*n*=42)	0.766*** (*n*=42)
	MCI-MCI (*n*=27)	-	-	0.472** (*n*=26)	- (*n*=26)	0.577***	0.651***
	MCI-AD (*n*=27)	-0.605***	-	0.640*** (*n*=23)	-0.457* (*n*=23)	0.736***	0.694***
	AD (*n*=27)	-	-	0.758*** (*n*=26)	- (*n*=26)	0.722***	0.738***
12-months	MCI-MCI (*n*=27)	-	- (*n*=15)	0.703*** (*n*=15)	-0.575* (*n*=15)	0.804*** (*n*=15)	0.849*** (*n*=15)
	MCI-AD (*n*=25)	-	- (*n*=19)	0.863*** (*n*=18)	-0.732*** (*n*=18)	0.913*** (*n*=19)	0.927*** (*n*=19)
	AD (*n*=24)	-	- (*n*=18)	0.691*** (*n*=18)	- (*n*=18)	0.804*** (*n*=18)	0.770*** (*n*=18)
24-months	MCI-MCI (*n*=24)	-	- (*n*=14)	0.925*** (*n*=14)	- (*n*=14)	0.782*** (*n*=14)	0.924*** (*n*=14)
	MCI-AD (*n*=24)	-	- (*n*=18)	0.716*** (*n*=17)	-0.556* (*n*=17)	0.902*** (*n*=18)	0.919*** (*n*=18)
	AD (*n*=18)	-	0.662** (*n*=16)	0.585* (*n*=16)	- (*n*=16)	0.741** (*n*=16)	0.650** (*n*=16)

[‡]All correlations calculated using the Spearman's rank correlation test
MCI-MCI= MCI patients who remained MCI at the 24-month follow up
MCI-AD= MCI patients who converted to Alzheimer's disease at the 24-month follow up
AD patients diagnosed with Alzheimer's disease at baseline, *MMSE* Mini-Mental State Examination score
*$p \leq 0.05$, **$p \leq 0.01$, ***$p \leq 0.001$

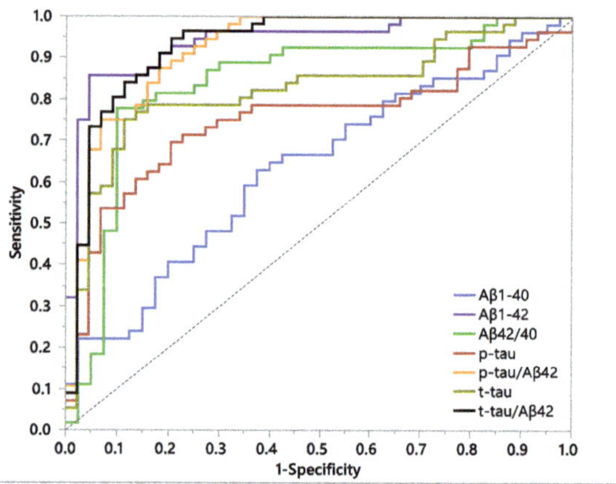

	Cutoff	Sensitivity (%)	Specificity (%)	AUC (95% CI)
$A\beta_{1\text{-}42}$	<579 pg/mL	86	96	0.94 (0.88-1.00)
t-tau/$A\beta_{1\text{-}42}$	>0.364	96	78	0.94 (0.89-1.00)
p-tau/$A\beta_{1\text{-}42}$	>0.126	75	96	0.93 (0.88-0.99)
$A\beta_{42/40}$	<0.041	78	90	0.85 (0.75-0.95)
t-tau	>470 pg/mL	75	91	0.84 (0.73-0.94)
p-tau	>71.6 pg/mL	71	80	0.77 (0.64-0.89)
$A\beta_{1\text{-}40}$	>15986 pg/mL	63	63	0.64 (0.51-0.78)

Fig. 6 Receiver operator characteristic (ROC) curves of AD CSF biomarkers. For each CSF biomarker analyte or ratio the table indicates the cutoff value, sensitivity (%), specificity (%), and area under the ROC curve (AUC) with the corresponding 95% confidence interval. A clinical diagnosis of healthy control versus AD was used as the dichotomous variable to define CSF cutoffs based on the best performing Youden index

and positive (CSF(+)) subjects within the diagnostic groups (Fig. 7). When utilizing the cutoffs for CSF t-tau (> 470 pg/mL) and p-tau (> 71.6 pg/mL) in the MCI and AD patients, we discovered significant differences where CSF(+) patients had elevated αSyn compared to CSF(−) patients (Fig. 7e-f). Additionally, the CSF p-tau/$A\beta_{42}$ cutoff (> 0.126) applied in the AD patient group showed that the CSF(+) group had higher CSF αSyn than the CSF(−) group (Fig. 7c).

Descriptive statistics of DIAN participants

In the ADAD mutation carrying participants from the DIAN study, $n = 6$ *APP*, $n = 33$ *PSEN1* and $n = 3$ *PSEN2* mutations were represented (data not shown). Of the $n = 92$ included ADAD mutation carriers 34.8% had cognitive symptoms (CDR ≥0.5). Specifically, 37.5% of *APP*, 42.0% of *PSEN1* and 11.1% of *PSEN2* mutation carriers were symptomatic (Table 3). More than half of the *PSEN2* mutation carriers carried the *APOEε4* allele, whereas a quarter of the *APP* mutation carriers and less than 20% of the *PSEN1* mutation carriers were *APOEε4* carriers (Table 3). In all three ADAD mutation groups, there was a higher percentage of *APOEε4* carriers amongst the symptomatic mutation carriers ($p < 0.01$). Further, we found significant variability in the EYO, MMSE and CDR scores, and CSF levels of $A\beta_{1-42}$, $A\beta_{42/40}$, t-tau, and p-tau between ADAD mutation carriers and non-mutation carriers (Table 3). When the three ADAD mutation carrier groups where subdivided into symptomatic (CDR ≥ 0.5) and asymptomatic (CDR < 0.5) individuals it was found that age of examination, EYO and MMSE scores as well as CSF $A\beta_{42/40}$ and CSF levels of t-tau, and p-tau significantly differed between symptomatic versus asymptomatic in *APP* mutation carriers (Table 3). Age of examination, EYO, MMSE and CDR scores, CSF $A\beta_{42/40}$, $A\beta_{1-42}$, t-tau, and p-tau significantly differed between symptomatic and asymptomatic *PSEN1* mutation carriers, while no differences could be observed in the *PSEN2* mutation carriers due to insufficient sample size (Table 3).

Cerebrospinal fluid αSyn levels in ADAD mutation carriers are related to onset of cognitive symptoms

After pooling all individual gene mutations into their respective groups (*APP*, *PSEN1* and *PSEN2* mutation carriers) we compared the CSF αSyn levels between the three groups and found no significant differences between the groups of ADAD mutation carriers or versus non-mutation carriers (Fig. 8a). Upon subdividing the ADAD mutation carriers into either asymptomatic or symptomatic individuals we found that symptomatic mutation carriers had higher CSF αSyn compared to non-mutation carriers (Fig. 8b); no further differences were found when *APOEε4* status was included as a covariate (data not shown). Importantly, the temporal trajectory of CSF αSyn versus EYO showed a weak-to-moderate increase in both ADAD mutation carriers (Pearson's $R = 0.20$; $p = 0.056$) and in non-carriers (Pearson's $R = 0.31$; $p = 0.032$), where the confidence bands around both trajectories showed a substantial overlap (Fig. 8c).

Cerebrospinal fluid αSyn and AD biomarker associations in the DIAN cohort

To assess links between CSF αSyn levels and AD brain pathology using surrogate CSF AD biomarkers in the DIAN participants we investigated potential correlations between CSF αSyn and CSF levels of $A\beta_{1-42}$, $A\beta_{1-40}$, $A\beta_{42/40}$, t-tau, and p-tau in ADAD non-mutation carriers and the three ADAD mutation carrier groups sorted into asymptomatic and symptomatic subgroups (the $n = 2$ *PSEN1* symptomatic patient group was excluded from analysis) (Table 4). We found no correlations between CSF αSyn and $A\beta_{1-42}$ when combining asymptomatic and symptomatic ADAD mutation carriers, but did find a positive correlation between CSF αSyn and $A\beta_{1-40}$ in non-mutation carriers, and both asymptomatic and symptomatic *APP* and *PSEN1* mutation carriers. Inverse correlations between CSF αSyn and $A\beta_{42/40}$ were found in symptomatic *APP* and *PSEN1* mutation carriers, but most notably in the symptomatic *APP* mutation carrier group (Table 4). There were strong positive associations between

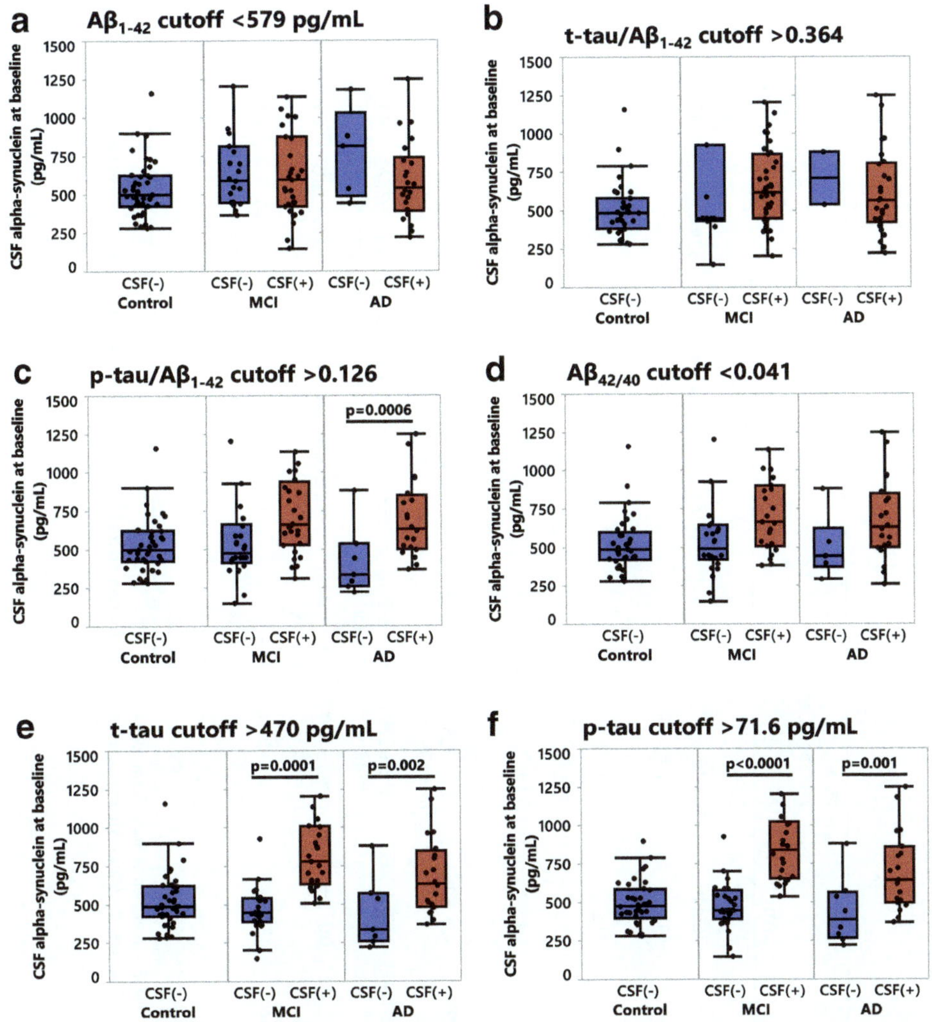

Fig. 7 Baseline levels of CSF αSyn in the longitudinal cohort based on ROC cutoff values. **a-f** Clinically diagnosed subjects subdivided into either CSF Alzheimer's disease (AD) biomarker negative (CSF(−)) or positive (CSF(+)) groups based on cutoff values. *P*-values were calculated using the one-way ANCOVA of log-transformed data, with age entered as a covariate with post-hoc testing by use of the student's *t*-test. Results are displayed without log-transformation

CSF αSyn and t-tau in ADAD non-mutation carriers and in all three mutation carrier groups, both asymptomatic and symptomatic, with the strongest correlations found in asymptomatic *PSEN1* and symptomatic *APP* mutation carriers (Table 4). Asymptomatic *PSEN1* mutation carriers also exhibited a significant positive relationship between CSF αSyn and p-tau while both non-mutation carriers and *PSEN2* mutation carriers demonstrated a significant positive correlation between CSF αSyn and p-tau levels.

Associations between brain Aβ plaque load and CSF αSyn in the DIAN cohort

Data from PiB-PET examinations reflecting in vivo amyloid plaque burden were available for $n = 132$ of the DIAN participants (dataset freeze 10). The frequency of PiB-positive ((SUVR/brainstem) ≥0.72) subjects was; 49.1% (28/57) of asymptomatic mutation carriers, 89.3% (25/28) of symptomatic mutation carriers, and 4.3% (2/47) of non-mutation carriers. As expected, PiB-retention increased with EYO in mutation carriers, while no significant increase was noted in non-mutation carriers (data not shown).

Potential associations between CSF αSyn levels and in vivo amyloid plaque burden in subjects from the DIAN cohort were investigated by analyzing the linear regressions of mean cortical PiB-retention versus CSF αSyn in ADAD mutation carriers and non-mutation carriers separately, as well as stratifying mutation carriers into asymptomatic and symptomatic groups, adding EYO as independent predictor in all regression models; these four models resulted in no significant associations between mean cortical PiB and αSyn (Table 5). When looking at the subset of PiB-positive asymptomatic ADAD

Table 3 DIAN cohort baseline characteristics

	Non-Mutation Carriers	All Mutation Carriers	*APP* Mutation Carriers			*PSEN1* Mutation Carriers			*PSEN2* Mutation Carriers			*p*-value[a]
			All	Asym	Sym	All	Asym	Sym	All	Asym	Sym	
N	50	92	24	15	9	50	29	21	18	16	2	
Sex (% f/m)	56/44	55/45	63/37	67/33	56/44	60/40	52/48	71/29	33/66	32/68	50/50	-
Age at examination (yrs)	40 ± 11	40 ± 10	42 ± 11	39 ± 12	49 ± 6	39 ± 10	34 ± 9	46 ± 8	40 ± 10	38 ± 9	57	***[c]
EYO	−9 ± 11	−7 ± 12	5 ± 11	−11 ± 11	4±3	−6 ± 12	−14 ± 10	5 ± 4	−13 ± 11	−15 ± 9	7	-
MMSE[§]	29 ± 1	27 ± 5	27 ± 5	29 ± 1	23 ± 7	26 ± 6	29 ± 1	21 ± 7	29 ± 3	30 ± 1	20	*[c]
CDR	0 ± 0	0.3 ± 0.6	0.3 ± 0.5	0 ± 0	1 ± 0.5	0.4 ± 0.7	0 ± 0	1 ± 0.8	0.1 ± 0.5	0 ± 0	1	-[d]
APOEε4 carrier (%)	36	27	25	20	33	18	14	24	56	50	100	**[e]
CSF $A\beta_{1-42}$ (pg/mL)	858 ±264	613 ± 320 (*n* = 89)	613 ± 353 (*n* = 23)	734 ± 393 (*n* = 14)	425 ± 164	621 ± 335 (*n* = 49)	770 ± 318 (*n* = 28)	422 ± 243	592 ± 233 (*n* = 17)	596 ± 249 (*n* = 15)	565	-[c]
CSF $A\beta_{1-40}$ (pg/mL)	8173 ± 2751	7477 ± 3146 (*n* = 89)	6639 ± 3278 (*n* = 23)	6302 ± 3494 (*n* = 14)	7162 ± 3033	7563 ± 2785 (*n* = 49)	8313 ± 2918 (*n* = 28)	6563 ± 2300	8363 ± 3810 (*n* = 17)	8811 ± 3776 (*n* = 15)	5005	-[c]
CSF $A\beta_{42/40}$	0.115 ± 0.045	0.092 ± 0.052 (*n* = 89)	0.106 ± 0.065 (*n* = 23)	0.128 ± 0.066 (*n* = 14)	0.072 ± 0.049	0.088 ± 0.048 (*n* = 49)	0.102 ± 0.047 (*n* = 28)	0.070 ± 0.045	0.082 ± 0.037 (*n* = 17)	0.076 ± 0.031 (*n* = 15)	0.129	-[c]
CSF t-tau (pg/mL)	56 ± 26 (*n* = 49)	106 ± 73 (*n* = 87)	90 ± 64 (*n* = 23)	60 ± 30 (*n* = 14)	137 ± 76	124 ± 82 (*n* = 47)	94 ± 56 (*n* = 26)	162 ± 94	75 ± 39 (*n* = 17)	72 ± 40 (*n* = 15)	98	*[c]
CSF p-tau (pg/mL)	29 ± 14 (*n* = 49)	54 ± 34 (*n* = 89)	46 ± 25 (*n* = 23)	32 ± 11 (*n* = 14)	69 ± 24	60 ± 37 (*n* = 49)	45 ± 28 (*n* = 28)	81 ± 38	45 ± 31 (*n* = 17)	48 ± 32 (*n* = 15)	29	-[c]

Asym = asymptomatic *Sym* = symptomatic *EYO* = estimated yeats from symptom onset *MMSE* = Mini-Mental State Exam scorre
CDR = Clinical Dementia Rating score *a* = *p*-value versus all mutation carriers *c* = Kruskal-Wallis test *d* = Mann-Whitney *U* test e = χ^2 test § = non-normally distributed data presented at mean ± standard deviation in order to maintain patient anonymity
- = non-significant * = $p \leq 0.05$ ** = $p \leq 0.01$ *** = $p \leq 0.001$

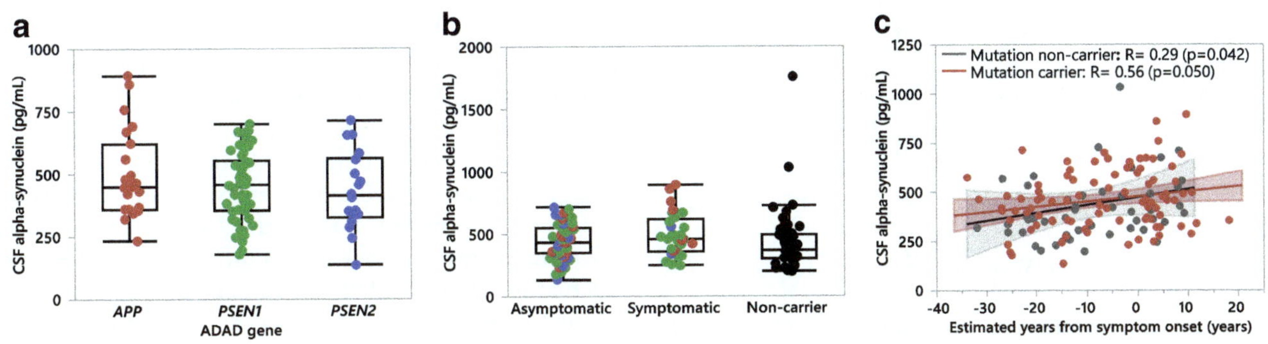

Fig. 8 CSF αSyn levels quantified in participants from the DIAN cohort. **a** CSF αSyn levels grouped by ADAD gene mutation. **b** CSF αSyn in ADAD non-mutation carriers versus asymptomatic and symptomatic ADAD mutation carriers. Data point coloring represents the three ADAD genes as shown in (**a**). *P*-values are presented first as uncorrected followed by Bonferroni corrected (a-b: $n = 3$ comparisons). *P*-values calculated using the non-parametric Kruskal-Wallis test and a Bonferroni correction was used to account for multiple comparisons. **c** Correlation between CSF αSyn and the estimated years from symptom onset (EYO) in ADAD mutation carriers and non-mutation carriers; correlations are calculated using Pearson's correlation test after log-transformation of the CSF αSyn data and results are displayed without log-transformation or age-correction (raw data)

mutation carriers, however, the same linear regression model showed a significant positive association between mean cortical PiB and αSyn ($\beta = 0.44$, $p = 0.010$); results from exploratory linear regression models in 42 regions of interest in PiB-positive asymptomatic mutation carriers are illustrated in Fig. 9a in which the regional regression β coefficients are displayed on a 3D brain surface. The strongest associations between CSF αSyn and amyloid plaque load in PiB-positive asymptomatic ADAD mutation carriers were found in regions including the posterior cingulate, superior temporal and frontal (Fig. 9a).

A potential effect of *APOEε4* on the association between PiB and αSyn in ADAD mutation carriers was assessed by the linear regression model: Mean cortical PiB ~ αSyn + *APOEε4* + EYO + αSyn**APOEε4*, and results are presented on Table 5. The interaction of αSyn**APOEε4* was significant, indicating that the relationship between PiB-retention and αSyn is different in *APOEε4*-positive than in *APOEε4*-negative ADAD mutation carriers. In *APOEε4*-positive ADAD mutation carriers, there was a strong positive association between αSyn and mean cortical PiB ($\beta = 0.39$, $p = 0.032$) (Table 5), while this association was not significant in *APOEε4*-negative mutation carriers ($\beta = -0.10$, $p = 0.348$) (data not shown). Exploratory positive associations between αSyn and regional PiB in *APOEε4*-positive ADAD mutation carriers in the 42 regions of interest are illustrated on the 3D brain display, showing strong associations in regions including the superior temporal, anterior and posterior cingulate, parietal and frontal regions (Fig. 9b). The CSF αSyn-*APOEε4*-PiB relationship was specific and could not be replicated by replacing CSF αSyn with CSF t-tau levels (data not shown).

The positive significant associations between CSF αSyn and regional amyloid plaque load in PiB-positive asymptomatic ADAD mutation carriers are illustrated in Fig. 10 in regions of early amyloid plaque deposition within the frontal and temporal lobes (Fig. 10a-b), contrasting with respective negative associations observed in few regions in the symptomatic mutation carriers, mostly restricted to areas of late amyloid plaque deposition including paracentral and post-central regions (Fig. 10c-d).

Discussion

The role of αSyn in the context of AD pathophysiology is unknown. In this study we set out to investigate the

Table 4 DIAN cohort CSF αSyn correlation analyses[a]

	Non- mutation carriers	*APP* mutation carriers		*PSEN1* mutation carriers		*PSEN2* mutation carriers	
		Asym ($n = 15$)	Sym ($n = 9$)	Asym ($n = 29$)	Sym ($n = 21$)	Asym ($n = 16$)	Sym ($n = 2$)
CSF $A\beta_{1-42}$	0.595***	–	–	–	–	–	¤
CSF $A\beta_{1-40}$	0.744***	0.610*	0.717*	0.526**	0.788***	–	¤
CSF $A\beta_{42/40}$	–	–	−0.800**	–	−0.464*	–	¤
CSF t-tau	0.716***	0.676*	0.917***	0.943***	0.664***	0.736**	¤
CSF p-tau	0.544***	–	–	0.809***	0.662***	–	¤

[a] = All correlations calculated using the Spearman's rank correlation test
Asym = asymptomatic Sym = symptomatic
¤ = not determined * = $p \leq 0.05$ ** = $p \leq 0.01$ *** = $p \leq 0.001$

Table 5 Linear regression models relating mean cortical ^{11}C-Pittsburgh Compound-B uptake and αSyn in ADAD mutation carriers

Model (PiB-positive asymptomatic mutation carriers): Mean cortical PiB ~ αSyn + EYO				
Independent predictors	Standardized β	Std. Error	*t* value	*p* value
αSyn	0.375	0.00047	2.348	0.027
EYO	0.444	0.00676	2.780	0.010
Model (all mutation carriers): Mean cortical PiB ~ αSyn + *APOEε4* + EYO + αSynAPOEε4***				
Independent predictors	Estimate	Std. Error	*t* value	*p* value
αSyn	−0.00038	0.00039	−0.970	0.335
APOEε4	−0.98698	0.44767	−2.205	0.030
EYO	0.02682	0.00435	6.164	0.000
αSyn**APOEε4*	0.00244	0.00099	2.468	0.016
Model (*APOEε4*-positive mutation carriers): Mean cortical PiB ~ αSyn + EYO				
Independent predictors	Standardized β	Std. Error	*t* value	*p* value
αSyn	0.393	0.00088	2.321	0.032
EYO	0.533	0.00864	3.147	0.005

PiB = ^{11}C-Pittsburgh Compound-B *EYO* = estimated years from symptom onset

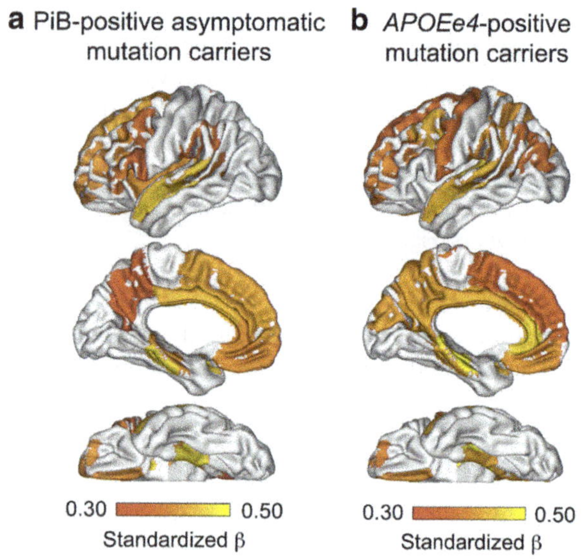

Fig. 9 Brain maps depicting significant associations between CSF αSyn and PiB-retention in ADAD mutation carrier subgroups. **a** PiB-positive asymptomatic (CDR < 0.5), **b** *APOEε4*-positive (CDR < 0.5). The significant associations between CSF αSyn and PiB are represented by the standardized β coefficient corresponding to independent predictor αSyn in the linear regression model for PiB ~ αSyn + EYO, where EYO is estimated years from symptom onset. Increasing positive associations (increasing β values) are labeled from red to yellow

relevance of αSyn to both sporadic and familial AD by examining CSF levels of αSyn and potential associations to various AD disease parameters including cognition, CSF AD biomarkers, and in familial AD patients we also assessed potential relationships between CSF αSyn and brain amyloid plaque burden. Although the significance of CSF αSyn levels to brain parenchymal αSyn levels are still unknown, mainly due to the lack of ante-mortem αSyn brain imaging tools, we anticipate that CSF αSyn fluctuations mirror αSyn alterations in the brain. We studied subjects with MCI, sporadic AD and ADAD in order to pinpoint any potential pre-clinical, prodromal, or mid- to late stage disease pathways linking αSyn to AD pathogenesis. We further explored the impact of the AD risk allele *APOEε4* on the identified associations.

Numerous studies have reported αSyn pathology in approximately 50–60% of autopsied AD patients [1, 16, 31]. Patients with mixed pathologies as well as animal models expressing combined AD and αSyn pathologies tend to exhibit amplified deterioration, typically enduring more severe symptoms and shorter survival rates [9, 41]. Interestingly, Wirths and colleagues described co-accumulation of αSyn in Aβ plaques and dystrophic neurites only in patients with the Lewy body variant of AD rather than in 'typical AD' cases [65]. Furthermore, results from another study of 147 neuropathologically confirmed AD cases suggested that the majority of AD patients who exhibited very few or no neocortical neurofibrillary tangles, termed 'plaque-only AD', exhibited Lewy body αSyn pathology, while patients with the Lewy body variant of AD were found to be 'plaque-only AD' cases, hence, 'plaque-only AD' and Lewy body variant AD were diagnostically indistingushable from one another [17]. Currently, the mechanisms driving comorbid αSyn and AD pathologies have yet to be elucidated making it impossible to predict comorbidity or to diagnose patients ante-mortem, however, we speculate that altered CSF αSyn may be a predictive feature of AD patients that already are harboring or that will develop αSyn co-pathology.

The identification of a molecular interaction between presenilin 1 (PS1), encoded by the *PSEN1* gene, and αSyn was suggested to in part explain the clinical and pathophysiological overlaps between AD and synucleinopathies like DLB [64]. The relevance of this molecular interaction was illustrated by two previous reports demonstrating a strong association between ADAD-causing *PSEN1* mutations and Lewy body pathology specifically in the amygdala [29, 31]. With a direct interaction between αSyn and PS1, the latter of which is a key player in Aβ peptide production [8], it is tempting to speculate that αSyn may have a modulatory role in Aβ production and/or deposition. In support, our current study revealed an association between higher CSF αSyn levels and amyloid plaque burden in several brain areas of PiB-positive asymptomatic ADAD mutation carriers.

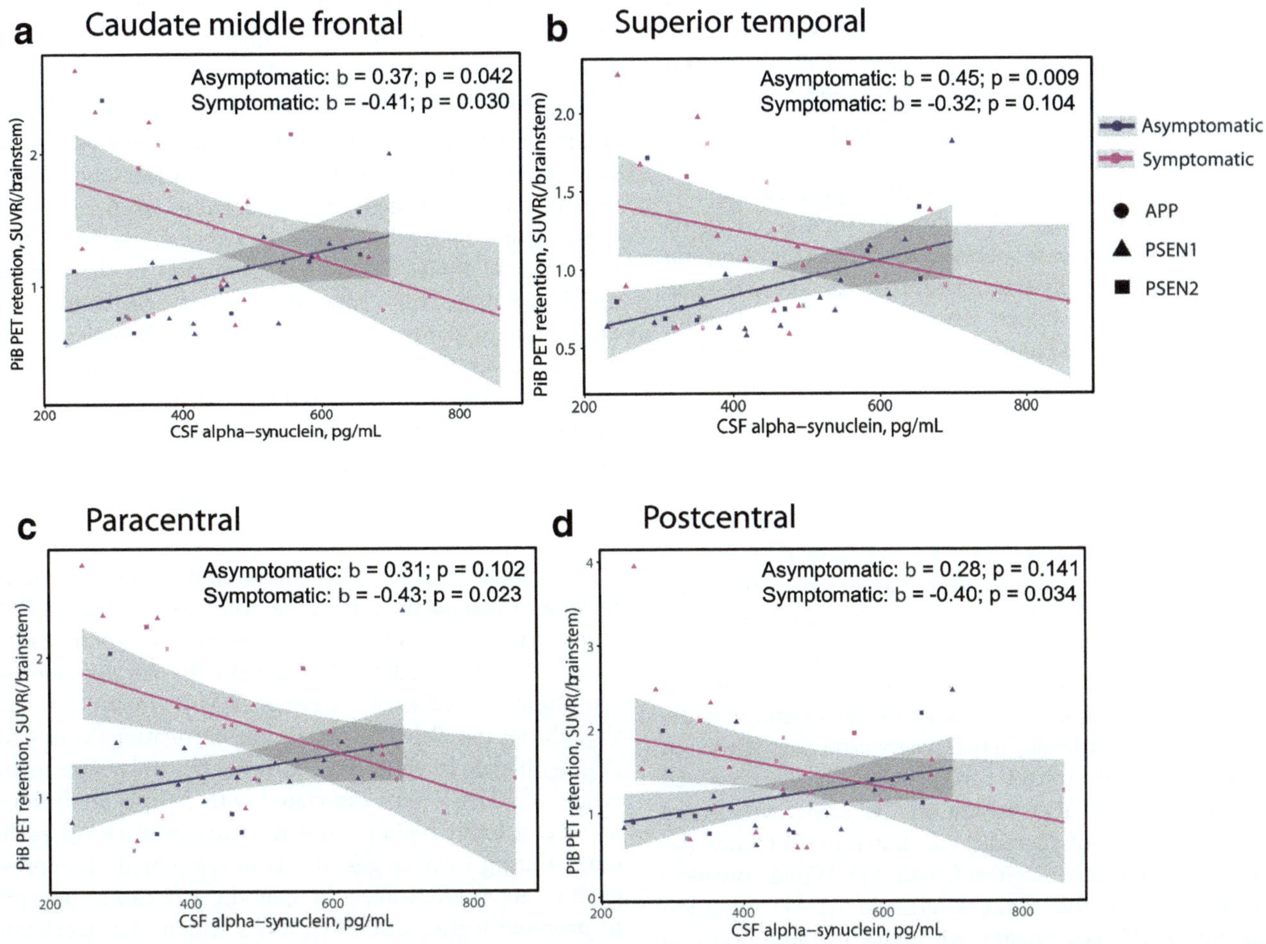

Fig. 10 Significant associations between CSF αSyn and regional PiB-retention in PiB-positive asymptomatic (CDR < 0.5, *n* = 28) ADAD mutation carriers (blue) and in symptomatic (CDR ≥0.5, *n* = 25) ADAD mutation carriers (purple) . The significant associations between CSF αSyn and PiB are represented by the standardized β coefficient corresponding to independent predictor αSyn in the linear regression model for PiB ~ αSyn + EYO, where EYO is estimated years from symptom onset, in four brain regions (**a-d**)

The brain regions exhibiting the highest rates of annual Aβ accumulation in PiB-negative (< 0.87 SUVR) individuals with pathological CSF $A\beta_{1-42}$ levels (< 192 ng/L), termed as 'early Aβ accumulators' by Palmqvist and colleagues in a recent publication [44], exhibited a striking overlap with the brain regions for which we found significant positive correlations between CSF αSyn and PiB-retention. Similarly, a recent study of cognitively normal individuals with subjective memory complaints documented a positive association between CSF αSyn concentrations and brain Aβ deposition [59]. Rather surprisingly, our identified relationship between CSF αSyn levels and Aβ deposition in presymptomatic ADAD patients was not paralleled by any significant correlations between CSF αSyn and $A\beta_{1-42}$ levels. Instead, CSF αSyn levels were consistently positively associated with CSF $A\beta_{1-40}$ levels levels across all studied groups. The relevance of CSF $A\beta_{1-40}$ in the presymptomatic stage of AD, during which Aβ deposition occurs, remains to be determined.

In support of our own results [61, 62] and those reported by other groups [60] we further identified strong positive correlations between CSF αSyn and CSF t-tau and p-tau in both studied cohorts. By stratifying the studied sporadic MCI and AD patients based on pathological CSF AD biomarker levels, we found elevated CSF αSyn levels in patients that exhibited pathological CSF tau but not Aβ levels. Similar to CSF levels of tau, CSF αSyn levels were previously proposed to potentially function as a marker of synapse loss and neurodegeneration [43], however the lack of significantly altered levels of CSF αSyn in AD does not support this notion. A direct link between αSyn and tau pathology was instead suggested by findings showing that αSyn promotes GSK-3β-mediated tau phosphorylation [25]. The authors of the same study further demonstrated that αSyn directly interacted with both tau and GSK-3β. Another in vitro study employing rat PC12 cells demonstrated that extracellular αSyn in fact was involved in GSK-3β-mediated hyperphosphorylation of tau [15]. In cognitively intact control subjects we found strong positive correlations between levels of CSF αSyn and both t-tau and p-tau, which in our view may challenge the notion that the relationship between αSyn and tau is indeed pathological. Due to our findings of a consistent correlation between CSF αSyn and tau levels across diagnostic groups including healthy controls, we speculate that the this association might be due to non-conventional exosome-related release mechanisms [12, 51] for both tau and αSyn, without any clear disease association. The exact relevance of the described seemingly robust relationship between levels of CSF tau and αSyn requires clarification, preferably in future studies assessing potential links between CSF αSyn and ante-mortem tau pathology using novel tau tracers and imaging techniques [50].

With the *APOEε4* allele as a common denominator in terms or risk of disease for both AD and DLB [7] we were interested in assessing potential effects of this gene variant on CSF αSyn levels in the investigated cohorts. In subjects from the MCI-AD diagnostic group who exhibited elevated CSF αSyn levels compared to controls at baseline, homozygous *APOEε4* carriers exhibited the highest CSF αSyn levels. This observation was absent in AD patients and control subjects. Hence, we observed an effect of the *APOEε4* variant on CSF αSyn levels in the prodromal phase of sporadic AD, but no effect once patients were clinically diagnosed with AD. When considering any effect of the *APOEε4* allele in ADAD mutation carrying DIAN participants, we found no differences in CSF αSyn between *APOEε4* positive versus *APOEε4* negative participants, or within the *APP*, *PSEN1* or *PSEN2* mutation carrying groups. However, presymptomatic Aβ deposition in ADAD mutation carriers was positively associated with CSF αSyn levels only in *APOEε4* positive subjects. We hypothesize that an association between CSF αSyn and Aβ deposition at the presymptomatic stage of AD may be further supported by the *APOEε4* variant which in previous studies has been shown to promote Aβ deposition even in cognitively intact individuals [40].

The regulatory mechanisms governing αSyn levels in brain parenchyma and CSF are unknown. However, there is a clear difference between CSF αSyn levels in AD patients and those with synucleinopathies, where patients afflicted with the latter disorders consistently exhibit reduced levels [18, 24, 38, 39, 56, 57] suggesting a disease-specific process that disrupts the balance between the intracellular and extracellular pools of αSyn. Kallikrein-6, also called neurosin, is one of few reported extracellular proteases shown to cleave αSyn [52, 55]. Increasing the expression of kallikrein-6 in the brains of a mouse model of Lewy body disease promoted αSyn clearance and reduced αSyn pathology [52]. Further, we have shown that patients with synucleinopathies not only exhibited low CSF αSyn levels but also reduced levels of kallikrein-6 [62]. Thus, our previous results combined with those from animal studies suggest that an imbalance between αSyn and kallikrein-6 may promote synucleinopathy. Recently we also reported that the AD and MCI-AD patients examined in the current study did not exhibit altered levels of CSF kallikrein-6 compared to controls, whereas MCI-MCI patients had slightly lower CSF kallikrein-6 levels compared to controls [45]. Hence, the elevated CSF αSyn levels observed in the MCI-AD group were not paralleled by increased kallikrein-6 levels suggesting a potential imbalance between kallikrein-6 and αSyn in this group.

We conclude, CSF αSyn levels alone appear unfit to serve as a diagnostic marker for AD however higher CSF αSyn concentrations were associated with the progression from MCI to sporadic AD, and with the development of symptoms in subjects carrying ADAD mutations. Although not paralleled by significant correlations with CSF $A\beta_{1-42}$, higher levels of CSF αSyn in the presymptomatic stages of ADAD were associated with Aβ plaque burden in several brain regions known to accumulate Aβ pathology during early stages of AD development. The presence of an *APOEε4* allele in sporadic AD cases appeared to promote higher CSF αSyn levels which may accelerate the processes linking αSyn to Aβ deposition in AD. The *APOEε4* allele may be involved in molecular processes governing CSF αSyn levels, which in turn appear associated with the presymptomatic build-up of Aβ plaque burden in the brain during AD development. Future studies assessing αSyn in paired CSF and autopsied brain tissues are needed in order to decipher the relevance of altered CSF αSyn levels in the pathophysiology of AD.

Abbreviations

3T-MRI: 3-Tesla magnetic resonance imaging; AD: Alzheimer's disease; ADAD: Early-onset autosomal dominant Alzheimer's disease; ADNI: Alzheimer's Disease Neuroimaging Initiative; *APOEε4*: Apolipoprotein E epsilon 4; *APP*: Amyloid precursor protein gene; AUC: Area under the curve; $A\beta_{1-40}$: Amyloid-β_{1-40}; $A\beta_{1-42}$: Amyloid-β_{1-42};; $A\beta_{42/40}$: Amyloid-$\beta_{1-42/1-40}$ ratio; CSF: Cerebrospinal fluid; DIAN: Dominantly Inherited Alzheimer's Network; ELISA: Enzyme-linked immunosorbent assay; EYO: Estimated years from symptom onset; FDG: Fluorodeoxyglucose; MCI: Mild cognitive impairment; MCI-AD: MCI patients who converted to Alzheimer's disease at the 24-month follow up; MCI-MCI: MCI patients who remained MCI at the 24-month follow up; MMSE: Mini Mental State Examination score; MPRAGE: Magnetization-prepared 180 degrees radio-frequency pulses and rapid gradient-echo sampling; PET: Positron emission tomography; PiB: ^{11}C-Pittsburgh Compound-B; PS1: Presenilin 1 protein; *PSEN1*: Presenilin 1 gene; *PSEN2*: Presenilin 2 gene; p-tau: Phosphorylated tau; ROC: Receiver operating characteristic; SUVR: Standardized Uptake Value Ratio; t-tau: Total tau

Acknowledgements
This manuscript has been reviewed by DIAN Study investigators for scientific content and consistency of data interpretation with previous DIAN Study publications. We acknowledge the altruism of the participants and their families and contributions of the DIAN research and support staff at each of the participating sites for their contributions to this study. The authors further would like to acknowledge the generous participation of the study subjects included in the sporadic cohort as well as all the clinical and laboratory staff involved.

Funding
Data collection and sharing for this project was supported by The Dominantly Inherited Alzheimer's Network (DIAN, NIH project U19AG032438, RJB) funded by the National Institute on Aging (NIA), the German Center for Neurodegenerative Diseases (DZNE), Raul Carrea Institute for Neurological Research (FLENI), Partial support by the Research and Development Grants for Dementia from Japan Agency for Medical Research and Development, AMED, and the Korea Health Technology R&D Project through the Korea Health Industry Development Institute (KHIDI). The current study was supported by grants from the Swedish Dementia Foundation (Demensfonden, HMN), the Alzheimer's Association (2015-NIRG-339824, HMN), Stiftelsen Olle Engkvist Byggmästare (184–333, HMN) and a VINNMER and Marie Curie Fellowship (2015–04905, HMN).

Authors' contributions
HMN conceived of the idea for the study. DT drafted the manuscript. HMN and DT quantified and statistically analyzed all CSF αSyn levels and correlations. HMN, KP and DT interpreted the non-neuroimaging data. AN and ERV analyzed and interpreted all DIAN PET imaging data. SBS and GRG performed the clinical assessments of subjects from the longitudinal cohort. SBS, GBe, GRG, CL, IM, GRG, GBr and LRW contributed to clinical data acquisition, including the CSF biomarker profiles for the longitudinal cohort. TLSB, CMK, AF, JCM and RJB contributed to DIAN participant enrollment and collection of the DIAN data provided for the current study. DT, ERV, SBS, GB, CL, IM, GRG, GBe, KP, GB, TLSB, CML, AF, JCM, RJB, AN, LRW and HMN provided critical input and revision of the manuscript for intellectual content. All authors have approved of the content of the final manuscript.

Competing interests
The authors declare that they have no competing interests.

Author details
[1]Department of Biochemistry and Biophysics, Stockholm University, Svante Arrhenius väg 16B, 106 91 Stockholm, Sweden. [2]Department of Neurobiology, Care Sciences and Society, Karolinska Institutet, Stockholm, Sweden. [3]Department of Neurology, University Hospital of Trondheim, Trondheim, Norway. [4]Department of Neuroscience, Norwegian University of Science and Technology, Trondheim, Norway. [5]Department of Neuroscience, Mayo Clinic College of Medicine, Jacksonville, FL, USA. [6]Department of Radiology, Washington University School of Medicine, St Louis, MO, USA. [7]Department of Psychiatry, Washington University School of Medicine, St Louis, MO, USA. [8]Department of Neurology, Washington University School of Medicine, St Louis, MO, USA. [9]The Aging Research Center, Karolinska University Hospital, Stockholm, Sweden.

References

1. Arai Y, Yamazaki M, Mori O, Muramatsu H, Asano G, Katayama Y (2001) α-Synuclein-positive structures in cases with sporadic Alzheimer's disease: morphology and its relationship to tau aggregation. Brain Res 888:287–296 Elsevier
2. Bachhuber T, Katzmarski N, McCarter JF, Loreth D, Tahirovic S, Kamp F et al (2015) Inhibition of amyloid-β plaque formation by α-synuclein. Nat Med 21: 802–807 Nature Publishing Group, a division of Macmillan Publishers Limited. All Rights Reserved
3. Bateman RJ, Aisen PS, De Strooper B, Fox NC, Lemere CA, Ringman JM et al (2011) Autosomal-dominant Alzheimer's disease: a review and proposal for the prevention of Alzheimer's disease. Alzheimers Res Ther 3:1 BioMed Central
4. Bateman RJ, Xiong C, Benzinger TLS, Fagan AM, Goate A, Fox NC et al (2012) Clinical and biomarker changes in dominantly inherited Alzheimer's disease. N Engl J Med 367:795–804 Massachusetts Medical Society
5. Berge G, Lauridsen C, Sando SB, Holder DJ, Møller I, Aasly JO et al (2015) Effect of Tween-20 on Core biomarkers measured in cerebrospinal fluid from patients with Alzheimer's disease, mild cognitive impairment, or healthy control individuals. J Alzheimer's Dis 49:493–502 IOS Press
6. Berge G, Sando SB, Albrektsen G, Lauridsen C, Møller I, Grøntvedt GR et al (2016) Alpha-synuclein measured in cerebrospinal fluid from patients with Alzheimer's disease, mild cognitive impairment, or healthy controls: a two year follow-up study. BMC Neurol BioMed Central 16:180
7. Berge G, Sando SB, Rongve A, Aarsland D, White LR (2014) Apolipoprotein E ε2 genotype delays onset of dementia with Lewy bodies in a Norwegian cohort. J Neurol Neurosurg Psychiatry 85:1227–1231 BMJ Publishing Group Ltd
8. Chow VW, Mattson MP, Wong PC, Gleichmann M (2010) An overview of APP processing enzymes and products. NeuroMolecular Med 12:1–12 Humana Press Inc
9. Clinton LK, Blurton-Jones M, Myczek K, Trojanowski JQ, LaFerla FM (2010) Synergistic interactions between Abeta, tau, and alpha-synuclein: acceleration of neuropathology and cognitive decline. J Neurosci 30:7281–7289
10. Corder EH, Saunders AM, Strittmatter WJ, Schmechel DE, Gaskell PC, Small GW et al (1993) Gene dose of apolipoprotein E type 4 allele and the risk of Alzheimer's disease in late onset families. Science 261:921–923 American Association for the Advancement of Science
11. Eisenstein SA, Koller JM, Piccirillo M, Kim A, Antenor-Dorsey JAV, Videen TO et al (2012) Characterization of extrastriatal D2 in vivo specific binding of [^{18}F](N-methyl)benperidol using PET. Synapse 66:770–780 NIH Public Access
12. Emmanouilidou E, Melachroinou K, Roumeliotis T, Garbis SD, Ntzouni M, Margaritis LH et al (2010) Cell-produced alpha-synuclein is secreted in a calcium-dependent manner by exosomes and impacts neuronal survival. J Neurosci 30:6838–6851 Society for Neuroscience
13. Fagan AM, Mintun MA, Mach RH, Lee S-Y, Dence CS, Shah AR et al (2006) Inverse relation between in vivo amyloid imaging load and cerebrospinal fluid $A\beta_{42}$ in humans. Ann Neurol 59:512–519 Wiley-Blackwell
14. Fischl B, Salat DH, Busa E, Albert M, Dieterich M, Haselgrove C et al (2002) Whole brain segmentation: automated labeling of neuroanatomical structures in the human brain. Neuron 33:341–355
15. Gąssowska M, Czapski GA, Pająk B, Cieślik M, Lenkiewicz AM, Adamczyk A. Extracellular α-Synuclein Leads to Microtubule Destabilization via GSK-3β-Dependent Tau Phosphorylation in PC12 Cells. Paudel HK, editor. PLoS One. 2014;9:e94259 Public Library of Science
16. Hamilton RL (2000) Lewy bodies in Alzheimer's disease: a neuropathological review of 145 cases using alpha-synuclein immunohistochemistry. Brain Pathol 10:378–384
17. Hansen LA, Masliah E, Galasko D, Terry RD (1993) Plaque-only Alzheimer disease is usually the lewy body variant, and vice versa. J Neuropathol Exp Neurol 52:648–654
18. Hong Z, Shi M, Chung KA, Quinn JF, Peskind ER, Galasko D et al (2010) DJ-1 and alpha-synuclein in human cerebrospinal fluid as biomarkers of Parkinson's disease. Brain 133:713–726 Oxford University Press
19. Irwin DJ, Lee VM-Y, Trojanowski JQ (2013) Parkinson's disease dementia: convergence of α-synuclein, tau and amyloid-β pathologies. Nat Rev Neurosci NIH Public Access 14:626–636
20. Iwai A, Masliah E, Yoshimoto M, Ge N, Flanagan L (1995) Rohan de Silva H., et al. the precursor protein of non-Aβ component of Alzheimer's disease amyloid is a presynaptic protein of the central nervous system. Neuron 14: 467–475
21. Jack CR, Bernstein MA, Borowski BJ, Gunter JL, Fox NC, Thompson PM et al (2010) Update on the magnetic resonance imaging core of the Alzheimer's disease neuroimaging initiative. Alzheimers Dement 6:212–220 NIH Public Access
22. Jack CR, Bernstein MA, Fox NC, Thompson P, Alexander G, Harvey D et al (2008) The Alzheimer's Disease Neuroimaging Initiative (ADNI): MRI methods. J Magn Reson Imaging 27:685–691 NIH Public Access
23. Kallhoff V, Peethumnongsin E, Zheng H (2007) Lack of α-synuclein increases amyloid plaque accumulation in a transgenic mouse model of Alzheimer's

disease. Mol Neurodegener BioMed Central 2:6
24. Kasuga K, Tokutake T, Ishikawa A, Uchiyama T, Tokuda T, Onodera O et al (2010) Differential levels of alpha-synuclein, beta-amyloid42 and tau in CSF between patients with dementia with Lewy bodies and Alzheimer's disease. J Neurol Neurosurg Psychiatry 81:608–610 BMJ Publishing Group Ltd
25. Kawakami F, Suzuki M, Shimada N, Kagiya G, Ohta E, Tamura K et al (2011) Stimulatory effect of α-synuclein on the tau-phosphorylation by GSK-3β. FEBS J 278(10.1111):4895–4904 Wiley/Blackwell
26. Korff A, Liu C, Ginghina C, Shi M, Zhang J (2013) α-Synuclein in cerebrospinal fluid of Alzheimer's disease and mild cognitive impairment. J Alzheimer's Dis 36:679–688 IOS Press
27. Larson ME, Greimel SJ, Amar F, LaCroix M, Boyle G, Sherman MA et al (2017) Selective lowering of synapsins induced by oligomeric α-synuclein exacerbates memory deficits. Proc Natl Acad Sci 114:E4648–E4657 National Academy of Sciences
28. Larson ME, Sherman MA, Greimel S, Kuskowski M, Schneider JA, Bennett DA et al (2012) Soluble alpha-Synuclein is a novel modulator of Alzheimer's disease pathophysiology. J Neurosci 32:10253–10266
29. Leverenz JB, Fishel MA, Peskind ER, Montine TJ, Nochlin D, Steinbart E et al (2006) Lewy Body Pathology in Familial Alzheimer Disease. Arch Neurol 63: 370 American Medical Association
30. Li YJ, Hauser MA, Scott WK, Martin ER, Booze MW, Qin XJ et al (2004) Apolipoprotein E controls the risk and age at onset of Parkinson disease. Neurology 62:2005–2009
31. Lippa CF, Fujiwara H, Mann DMA, Giasson B, Baba M, Schmidt ML et al (1998) Lewy bodies contain altered α-Synuclein in brains of many familial Alzheimer's disease patients with mutations in Presenilin and amyloid precursor protein genes. Am J Pathol 153:1365–1370
32. Liu C-C, Liu C-C, Kanekiyo T, Xu H, Bu G (2013) Apolipoprotein E and Alzheimer disease: risk, mechanisms and therapy. Nat Rev Neurol 9:106–118 NIH Public Access
33. Masliah E, Iwai A, Mallory M, Uéda K, Saitoh T (1996) Altered presynaptic protein NACP is associated with plaque formation and neurodegeneration in Alzheimer's disease. Am J Pathol 148:201–210 American Society for Investigative Pathology
34. McKhann G, Drachman D, Folstein M, Katzman R, Price D, Stadlan EM (1984) Clinical diagnosis of Alzheimer's disease: report of the NINCDS-ADRDA work group under the auspices of Department of Health and Human Services Task Force on Alzheimer's disease. Neurology 34:939–944 American Academy of Neurology
35. McKhann GM, Knopman DS, Chertkow H, Hyman BT, Jack CR, Kawas CH et al (2011) The diagnosis of dementia due to Alzheimer's disease: recommendations from the National Institute on Aging-Alzheimer's association workgroups on diagnostic guidelines for Alzheimer's disease. Alzheimers Dement 7:263–269 Elsevier
36. Mintun MA, LaRossa GN, Sheline YI, Dence CS, Lee SY, Mach RH et al (2006) [11C]PIB in a nondemented population: Potential antecedent marker of Alzheimer disease. Neurology 67:446–452
37. Mollenhauer B, Cullen V, Kahn I, Krastins B, Outeiro TF, Pepivani I et al (2008) Direct quantification of CSF α-synuclein by ELISA and first cross-sectional study in patients with neurodegeneration. Exp Neurol 213:315–325 Academic Press
38. Mollenhauer B, Cullen V, Kahn I, Krastins B, Outeiro TF, Pepivani I et al (2008) Direct quantification of CSF α-synuclein by ELISA and first cross-sectional study in patients with neurodegeneration. Exp Neurol 213:315–325
39. Mollenhauer B, Locascio JJ, Schulz-Schaeffer W, Sixel-Döring F, Trenkwalder C, Schlossmacher MG (2011) α-Synuclein and tau concentrations in cerebrospinal fluid of patients presenting with parkinsonism: a cohort study. Lancet Neurol. Elsevier 10:230–240
40. Morris JC, Roe CM, Xiong C, Fagan AM, Goate AM, Holtzman DM et al (2010) APOE predicts amyloid-beta but not tau Alzheimer pathology in cognitively normal aging. Ann Neurol 67:122–131 NIH Public Access
41. Mueller C, Ballard C, Corbett A, Aarsland D (2017) The prognosis of dementia with Lewy bodies. Lancet Neurol 16:390–398
42. Ogaki K, Martens YA, Heckman MG, Koga S, Labbé C, Lorenzo-Betancor O et al (2018) Multiple system atrophy and apolipoprotein E. Mov Disord 33(4):647–650
43. Öhrfelt A, Grognet P, Andreasen N, Wallin A, Vanmechelen E, Blennow K et al (2009) Cerebrospinal fluid α-synuclein in neurodegenerative disorders—a marker of synapse loss? Neurosci Lett 450:332–335 Elsevier
44. Palmqvist S, Schöll M, Strandberg O, Mattsson N, Stomrud E, Zetterberg H et al (2017) Earliest accumulation of β-amyloid occurs within the default-mode network and concurrently affects brain connectivity. Nat Commun 8: 1214 Nat Nature Publishing Group
45. Patra K, Soosaipillai A, Sando SB, Lauridsen C, Berge G, Møller I et al (2018) Assessment of kallikrein 6 as a cross-sectional and longitudinal biomarker for Alzheimer's disease. Alzheimers Res Ther 10:9 BioMed Central
46. Pletnikova O, West N, Lee MK, Rudow GL, Skolasky RL, Dawson TM et al (2005) Aβ deposition is associated with enhanced cortical α-synuclein lesions in Lewy body diseases. Neurobiol Aging 26:1183–1192
47. Rosén C, Hansson O, Blennow K, Zetterberg H (2013) Fluid biomarkers in Alzheimer's disease – current concepts. Mol Neurodegener 8:20
48. Rousset OG, Ma Y, Evans AC (1998) Correction for partial volume effects in PET: principle and validation. J Nucl Med 39:904–911
49. Rowland DJ, Garbow JR, Laforest R, Snyder AZ, Xu S, Zanzonico PB et al (2005) Registration of [18F]FDG microPET and small-animal MRI. Nucl Med Biol 32:567–572 Elsevier
50. Saint-Aubert L, Lemoine L, Chiotis K, Leuzy A, Rodriguez-Vieitez E, Nordberg A (2017) Tau PET imaging: present and future directions. Mol Neurodegener 12:19 BioMed Central
51. Saman S, Kim W, Raya M, Visnick Y, Miro S, Saman S et al (2012) Exosome-associated tau is secreted in tauopathy models and is selectively phosphorylated in cerebrospinal fluid in early Alzheimer disease. J Biol Chem 287:3842–3849 American Society for Biochemistry and Molecular Biology
52. Spencer B, Michael S, Shen J, Kosberg K, Rockenstein E, Patrick C et al (2013) Lentivirus mediated delivery of neurosin promotes clearance of wild-type α-synuclein and reduces the pathology in an α-synuclein model of LBD. Mol Ther 21:31–41 Elsevier
53. Su Y, Blazey TM, Owen CJ, Christensen JJ, Friedrichsen K, Joseph-Mathurin N et al (2016) Quantitative Amyloid Imaging in Autosomal Dominant Alzheimer's Disease: Results from the DIAN Study Group. PLoS One 11: e0152082 Herholz K, editor. Public Library of Science
54. Su Y, D'Angelo GM, Vlassenko AG, Zhou G, Snyder AZ, Marcus DS et al (2013) Quantitative Analysis of PiB-PET with FreeSurfer ROIs. PLoS One 8: e73377 Chen K, editor. Public Library of Science
55. Tatebe H, Watanabe Y, Kasai T, Mizuno T, Nakagawa M, Tanaka M et al (2010) Extracellular neurosin degrades α-synuclein in cultured cells. Neurosci Res 67:341–346 Elsevier
56. Tateno F, Sakakibara R, Kawai T, Kishi M, Murano T (2012) Alpha-synuclein in the cerebrospinal fluid differentiates Synucleinopathies (Parkinson disease, dementia with Lewy bodies, multiple system atrophy) from Alzheimer disease. Alzheimer Dis Assoc Disord 26:213–216
57. Tokuda T, Salem SA, Allsop D, Mizuno T, Nakagawa M, Qureshi MM et al (2006) Decreased α-synuclein in cerebrospinal fluid of aged individuals and subjects with Parkinson's disease. Biochem Biophys Res Commun 349:162–166 Academic Press
58. Uéda K, Fukushima H, Masliah E, Xia Y, Iwai A, Yoshimoto M et al (1993) Molecular cloning of cDNA encoding an unrecognized component of amyloid in Alzheimer disease. Proc Natl Acad Sci U S A 90:11282–11286
59. Vergallo A, Bun R-S, Toschi N, Baldacci F, Zetterberg H, Blennow K et al (2018) Association of cerebrospinal fluid α-synuclein with total and phospho-tau 181 protein concentrations and brain amyloid load in cognitively normal subjective memory complainers stratified by Alzheimer's disease biomarkers. Alzheimers Dement:1–9. https://doi.org/10.1016/j.jalz.2018.06.3053.
60. Wang H, Stewart T, Toledo JB, Ginghina C, Tang L, Atik A et al (2018) A longitudinal study of Total and phosphorylated α-Synuclein with other biomarkers in cerebrospinal fluid of Alzheimer's disease and mild cognitive impairment. J Alzheimer's Dis 61:1541–1553 IOS Press
61. Wennström M, Londos E, Minthon L, Nielsen HM (2012) Altered CSF orexin and α-synuclein levels in dementia patients. J Alzheimers Dis 29:125–132
62. Wennström M, Surova Y, Hall S, Nilsson C, Minthon L, Boström F et al (2013) Low CSF levels of both α-synuclein and the α-synuclein cleaving enzyme neurosin in patients with synucleinopathy. PLoS One 8:e53250 Public Library of Science
63. Winblad B, Palmer K, Kivipelto M, Jelic V, Fratiglioni L (2004) Mild cognitive impairment – beyond controversies, towards a consensus : report of the

International Working Group on Mild Cognitive Impairment, pp 240–246
64. Winslow AR, Moussaud S, Zhu L, Post KL, Post KL, Dickson DW et al (2014) Convergence of pathology in dementia with Lewy bodies and Alzheimer's disease: a role for the novel interaction of alpha-synuclein and presenilin 1 in disease. Brain Oxford 137:1958–1970 University Press
65. Wirths O, Weickert S, Majtenyi K, Havas L, Kahle PJ, Okochi M et al (2000) Lewy body variant of Alzheimer's disease: alpha-synuclein in dystrophic neurites of a beta plaques. Neuroreport 11:3737–3741
66. Wong YC, Krainc D (2017) A-Synuclein Toxicity in Neurodegeneration: Mechanism and Therapeutic Strategies. Nat Med 23:1–13 Nature Publishing Group
67. Xia M, Wang J, He Y (2013) BrainNet Viewer: A Network Visualization Tool for Human Brain Connectomics. PLoS One 8:e68910 Csermely P, editor. Public Library of Science
68. Zetterberg H, Petzold M, Magdalinou N (2014) Cerebrospinal fluid α-synuclein levels in Parkinson's disease - changed or unchanged? Eur J Neurol 21(10.1111):365–367 Wiley/Blackwell

8

In vivo induction of membrane damage by β-amyloid peptide oligomers

Carl Julien[1,4], Colson Tomberlin[1], Christine M. Roberts[1], Aumbreen Akram[2], Gretchen H. Stein[3], Michael A. Silverman[2,5] and Christopher D. Link[1*]

Abstract

Exposure to the β-amyloid peptide (Aβ) is toxic to neurons and other cell types, but the mechanism(s) involved are still unresolved. Synthetic Aβ oligomers can induce ion-permeable pores in synthetic membranes, but whether this ability to damage membranes plays a role in the ability of Aβ oligomers to induce tau hyperphosphorylation, or other disease-relevant pathological changes, is unclear. To examine the cellular responses to Aβ exposure independent of possible receptor interactions, we have developed an in vivo *C. elegans* model that allows us to visualize these cellular responses in living animals. We find that feeding *C. elegans E. coli* expressing human Aβ induces a membrane repair response similar to that induced by exposure to the CRY5B, a known pore-forming toxin produced by *B. thuringensis*. This repair response does not occur when *C. elegans* is exposed to an Aβ Gly37Leu variant, which we have previously shown to be incapable of inducing tau phosphorylation in hippocampal neurons. The repair response is also blocked by loss of calpain function, and is altered by loss-of-function mutations in the *C. elegans* orthologs of BIN1 and PICALM, well-established risk genes for late onset Alzheimer's disease. To investigate the role of membrane repair on tau phosphorylation directly, we exposed hippocampal neurons to streptolysin O (SLO), a pore-forming toxin that induces a well-characterized membrane repair response. We find that SLO induces tau hyperphosphorylation, which is blocked by calpain inhibition. Finally, we use a novel biarsenical dye-tagging approach to show that the Gly37Leu substitution interferes with Aβ multimerization and thus the formation of potentially pore-forming oligomers. We propose that Aβ-induced tau hyperphosphorylation may be a downstream consequence of induction of a membrane repair process.

Keywords: Alzheimer's disease, β-amyloid, Tau, *Caenorhabditis elegans*, Pore-forming toxin

Introduction

Alzheimer's disease is characterized by the deposition in the brain of senile plaques, composed largely of the β-amyloid peptide (Aβ). The amyloid cascade hypothesis posits that in Alzheimer's disease, accumulation of Aβ in the brain ultimately leads to changes in tau metabolism, which leads to the deposition of tau in neurofibrillary tangles (NFTs) that are the likely proximal cause of the neuronal loss observed in this disease [25]. While there is still significant controversy as to whether the amyloid cascade hypothesis explains Alzheimer's pathology [28, 39, 72], there is extensive evidence (in vitro and in vivo) that exposure of neurons to Aβ can lead to tau hyperphosphorylation, a potential driver of tau deposition [23, 51, 87]. Multiple tau kinases have been identified (reviewed in [57]), as have pathways by which these kinases could be activated to increase tau phosphorylation [15, 55, 89]. CDK5 and GSK3β have emerged as strong candidates for disease-relevant tau kinases [27, 50], and activation of both these kinases can occur downstream of calpain activation. Calpains are calcium-dependent proteases that can cleave P35 to produce P25, a strong activator of CDK5 [81] or directly truncate GSK3β, leading to its activation [34]. Many studies have also demonstrated that exposure of neurons to Aβ can lead to calcium influx [30, 77, 78] and thus there exists a well-supported molecular mechanism explaining how Aβ exposure can lead to tau hyperphosphorylation. Less clear is the biological rationale for Aβ-induced tau phosphorylation. Is this a selected function of Aβ, an indirect consequence of an evolved interaction of Aβ with cells, or an essentially

* Correspondence: linkc@colorado.edu
[1]Department of Integrative Physiology, University of Colorado at Boulder, Boulder, CO, USA
Full list of author information is available at the end of the article

random interaction with potentially deleterious outcomes?

Addressing "why" Aβ induces tau phosphorylation is complicated by uncertainty as to what, if any, selected function the Aβ peptide has. Similarly unresolved is how extracellular Aβ induces intracellular increases in calcium. Aβ toxicity in numerous models can be attenuated by blocking the NMDA-type glutamate receptor [5, 11, 24], suggesting that this calcium channel could be responsible for Aβ-dependent calcium influx. Similarly, the prion protein has been claimed to be a receptor for Aβ oligomers [7, 44, 74] and to moderate Aβ-dependent calcium influx [67]. Other studies have implicated AMPA glutamate receptors [2, 73, 88] or L-type voltage gated calcium channels [4] in cytoplasmic calcium increases resulting from Aβ exposure. While multiple calcium channels could be involved in Aβ-dependent calcium influx, other studies suggest that Aβ oligomers could act independently of endogenous calcium channels by directly forming a calcium-permeable pore. It has long been known that synthetic Aβ can form ion-permeable channels in synthetic membranes [6], a finding that has often been replicated [29, 41, 71]. Although ring-like structures of Aβ oligomers have been visualized by atomic force microscopy in synthetic membranes [46], Aβ membrane pores in pathologically-relevant cells have not been directly visualized or assayed.

One approach to determining if Aβ pores are relevant to Aβ neurotoxicity is to identify substitutions in the Aβ sequence that block pore formation in synthetic membranes, and then determine if these substitutions alter Aβ toxicity. Kim et al. [41] noted a sequence motif (Gly-XXX-Gly-XXX-Gly) present in both the C terminus of Aβ and the transmembrane domains of bacterial channel proteins, and they proposed that this motif (the "glycine zipper") could facilitate α-helical interactions driving Aβ oligomer assembly in membranes and subsequent pore formation. These researchers demonstrated that leucine substitutions at these glycine residues (particularly $Gly^{37}Leu$) prevented synthetic Aβ 1–42 from forming ion channels in synthetic membranes and also significantly reduced Aβ toxicity in Neuro 2A cells. Fonte et al. [20] extended these results and demonstrated that the $Gly^{37}Leu$ substitution both reduced Aβ toxicity in an in vivo transgenic *C. elegans* model that expresses human Aβ and prevented synthetic Aβ oligomers from inducing tau hyperphosphorylation in hippocampal neurons. Peters et al. [68] subsequently demonstrated that a pentapeptide derived from the glycine zipper sequence (GLMVG) inhibited Aβ synaptotoxicity. These observations are consistent with pore formation underlying Aβ neurotoxicity, but they do not directly demonstrate pore formation, and they cannot exclude other interpretations, such as the possibility that the glycine zipper substitutions interfere with interactions with specific cell surface receptors.

An alternative approach to inferring pore formation by Aβ is to look for its ability to induce a known biological response to exogenously induced pores, rather than attempting to image the pores themselves. Andrews and colleagues have defined a membrane repair pathway that occurs in response to exposure to streptolysin O, (SLO), a well-characterized bacterial pore-forming toxin [32]. As outlined in Fig. 1, this multi-step repair pathway involves: 1) calcium influx through the SLO pore, 2) fusion of local lysosomes with the plasma membrane, releasing

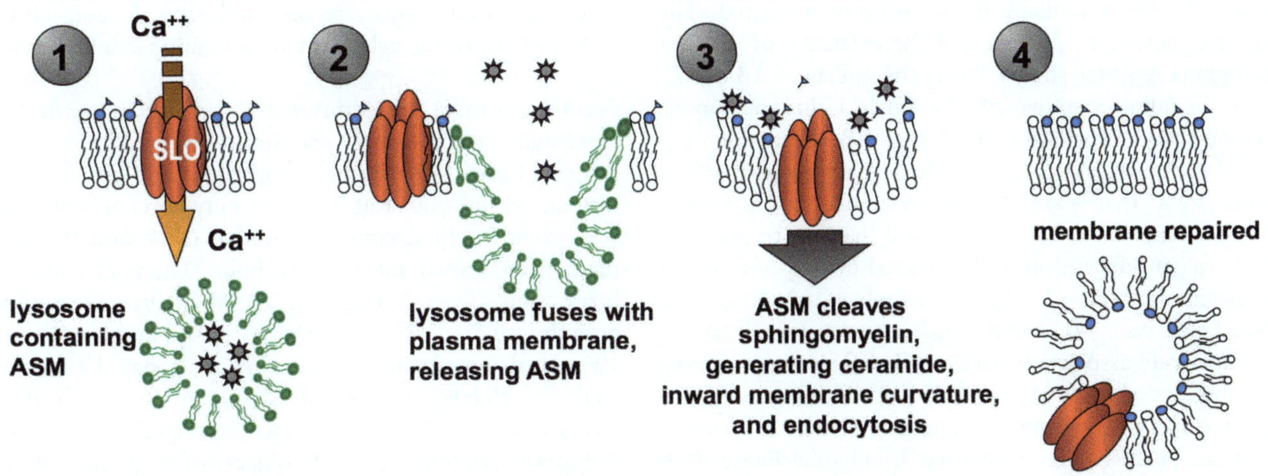

Fig. 1 Membrane repair model. This model is based on the mechanism for repair of plasma membrane (PM) pores created in mammalian cells by exposure to the bacterial pore-forming toxin SLO [32]. ASM is acidic spingomyelinase that is stored in lysosomes, which fuse to the PM in response to an influx of Ca2+ through the pore created by SLO (Step 1), thereby releasing ASM at the cell surface (Step 2). The localized release of ASM cleaves off the phosphoryl head group of sphingomyelinase in the vicinity of the pore to generate ceramide in that area (Step 3). Consequently, the PM around the pore undergoes inward curvature and endocytosis such that the pore is removed from the PM (Step 4)

lysosomal contents, 3) cleavage of nearby sphingolipids in the outer leaflet of the plasma membrane by lysosomally-derived acid sphingomyelinase, thereby inducing localized inward membrane curvature, and 4) removal of the pore-containing plasma membrane by endocytosis. While different membrane repair mechanisms are apparently employed for different classes of pore-forming toxins [19], if Aβ-induced calcium influx results from an SLO-analogous pore, three strong predictions can be made: 1) Aβ exposure will induce sphingomyelinase-dependent endocytosis, 2) non-toxic Aβ variants (e,g., $A\beta_{42}$ $Gly^{37}Leu$) will be incapable of inducing membrane repair because they cannot appropriately oligomerize to form membrane pores, and 3) exposure to pore-forming toxins will mimic the effects of Aβ oligomers, specifically the hyperphosphorylation of tau. Here we test these predictions using a novel *C. elegans* model and primary cultures of rat hippocampal neurons.

Transgenic *C. elegans* strains have been constructed that express human $A\beta_{42}$ [16–18, 47, 82], and these strains have a variety of phenotypes, depending on where the transgene is expressed. A confound of these models is that the detectable Aβ is intracellular when assayed by immunohistochemistry, so the degree of outside-in Aβ toxicity (i.e., extracellular Aβ affecting neighboring cells) is unclear. To circumvent this limitation, we have developed a "feeding" model, where *C. elegans* is fed *E. coli* engineered to secrete human Aβ, and the cellular effects of this exogenous peptide are assayed in intestinal cells. One rationale for this model is that the *C. elegans* intestine does not express candidate Aβ receptors (e.g., prion protein, NMDA glutamate receptors, α7nAChR, etc.), and thus any physiological response to Aβ is unlikely to be receptor-mediated. The transparency of *C. elegans* and the existence of relevant transgenic reporter strains allows the effects of Aβ exposure to be followed in live intact animals. Using this model, we show that exposure to wild type $A\beta_{42}$ (but not $A\beta_{42}$ $Gly^{37}Leu$) induces acid-sphingomyelinase-dependent endocytosis that parallels the response to a known pore-forming toxin, CRY5B. We find that this response to Aβ is calpain-dependent and is altered by loss-of-function mutations in the *C. elegans* orthologs of BIN1 and PICALM, two Alzheimer risk genes identified in genome-wide association studies [61, 84]. In hippocampal cultures, we show that the SLO pore-forming toxin induces calpain-dependent tau phosphorylation in primary neurons. Furthermore, exogenous sphingomyelinase itself can induce increased tau phosphorylation in these neurons. Finally, we use a novel tagging method to show that the $Gly^{37}Leu$ substitution does inhibit Aβ multimerization in a cellular context, thereby rationalizing why this Aβ variant is incapable of inducing membrane repair or tau phosphorylation. Taken together, these results support the view that tau hyperphosphorylation may be a downstream consequence of a membrane repair process, and that exogenous Aβ can induce membrane repair because of its ability to oligomerize into membrane pores.

Materials and methods

C. elegans strains and maintenance

C. elegans strains were generally maintained at 20 °C using standard methods [8]. Strains containing the *rynEx60* transgene were passaged at 25 °C to select for retention of this *pha-1* rescuing extrachromosomal transgene.

C. elegans strains used in this study.

Strain	Genotype
KWN117	*pha-1*(e2123) III; *him-5*(e1490) V; ***rnyEX60*** (*vha-6*::mCherry, *Pmyo-3*:: GFP, *pha-1* rescuing fragment)
GK280	*unc-119*(ed3) III; ***dkIs166***(*Popt-2*::GFP::*pgp-1*; *unc-119* rescuing fragment)
CL6651	*asm-1*(tm5267) II; *pha-1*(e2123) III; *him-5*(e1490) V; ***rnyEX60***
CL6626	*pha-1*(e2123) III(?); *amph-1*(ok3443); *him-5*(e1490) V (?); ***rnyEX60***
CL6605	*unc-11*(e47) I; *pha-1*(e2123) III (?); *him-5*(e1490) V (?); ***rnyEX60***
CL6623	*clp-4*(ok2808) III; *him-5*(e1490) V (?); ***rnyEX60***
RT327	*unc-119*(ed3) III; ***pwIS72***[*Pvha-6*::GFP::*rab-5*, *unc-119* rescuing fragment)

Construction of *E. coli* feeding strains

To engineer *E. coli* strains capable of secreting human Aβ, plasmids were constructed by inserting the wild type or $Gly^{37}Leu$ Aβ sequence between the NcoI and SacI sites of the arabinose-inducible pBAD gIII vector (Invitrogen), yielding pCL241 and pCL242, respectively. These plasmids were transformed into *E. coli* strain LMG194, allowing tight control of arabinose induction.

Feeding protocol and endosome scoring in *C. elegans*

Overnight cultures of *E. coli* strains were diluted 1:100 in Luria broth (+ 100 μg/mL ampicillin) and grown for 2 h in an orbital shaker at 37 °C. Twenty percent arabinose was added to final concentration of 0.2%, and the cultures were grown for another hour. These *E. coli* cultures were then spread on NGM plates (nematode growth media, [8]) containing 0.2% arabinose and 100 μg/mL ampicillin and used the next day. L4 worms were transferred onto the fresh plates and allowed to feed for 4 h at 25 °C. Worms were imaged with a Zeiss Axiophot microscope (40X objective) and endosomes were counted in the anterior 50 μm of the intestine.

Hippocampal neuronal culture

Different protocols were used for the preparation of hippocampal neurons in the Stein and Silverman labs.

Results obtained in replicate experiments using the two different protocols were highly reproducible.

(Stein lab) Hippocampal neurons were isolated from rat E18 hippocampus (BrainBits, LLC) according to the supplier's protocol. The neurons were plated at 10–20,000 cells/cm^2 on Lab-Tek CC2 4-well chamber slides coated with poly-D-lysine (Sigma P0899) and grown at 37 °C. The medium, Neurobasal + 2% B27 and 0.5 mM GlutaMax (Invitrogen), was supplemented with 25 mM glutamate for the first 4 days; thereafter the neurons were fed twice a week with medium lacking glutamate and were used for experiments after 9–16 days in vitro (DIV).

(Silverman lab) Primary hippocampal neuronal cultures from E18 embryonic rats (Charles River, USA) of either sex were prepared as described by Kaech and Banker [35] and kept in PNGM primary neuron growth media (Lonza, Basel, Switzerland). The glial feeder layer was derived from murine neural stem cells as described by [59]. All experiments with animals were approved by and followed the guidelines set out by the Simon Fraser University Animal Care Committee, Protocol 943-B05.

SLO activation and treatment of cells

SLO (Aalto Bio Reagents), a thiol-activated pore-forming toxin that loses activity upon storage at − 70^0 C, was reactivated before each experiment by incubating an aliquot of 3 mg/ml SLO (in Tris-buffered saline, pH 8.5) with an equal volume of 20 mM DTT for 10–15 min at room temperature. Activated SLO and vehicle were diluted with medium and added directly to the cells as 6X concentrates, or half the old medium on the cells was removed and replaced by a 2X concentration of SLO or its vehicle control. The toxicity of each aliquot of activated SLO was evaluated by exposing rat insulinoma cells (RIN5F in RPMI 1640 with 10% Hyclone FBS) to 0–2000 ng/ml SLO for 2 h and measuring viability by the reduction of MTT (Sigma M5655). The results indicated that SLO had little or no gross toxicity at $\leq$ 200 ng/ml and killed half the cells at 600–800 ng/ml. We treated neurons with 50–100 ng/ml SLO for 1–2 h.

Treatment with calpain inhibitor (PD 150606)

PD150606 (Sigma D5946) was dissolved in DMSO, aliquoted and stored at -20° C. It was added to cells at a final concentration of 30 μg.

Treatment with sphingomyelinase

Ten units of lyophilized *B. cereus* sphingomyelinase (SMase, Sigma S7651) were dissolved in 200 μl of cold 50% glycerol in sterile Ca^{2+}/Mg^{2+} free PBS, yielding 50 mU/μl. Because SMase loses activity even when stored at − 70^0 C, it was best used fresh. Neurons were treated with vehicle control and SMase diluted to 2.5 mU/ml in Neurobasal/B27 for 2 h.

Antibodies for immunostaining neurons or *C. elegans*

Primary antibodies used to stain neurons were mouse monoclonal PHF1 to p-Ser396/p-Ser404 tau (kind gift from Dr. Peter Davies), mouse monoclonal AT8 to p-Ser202/p-Thr205 tau (Pierce, Thermo) and rabbit polyclonal K9JA to total tau (Dako A0024). Secondary antibodies (goat anti-rabbit IgG Alexa Fluor 488, goat anti-mouse Alexa Fluor 594 and donkey anti-mouse IgG Alexa Fluor 555) and Prolong Gold Antifade mounting medium were from Invitrogen. The biarsenical dye used was from the TC-FlAsH™ II In-Cell Tetracysteine Tag Detection Kit (Molecular Probes, Eugene, OR). The anti-RME-1 antibody was obtained from the Developmental Studies Hybridoma Bank, and anti-Aβ mouse monoclonal 6E10 was purchased from BioLegend.

Immunocytochemistry

Neurons grown on chamber slides (Stein lab) were fixed for 15 min in 4% p-formaldehyde/4% sucrose, permeabilized 7 min in 0.25% Triton X-100, and blocked 1–2 h with 5% goat or sheep serum (all solutions in PBS). Slides were stained with primary antibodies (K9JA at 1:500, and AT8 or PHF1 at 1:50), secondary antibodies at 1:500, and mounted with ProLong Gold AntiFade. Images were obtained by using a defined scanning pattern to view the total tau fluorescence in sequential fields in each well, capturing images of fields with an average number of cells (typically 5–20 cells per field) using a 10X or 20X objective on a Zeiss Axioskop epifluorescence microscope equipped with 3i Slidebook image analysis software. Identical conditions and exposure times were used to capture an average of 18.5 fields of cells for each treatment presented in our figures, and the ratio of p-tau to t-tau was determined for each image. The mean for each treatment was calculated and compared to its control, and Student's t-test was used to determine if the difference of the means was significant. Similarly, neurons grown on coverslips (Silverman lab) were fixed in 4% paraformaldehyde for 15 min, blocked in 0.5% fish skin gelatin and 0.1%Triton X-100 in PBS, and immunostained with K9JA and AT8 or PHF1. In this case, to quantify tau phosphorylation, histograms were generated using ImageJ from the fluorescence intensity of each pixel across several images, and the average intensity was calculated [20]. Appropriate thresholds were applied to eliminate background signal before histogram analysis, and 17 images per experimental condition were analyzed from at least three independent neuronal cultures.

Preparation of Aβ oligomers

Aβ oligomers were prepared from synthetic Aβ (BioSynthesis) using the "ADDL" preparation originally described by Lambert et al. [43], and subsequently used by us to characterize wild type and Gly^{37}Leu oligomers [20]. Briefly, peptides were solubilized in hexafluoroisopropanol (HFIP) and desiccated in microfuge tubes, then dissolved in fresh, anhydrous DMSO (Sigma Hybri-Max D-2650) to make a ~5 mM solution. This solution was then diluted to ~100 µM with cold F12 media without phenol red (Biosource) and aged 24 h at 4 °C. The samples were centrifuged at 14,000 g for 10 min at 4 °C to remove any insoluble material, and the supernatants stored at 4 °C.

Immunoblotting

Neurons were treated with vehicle, SMase, SLO, and PD 150606 as described above and lysed in RIPA buffer containing Complete Protease Inhibitor Cocktail (Roche) and Halt Phosphatase Inhibitor Cocktail (Thermo Fisher). Samples (10 µg) were resolved on 10% SDS–polyacrylamide gels and transferred to polyvinylidene fluoride (PVDF) membranes. Membranes were incubated with the following primary antibodies overnight at 4 °C: PHF-1 (1:1000), AT8 (1:250), K9JA (1:2000), and anti–tubulin (1:1000). The membranes were imaged using Fujifilm LAS4000 Luminescent Imager. Densitometric scanning and quantitative analysis were carried out using ImageJ.

For assaying Aβ expression in engineered *E. coli* strains, bacterial cultures were spun down in a tabletop centrifuge (3 min at 10,000 rpm), and pellets were frozen at − 80 °C until use. Pellets were solubilized in RIPA buffer supplemented with AEBSF (Sigma P2714) and quantitated using Pierce BCA Protein Assay (Thermo Fisher 23,225). Protein samples were boiled in sample buffer (4% BME in NuPAGE LDS sample buffer (Invitrogen NP0007) and run at 180 V on NuPAGE 4–12%Bis-Tris Gels (Invitrogen, NP0321) using MES SDS Running Buffer (Invitrogen NP0002). Gels were transferred to 0.45 µm supported nitrocellulose (GE Osmonics WP4HY00010) using 20% methanol, 39 mM glycine, 48 mM Tris base at 21 V for 108 min. Prestained Rainbow size markers (Amersham Biosciences RPN755) were used to size bands. Blots were visualized by Ponceau stain, boiled for 3 min in PBS, blocked in TBS-Tween + 5% milk (100 mM Tris7.5, 150 mM NaCl, 0.1% Tween-20) and probed with Aβ antibody 6E10 (BioLegend 803,002) at a 1:1000 dilution. Secondary HRP-conjugated antibody (Sigma A5906 mouse) was used, and the blot was developed in ECL Plus (Amersham RPN2132).

Treating hippocampal neurons with Cys-tagged Aβ peptides

Rat hippocampal neurons (cultured in the Stein lab) were treated with wild-type and Cys-tagged Aβ peptides (synthesized by Bio Synthesis) and exposed to the FlAsH dye according to manufacturer's instructions with modifications [TC-FlAsH™ II In-Cell Tetracysteine Tag Detection Kit (Molecular Probes)]. Briefly, rat hippocampal neurons were treated to 10 µM EDTA to suppress background fluorescence for 10 min, and were then exposed to 2.5 µM Cys-tagged Aβ peptides or wild-type Aβ peptides for 1 h at 37 °C. FlAsH dye was then applied to neurons at 1:800 for 30 min. BAL wash buffer was used to remove the excess of the FlAsH dye and was replaced by warm HBSS after 15 min. Neurons were then immediately fixed and prepared for immunohistochemistry.

Microscopy

A Zeiss Axiophot microscope equipped with digital deconvolution optics and a Nikon Structured Illumination Super-resolution (Light Microscopy Core Facility in Department of Molecular, Cellular and Developmental Biology, University of Colorado Boulder) were used for imaging *C. elegans* and rat hippocampal neurons. Images generated in the Silverman lab were acquired using a Leica DMI6000B inverted epifluorescence microscope using a 63 X 1.4 N.A. oil-immersion objective equipped with a cooled CCD camera controlled by MetaMorph (Molecular Devices).

Statistical analysis

A one-tailed paired two sample for mean t-test, was used to determine significance between pair wise comparisons of control and experimental conditions. Log rank statistics were used to analyze survival curves in the CRY5B exposure experiments (Fig. 5b). Significant differences between the treatments were analyzed by t-tests with equal or unequal variance at a 95% confidence interval. The experiments were carried out in three different cultures and at least 60–70 cells were analyzed per condition. For the blots, cell lysates were collected from at least 3 cultures.

Results

Aβ induces a membrane repair process in *C. elegans*

To assay the ability of human Aβ to induce membrane repair in an in vivo model, we took advantage of previous work by Aroian and colleagues, who studied the effects of the CRY5B toxin on *C. elegans* [31, 38, 85]. The CRY5B protein is a member of the "Bt crystal" collection of pore-forming toxins secreted by *Bacillus thuringiensis*. To visualize the effects of CRY5B on *C. elegans* intestinal cells, these researchers used a transgenic reporter strain that expresses a fluorescent fusion protein, specifically localized to the lumen of the intestine, to monitor endocytosis following exposure to an *E. coli* strain engineered to express CRY5B. We have replicated these experiments (Fig. 2a, right top panel), and as previously reported [52], exposure to CRY5B results in a significant induction of endocytosis, a component of the intestinal membrane repair process. In parallel, we engineered *E. coli* to secrete Aβ 1–42, using the arabinose-inducible vector pBAD gIII,

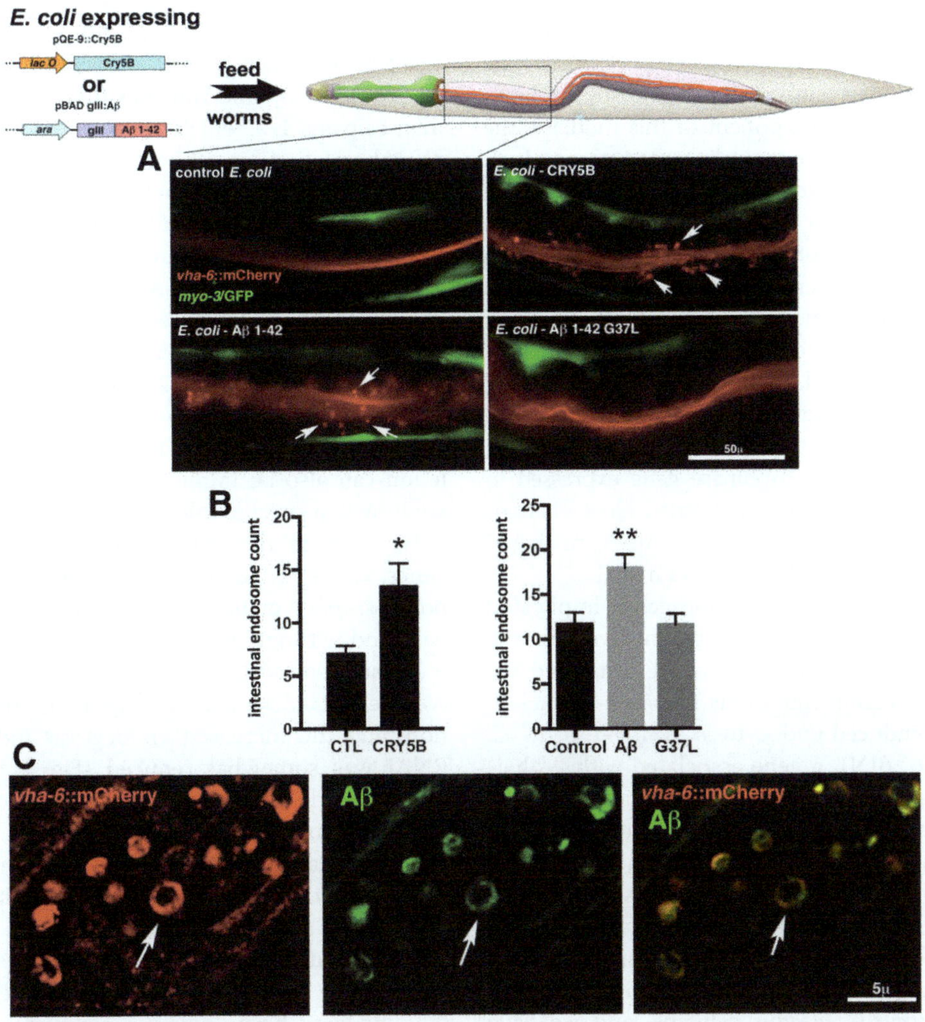

Fig. 2 Induction of intestinal endosomes in *C. elegans* fed *E. coli* that express human Aβ peptide. **a** Live images of *C. elegans* reporter strain KWN117 (*vha-6*::mCherry) fed *E. coli* expressing either vector control, CRY5B toxin, wild type $A\beta_{1-42}$, or $A\beta_{1-42}$ Gly^{37}Leu. Note that both CRY5B and wild type $A\beta_{1-42}$ induce intestinal endosomes (arrows) containing the *vha-6*::mCherry reporter usually associated with the lumenal membrane of the intestine. This induction does not occur in worms fed *E. coli* expressing the non-pore-forming $A\beta_{1-42}$ Gly^{37}Leu variant. **b** Quantification of endosome induction. **c** Super resolution image of intestinal endosomes induced in strain KWN117 by wild type $A\beta_{1-42}$. The treated worm was subsequently fixed, permeabilized, and probed using anti-Aβ antibody 6E10. Note that the Aβ staining co-localizes with the membrane-associated *vha-6*::mCherry reporter (arrow), not the endosomal lumen. $^{*}p < 0.05$ and $^{**}p < 0.01$ when compared with the vehicle control

and exposed reporter strain KWN117 (*vha-6*::mCherry) to those induced *E. coli* cultures. As shown in Fig. 2a, this resulted in an induction of endocytosis similar to that observed in the CRY5B-exposed worms. The increased intestinal vesicles observed in these experiments could be labeled with endosomal marker RAB-5::GFP or immunostained with antibodies against the endosomal RME-1 protein, demonstrating that these vesicles are indeed endosomes (Additional file 1: Figure S1 A, B). Importantly, this induction of endocytosis was not observed in KWN117 worms exposed to an *E. coli* strain expressing the Gly^{37}Leu variant of Aβ, which we have previously shown to be non-toxic in mammalian neurons [20]. (The engineered *E. coli* strain expressing the $A\beta_{42}$ Gly^{37}Leu variant actually produces higher levels of Aβ peptide than the $A\beta_{42}$ wild type strain when assayed by immunoblot, Additional file 1: Figure S1C).

If the intestinal endosomes induced by exposure to Aβ-expressing *E. coli* were in fact a result of a membrane repair process similar to that proposed by Andrews and colleagues (Fig. 1), our expectation was that these endosomes would contain membrane-associated Aβ that had been endocytosed from the luminal plasma membrane. We therefore fixed and permeabilized KWN117 reporter worms fed *E. coli* expressing Aβ, then immunostained with an Aβ-specific antibody (mAB 6E10). Super-resolution microscopy revealed cytoplasmic rings in intestinal cells that contained both the VHA-6::mCherry luminal marker and

Aβ (Fig. 2b). We interpret these rings as cross-sections of endosomal vesicles containing both membrane-associated Aβ and VHA-6::mCherry, consistent with the pore-removal-by-endocytosis model proposed by Andrews and colleagues. A critical component of this model is the initial exocytosis of acid sphingomyelinase, a lysosomal enzyme that when released extracellularly can cleave sphingomyelin in the outer leaflet of the plasma membrane lipid bilayer. This conversion of sphingomyelin to ceramide can lead to inward membrane curvature, which is thought to promote local endocytosis (Fig. 1). To test further whether the *C. elegans* response to Aβ exposure paralleled the Andrews model, we repeated the feeding experiments using a derivative of the *vha-6*::mCherry reporter strain that contained the *asm-1 (tm5627)* mutation, a deletion allele of a sphingomyelinase gene expressed in the intestine. We observed that loss of *asm-1* blocked endosome induction by feeding with CRY5B or Aβ-expressing *E. coli* (Fig. 3), again supporting the Andrews model.

To explore the possibility that some genes implicated in Alzheimer's disease might also be involved in the Aβ-induced membrane repair process, we examined the effects of loss-of-function mutations in *amph-1, unc-11,* and *clp-4* on Aβ-induced endocytosis. *Amph-1* is the *C. elegans* ortholog of BIN1, a gene associated with Alzheimer's risk in multiple genome-wide association studies [10, 75]. BIN1 is a member of the Bin1/amphiphysin/RVS167 (BAR) family of proteins that are involved in diverse cellular processes, including endocytosis. *Unc-11* is orthologous to PICALM, another well-established late onset Alzheimer's disease (LOAD) risk gene known to play a role in clathrin-mediated endocytosis [40]. *clp-4* encodes one of nine *C. elegans* calpains, and we have previously noted that this specific calpain is upregulated both in a transgenic *C. elegans* model of Aβ toxicity [26] and in worms exposed to CRY5B [31]. The CLP-4 calpain also activates the *C. elegans* CDK5 ortholog via cleavage of p35 to p25 [58], and thus has functions analogous to human calpains 1/2, which are required for Aβ-induced tau phosphorylation via p25 activation of CDK5 [50, 81]. As shown in Fig. 4a, introduction of either an *amph-1* deletion allele or an *unc-11* null allele (*e47*) into the *vha-6*::mCherry reporter background led to a significant increase in endosomes induced by Aβ feeding. (Similar results were observed when these genes were knocked down by RNAi, or if CRY5B exposure was used to induce endocytosis, Additional file 1: Figure S2 A, B.) In contrast, introduction of a *clp-4* deletion allele blocked the increase of endosomes induced by Aβ feeding. The effects of the *clp-4* deletion can also be mimicked by pre-treating the reporter strain with a specific calpain inhibitor, PD150606 (Fig. 4b).

To examine the epistatic relationship between these genes, we repeated the Aβ feeding assay after knocking down *amph-1* or *unc-11* by RNAi in the reporter strain, with and without the *clp-4* mutation. We found that the increased endocytosis induced by *amph-1* RNAi was completely blocked by the *clp-4* loss-of-function mutation, and the increased endocytosis induced by *unc-11* RNAi was somewhat reduced (Fig. 4c). Thus, *clp-4* is fully epistatic to *amph-1* and may moderate *unc-11* in the membrane repair process.

If *clp-4* plays a general upstream role in membrane repair, loss of *clp-4* should also block induction of endocytosis by CRY5B, and should be deleterious when *C. elegans* is challenged by this pore-forming toxin. As observed for Aβ feeding, introduction of the *clp-4* deletion allele into the *vha-6*::mCherry reporter completely blocked endosome induction by CRY5B feeding (Fig. 5a). CRY5B

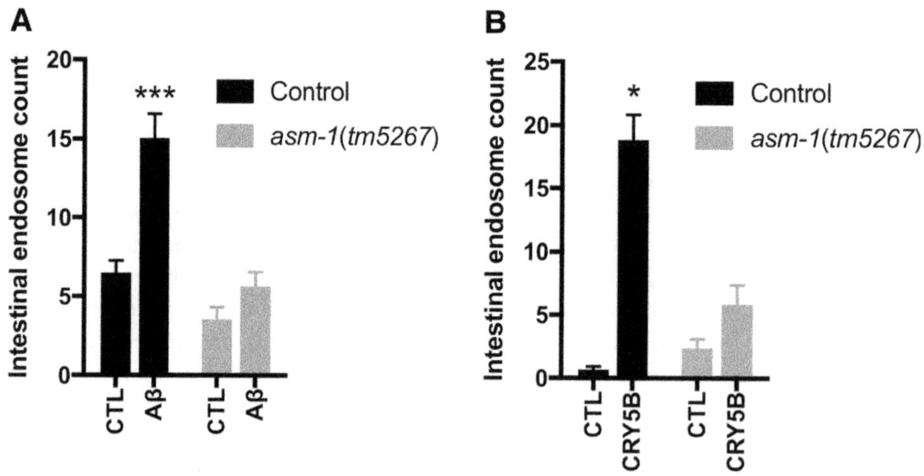

Fig. 3 Induction of intestinal endosomes is blocked by a mutation in the acid sphingomyelinase gene *asm-1*. **a** Introduction of the *asm-1*(tm5267) deletion allele into reporter strain KWN117 blocks the ability of *E. coli* expressing wild type $A\beta_{1-42}$ to induce intestinal endosomes. **b** The *asm-1*(tm5267) mutation also blocks endosome induction by the CRY5B toxin. *$p < 0.05$ and ***$p < 0.001$ when compared with the vehicle control

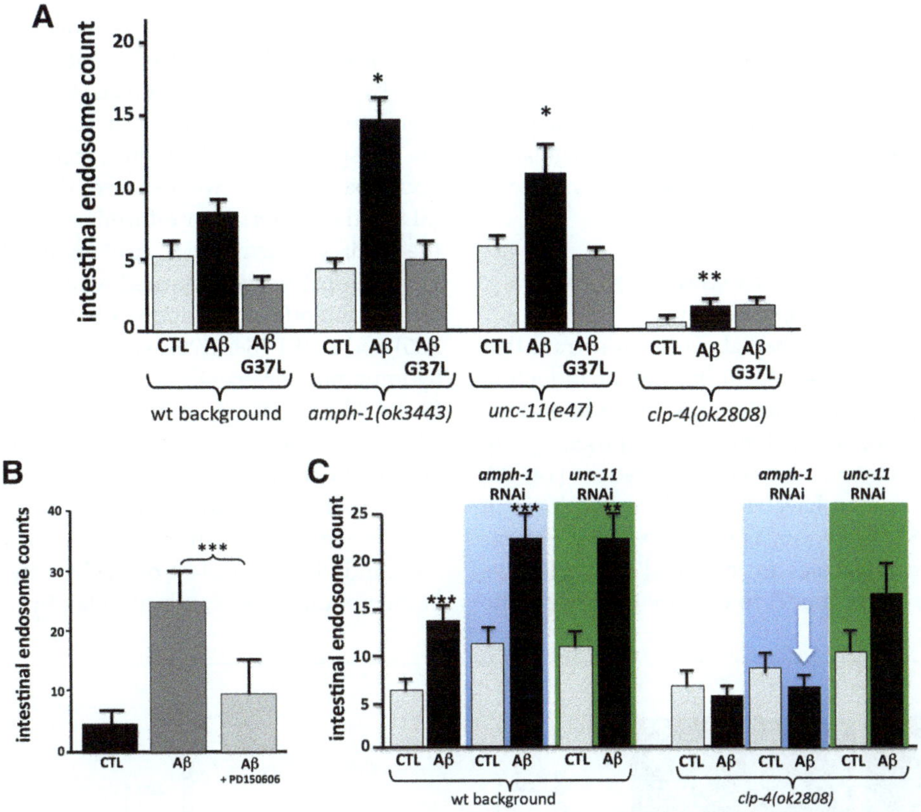

Fig. 4 Effects of *amph-1*, *unc-11* and *clp-4* mutations on Aβ-induced intestinal endosomes. **a** Strains containing the *vha-6*::mCherry reporter and deletion mutations in *amph-1*, *unc-11*, or *clp-4* were fed *E. coli* expressing wild type $A\beta_{1-42}$. Note that the *amph-1* and *unc-11* mutations increase, while the *clp-4* mutation decreases, the accumulation of intestinal endosomes. **b** Treatment with the calpain inhibitor PD150606 replicates the ability of the *clp-4* mutation to block Aβ-induced increases in intestinal endosomes. **c** RNAi treatments indicate the *clp-4* deletion mutation is epistatic to the effects of *amph-1* and *unc-11* knockdown. *$p < 0.05$, **$p < 0.01$ and ***$p < 0.001$ when compared with the vehicle control

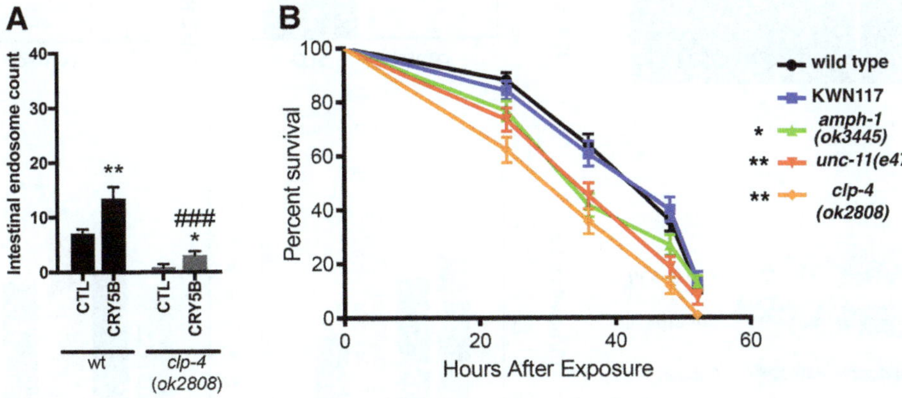

Fig. 5 *clp-4* deletion blocks endosome induction by CRY5B and sensitizes worms to CRY5B toxicity. **a** Induction of endosomes induced by CRY5B is blocked by the *clp-4*(ok2808) deletion mutation. *$p < 0.05$ when compared with the vehicle control. ###$p < 0.001$ when compared with the wt background. **b** Survival curves of *C. elegans* strains exposed to *E. coli* expressing CRY5B. Note that all three mutations that alter endosome accumulation also sensitize worms to CRY5B toxicity. *$p < 0.05$, **$p < 0.01$ compared to wild type control. NS = not significant

exposure can be lethal in *C. elegans*, with most worms dying within 2 days [85]. We found that loss of *clp-4* significantly reduced the survival rate in worms exposed to CRY5B (Fig. 5b). Interestingly, both the *amph-1* deletion and the *unc-11* mutation also significantly decreased survival after CRY5B exposure, suggesting that dysregulation of the membrane repair process may generally sensitize *C. elegans* to this toxin.

Induction of membrane repair increases tau phosphorylation in hippocampal neurons

The *C. elegans* studies described above demonstrate that Aβ exposure can lead to a calpain-dependent induction of membrane repair. Multiple groups, including ours, have shown that hippocampal neurons exposed to Aβ oligomers show increased tau phosphorylation [14, 20, 76], and previous studies have also implicated calpain in this process [42, 62]. However, as *C. elegans* does not express any tau-like molecules in intestinal cells [21], we could not use the worm model to directly investigate the possibility that tau hyperphosphorylation is a downstream consequence of Aβ-induced membrane repair. We thus turned to cultured hippocampal neurons to test this hypothesis, which predicts that insults that induce membrane repair should also induce calpain-dependent tau phosphorylation. To induce membrane damage in hippocampal neurons we exposed them to streptolysin O (SLO), the bacterial pore-forming toxin used to establish the Andrews membrane-repair model. We observed that hippocampal neurons treated with 50 ng/ml SLO had increased tau phosphorylation at both the AT8 (Ser 202/Thr 205) and PHF1 (Ser 396/Ser 404) epitopes, assayed either by immunofluorescence (Fig. 6a) or immunoblot (Fig. 6b) (See Additional file 2, Individual Experiment Data). Critically, SLO induction of tau hyperphosphorylation was blocked by treatment with the calpain inhibitor PD150606 (Fig. 6b), demonstrating that tau hyperphosphorylation in response to a pore-forming toxin was calpain-dependent.

As an independent test of the hypothesis that activation of membrane repair can lead to tau hyperphosphorylation,

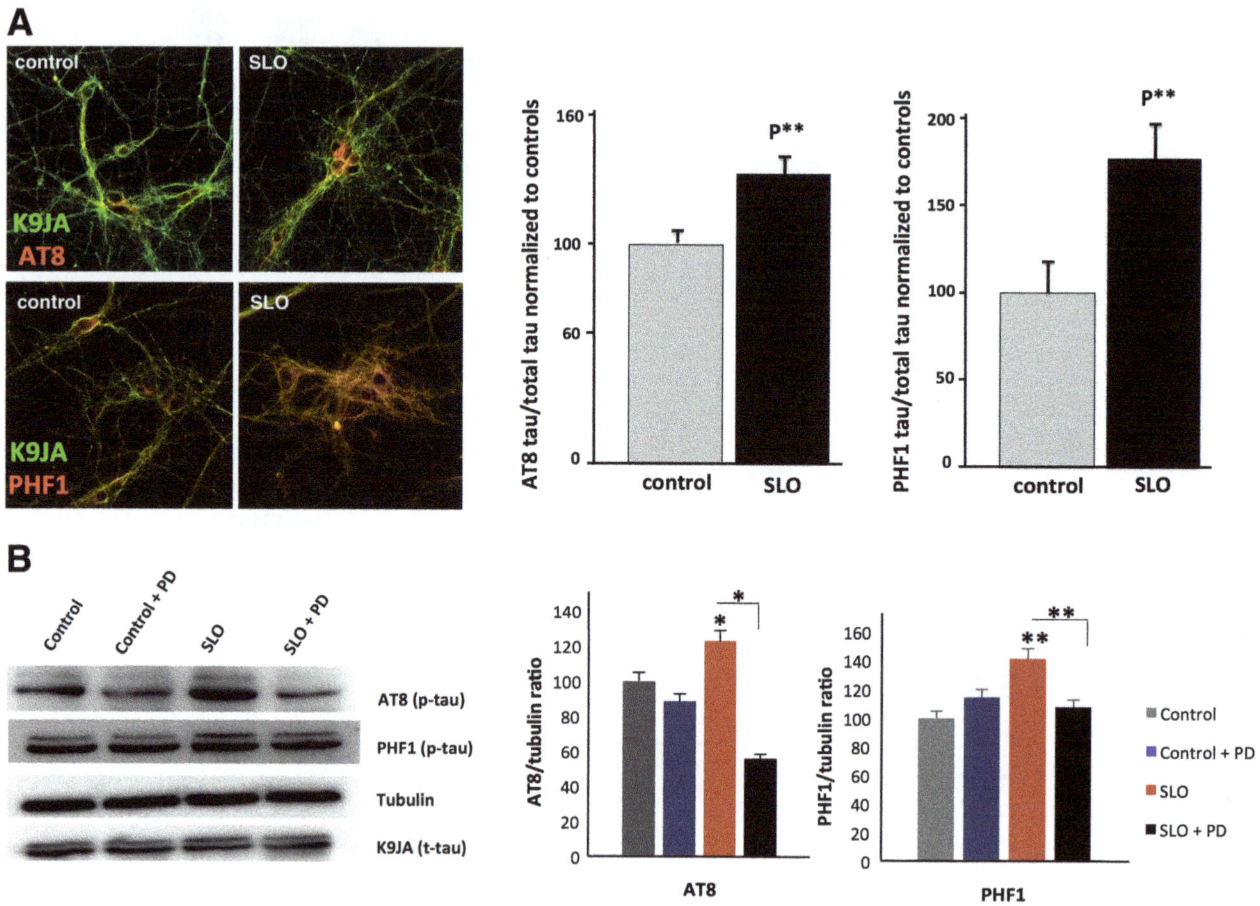

Fig. 6 Treatment of hippocampal neurons with streptolysin O (SLO) increases tau phosphorylation. **a** Treatment of hippocampal neurons with SLO increases tau immunoreactivity at the AT8 and PHF1 phospho-epitopes, assayed by quantification of immunofluoresence images and normalized to total tau. **b** Treatment of hippocampal neurons with SLO increases tau immunoreactivity at the AT8 and PHF1 phospho-epitopes, assayed by quantification of immunoblots. *$p < 0.05$, **$p < 0.01$ and ***$p < 0.001$ when compared with the vehicle control

we sought to artificially stimulate this process by exposing hippocampal neurons to exogenous sphingomyelinase, a treatment that has previously been shown to promote the membrane repair process in HeLa cells [79]. We observed that hippocampal neurons treated with 2.5 mU *B. cereus* sphingomyelinase had increased tau phosphorylation at both the AT8 and PHF1 epitopes, assayed either by immunofluorescence (Fig. 7a) or immunoblot (Fig. 7b).

Visualization of toxic Aβ oligomers using fluorescent biarsenical labeling

The ability of Aβ to form ion-permeable pores in synthetic membranes is blocked by the Gly[37]Leu substitution, an observation that supports the "glycine zipper" model of Aβ multimerization [41]. We have previously shown that Aβ Gly[37]Leu oligomers cannot induce tau hyperphosphorylation in hippocampal neurons [20], and we show in the present study that this substitution prevents the induction of membrane repair in a *C. elegans* model. These observations are consistent with the idea that the Gly[37]Leu substitution interferes with the assembly of Aβ multimers capable of inducing membrane repair and subsequent downstream tau hyperphosphorylation, even though they do not directly demonstrate that this substitution interferes with Aβ multimerization per se. To demonstrate this directly, we sought a means to label multimeric Aβ specifically by taking advantage of the way that biarsenical compounds interact with appropriately positioned Cys residues. The ability of pairs of di-cysteine residues to coordinate the binding of biarsenical small molecules can be used to fluorescently label proteins tagged with a tetra-cysteine motif (FlAsH and ReAsH labeling, [1]). Subsequent studies demonstrated that di-cysteine residues located in different protein domains

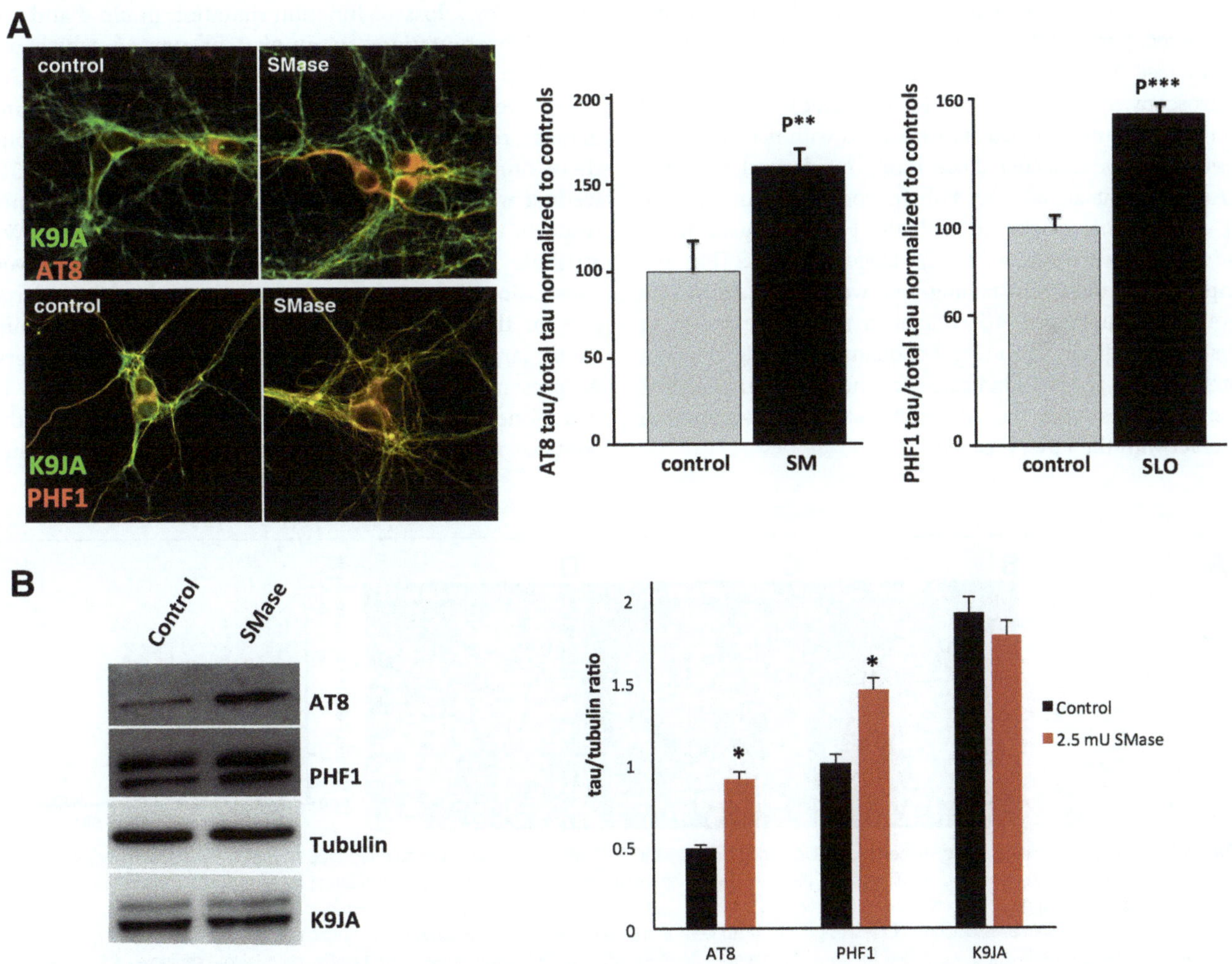

Fig. 7 Treatment of hippocampal neurons with exogenous bacterial sphingomyelinase increases tau phosphorylation. **a** Cultured hippocampal neurons were treated with 2.5 mU/ml *B. cereus* sphingomyelinase, and tau immunoreactivity at the AT8 and PHF1 phospho-epitopes was quantified by immunofluorescence (normalized to total tau). **b** Cultured hippocampal neurons were treated with 2.5 mU *B. cereus* sphingomyelinase, and tau immunoreactivity at the AT8 and PHF1 phospho-epitopes was quantified by immunoblot (normalized to total tau). $^{*}p < 0.05$, $^{**}p < 0.01$ and $^{***}p < 0.001$ when compared with the vehicle control

can also bind biarsenical fluorescent compounds if the di-cysteine residues are appropriately positioned in the protein [53]. The glycine zipper model predicts that α-helical interactions in a lipid environment will pair Aβ monomers such that their C-termini will be closely opposed, potentially allowing binding of biarsenical fluorescent probes (Fig. 8a).

We synthesized both wild type and Gly^{37}Leu Aβ 1–42 tagged with two cysteine residues at the C-terminus. The addition of the C-terminal cysteines did not prevent wild type Aβ from inducing tau phosphorylation in hippocampal neurons (Additional file 1: Figure S3 A). Furthermore, *E. coli* engineered to express the two-Cys version of wild type, but not Gly^{37}Leu, Aβ, induced endocytosis in the *C. elegans* model (Additional file 1: Figure S3B). We exposed hippocampal neurons to these Cys-tagged peptides using the same protocols we used previously to assay Aβ induction of tau. After 1 h exposure to Aβ oligomer preps, the cultured neurons were stained with 1:800 FlAsH dye for 30 min, then fixed, permeabilized and immunostained for Aβ (mAb 6E10) (see Methods). When imaged by super-resolution microscopy, neurons exposed to (untagged) Aβ 1–42 had foci of Aβ immuno-reactivity associated with neuronal processes, but as expected these rarely overlapped with the FlAsH reagent signal (Fig. 8b). In contrast, a number of Aβ-positive foci co-stained with the FlAsH reagent when the neurons were exposed to Cys-tagged Aβ 1–42 (Fig. 8c). Importantly, these co-staining-foci were reduced to the level of the untagged Aβ control in neurons exposed to Cys-tagged Aβ Gly^{37}Leu (Fig. 8d, quantified in Fig. 8e). We interpret this result to indicate that the Gly^{37}Leu substitution interferes with the Aβ multimerization required to interact with the FlAsH reagent.

Discussion

We have developed a novel *C. elegans* model to investigate the cellular response to exogenous Aβ, and we observe an induction of endocytosis similar to that associated with membrane repair in response to exposure to the CRY5B pore-forming toxin [31, 38, 52, 85]. The observations that Aβ induction of endocytosis is blocked by a mutation in a gene encoding acid sphingomyelinase and the apparent membrane association of Aβ in induced endosomes also support the view that exogenous Aβ induces a membrane repair process as described in Fig. 1. To link membrane damage/repair with a classic marker of AD pathology, we also show that hippocampal neurons exposed to either a pore-forming toxin (streptolysin O) or exogenous sphingomyelinase display increased tau hyperphosphorylation comparable to that seen after exposure to Aβ. In addition, the increase in endosomes induced by either Aβ or CRY5B is inhibited by a loss-of-function mutation in *clp-4* and enhanced by mutations in *amph-1* or *unc-11*, which are orthologs of human LOAD genes BIN1 and PICALM, respectively. These results further tie the membrane damage/repair process to Alzheimer's disease. Despite these similarities between the effects of Aβ-and CRY5B, feeding wild type worms Aβ-expressing *E. coli* does not result in the lethality observed in worms fed CRY5B. We attribute this difference to CRY5B being an evolved "professional" pore-forming toxin, assembled by a large protein that produces a pore size likely to be significantly larger [70] than a pore formed by Aβ oligomers. At present we cannot determine if the effects of Aβ in this model are due solely to the direct interaction of the peptide with the intestinal membrane, or whether

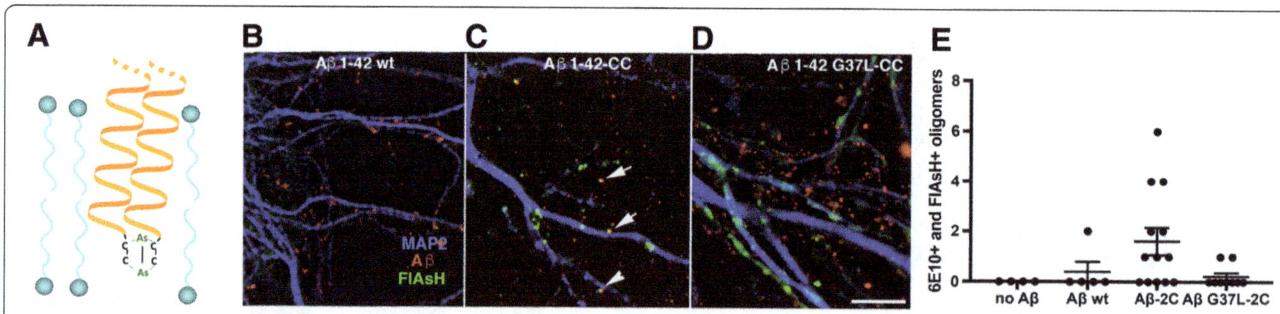

Fig. 8 Biarsenical dye staining of dicysteine-tagged synthetic wild type and Gly^{37}Leu variant Aβ$_{1–42}$ in cultured hippocampal neurons. **a** Schematic model of how membrane-associated Aβ dimers in a parallel α-helical arrangement could bind the biarsenical FlAsH reagent. **b** Super-resolution image of cultured hippocampal neurons exposed to synthetic wild type Aβ$_{1–42}$, treated with FlAsH reagent, fixed, permeabilized, and probed with anti-Aβ antibody 6E10. Note minimal association of FlAsH signal with Aβ immunoreactivity, expected because this synthetic peptide does not contain dicysteines. **c** Same experiment as described for panel "B", except treatment with synthetic dicysteine-tagged wild type Aβ. Note multiple foci of co-localized FlAsH and Aβ staining (arrows). **d** Same experiment as described for panel "B", except treatment with synthetic dicysteine-tagged Gly^{37}Leu Aβ. Note this substitution prevents the formation of co-staining foci, supporting the hypothesis that the Gly^{37}Leu substitution inhibits the assembly of (potentially pore-forming) multimers. (The increased neuronal process-associated FlAsH signal in the Gly^{37}Leu Aβ -treated cultures may reflect increased non-specific uptake of the FlAsH dye, because neurons treated with the non-toxic Gly^{37}Leu Aβ are healthier than neurons treated with wild type Aβ peptides.) **e** Quantification of co-labeling foci from multiple image fields acquired in the experiments described in **b-d**

proteins in the lumenal membrane are playing a role. We can conclude that suggested Aβ receptors (e.g., prion protein, NMDA glutamate receptors, α7nAChR) are highly unlikely to be involved, as there is no evidence these proteins are expressed in the *C. elegans* intestine.

The ability of the *clp-4* deletion to prevent endocytosis induced by Aβ or CRY5B implicates calpain activity in the membrane repair process. Calpains are known to target a number of cytoskeletal proteins [22], and to moderate cytoskeletal rearrangement, for example in focal adhesion remodeling [45]. Interestingly, in macrophages, calpain 2 cleavage of talin 1 is critical for endocytosis of the pore-forming "protective antigen" toxin produced by *Bacillus anthracis* [33]. We envision calpain activation as a conserved process required to induce local cytoskeletal changes that promote endocytosis in response to membrane damage.

The increased accumulation of Aβ-induced intestinal endosomes in the *amph-1* and *unc-11* mutants could result from either increased endosome formation or reduced endosome disassembly (or both). Loss of *amph-1* alters the distribution of the early endosome marker RAB-5 and leads to increases in the size of steady-state endosomes in the *C. elegans* intestine [49]. We note that an increase in the number and size of RAB-5 endosomes in AD brains has long been recognized [12]. BIN1, the ortholog of *amph-1*, has been implicated in the regulation of endocytosis in multiple contexts [65, 83, 86], even though its specific role in Alzheimer's disease is unclear. The rs59335482 risk allele, a 3 bp insertion upstream of the BIN1 coding sequence, increases BIN1 transcription and is correlated with tau but not β-amyloid loads in AD brains [13]. This study also reported that loss of *Amph* (the Drosophila ortholog of BIN1) suppressed tau pathology, but not Aβ pathology, in transgenic fly models based on ectopic expression of these AD-associated proteins in the fly eye. However, studies in mammalian neurons have indicated that loss of BIN1 promoted the propagation of tau pathology [9]. Our studies cannot resolve the discrepancies between these reports, but they do establish that *amph-1* in *C. elegans* (and potentially BIN1 in mammals) play a role in the membrane repair response to pore-forming toxins.

unc-11 encodes a *C. elegans* clathrin adaptor protein orthologous to AP180 and PICALM, two closely related mammalian proteins. The *unc-11* gene is expressed at high levels in neurons and at lower levels in other tissues, including the intestine. Interestingly, null alleles of *unc-11* do not prevent endocytosis of synaptic membrane, but do result in an increased size of synaptic endosomes [63]. Similarly, RNAi knockdown of PICALM in mammalian neurons leads to synaptic vesicles with variably increased size [69], suggesting that these proteins primarily regulate the sorting and recycling of endosomes rather than endocytosis per se. Thus, the increase of endosome number (and size) in the intestines of *unc-11* mutants exposed to CRY5B or Aβ likely results from altered endosome sorting or disassembly rather than upregulated endocytosis. Most studies support the view that PICALM expression is protective in AD, as full-length PICALM has been reported to be reduced in AD brain [3], and the protective allele of the primary AD-related SNP (rs3851179) is reported to increase PICALM mRNA levels in the brain [66].

The involvement of BIN1 and PICALM in endocytic processes has led multiple research groups to investigate their possible role in the production of Aβ, which largely occurs during the endocytic recycling of Amyloid Precursor Protein (APP) [56]. Indeed, loss of BIN1 has been reported to increase the production of Aβ [3, 60], whereas knockdown of PICALM is reported to reduce Aβ production [37, 80]. If, as described above, the risk alleles of BIN1 and PICALM lead to increased and decreased expression of these two genes, respectively, then the reported effects of these genes on Aβ production are actually the opposite of what might be expected. Our studies explore an alternative (but not mutually exclusive) explanation for the association of these genes with AD risk: a role in the cellular response to extracellular Aβ. We find that loss of function mutations in *amph-1* or *unc-11* sensitize worms to the CRY5B pore-forming toxin (Fig. 5b), and lead to similar dysregulation of the endocytosis induced by CRY5B or Aβ. These results are consistent with BIN1 and PICALM risk alleles acting by modulating cellular responses to extracellular Aβ.

The ability of *asm-1* mutations to suppress Aβ- or CRY5B-induced endocytosis supports the view that Aβ can instigate a membrane repair process analogous to that occurring in mammalian cells challenged by a pore-forming toxin such as SLO [32]. We reasoned that if Aβ acts as a pore-forming toxin in the AD brain, known effects of Aβ exposure such as tau hyperphosphorylation might be replicated using pore-forming toxins. We find that exposure of rat hippocampal neurons to SLO results in calpain-dependent tau hyperphosphorylation at two epitopes classically associated with AD pathology. Although we cannot exclude the possibility that SLO mimics Aβ-induced tau phosphorylation by an unrelated pathway, the ability of exogenous sphingomeylinase to induce tau hyperphosphorylation supports the view that tau hyperphosphorylation can be a downstream consequence of membrane damage/repair. Attempts to induce endocytosis in *C. elegans* intestines by feeding worms *B. cereus* sphingomeylinase were unsuccessful, possibly due to reduced activity of the enzyme at the lower temperature required for worm maintenance (20 °C for *C. elegans* vs. 37 °C for cultured neurons) and/or degradation of the enzyme in the intestinal lumen. Our experiments examining the effects of exogenous sphingomeylinase on tau hyperphosphorylation also cannot determine if sphingomeylin-based secondary messengers are playing a role. Similarly, determining whether tau phosphorylation is a functional component of membrane repair or simply an incidental consequence will require additional investigation.

A major observation supporting the possible relevance of toxic Aβ pores is the dramatically reduced toxicity of the Aβ Gly^{37}Leu variant in both transgenic *C. elegans* models and mammalian neurons [20]. This variant, which unlike wild type Aβ cannot induce ion-permeable channels in synthetic membranes, was investigated by the Bowie lab based on modeling studies that suggested it could not assemble pore-forming oligomers due to interference with a "glycine zipper" motif [41]. However, it had not been demonstrated previously that the Gly substitution in this critical variant actually alters Aβ multimerization, and in fact the Gly^{37}Leu substitution does not reduce the stable oligomer species assayed by SDS-PAGE (see Additional file 1: Figure S1C). We therefore sought an approach to assay Aβ multimerization in vivo that could capture potentially less stable, membrane-associated oligomers. Using hippocampal neurons exposed to dicysteine-tagged synthetic Aβ and detection of closely associated pairs of dicysteine tags by means of a membrane-permeant biarsenical dye, we provide evidence that the Gly^{37}Leu substitution does indeed inhibit Aβ multimerization in a cellular context. We suggest that the Aβ Gly^{37}Leu variant can be used as a control peptide for investigating Aβ oligomer effects and could be superior to typical Aβ sequence-scrambled peptides used for this purpose.

More than 20 years ago, the discovery by Arispe and colleagues that Aβ could form ion channels in synthetic membranes led them to propose that Aβ membrane pores could contribute to Alzheimer pathology [6]. While multiple studies since have been supportive [36], this hypothesis has neither been directly supported (e.g., by the visualization of Aβ pores in AD pathological tissue), nor convincingly discounted. The failure to visualize Aβ pores in AD brains may not be surprising given the rapidity with which membrane pores are removed in mammalian cells (< 1 min in HEK293 cells; [32]), whereas the consequences of the repair process, such as hyperphosphorylation of tau, might accumulate. Our studies cannot directly prove the disease relevance of the pore-forming capacity of Aβ, but they do suggest a possible re-interpretation of existing human data. In particular, the demonstration of tau hyperphosphorylation as a downstream consequence of membrane damage/repair suggests mechanisms that may also be in play in non-AD tauopathies such as Chronic Traumatic Encephalopathy (CTE) and Niemann Pick type C disease [64]. Multiple explanations have been advanced to explain the dramatically increased AD risk for ApoE4 allele carriers (reviewed in [48]); our results support the idea that this risk could be due to altered membrane repair resulting from the reduced cholesterol and phospholipid secretion observed in APOE4 glia and neurons [54].

Acknowledgements

This work was supported by US National Institutes of Health award R21 AG049693 to CDL and Natural Science and Engineering Research Council of Canada (327100) and the Canadian Institutes of Health Research (90396) awards to MAS. We would like to thank Andrew Gaines and Josh Morgenstern for assistance with pilot experiments. Some nematode strains were provided by the *Caenorhabditis* Genetics Center, funded by the NIH National Center for Research Resources. We would also like to thank Dr. Raffi Aroian for providing the CRY5B expression vector.

Authors' contributions

CJ designed and performed both *C. elegans* and hippocampal neuron experiments, and contributed to the writing of this manuscript. CT performed *C. elegans* experiments, including the CRY5B survival studies. CMR constructed and characterized the E. coli strains expressing human Aβ. AA performed hippocampal neuron experiments, including tau immunoblots. GHS designed and performed hippocampal neuron experiments and contributed to the writing of this manuscript. MAS and CDL contributed to experimental design and the writing of the manuscript. All authors have read and approved of this manuscript.

Competing interests

The authors declare that they have no competing interests.

Author details

[1]Department of Integrative Physiology, University of Colorado at Boulder, Boulder, CO, USA. [2]Department of Biological Sciences, and the Centre for Cell Biology, Development, and Disease, Simon Fraser University, Burnaby, British Columbia, Canada. [3]Department of Molecular, Cellular, and Developmental Biology, University of Colorado, Boulder, Colorado 80309-0347, USA. [4]Centre de Recherche en Sciences Animales de Deschambault (CRSAD), Deschambault, Quebec G0A 1S0, Canada. [5]Centre for Cell Biology, Development, and Disease), Simon Fraser University, British Columbia V5A 1S6, Burnaby, Canada.

References

1. Adams SR, Campbell RE, Gross LA, Martin BR, Walkup GK, Yao Y, Llopis J, Tsien RY (2002) New biarsenical ligands and tetracysteine motifs for protein labeling in vitro and in vivo: synthesis and biological applications. J Am Chem Soc 124:6063–6076
2. Alberdi E, Sanchez-Gomez MV, Cavaliere F, Perez-Samartin A, Zugaza JL, Trullas R, Domercq M, Matute C (2010) Amyloid beta oligomers induce Ca2+ dysregulation and neuronal death through activation of ionotropic glutamate receptors. Cell Calcium 47:264–272. https://doi.org/10.1016/j.ceca.2009.12.010
3. Ando K, Brion JP, Stygelbout V, Suain V, Authelet M, Dedecker R, Chanut A, Lacor P, Lavaur J, Sazdovitch V et al (2013) Clathrin adaptor CALM/PICALM is associated with neurofibrillary tangles and is cleaved in Alzheimer's brains. Acta Neuropathol 125:861–878. https://doi.org/10.1007/s00401-013-1111-z
4. Anekonda TS, Quinn JF, Harris C, Frahler K, Wadsworth TL, Woltjer RL (2011) L-type voltage-gated calcium channel blockade with isradipine as a therapeutic strategy for Alzheimer's disease. Neurobiol Dis 41:62–70. https://doi.org/10.1016/j.nbd.2010.08.020

5. Arbel-Ornath M, Hudry E, Boivin JR, Hashimoto T, Takeda S, Kuchibhotla KV, Hou S, Lattarulo CR, Belcher AM, Shakerdge N et al (2017) Soluble oligomeric amyloid-beta induces calcium dyshomeostasis that precedes synapse loss in the living mouse brain. Mol Neurodegener 12:27. https://doi.org/10.1186/s13024-017-0169-9
6. Arispe N, Pollard HB, Rojas E (1993) Giant multilevel cation channels formed by Alzheimer disease amyloid beta-protein [A beta P-(1-40)] in bilayer membranes. Proc Natl Acad Sci U S A 90:10573–10577
7. Bate C, Williams A (2011) Amyloid-beta-induced synapse damage is mediated via cross-linkage of cellular prion proteins. J Biol Chem 286: 37955–37963. https://doi.org/10.1074/jbc.M111.248724
8. Brenner S (1974) The genetics of Caenorhabditis elegans. Genetics 77:71–94
9. Calafate S, Flavin W, Verstreken P, Moechars D (2016) Loss of Bin1 promotes the propagation of tau pathology. Cell Rep 17:931–940. https://doi.org/10.1016/j.celrep.2016.09.063
10. Carrasquillo MM, Belbin O, Hunter TA, Ma L, Bisceglio GD, Zou F, Crook JE, Pankratz VS, Sando SB, Aasly JO et al (2011) Replication of BIN1 association with Alzheimer's disease and evaluation of genetic interactions. J Alzheimers Dis 24:751–758. https://doi.org/10.3233/JAD-2011-101932
11. Castaneda MT, Lopez ED, Touhami A, Tovar R, Ortega MR, Rodriguez JM (2015) Neuroprotection of medial septal cholinergic neurons by memantine after intralateral septal injection of Abeta1-40. Neuroreport 26:450–454. https://doi.org/10.1097/WNR.0000000000000364
12. Cataldo AM, Barnett JL, Pieroni C, Nixon RA (1997) Increased neuronal endocytosis and protease delivery to early endosomes in sporadic Alzheimer's disease: neuropathologic evidence for a mechanism of increased beta-amyloidogenesis. J Neurosci 17:6142–6151
13. Chapuis J, Hansmannel F, Gistelinck M, Mounier A, Van Cauwenberghe C, Kolen KV, Geller F, Sottejeau Y, Harold D, Dourlen P et al (2013) Increased expression of BIN1 mediates Alzheimer genetic risk by modulating tau pathology. Mol Psychiatry 18:1225–1234. https://doi.org/10.1038/mp.2013.1
14. De Felice FG, Wu D, Lambert MP, Fernandez SJ, Velasco PT, Lacor PN, Bigio EH, Jerecic J, Acton PJ, Shughrue PJ et al (2008) Alzheimer's disease-type neuronal tau hyperphosphorylation induced by A beta oligomers. Neurobiol Aging 29:1334–1347. https://doi.org/10.1016/j.neurobiolaging.2007.02.029
15. Deng J, Habib A, Obregon DF, Barger SW, Giunta B, Wang YJ, Hou H, Sawmiller D, Tan J (2015) Soluble amyloid precursor protein alpha inhibits tau phosphorylation through modulation of GSK3beta signaling pathway. J Neurochem 135:630–637. https://doi.org/10.1111/jnc.13351
16. Diomede L, Di Fede G, Romeo M, Bagnati R, Ghidoni R, Fiordaliso F, Salio M, Rossi A, Catania M, Paterlini A et al (2014) Expression of A2V-mutated Abeta in *Caenorhabditis elegans* results in oligomer formation and toxicity. Neurobiol Dis 62:521–532. https://doi.org/10.1016/j.nbd.2013.10.024
17. Dosanjh LE, Brown MK, Rao G, Link CD, Luo Y (2010) Behavioral phenotyping of a transgenic Caenorhabditis elegans expressing neuronal amyloid-beta. J Alzheimers Dis 19:681–690. https://doi.org/10.3233/JAD-2010-1267
18. Duan Z, Sesti F (2013) A Caenorhabditis elegans model system for amylopathy study. J Vis Exp:e50435. https://doi.org/10.3791/50435
19. Etxaniz A, Gonzalez-Bullon D, Martin C, Ostolaza H (2018) Membrane repair mechanisms against Permeabilization by pore-forming toxins. Toxins (Basel) 10. https://doi.org/10.3390/toxins10060234
20. Fonte V, Dostal V, Roberts CM, Gonzales P, Lacor PN, Velasco PT, Magrane J, Dingwell N, Fan EY, Silverman MA et al (2011) A glycine zipper motif mediates the formation of toxic beta-amyloid oligomers in vitro and in vivo. Mol Neurodegener 6:61. https://doi.org/10.1186/1750-1326-6-61
21. Goedert M, Baur CP, Ahringer J, Jakes R, Hasegawa M, Spillantini MG, Smith MJ, Hill F (1996) PTL-1, a microtubule-associated protein with tau-like repeats from the nematode Caenorhabditis elegans. J Cell Sci 109(Pt 11): 2661–2672
22. Goll DE, Thompson VF, Li H, Wei W, Cong J (2003) The calpain system. Physiol Rev 83:731–801. https://doi.org/10.1152/physrev.00029.2002
23. Grueninger F, Bohrmann B, Czech C, Ballard TM, Frey JR, Weidensteiner C, von Kienlin M, Ozmen L (2010) Phosphorylation of Tau at S422 is enhanced by Abeta in TauPS2APP triple transgenic mice. Neurobiol Dis 37:294–306. https://doi.org/10.1016/j.nbd.2009.09.004
24. Guivernau B, Bonet J, Valls-Comamala V, Bosch-Morato M, Godoy JA, Inestrosa NC, Peralvarez-Marin A, Fernandez-Busquets X, Andreu D, Oliva B et al (2016) Amyloid-beta Peptide Nitrotyrosination Stabilizes Oligomers and Enhances NMDAR-Mediated Toxicity. J Neurosci 36:11693–11703. https://doi.org/10.1523/JNEUROSCI.1081-16.2016
25. Hardy J, Selkoe DJ (2002) The amyloid hypothesis of Alzheimer's disease: progress and problems on the road to therapeutics. Science 297:353–356. https://doi.org/10.1126/science.1072994
26. Hassan WM, Dostal V, Huemann BN, Yerg JE, Link CD (2015) Identifying Abeta-specific pathogenic mechanisms using a nematode model of Alzheimer's disease. Neurobiol Aging 36:857–866. https://doi.org/10.1016/j.neurobiolaging.2014.10.016
27. Hernandez F, Lucas JJ, Avila J (2013) GSK3 and tau: two convergence points in Alzheimer's disease. J Alzheimers Dis 33(Suppl 1):S141–S144. https://doi.org/10.3233/JAD-2012-129025
28. Herrup K (2015) The case for rejecting the amyloid cascade hypothesis. Nat Neurosci 18:794–799. https://doi.org/10.1038/nn.4017
29. Hirakura Y, Lin MC, Kagan BL (1999) Alzheimer amyloid abeta1-42 channels: effects of solvent, pH, and Congo red. J Neurosci Res 57:458–466
30. Ho R, Ortiz D, Shea TB (2001) Amyloid-beta promotes calcium influx and neurodegeneration via stimulation of L voltage-sensitive calcium channels rather than NMDA channels in cultured neurons. J Alzheimers Dis 3:479–483
31. Huffman DL, Abrami L, Sasik R, Corbeil J, van der Goot FG, Aroian RV (2004) Mitogen-activated protein kinase pathways defend against bacterial pore-forming toxins. Proc Natl Acad Sci U S A 101:10995–11000. https://doi.org/10.1073/pnas.0404073101
32. Idone V, Tam C, Goss JW, Toomre D, Pypaert M, Andrews NW (2008) Repair of injured plasma membrane by rapid Ca2+–dependent endocytosis. J Cell Biol 180:905–914. https://doi.org/10.1083/jcb.200708010
33. Jeong SY, Martchenko M, Cohen SN (2013) Calpain-dependent cytoskeletal rearrangement exploited for anthrax toxin endocytosis. Proc Natl Acad Sci U S A 110:E4007–E4015. https://doi.org/10.1073/pnas.1316852110
34. Jin N, Yin X, Yu D, Cao M, Gong CX, Iqbal K, Ding F, Gu X, Liu F (2015) Truncation and activation of GSK-3beta by calpain I: a molecular mechanism links to tau hyperphosphorylation in Alzheimer's disease. Sci Rep 5:8187. https://doi.org/10.1038/srep08187
35. Kaech S, Banker G (2006) Culturing hippocampal neurons. Nat Protoc 1: 2406–2415. https://doi.org/10.1038/nprot.2006.356
36. Kagan BL (2012) Membrane pores in the pathogenesis of neurodegenerative disease. Prog Mol Biol Transl Sci 107:295–325. https://doi.org/10.1016/B978-0-12-385883-2.00001-1
37. Kanatsu K, Hori Y, Takatori S, Watanabe T, Iwatsubo T, Tomita T (2016) Partial loss of CALM function reduces Abeta42 production and amyloid deposition in vivo. Hum Mol Genet 25:3988–3997. https://doi.org/10.1093/hmg/ddw239
38. Kao CY, Los FC, Huffman DL, Wachi S, Kloft N, Husmann M, Karabrahimi V, Schwartz JL, Bellier A, Ha C et al (2011) Global functional analyses of cellular responses to pore-forming toxins. PLoS Pathog 7:e1001314. https://doi.org/10.1371/journal.ppat.1001314
39. Karran E, De Strooper B (2016) The amyloid cascade hypothesis: are we poised for success or failure? J Neurochem 139(Suppl 2):237–252. https://doi.org/10.1111/jnc.13632
40. Kim JA, Kim HL (2001) Cell-free expression and functional reconstitution of CALM in clathrin assembly. Exp Mol Med 33:89–94. https://doi.org/10.1038/emm.2001.16
41. Kim S, Jeon TJ, Oberai A, Yang D, Schmidt JJ, Bowie JU (2005) Transmembrane glycine zippers: physiological and pathological roles in membrane proteins. Proc Natl Acad Sci U S A 102:14278–14283. https://doi.org/10.1073/pnas.0501234102
42. Kurbatskaya K, Phillips EC, Croft CL, Dentoni G, Hughes MM, Wade MA, Al-Sarraj S, Troakes C, O'Neill MJ, Perez-Nievas BG et al (2016) Upregulation of calpain activity precedes tau phosphorylation and loss of synaptic proteins in Alzheimer's disease brain. Acta Neuropathol Commun 4:34. https://doi.org/10.1186/s40478-016-0299-2
43. Lambert MP, Viola KL, Chromy BA, Chang L, Morgan TE, Yu J, Venton DL, Krafft GA, Finch CE, Klein WL (2001) Vaccination with soluble Abeta oligomers generates toxicity-neutralizing antibodies. J Neurochem 79:595–605
44. Lauren J, Gimbel DA, Nygaard HB, Gilbert JW, Strittmatter SM (2009) Cellular prion protein mediates impairment of synaptic plasticity by amyloid-beta oligomers. Nature 457:1128–1132. https://doi.org/10.1038/nature07761
45. Lebart MC, Benyamin Y (2006) Calpain involvement in the remodeling of cytoskeletal anchorage complexes. FEBS J 273:3415–3426. https://doi.org/10.1111/j.1742-4658.2006.05350.x
46. Lin H, Bhatia R, Lal R (2001) Amyloid beta protein forms ion channels: implications for Alzheimer's disease pathophysiology. FASEB J 15:2433–2444. https://doi.org/10.1096/fj.01-0377com
47. Link CD (1995) Expression of human beta-amyloid peptide in transgenic Caenorhabditis elegans. Proc Natl Acad Sci U S A 92:9368–9372

48. Liu CC, Liu CC, Kanekiyo T, Xu H, Bu G (2013) Apolipoprotein E and Alzheimer disease: risk, mechanisms and therapy. Nat Rev Neurol 9:106–118. https://doi.org/10.1038/nrneurol.2012.263
49. Liu O, Grant BD (2015) Basolateral endocytic recycling requires RAB-10 and AMPH-1 mediated recruitment of RAB-5 GAP TBC-2 to endosomes. PLoS Genet 11:e1005514. https://doi.org/10.1371/journal.pgen.1005514
50. Liu SL, Wang C, Jiang T, Tan L, Xing A, Yu JT (2016) The role of Cdk5 in Alzheimer's disease. Mol Neurobiol 53:4328–4342. https://doi.org/10.1007/s12035-015-9369-x
51. Lopes JP, Oliveira CR, Agostinho P (2010) Neurodegeneration in an Abeta-induced model of Alzheimer's disease: the role of Cdk5. Aging Cell 9:64–77. https://doi.org/10.1111/j.1474-9726.2009.00536.x
52. Los FC, Kao CY, Smitham J, McDonald KL, Ha C, Peixoto CA, Aroian RV (2011) RAB-5- and RAB-11-dependent vesicle-trafficking pathways are required for plasma membrane repair after attack by bacterial pore-forming toxin. Cell Host Microbe 9:147–157. https://doi.org/10.1016/j.chom.2011.01.005
53. Luedtke NW, Dexter RJ, Fried DB, Schepartz A (2007) Surveying polypeptide and protein domain conformation and association with FlAsH and ReAsH. Nat Chem Biol 3:779–784. https://doi.org/10.1038/nchembio.2007.49
54. Mahley RW (2016) Central nervous system lipoproteins: ApoE and regulation of cholesterol metabolism. Arterioscler Thromb Vasc Biol 36:1305–1315. https://doi.org/10.1161/ATVBAHA.116.307023
55. Mairet-Coello G, Courchet J, Pieraut S, Courchet V, Maximov A, Polleux F (2013) The CAMKK2-AMPK kinase pathway mediates the synaptotoxic effects of Abeta oligomers through Tau phosphorylation. Neuron 78:94–108. https://doi.org/10.1016/j.neuron.2013.02.003
56. Manucat-Tan NB, Saadipour K, Wang YJ, Bobrovskaya L, Zhou XF (2018) Cellular trafficking of amyloid precursor protein in Amyloidogenesis physiological and pathological significance. Mol Neurobiol. https://doi.org/10.1007/s12035-018-1106-9
57. Martin L, Latypova X, Wilson CM, Magnaudeix A, Perrin ML, Yardin C, Terro F (2013) Tau protein kinases: involvement in Alzheimer's disease. Ageing Res Rev 12:289–309. https://doi.org/10.1016/j.arr.2012.06.003
58. Meng L, Zhang A, Jin Y, Yan D (2016) Regulation of neuronal axon specification by glia-neuron gap junctions in *C. elegans*. Elife 5. https://doi.org/10.7554/eLife.19510
59. Miranda CJ, Braun L, Jiang Y, Hester ME, Zhang L, Riolo M, Wang H, Rao M, Altura RA, Kaspar BK (2012) Aging brain microenvironment decreases hippocampal neurogenesis through Wnt-mediated survivin signaling. Aging Cell 11:542–552. https://doi.org/10.1111/j.1474-9726.2012.00816.x
60. Miyagawa T, Ebinuma I, Morohashi Y, Hori Y, Young Chang M, Hattori H, Maehara T, Yokoshima S, Fukuyama T, Tsuji S et al (2016) BIN1 regulates BACE1 intracellular trafficking and amyloid-beta production. Hum Mol Genet 25:2948–2958. https://doi.org/10.1093/hmg/ddw146
61. Naj AC, Jun G, Reitz C, Kunkle BW, Perry W, Park YS, Beecham GW, Rajbhandary RA, Hamilton-Nelson KL, Wang LS et al (2014) Effects of multiple genetic loci on age at onset in late-onset Alzheimer disease: a genome-wide association study. JAMA Neurol 71:1394–1404. https://doi.org/10.1001/jamaneurol.2014.1491
62. Nikkel AL, Martino B, Markosyan S, Brederson JD, Medeiros R, Moeller A, Bitner RS (2012) The novel calpain inhibitor A-705253 prevents stress-induced tau hyperphosphorylation in vitro and in vivo. Neuropharmacology 63:606–612. https://doi.org/10.1016/j.neuropharm.2012.05.011
63. Nonet ML, Holgado AM, Brewer F, Serpe CJ, Norbeck BA, Holleran J, Wei L, Hartwieg E, Jorgensen EM, Alfonso A (1999) UNC-11, a Caenorhabditis elegans AP180 homologue, regulates the size and protein composition of synaptic vesicles. Mol Biol Cell 10:2343–2360. https://doi.org/10.1091/mbc.10.7.2343
64. Padamsey Z, McGuinness L, Emptage NJ (2017) Inhibition of lysosomal Ca(2+) signalling disrupts dendritic spine structure and impairs wound healing in neurons. Commun Integr Biol 10:e1344802. https://doi.org/10.1080/19420889.2017.1344802
65. Pant S, Sharma M, Patel K, Caplan S, Carr CM, Grant BD (2009) AMPH-1/Amphiphysin/Bin1 functions with RME-1/Ehd1 in endocytic recycling. Nat Cell Biol 11:1399–1410. https://doi.org/10.1038/ncb1986
66. Parikh I, Fardo DW, Estus S (2014) Genetics of PICALM expression and Alzheimer's disease. PLoS One 9:e91242. https://doi.org/10.1371/journal.pone.0091242
67. Peters C, Espinoza MP, Gallegos S, Opazo C, Aguayo LG (2015) Alzheimer's Abeta interacts with cellular prion protein inducing neuronal membrane damage and synaptotoxicity. Neurobiol Aging 36:1369–1377. https://doi.org/10.1016/j.neurobiolaging.2014.11.019
68. Peters C, Fernandez-Perez EJ, Burgos CF, Espinoza MP, Castillo C, Urrutia JC, Streltsov VA, Opazo C, Aguayo LG (2013) Inhibition of amyloid beta-induced synaptotoxicity by a pentapeptide derived from the glycine zipper region of the neurotoxic peptide. Neurobiol Aging 34:2805–2814. https://doi.org/10.1016/j.neurobiolaging.2013.06.001
69. Petralia RS, Wang YX, Indig FE, Bushlin I, Wu F, Mattson MP, Yao PJ (2013) Reduction of AP180 and CALM produces defects in synaptic vesicle size and density. NeuroMolecular Med 15:49–60. https://doi.org/10.1007/s12017-012-8194-x
70. Peyronnet O, Vachon V, Schwartz JL, Laprade R (2001) Ion channels induced in planar lipid bilayers by the bacillus thuringiensis toxin Cry1Aa in the presence of gypsy moth (Lymantria dispar) brush border membrane. J Membr Biol 184:45–54
71. de Planque MR, Raussens V, Contera SA, Rijkers DT, Liskamp RM, Ruysschaert JM, Ryan JF, Separovic F, Watts A (2007) Beta-sheet structured beta-amyloid(1-40) perturbs phosphatidylcholine model membranes. J Mol Biol 368: 982–997 Doi https://doi.org/10.1016/j.jmb.2007.02.063
72. Reitz C Alzheimer's disease and the amyloid cascade hypothesis: a critical review. Int J Alzheimers Dis 2012, 2012:369808. https://doi.org/10.1155/2012/369808
73. Resende R, Pereira C, Agostinho P, Vieira AP, Malva JO, Oliveira CR (2007) Susceptibility of hippocampal neurons to Abeta peptide toxicity is associated with perturbation of Ca2+ homeostasis. Brain Res 1143:11–21. https://doi.org/10.1016/j.brainres.2007.01.071
74. Rushworth JV, Griffiths HH, Watt NT, Hooper NM (2013) Prion protein-mediated toxicity of amyloid-beta oligomers requires lipid rafts and the transmembrane LRP1. J Biol Chem 288:8935–8951. https://doi.org/10.1074/jbc.M112.400358
75. Seshadri S, Fitzpatrick AL, Ikram MA, DeStefano AL, Gudnason V, Boada M, Bis JC, Smith AV, Carassquillo MM, Lambert JC et al (2010) Genome-wide analysis of genetic loci associated with Alzheimer disease. JAMA 303:1832–1840. https://doi.org/10.1001/jama.2010.574
76. Seward ME, Swanson E, Norambuena A, Reimann A, Cochran JN, Li R, Roberson ED, Bloom GS (2013) Amyloid-beta signals through tau to drive ectopic neuronal cell cycle re-entry in Alzheimer's disease. J Cell Sci 126: 1278–1286. https://doi.org/10.1242/jcs.1125880
77. Small DH, Gasperini R, Vincent AJ, Hung AC, Foa L (2009) The role of Abeta-induced calcium dysregulation in the pathogenesis of Alzheimer's disease. J Alzheimers Dis 16:225–233. https://doi.org/10.3233/JAD-2009-0951
78. Sul D, Kim HS, Lee D, Joo SS, Hwang KW, Park SY (2009) Protective effect of caffeic acid against beta-amyloid-induced neurotoxicity by the inhibition of calcium influx and tau phosphorylation. Life Sci 84:257–262. https://doi.org/10.1016/j.lfs.2008.12.001
79. Tam C, Idone V, Devlin C, Fernandes MC, Flannery A, He X, Schuchman E, Tabas I, Andrews NW (2010) Exocytosis of acid sphingomyelinase by wounded cells promotes endocytosis and plasma membrane repair. J Cell Biol 189:1027–1038. https://doi.org/10.1083/jcb.201003053
80. Thomas RS, Henson A, Gerrish A, Jones L, Williams J, Kidd EJ (2016) Decreasing the expression of PICALM reduces endocytosis and the activity of beta-secretase: implications for Alzheimer's disease. BMC Neurosci 17:50. https://doi.org/10.1186/s12868-016-0288-1
81. Town T, Zolton J, Shaffner R, Schnell B, Crescentini R, Wu Y, Zeng J, DelleDonne A, Obregon D, Tan J et al (2002) p35/Cdk5 pathway mediates soluble amyloid-beta peptide-induced tau phosphorylation in vitro. J Neurosci Res 69:362–372. https://doi.org/10.1002/jnr.10299
82. Treusch S, Hamamichi S, Goodman JL, Matlack KE, Chung CY, Baru V, Shulman JM, Parrado A, Bevis BJ, Valastyan JS et al (2011) Functional links between Abeta toxicity, endocytic trafficking, and Alzheimer's disease risk factors in yeast. Science 334:1241–1245. https://doi.org/10.1126/science.1213210
83. Ubelmann F, Burrinha T, Salavessa L, Gomes R, Ferreira C, Moreno N, Guimas Almeida C (2017) Bin1 and CD2AP polarise the endocytic generation of beta-amyloid. EMBO Rep 18:102–122. https://doi.org/10.15252/embr.201642738
84. Wang Z, Lei H, Zheng M, Li Y, Cui Y, Hao F (2016) Meta-analysis of the association between Alzheimer disease and variants in GAB2, PICALM, and SORL1. Mol Neurobiol 53:6501–6510. https://doi.org/10.1007/s12035-015-9546-y
85. Wei JZ, Hale K, Carta L, Platzer E, Wong C, Fang SC, Aroian RV (2003) Bacillus thuringiensis crystal proteins that target nematodes. Proc Natl Acad Sci U S A 100:2760–2765. https://doi.org/10.1073/pnas.0538072100
86. Yao J, Nowack A, Kensel-Hammes P, Gardner RG, Bajjalieh SM (2010) Cotrafficking of SV2 and synaptotagmin at the synapse. J Neurosci 30:5569–5578. https://doi.org/10.1523/JNEUROSCI.4781-09.2010

87. Zempel H, Thies E, Mandelkow E, Mandelkow EM (2010) Abeta oligomers cause localized Ca(2+) elevation, missorting of endogenous Tau into dendrites, Tau phosphorylation, and destruction of microtubules and spines. J Neurosci 30:11938–11950. https://doi.org/10.1523/JNEUROSCI.2357-10.2010
88. Zhao WQ, Santini F, Breese R, Ross D, Zhang XD, Stone DJ, Ferrer M, Townsend M, Wolfe AL, Seager MA et al (2010) Inhibition of calcineurin-mediated endocytosis and alpha-amino-3-hydroxy-5-methyl-4-isoxazolepropionic acid (AMPA) receptors prevents amyloid beta oligomer-induced synaptic disruption. J Biol Chem 285:7619–7632. https://doi.org/10.1074/jbc.M109.057182
89. Zhou M, Huang T, Collins N, Zhang J, Shen H, Dai X, Xiao N, Wu X, Wei Z, York J et al (2016) APOE4 induces site-specific Tau phosphorylation through Calpain-CDK5 signaling pathway in EFAD-Tg mice. Curr Alzheimer Res 13: 1048–1055

9

Deficit of corpus callosum axons, reduced axon diameter and decreased area are markers of abnormal development of interhemispheric connections in autistic subjects

Jarek Wegiel[1*], Wojciech Kaczmarski[1], Michael Flory[2], Veronica Martinez-Cerdeno[3], Thomas Wisniewski[4], Krzysztof Nowicki[1], Izabela Kuchna[1] and Jerzy Wegiel[1]

Abstract

Introduction: In autism spectrum disorder, lack of coherence and of complex information processing, and narrowly focused interests and repetitive behaviors are considered a sign of long-range underconnectivity and short-range overconnectivity. The goal of this morphometric study of five anatomically and functionally different segments of the corpus callosum (CC) was to establish patterns of differences between long-range interhemispheric connections in nine neurotypical and nine autistic subjects.

Results: Electron microscopy revealed a significant reduction in average axon diameter and axon cross-sectional area in autistic subjects, and reduction in CC segment–specific diversification of connections of functionally different cortical regions. The study shows an increase in the percentage of small diameter axons (< 0.651 μm) and a decrease in the percentage of axons with large diameter (> 1.051 μm). The total number of small-diameter axons is reduced in segment I and III by 43% on average. The number of medium- and large-diameter axons is reduced in all five CC segments by an average of 49 and 72%, respectively.

Conclusions: The detected pattern of pathology suggests a failure of mechanisms controlling guidance of axons during development leading to axonal deficit, and failure of mechanisms controlling axon structure. A reduction in axon diameter may affect the velocity and volume of signal transmission, and distort functional specialization of CC segments. Significant deficits in axon number and reduction in axon size in all five CC segments appear to be substantial components of brain connectome integrity distortion which may contribute to the autism phenotype.

Keywords: Autism, Neuropathology, Corpus callosum, Axon, Electron microscopy, Morphometry

Introduction

Developmental abnormalities of brain connectivity in autism

Autism spectrum disorders (ASDs) are debilitating neurodevelopmental conditions characterized by impairments in social interactions and communication combined with restricted interests and repetitive behaviors [3]. Despite the genetic and clinical heterogeneity of ASDs, Geschwind and Levitt [16] hypothesized that a common feature of the brain across the many forms of autism is developmental disconnection, resulting from the failure to develop normal connections between higher-order association areas of the temporal, parietal, and frontal cortices. Just et al. [28] formulated an underconnectivity theory, whereas Belmonte et al. [6] suggested a combination of reduced long-distance, and increased local connectivity in ASD. Functional [27, 30, 34] and structural [5, 31] neuroimaging findings support the concept of disconnection [12, 15]

* Correspondence: Jarek.Wegiel@opwdd.ny.gov
[1]Department of Developmental Neurobiology, New York State Institute for Basic Research in Developmental Disabilities, 1050 Forest Hill Road, Staten Island, NY 10314, USA
Full list of author information is available at the end of the article

and underconnectivity [25, 28]. The first postmortem study of axonal connections in autism published by Zikopoulos and Barbas [61] revealed disconnection of long-distance pathways and excessive connections between neighboring areas in the white matter adjacent to the anterior cingulate cortex, orbitofrontal cortex and lateral prefrontal cortex suggesting contribution of these developmental defects to diagnostic behavioral alterations in autism. The combined effect of developmental abnormalities reported in imaging and postmortem studies is loss of integrity in brain connectivity, which appears to be essential for deficits of higher-order social, emotional, and communication activities in autistic subjects [46, 61].

Corpus callosum (CC) developmental abnormalities in autism

The CC, the largest fiber tract in the human brain [23, 45, 55], is a primary target for testing the hypothesis of long-range underconnectivity. The CC connects brain hemispheres and facilitates integration of sensory and motor information with executive functions, social interactions, and language [38]. Structural MRI of the brain of autistic subjects shows a thinner CC and reduced CC midsagittal area, suggesting a deficit of long-range interhemispheric axonal connections [9, 21, 27, 36, 41, 42]. Brain fiber tractography based on diffusion tensor imaging divided the human CC into five segments, with axons projected through the brain midline by neurons of the prefrontal (S I), premotor and supplementary motor (S II), motor (S III), sensory (S IV), and parietal, occipital, and temporal cortices (S V) [24]. Morphometry of the CC of neurotypical subjects reveals a segment-specific ratio of axons of different diameter. In the brains of control subjects, the percentage of small-diameter axons is the largest in S1 and decreases in posterior segments, whereas the percentage of medium- and large-diameter axons increases in the middle and caudal portions of the CC [45]. Axon diameter, cross-sectional area, and myelin sheath thickness correlate with the velocity and volume of signal transmission [39, 47, 48]. Therefore, a CC segment–specific combination of these structural characteristics reflects the function of axons connecting specialized cortical regions.

To explain clinical MRI characterization of developmental anomalies of interhemispheric connections, we performed the first postmortem morphometric light microscopy study of the CC of autistic individuals [56]. The study revealed a significant deficit in interhemispheric axonal connections and two types of developmental abnormalities contributing to this deficit: a topographically selective partial CC agenesis and topographically non-selective hypoplasia. The detected regional agenesis and hypoplasia in CC segments are signs of focal interhemispheric disconnection and diffuse underconnectivity in autism.

CC segmentation

There are no macroscopic anatomical landmarks that delineate distinct callosal areas in a midsaggital CC cross-section of the human corpus callosum. Several geometric partitioning schemes have been designed including the commonly used Witelson [60] CC division into five vertical segments. However, this segmentation was based predominantly on non-human primate studies. In this study, Hofer and Frahm [24] CC segmentation based on human brain diffusion tensor imaging (DTI) was applied. DTI fiber tracking identifies axonal bundles that cross the CC midline and separates transcallosal fiber tracts with respect to their specific cortical projections. Hofer and Frahm classification distinguishes segment (S) I with prefrontal cortex projections, SII with premotor and supplementary connections, SIII with motor connections, SIV with sensory cortex connections, and SV with parietal, temporal and occipital connections. Segmentation of the CC from the most rostral to the most caudal end of the CC defines the rostro-caudal extent of each segment with SI - 17%; SII - 33%, SIII -17%; SIV -8%; and SV – 25%.

Aims

The aims of this electron microscopy study of axons in all five structurally and functionally different segments in the CC of autistic and control subjects were (a) to define the range of deficits in the number of small-, medium-, and large-diameter axons in CC segments connecting functionally diverse cortical regions; and (b) to define distortion in the segment-specific ratio of axons of different diameters as a marker of loss of specialization of interhemispheric connectivity in autism.

Materials and methods

Cohort

This study of the corpus callosum in autistic subjects is a component of a project designed by several research groups and supported by Autism Speaks and the Autism Tissue Program. The overall goal was to overcome problems associated with scarcity of material for postmortem studies resulting in studies of 3–4 cases by different groups applying different clinical criteria and different methods of examination of a few brain regions. Screening of 18 pairs of autistic and control subjects identified 12 age- and sex-matched pairs corresponding to inclusion criteria. One brain hemisphere of these subjects was processed and embedded in celloidin for detection of global patterns of anatomical, cellular and subcellular changes in autism.

This electron microscopic study of the CC was performed on 9 pairs of autistic and control subjects with well-preserved CC ultrastructure. The demographics, information about intellectual deficits, seizures, and cause of death are summarized in Table 1. In both the autistic and control group, the same number of males (6) and females (3) was examined. The age of subjects with autism was 5 to 60 years (25.4 on average) and of control subjects 4 to 52 years (26.3 on average). The diagnosis of autism was confirmed with the Autism Diagnostic Interview, Revised [33]. Intellectual deficit ranging from mild to severe was reported in six autistic subjects, and seizures in five cases. The average postmortem interval (PMI) was 21.2 and 15 h in the autistic and control groups, respectively. Differences in age, PMI, brain weight, and brain weight loss during dehydration were not significant.

Tissue sampling for estimation of CC area, number of axons, and EM-based axon morphometry

The brain hemisphere was fixed in 10% buffered formalin for at least 3 months, washed for 24 h in water to remove fixer, dehydrated in increasing concentrations of ETOH, embedded in celloidin [22] and cut into 200-μm-thick frontal hemispheric sections. For morphometric study, a 2-mm wide CC strip was cut off from every twelfth coronal hemispheric section, impregnated with 2% osmium tetroxide, dehydrated and embedded in Epon. Resin blocks were cut into 1-μm-thick sections. Samples were oriented to cut axons perpendicularly to axon long axis and stained with a 2% solution of p-phenylenediamine (PPD). Every 60th section was used to determine the entire midsagittal area of the CC and each of the five segments, whereas 21 equidistant sections were used for estimation of the total number and numerical density of axons under a light microscope. The results were summarized in our previous paper [56].

Electron microscopy study

Thirteen Epon blocks represented the entire rostro-caudal extent of the CC in each brain of nine control subjects and nine autistic subjects, including three Epon blocks for S I, S II, and S V and two blocks for the shorter segments, III and IV (Fig. 1). For electron microscopy, Epon blocks were cut into 60-nm-thick sections using a Reichert Ultracut S (Leica, Austria) ultramicrotome and Diatome 45° knife (Diatome, U.S., Fort Washington, PA, USA), and collected on formvar-coated copper grids. Sections were stained with uranyl acetate and photographed at a magnification of 15,000x using a Hitachi H7500 transmission electron microscope with a CCD camera and Advanced Microscopy Techniques' (AMT) Image Capture Engine (Danvers, MA). For each case, 12 electron micrographs were used for measurements of axon diameter, cross-sectional area, and myelin thickness. Background

Table 1 Preserved tissue and clinical and postmortem records

Group and case ID	Sex	Age (y)	Intellectual deficit	Seizures age of onset	Cause of death	PMI (h)	BW (g)	Weight loss (%)
A1	F	5	–	–	Traumatic injuries	13	1390	52
A2	M	8	–	–	Rhabdomysarcoma	22	1570	45
A3	F	11	Mild	4.5 m	Seizure-related drowning	13	1460	52
A4	M	13	Severe	2 y	Seizure-related	8	1470	39
A5	F	21	Moderate	5 y	Seizure related resp. failure	50	1108	43
A6	M	23	Severe	23 y	Seizure-related resp. failure	14	1610	60
A7	M	36	Severe	–	Cardiac arrest	24	1480	44
A8	M	52	–	–	Heart failure	11	1324	53
A9	M	60	Moderate	3 y	Pancreatic cancer	26	1210	38
Mean ± SD		25.4 ± 19.8				21.2 ± 13.6	1402 ± 164	47.3 ± 7.3
C1	F	4	–	–	Acute bronchopneumonia	17	1530	49
C2	M	14	–	–	Electrocution	20	1464	44
C3	F	15	–	–	Traumatic injuries	9	1250	49
C4	F	20	–	–	Traumatic injuries	9	1340	52
C5	M	23	–	–	Ruptured spleen	6	1520	41
C6	M	29	–	–	Traumatic injuries	13	1514	49
C7	M	32	–	–	Asphyxia	24	1364	42
C8	M	48	–	–	Heart atherosclerosis	24	1412	39
C9	M	52	–	–	Heart atherosclerosis	13	1430	48
Mean ± SD		26.3 ± 15.8				15 ± 6.6	1424 ± 94	45.9 ± 4.5

A autism, *C* control, *PMI* postmortem interval, *h* hours; Weight loss, decrease (%) of hemispheric brain sample weight during dehydration in ethyl alcohol; *SD* standard deviation

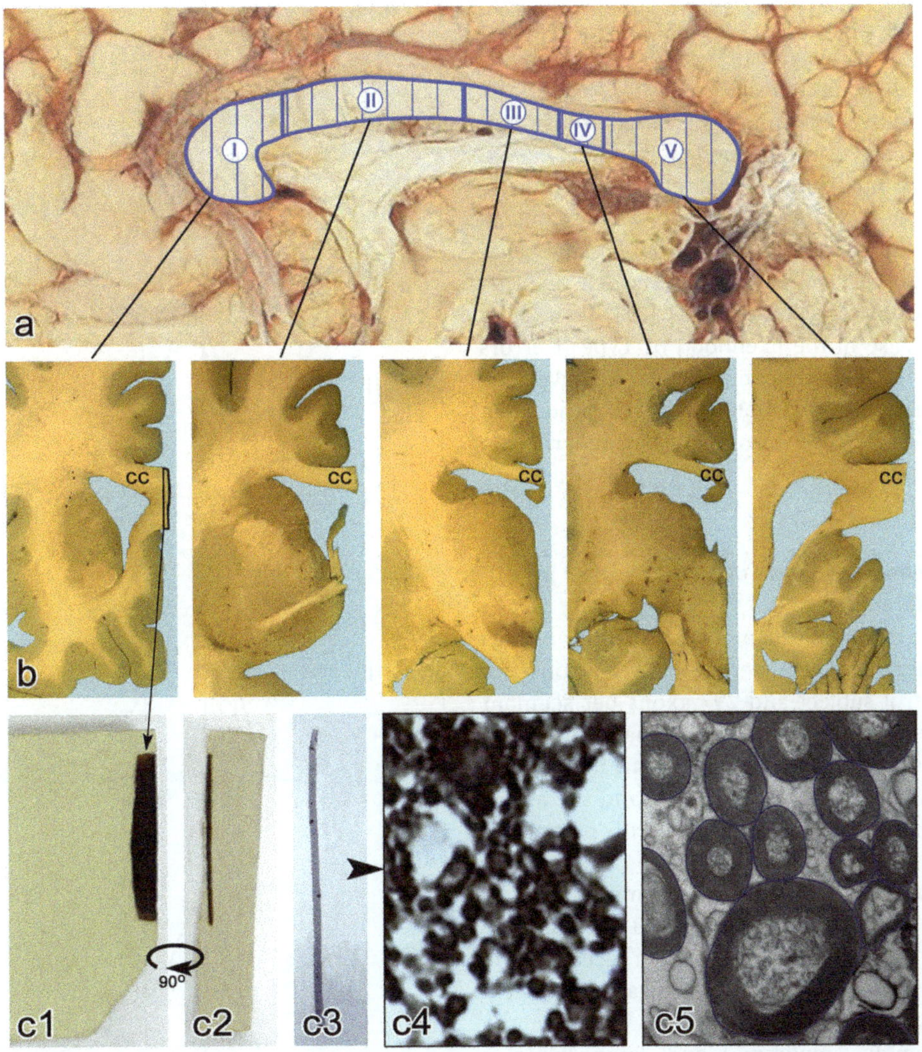

Fig. 1 CC sampling for light and electron microscopy study. Diffusion tensor imaging and fiber tractography-based vertical partitioning of the CC into five segments corresponding to interhemispheric cortical connections (**a**). Celloidin embedded brain hemisphere was cut into 200-μm-thick gapless serial sections (**b**). From thirteen equidistant CC samples representing all segments (**b**) a 2-mm-wide CC sample was cut off and embedded in Epon (**c1**). Semithin sections were cut (**c2**) and stained with PPD (**c3**) for light microscopy morphometric studies [56]. For the EM morphometric study of axons Epon blocks were trimmed and 60-nm-thick sections were cut and stained with uranyl acetate. The calibrated electron micrographs obtained at 15,000x magnification were used for manual delineation of the axon border (**c4**) to estimate the diameter and crossectional area of the axon, as well as to estimate the thickness of the myelin sheath measured at four equidistant points. The measurements were performed with Image J software

correction was applied to reduce risk of distortions during image analysis. For morphometric analysis, the images were saved in the jpeg format.

Axon measurements

The study of CC axons was limited to myelinated axons, which are more common (comprising approximately 90% of all axons [32] and are better preserved than non-myelinated axons [45]. On average in the CC of each control and autistic subject, 1676 and 1486 axons, respectively, were measured. The total number of axons measured in nine control subjects was 15,085, and in nine autistic subjects 13,376. Oversampling was necessary to detect significant differences in all CC segments, including the smallest sector IV representing only 8% of the linear rostro-caudal CC extent [24]. Axons were measured manually by delineation of the outer axon myelin contour (Fig. 1). Image J software provided measurements of axon diameter (Feret's diameter, μm) and area (μm^2). Average myelin thickness of axons was estimated by measuring the myelin sheath thickness at four equidistant points.

Statistical analysis

Analyses of axonal diameter, cross-sectional area, and myelin thickness across diagnostic groups and segments of the CC were done in repeated-measures ANOVAs, using the conservative Greenhouse-Geisser adjustment for the non-sphericity of the dependent variables across segments [19]. Comparisons of percentages of small-, medium-, and large-diameter axons across five segments were done using the wsanova routine [17]. *Post-hoc* analyses used one-way ANOVAs and T-tests. All analyses were performed using Stata version 15.1 [51]. To address the potential increase in the likelihood of Type II errors (false positives) with multiple comparisons, we have used control of the False Discovery Rate (FDR) [7] rather than performing Bonferroni-type corrections of our criterion for significance [18].

Results

Differences between distribution of axon diameter in control and autistic subjects

Characterization of axon diameter distribution was based on comparison of an average of 335 axons per segment per control subject and 297 axons per segment per autistic subject. The reason for examination of fewer axons in the autistic group was lack of axons in S IV and a partial deficit in the caudal part of S III and the anterior part of S V in three autistic subjects with partial CC agenesis.

The distribution curves for all five segments (Fig. 2) suggested that three intervals (< 0.651 µm; 0.651–1.051 µm and > 1.051 µm) identify differences in the prevalence of small-, medium-, and large-diameter axons. This classification was applied to compare the percentage of small-, medium-, and large-diameter axons in control and autistic groups. The axon diameter distribution curves for each of the five CC segments revealed an increased percentage of small-diameter axons and a decreased percentage of large-diameter axons in autistic subjects. The prevalence of small- and the deficit of large-diameter axons was relatively low in S I, but the difference became larger in the four posterior segments.

The detected differences among autistic subjects in the ratio of varying diameter axons in the CC segments reflected a distortion in structural and functional specialization of individual CC segments connecting functionally different cortical regions. To define inter-hemispheric connection abnormalities, inter-segmental differences in average axon diameter, cross-sectional area and myelin thickness, and the percentage and total number of small-, medium-, and large-diameter axons were characterized within the control and autistic groups.

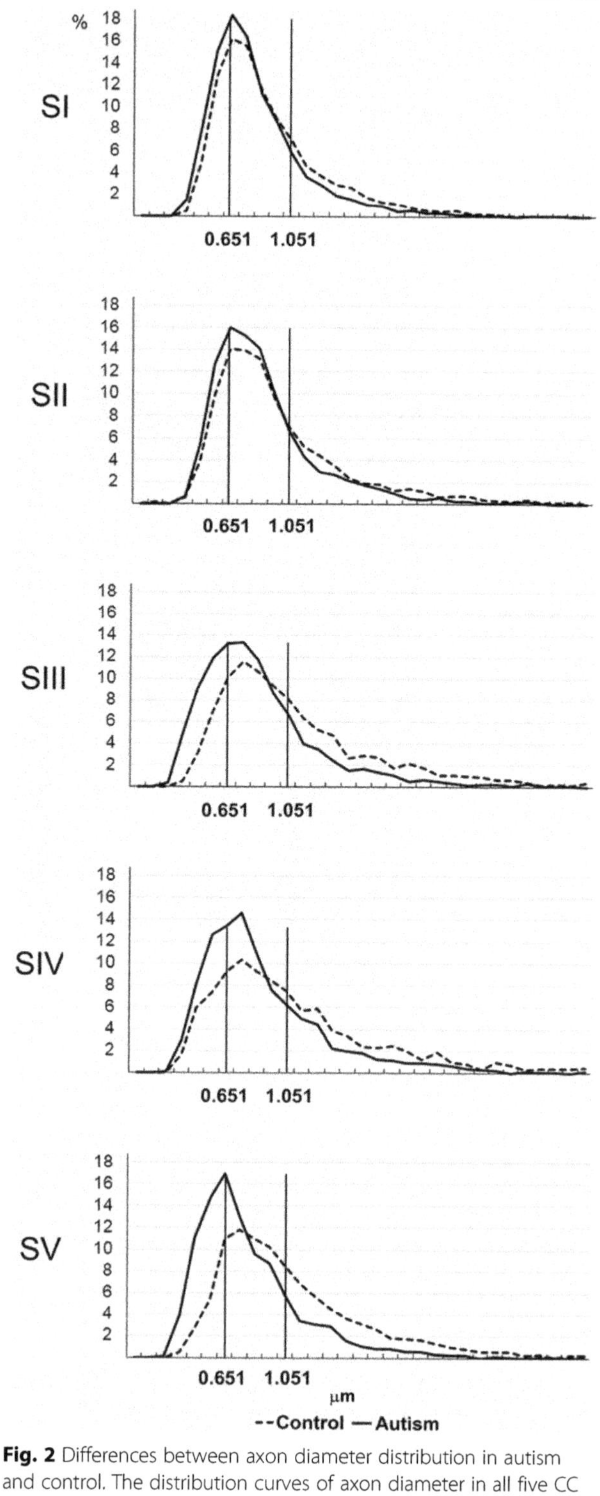

Fig. 2 Differences between axon diameter distribution in autism and control. The distribution curves of axon diameter in all five CC segments in autistic and control subjects revealed a higher percentage of small-diameter axons and a lower percentage of large-diameter axons in autistic subjects, with more prominent differences in posterior S III– S V than in anterior S I and S II

Differences in average axon diameter, cross-sectional area and myelin thickness in CC segments in neurotypical control and autistic subjects

An overall ANOVA analysis of axon measurements showed that in the autistic subjects, average axon diameter was significantly less than in the control group (F [3] = 18.31, $p = 0.0001$) and diameter differed across segments (F [6] = 7.32, Greenhouse-Geisser $p = 0.0012$) [19].

Within-segment ANOVA-based comparison of average axon diameter in autistic and control subjects revealed a significantly smaller diameter in autistic subjects, with the smallest difference in S I (11.2%) and a linear increase to a 28.6% deficit in S V (Table 2, Fig. 3). Reduced axon diameter is paralleled by a very significant deficit in axon cross-sectional area ranging from 24.6% in S 1 to 51.6% in S V in autistic subjects. Measurements of the thickness of the axon myelin sheath revealed insignificant deficits in S I–IV and a significant 15.8% ($p < 0.018$) deficit limited to SV.

The second analysis, focused on detection of differences between axon diameters in individual segments of the CC in neurotypical control subjects, revealed significant intersegmental differences, with an increase of average axon diameter from 0.869 μm in S I to 1.108 μm in S III, 1.120 μm in S IV, and 1.075 μm in S V (Fig. 3, Table 2). Analysis showed significant increase of axon diameter in three posterior segments of control subjects (Table 3). The average axon area increased from 0.613 um^2 in S I to 1.157 um^2 in S IV and 0.968 μm^2 in S V, and the increase was significant in these segments. The increase in axon myelin thickness was small and was significant only in S III. This pattern of significant inter-segmental differences reflects a specialization in CC connections of functionally different cortices in the human brain.

In contrast to the broad range of intersegmental differences in the control group, in autistic subjects axon diameter and area varied in a narrow range, and differences between segments were not significant (Table 3). The flat trend line reflecting a rather uniform diameter of axons and axon area in all segments in autistic subjects (Fig. 3) was a sign of lack of structural differentiation of axons in CC segments.

Autism was a less significant factor in myelin thickness than it was in diameter or area (F [3] = 5.42, $p = 0.0233$), while neither the segment nor its interaction with the diagnosis had a significant effect.

Decrease in the percentage of small-diameter axons and increase in percentage of large-diameter axons in posterior segments of the CC in neurotypical control subjects

To learn which factors define the higher, segment-specific increase in average axon diameter in S II–V of control subjects, the percentage of small-, medium-, and large-diameter axons was determined in the five segments. The percentages of axons falling within each size range differed strongly by segment for controls (F [6] = 6.61, Greenhouse-Geisser $p = 0.0005$).

In control subjects, the percentage of small-diameter axons (< 0.651 μm) decreased from 34.6% in S I to 19.1% in S V ($p < 0.002$) (Tables 4, 5; Fig. 4). The opposite trajectory was detected for large-diameter (> 1.05 μm) axons, with an increase in the percentage of these axons from

Table 2 The difference in average axon diameter, midsagittal area, and myelin thickness in five CC segments in autistic (A) and control (C) subjects

Parameter	Group	CC Segment				
		S I	S II	S III	S IV	S V
Axon diameter (μm) (SE)	C	0.869 (0.025)	0.945 (0.026)	1.108 (0.036)	1.120 (0.058)	1.075 (0.043)
	A	0.772 (0.016)	0.836 (0.020)	0.837 (0.059)	0.858 (0.061)	0.768 (0.046)
	$p <$	0.005**	0.004**	0.001***	0.010**	0.0002***
	Difference	−11.2%	−11.5%	−24.4%	−23.4%	−28.6%
Axon area (μm^2) (SE)	C	0.613 (0.042)	0.739 (0.039)	1.075 (0.068)	1.157 (0.117)	0.968 (0.086)
	A	0.462 (0.024)	0.543 (0.032)	0.594 (0.071)	0.620 (0.081)	0.469 (0.056)
	$p <$	0.006**	0.001***	0.0002***	0.005**	0.0002***
	Difference	−24.6%	−26.5%	−44.8%	−46.9%	−51.6%
Myelin thickness (μm) (SE)	C	0.149 (0.006)	0.153 (0.006)	0.162 (0.005)	0.163 (0.008)	0.157 (0.004)
	A	0.139 (0.003)	0.149 (0.003)	0.142 (0.011)	0.145 (0.007)	0.132 (0.008)
	$p <$	0.197	0.538	0.092	0.141	0.018*
	Difference	ns	ns	ns	ns	−15.8%

CC Corpus callosum, *SE* standard error, *ns* not significant
$^*p \leq 0.05$; $^{**}p \leq 0.01$; $^{***}p \leq 0.001$

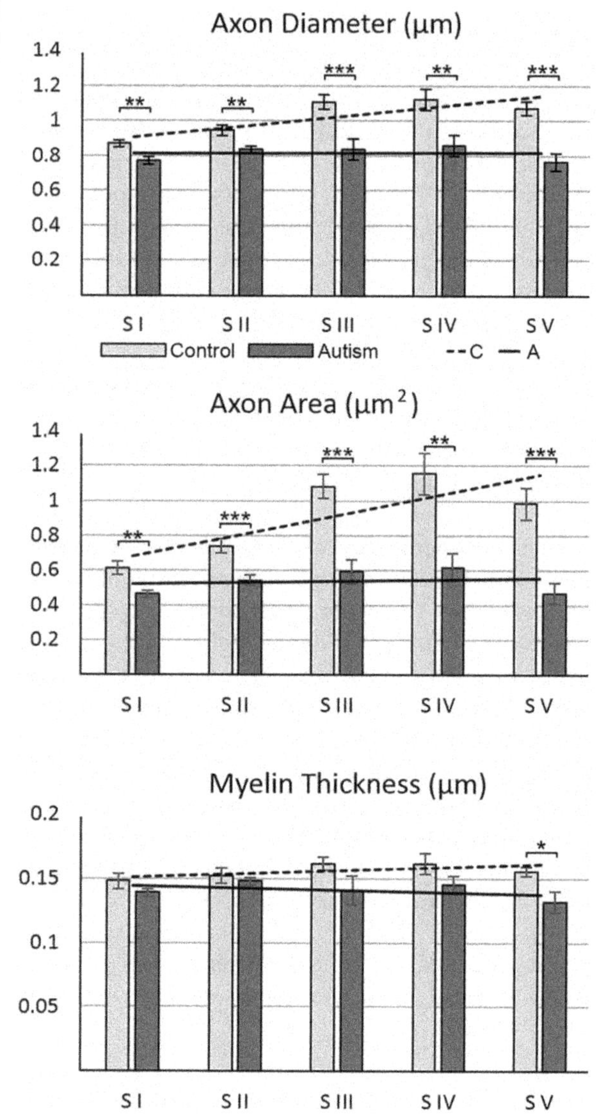

Fig. 3 Axon diameter, cross-sectional area, and myelin thickness. In the CC of control subjects, axon diameter and cross-sectional area were the lowest in S 1 but increased significantly in S II–V, and the increment in segments varied in a broad range. In autistic subjects, the average axon diameter and cross-sectional area were significantly less than in the control group, and differences between segments became insignificant. In contrast to variations of axon diameter and cross-sectional area, the average thickness of myelin sheath was comparable in five segments of control subjects. In autistic subjects, significant reduction in myelin thickness was detectable only in S V

22.2% in S I to 40.2% in S IV ($p < 0.0015$). In contrast to the distinct trajectories of small- and large-diameter axons, the percentage of medium-diameter axons varied in a narrow range from 35.4% (S IV) to 43.8% (S II). These differences were not significant, and there was no detectable trend in the rostro-caudal extent of the CC (Table 5).

Table 3 Significance of difference of axon diameter, cross-sectional area, and myelin thickness between S I and other segments in the CC in autistic and control subjects (test for diversity of axons within segments)

Parameter	Group	S I/S II	S I/S III	S I/S IV	S I/S V
Axon diameter	C	0.055	0.0001***	0.003**	0.0004***
Axon cross-sectional area	C	0.049*	0.0004***	0.003**	0.0014***
Myelin sheath thickness	C	0.483	0.042*	0.101	0.200
Axon diameter	A	0.218	0.440	0.440	0.947
Axon cross-sectional area	A	0.174	0.174	0.174	0.931
Myelin sheath thickness	A	0.055	0.964	0.929	0.891

P values incorporate multiple-comparison adjustments to maintain a false discovery rate (FDR) of 0.05. $*p \leq 0.05$; $**p \leq 0.01$; $***p \leq 0.001$. *S* Segment, *CC* corpus callosum, *C* control, *A* autism

Reverse trajectory in the CC of autistic subjects with an increase in the percentage of small-diameter axons and decrease of large-diameter axons in the posterior CC segments (Fig. 4, Tables 4, 5)

A comparison of the percentage of small-diameter axons in control and autistic subjects revealed an increased contribution of small-diameter axons in all segments of autistic subjects, with a significant increase limited to S I (+ 24%; ($p < 0.027$) and S V (+ 143%; $p < 0.0014$). Comparison of large-diameter axons revealed a reverse trend, with a deficit of large axons in all CC segments of autistic subjects, ranging from − 34.02% in S I ($p < 0.005$) and − 58.2% in S V ($p < 0.0001$).

Table 5 summarizes the results of analyses focused on intersegmental differences in the ratio of axons with different diameters. In neurotypical subjects, a decrease in percentage of small-diameter axons was observed in all segments, but a significant decrease was detected in S III and S V. The increase in percentage of large-diameter axons was significant in all posterior segments in the control group. In the autistic group, the percentages of axons falling within each size range do not show a significant difference (F [6] = 0.99, $p = 0.4310$). Table 5 illustrates the absence of intersegmental differences in all CC segments in autistic subjects.

This pattern was the product of a significant developmental pathology of the CC of autistic subjects documented with an almost flat trendline in the percentage of small-, medium-, and large-diameter axons (Fig. 4). These flat trendlines distinguished the CC of autistic subjects from a very consistent and significant decrease in the percentage of small-diameter axons and consistent and significant increase in the percentage of large-diameter axons in the posterior segments of the CC of neurotypical subjects. The results reflected an almost total loss of structural and functional differentiation of individual segment connections critical for normal human brain function.

Table 4 Different patterns of distribution of the percentage (± SE) of small-, medium-, and large-diameter axons in segments I–V in autistic and control subjects

Axon diameter (µm)	Group	Percentage of axons in CC segments				
		S I	S II	S III	S IV	S V
< 0.651	C	34.6 (2.6)	28.0 (3.0)	20.5 (2.4)	24.3 (5.6)	19.1 (3.4)
	A	43.1 (2.2)	35.0 (2.4)	38.9 (7.9)	37.8 (7.8)	46.6 (6.2)
	$p <$	0.027*	0.093	0.058	0.175	0.0014**
	Diff.	+ 24.42%	ns	ns	ns	+ 143.7%
0.651–1.051	C	43.1 (1.3)	43.8 (1.3)	39.8 (1.2)	35.4 (2.8)	41.6 (2.0)
	A	42.2 (1.8)	45.8 (2.1)	40.8 (5.3)	39.0 (4.2)	36.9 (3.4)
	$p <$	0.696	0.429	0.858	0.472	0.261
	Diff.	ns	ns	ns	ns	ns
> 1.051	C	22.2 (1.9)	28.2 (2.1)	39.6 (3.3)	40.2 (3.8)	39.2 (3.2)
	A	14.7 (1.3)	19.0 (2.1)	20.3 (3.0)	23.1 (3.8)	16.4 (3.1)
	$p <$	0.005**	0.008**	0.007***	0.010**	0.0001***
	Diff.	−34.02%	−32.52%	−48.82%	−42.52	−58.2%

SE Standard error, *CC* corpus callosum, *S* segment, *C* control, *A* autism; Diff., difference; ns, not significant. Standard errors in parentheses

The dominant role of axon number developmental deficits in a significant distortion of connectivity in all five CC segments in autistic subjects

The major factor defining structure and function of interhemispheric connectivity was the developmental focal and diffuse deficit of axons in all segments and in the entire CC of autistic subjects. The total number of myelinated axons was 49.42% less in the autistic (44.304 million) than in the control group (89.655 million) (Table 6).

The estimates of the total number of small-diameter (< 0.651 µm) axons revealed a significant reduction in S I and S III by 46.5 and 40.1%, respectively (Table 7). Evaluation of the total number of large-diameter (> 1.05 µm) axons revealed a very consistent and significant deficit in all five segments in autistic subjects, by 72.6% on average. The number of medium-size axons (0.651 to 1.051 µm in diameter) was also reduced significantly in all five segments, by 49.0% on average (Fig. 5).

Table 5 Significance of difference in the percentage of small-, medium-, and large-diameter axons in segment I compared to four other segments in the CC in autistic and control subjects

Axon diameter (µm)	Group	S I/S II	S I/S III	S I/S IV	S I/S V
< 0.651	C	0.073	0.002**	0.073	0.002**
0.651–1.051	C	0.646	0.224	0.052	0.616
> 1.051	C	0.048*	0.0002***	0.0015**	0.0002***
< 0.651	A	0.028*	0.663	0.663	0.663
0.651–1.051	A	0.232	0.745	0.404	0.232
> 1.051	A	0.221	0.221	0.221	0.684

CC corpus callosum, *S* segment, *C* control, *A* autism. *P* values incorporate multiple-comparison adjustments to maintain a false discovery rate (FDR) of 0.05. *$p \leq 0.05$; **$p \leq 0.01$; ***$p \leq 0.001$

The study revealed multilevel consequences of two pathological developmental processes: those reducing the number of axons, and those reducing the diameter of axons in autistic subjects. The final product is a segment-specific deficit of small-, medium-, and large-diameter axons, and loss of the integrity of interhemispheric connections in autistic subjects.

Discussion

Processing deficits in high-functioning autistic individuals are detected as shortages in integration of information and coordination of multiple neuronal networks [35, 36]. The CC, the largest human brain commissure, provides long-range interhemispheric connections integrating and synchronizing anatomically and functionally specialized cortices. The pattern of detected pathology of CC connections in autistic subjects indicates that these abnormalities are a product of failure of two types of mechanisms: (a) those controlling the number of interhemispheric connections, and (b) those controlling axon structure, including axon diameter and cross-sectional area. Failure of the first type of mechanism results in loss of, or reduction in the number of axons and underconnectivity affecting all major anatomical and functional cortical regions. Failure of the second type of mechanisms results in reduction of axon diameter and cross-sectional area in all five CC segments already affected by an axon deficit. Functional studies have revealed that decrease of axon diameter correlates with reduction in the velocity and volume of signal transmission [8, 40, 47, 48]. Our study of autistic subjects provides evidence for the: (a) reduction in the number of long-range axonal connections in the CC, (b) abnormal structure and function of preserved axons,

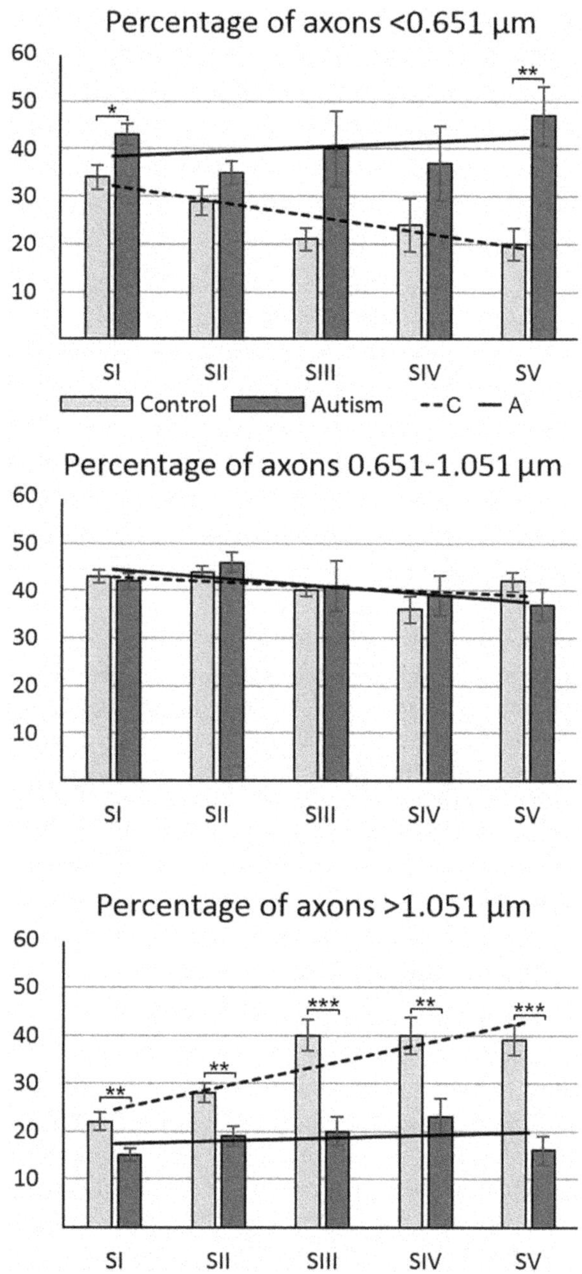

Fig. 4 Changes of the percentage of small-, medium-, and large-diameter axons in autism. Estimation of the percentage of small- (< 0.651 μm), medium- (> 0.651–1.051 μm), and large-diameter (> 1.051 μm) axons revealed that three features characterized CC connectivity in neurotypical control subjects: significant decrease in the percentage of small-diameter axons in posterior segments, especially in S III–S V; increase in the percentage of large-diameter axons, especially prominent in S III–S V, and a broad heterogeneity of axon size in all segments. The sign of pathological remodeling of interhemispheric connections in CC segments in autistic subjects was stabilization of the percentage of small-, medium-, and large-diameter axons close to approximately 40, 40, and 20%, respectively. The effect of these changes in autistic subjects is the loss of structural diversity of axons typical for a normal brain, resulting in almost flat trend lines for the percentages of small-, medium-, and large-diameter axons in CC segments in autistic subjects

and (c) partial loss of structural and functional specialization of CC segments and integrity of interhemispheric connections.

Interhemispheric underconnectivity

Numerous structural MRI studies have revealed in autistic subjects a reduced size of the CC, and abnormal shape, including thinning of the CC, considered to be markers of abnormal interhemispheric connectivity [10, 14, 20, 21, 27, 29, 41, 50, 54]. The first postmortem study of 11 brains of autistic subjects, including nine brains examined in this ultrastructural study, expanded on MRI observations with light microscopy resolution and counts of axons in five CC segments [56]. In the examined brains of autistic subjects, two types of CC pathology contributed to interhemispheric underconnectivity: focal agenesis and diffuse hypoplasia. In three autistic subjects, total agenesis of S IV disconnected the sensory cortex, whereas partial agenesis of S III and S V partially disconnected the motor and parietal cortex. Partial agenesis of S I reduced interhemispheric connections of the prefrontal cortex. This topography revealed several CC segments with a high susceptibility to misguidance of axons and regional interhemispheric disconnection in autistic subjects. The second major cause of the deficit of interhemispheric connections detected in all examined autistic subjects was hypoplasia, defined by CC thinning, reduction of CC midsagittal area, and diffuse 62% reduction in the number of axons affecting all CC segments. The effect is loss of structural and functional integrity of interhemispheric connections, which may result in dysregulation of the velocity and volume of information transferred between hemispheres, loss of coherence and deficits in complex information processing in autistic subjects.

Mechanisms contributing to underconnectivity and abnormal structure/function of axons

Two independent and methodologically different studies of genes involved in neurodevelopmental disorders identified robust, statistically significant evidence for convergence of the input of ASD risk genes in glutamatergic projection neurons in cerebral cortex layers 2/3 [37] and 5/6 [59]. Neuropathological studies of cortical neurons revealed that the most common neuron developmental anomalies reported in the cerebral cortex of autistic subjects are multiregional reduction of neuron soma size [26, 52, 53], minicolumn pathology always associated with a decrease in neuron size [11], and cortical dysplasia with smaller neurons [10, 57]. The link between reduced size of neurons and reduced axon diameter is strengthened by the fact that these reports were based

Table 6 The difference between number (million ± SE) of axons in five CC segments in autistic and control subjects

Group	S I	S II	S III	S IV	S V	Total number
C	25.900 (1.891)	24.944 (1.175)	9.878 (1.417)	3.711 (0.326)	25.222 (2.518)	89.655 (4.186)
A	10.933 (2.010)	15.500 (1.642)	3.211 (0.644)	1.360 (0.441)	13.300 (1.618)	44.304 (4.885)
p <	0.0001***	0.0003***	0.0012**	0.0006***	0.0011**	0.0001***
Differ.	−57.79%	−37.87%	67.49%	−63.35%	−47.27%	−49.42%

SE standard error, *CC* corpus callosum, *S* segment, *C* control, *A* autism; Differ., difference. Standard errors in parentheses. $^{**}p \leq 0.01$; $^{***}p \leq 0.001$

on examination of brains described in the current CC study. Energy capacity estimated by percentage of axon volume occupied with mitochondria increases with axon diameter [39]. One may assume that the energy capacity of smaller-diameter axons is proportional to the energy capacity of a reduced volume of neurons. CC connection pathology might be an effect of developmental anomalies of both neurons and glia. Long-range interhemispheric connectivity is controlled in utero by brain midline glial cells at six decision points. Misguidance of axons at any of these points by factors independent of the axon may change the trajectory and target of the axon [4, 43, 44] and result in partial CC agenesis and hypoplasia in autistic subjects.

Broad range of differences in axon diameter in CC segments in neurotypical subjects

Precision in synchronization of information transfer between hemispheres became more challenging with an evolutionary increase of brain size in apes and humans, which required both an increase in the velocity of long-distance interhemispheric conductivity and development of a variety of conduction velocities necessary for the functional diversity of human cortices and the brain [40].

Our study of neurotypical controls revealed a large range of axon diameters, reflecting a variety of conduction velocities, in each of the five CC segments, essential for cortices with different functions. Average axon diameter and cross-sectional area were the smallest in the first segment and increased in the posterior segments. The average diameter and cross-sectional area of axons in interhemispheric connections of the sensory cortex (S IV) were increased by 29 and 88%, respectively, in comparison to axons of the prefrontal cortex (S I).

Another feature of the CC in neurotypical controls was an increase in the percentage of large-diameter (> 1.051 μm) axons, known as high-velocity axons, from 22% in connections of the prefrontal cortex (S I) to 40–39% in connections of the motor (S III), sensory (S IV), and parietal, occipital, and temporal cortices (S V). High-transmission velocity across distant regions is essential for synchronization of the activity of remote neuronal circuits [49]. The percentage of small-diameter axons (< 0.651 μm) decreased from 34 to 29% in S I and II to approximately 20–24% in S III–S V. The axon composition detected in the examined control group resembled a higher percentage of small-diameter axons in the anterior CC and a higher percentage of large-diameter axons in the posterior portion of the CC reported in other studies of neurotypical subjects. This pattern reflected the structural and functional specialization of each segment of

Table 7 The difference between number (± SE) of small-, medium-, and large-diameter axons in five CC segments in autistic and control subjects

Axon diameter	Group	S I	S II	S III	S IV	S V
< 0.651	C	8.789 (0.711)	6.955 (0.745)	1.924 (0.259)	0.862 (0.187)	4.427 (0.568)
	A	4.701 (0.890)	5.422 (0.742)	1.153 (0.150)	0.817 (0.291)	5.881 (0.885)
	p <	0.0025	0.164 (ns)	0.0253	0.8916 (ns)	0.1860 (ns)
	Diff.	−46.5%		−40.1%		
> 0.651 < 1.051	C	11.177 (0.930)	10.889 (0.530)	3.967 (0.639)	1.319 (0.144)	10.405 (0.934)
	A	4.704 (0.972)	7.083 (0.836)	1.633 (0.332)	0.757 (0.178)	5.084 (0.974)
	p <	0.0002	0.0014	0.0071	0.0294	0.0012
	Diff.	−57.9%	−35.0%	−58.8%	−42.6%	−51.2%
> 1.051	C	5.933 (0.951)	7.099 (0.735)	3.985 (0.679)	1.529 (0.231)	10.389 (1.749)
	A	1.527 (0.275)	2.993 (0.438)	0.825 (0.183)	0.465 (0.159)	2.334 (0.644)
	p <	0.0015	0.0002	0.0014	0.0049	0.0015
	Diff.	−74.2%	−57.8%	−79.3%	−69.6%	−77.6%

SE standard error, *CC* corpus callosum, *S* segment, *C* control, *A* autism; Differ., difference. Standard errors in parentheses

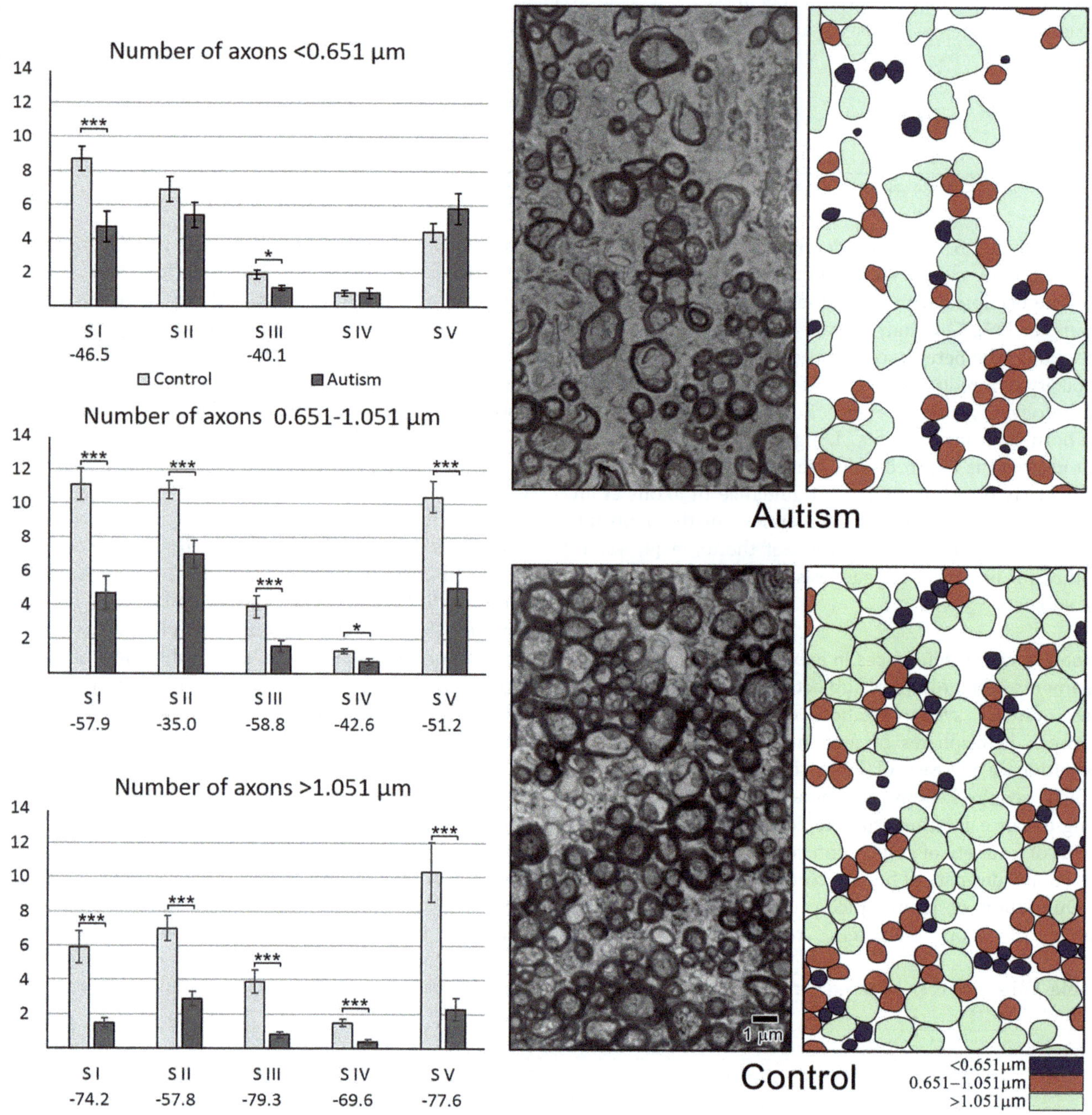

Fig. 5 Segment-specific deficits in the number of small-, medium-, and large-diameter axons in autism. Estimation of total number (millions) of axons in the CC of control and autistic subjects revealed significant deficit in total number of small-diameter axons in S I (46.5%) and S III (40.1%) in autistic subjects. The dominant feature of all five segments was 49% deficit of medium-diameter axons and 72.6% deficit of large-diameter axons in autistic subjects. Electron micrographs and respective plots of axons illustrate deficit of small-, medium-, and large-diameter axons in CC genu (S I) in autistic subject compared to control subject

interhemispheric connections being critical for normal human brain function [1, 2, 45].

Loss of diversity of axonal connections in five CC segments in autistic subjects

Velocity of signal transmission and firing rate correlate with axon diameter. Distribution of axon diameter, narrow or broad, symmetric or skewed, reflects the heterogeneity of information rates conveyed by the axon. A wide velocity range provides the wide range of timing necessary for synchronization of information delivery, processing, and response in an unaffected brain [39]. EM-based measurements of axons in autistic subjects revealed a reduction of average axon diameter by 12–29%

and of average axon cross-sectional area by 25–52% in the rostro-caudal CC extent. These structural abnormalities resulted in a loss of the axon structural and functional diversification critical for regional and global cortical functionality.

Posterior CC segments are involved in the transfer of motor [55], somatosensory [13], and auditory and visual information [58] with a high demand for fast conduction and processing. However, reduction in the total number of axons with a diameter of 0.651–1.051 μm by an average of 49%, and of axons with a diameter > 1.051 μm by an average of 72.6%, is an indicator of a significant reduction in conductivity by both medium- and large-diameter axons. Moreover, the deficit of medium- and large-diameter axons affects not only posterior CC segments but also S I and S II connecting the prefrontal, premotor, and supplementary motor cortices. These deficits and the decreased diversification of interhemispheric connections may make a major contribution to the autism phenotype and comorbidities including intellectual deficits.

Conclusions

This study revealed the complex nature of developmental defects affecting the structure and function of interhemispheric connections in autistic subjects. Agenesis and hypoplasia caused a severe reduction in the number of CC axons and loss of integrity of interhemispheric connectivity. Developmental anomalies of axons included a significant increase in the percentage of small-diameter and low-velocity axons, and a decrease in the percentage of large-diameter high-velocity axons in the CC of autistic subjects. The loss of structural and functional diversity in CC segments in autistic subjects reflected a loss of specialization of interhemispheric connections essential for higher-order functions and for integration of multisensory perception. In general, light and electron microscopy studies indicate that autism is associated with extensive developmental distortion of structure and function of interhemispheric connectivity.

Acknowledgments

Hemispheric brain samples were obtained from the Harvard Brain Tissue Resource Center, Belmont, MA, supported in part by PHS grant number R24-MH 068855; the Brain and Tissue Bank for Developmental Disorders of the National Institute of Child Health and Human Development at the University at Maryland; and the New York State Brain and Tissue Bank at the Institute for Basic Research in Developmental Disabilities (IBR), Staten Island, NY. Tissue and clinical record acquisition was coordinated by the Autism Tissue Program (ATP; Director: Jane Pickett, Ph.D.) of Autism Speaks (AS) (currently Autism BrainNet, sponsored by the Simons Foundation and Autism Speaks). Tissue was processed, cut, and stained at IBR and shared with partnering institutions consistently with AS/ATP recommendations.
We thank Mrs. Maureen Marlow for manuscript editing.

Funding

This study was supported in part by funds from the New York State Office for People with Developmental Disabilities, a grant from the Department of Defense Autism Spectrum Disorders Research Program (AS073234, J.W., T.W.), and a grant from Autism Speaks (Princeton, NJ; J.W.).

Authors' contributions

JW designed the project, drafted the manuscript, and together with WK performed the study. MF analyzed data. IK and KN assisted in tissue preservation and morphometric studies. VM-C. contributed to research design and writing manuscript. TW contributed to neuropathological evaluation and interpretation of clinicopathological data. JW. assisted in project design, coordination of tissue preservation and examination. All authors read and approved the final manuscript.

Competing interests

The authors declare that they have no competing interests.

Author details

[1]Department of Developmental Neurobiology, New York State Institute for Basic Research in Developmental Disabilities, 1050 Forest Hill Road, Staten Island, NY 10314, USA. [2]Research Design and Analysis Services, New York State Institute for Basic Research in Developmental Disabilities, Staten Island, NY 10314, USA. [3]Pathology and Laboratory Medicine, Institute for Pediatric Regenerative Medicine, MIND Institute, University of California, Davis, CA, USA. [4]Departments of Neurology, Pathology and Psychiatry, NYU Langone Medical Center, New York, NY, USA.

References

1. Aboitiz F, Montiel J (2003) one hundred million years of interhemispheric communication: the history of the corpus callosum. Braz J Med Biol Res 2003;36:409–420
2. Aboitiz F, Scheibel AB, Fisher RS, Zaidel E (1992) Fiber composition of the human corpus callosum. Brain Res 598:143–153
3. American Psychiatric Association (2013) Diagnostic and statistical manual of mental disorders, 5th edn. VA, Arlington
4. Bagri A, Marin O, Plump AS, Mak J, Pleasure SJ, Rubenstein JL et al (2002) Slit proteins prevent midline crossing and determine the dorsoventral position of major axonal pathways in mammalian forebrain. Neuron 33:233–248
5. Barnea-Goraly N, Kwon H, Menon V, Eliez S, Lotspeich L, Reiss AL (2004) White matter structure in autism: preliminary evidence from diffusion tensor imaging. Biol Psychiatry 55:323–326
6. Belmonte MK, Allen G, Beckel-Mitchener A, Boulanger LM, Carper RA, Webb SJ (2004) Autism and abnormal development of brain connectivity. J Neurosci 24:9228–9231
7. Benjamini Y, Hochberg Y (1995) Controlling the false discover rate: a practical and powerful approach to multiple testing. J Royal Statistical Society, Series B 57:289–300
8. Caminiti R, Carducci F, Piervincenzi C, Battaglia-Mayer A, Confalone G, Visco-Comandini F et al (2013) Diameter, length, speed, and conduction delay of callosal axons in macaque monkeys and humans: comparing data from histology and magnetic resonance imaging diffusion tractography. J Neurosci 33:14501–14511
9. Casanova FM, El-Baz A, Elnakib A, Switala AE, Williams EL, Williams DL et al (2011) Quantitative analysis of the shape of the corpus callosum in autistic individuals. Autism 15:223–238
10. Casanova MF, El-Baz A, Kamat SS, Dombroski BA, Khalifa F, Elnakib A et al (2013) Focal cortical dysplasias in autism spectrum disorders. Acta Neuropathol Commun 1:67
11. Casanova MF, Van Kooten IAJ, Switala AE, Van Engeland H, Heinsen H, Steinbusch HWM et al (2006) Minicolumnar abnormalities in autism. Acta Neuropathol 112:287 303

12. Courchesne E, Pierce K (2005) Why the frontal cortex in autism might be talking only to itself: local overconnectivity but long-distance disconnection. Curr Opin Neurobiol 23:225–230
13. Fabri M, Polonara G, Del Pesce M, Quattrini A, Salvolini U, Manzoni T (2001) Posterior corpus callosum and interhemispheric transfer of somatosensory information: an fMRI and neuropsychological study of a partially callosotomized patient. J Cogn Neurosci 13:1071–1079
14. Freitag CM, Luders E, Hulst HE, Narr KL, Thompson PM, Toga AW et al (2009) Total brain volume and corpus callosum size in medication-naïve adolescents and young adults with autism spectrum disorder. Biol Psychiatry 66:316–319
15. Frith CD (2004) Is autism a disconnection disorder? Lancet Neurol 3:577
16. Geschwind DH, Levitt P (2007) Autism spectrum disorders: developmental disconnection syndromes. Curr Opin Neurobiol 17:103–111
17. Gleason JR. Within subjects (repeated measures) ANOVA, including between subjects factors. Stata Technical Bulletin 1999;47:40–45
18. Glickman ME, Rao SR, Schultz MR (2014) False discovery rate control is a recommended alternative to Bonferroni-type adjustments in health studies. J Clin Epidemiol 67:850–857
19. Greenhouse SW, Geisser S (1959) On methods in the analysis of profile data. Psychometrika 24:95–112
20. Hardan AY, Minshew NJ, Keshavan MS (2000) Corpus callosum size in autism. Neurology 55:1033–1036
21. Hardan AY, Pabalan M, Gupta N, Bansal R, Melhem NM, Fedorov S et al (2009) Corpus callosum volume in children with autism. Psychiatry Res Neuroimaging 174:57–61
22. Heinsen H, Arzberger T, Schmitz C (2000) Celloidin mounting (embedding without infiltration) – a new, simple and reliable method for producing serial sections of high thickness through complete human brains and its application to stereological and immunohistochemical investigation. J Chem Neuroanat 20:49–59
23. Highley JR, Esiri MM, McDonald B, Cortina-Borja M, Herron BM, Crow TJ (1999) The size and fiber composition of the corpus callosum with respect to gender and schizophrenia: a post-mortem study. Brain 122:99–110
24. Hofer S, Frahm J (2006) Topography of the human corpus callosum revisited–comprehensive fiber tractography using diffusion tensor magnetic resonance imaging. NeuroImage 32:989–994
25. Hughes JR (2007) Autism: the first firm finding = underconnectivity? Epilepsy Behav 11:20–24
26. Jacot-Descombes S, Uppal N, Wicinski B, Santos M, Schmeidler J, Giannakopoulos P et al (2012) Decreased pyramidal neuron size in Brodmann areas 44 and 45 in patients with autism. Acta Neuropathol 124:67–79
27. Just MA, Cherkassky VL, Keller TA, Kana RK, Minshew NJ (2007) Functional and anatomical cortical underconnectivity in autism: evidence from an fMRI study of an executive function task and corpus callosum morphometry. Cereb Cortex 17:951–961
28. Just MA, Cherkassky VL, Keller TA, Minshew NJ (2004) Cortical activation and synchronization during sentence comprehension in high-functioning autism: evidence of underconnectivity. Brain 127:1811–1821
29. Just MA, Keller TA, Malave VL, Kana RK, Varma S (2012) Autism as a neural systems disorder: a theory of frontal posterior underconnectivity. Neurosci Biobehav Rev 36:1292–1313
30. Kana RK, Keller TA, Minshew NJ, Just MA (2007) Inhibitory control in high functioning autism: decreased activation and underconnectivity in inhibition networks. Biol Psychiatry 62:198–206
31. Kumar A, Sundaram SK, Sivaswamy L, Behen ME, Makki MI, Ager J et al (2010) Alterations in frontal lobe tracts and corpus callosum in young children with autism spectrum disorder. Cereb Cortex 20:2103–2113
32. LaMantia AS, Rakic P (1990) Cytological and quantitative characteristics of four cerebral commissures in the rhesus monkey. J Comp Neurol 291:520–537
33. Lord C, Risi S, Lambrecht L, Cook EH Jr, Leventhal BL, DiLavore PC et al (2000) The autism diagnostic observation schedule-generic: a standard measure of social and communication deficits associated with the spectrum of autism. J Autism Dev Disord 30:205–223
34. Minshew NJ, Keller TA (2010) The nature of brain dysfunction in autism: functional brain imaging studies. Curr Opin Neurol 23:124–130
35. Minshew NJ, Sweeney J, Luna B (2002) Autism as a selective disorder of complex information processing and underdevelopment of neocortical systems. Mol Psychiatry 7(Suppl 2):S14–S15
36. Minshew NJ, Williams DL (2007) The new neurobiology of autism: cortex, connectivity, and neuronal organization. Arch Neurol 65:945–950
37. Parikshak NN, Luo R, Zhang A, Won H, Lowe JK, Chandran V et al (2013) Integrative functional genomic analyses implicate specific molecular pathways and circuits in autism. Cell 155:1008–1021
38. Paul LK, Van Lancker D, Schieffer B, Dietrich R, Brown WS (2003) Communicative deficits in individuals with agenesis of the corpus callosum: nonliteral language and affective prosody. Brain Lang 85:313–324
39. Perge JA, Niven JE, Mugnaini E, Balasubramanian V, Sterling P (2012) Why do axons differ in caliber? J Neurosci 32:626–638
40. Phillips KA, Stimpson CD, Smaers JB, Raghanti MA, Jacobs B, Popratiloff A et al (2015) The corpus callosum in primates: processing speed of axons and the evolution of hemispheric asymmetry. Proc R Soc B 282:1535
41. Piven J, Bailey J, Ranson BJ, Arndt S (1997) An MRI study of the corpus callosum in autism. Am J Psychiatry 154:1051–1056
42. Prigge MBD, Lange N, Bigler ED, Merkley TL, Neeley ES, Abildskov TJ et al (2013) Corpus callosum area in children and adults with autism. Res Autism Spectr Disord 7:221–234
43. Ren T, Anderson A, Shen W-B, Huang H, Plachez C, Zhang J et al (2006) Imaging, anatomical, molecular analysis of callosal formation in the developing human brain. Anat Record A 288A:191–204
44. Richards LJ, Plachez C, Ren T (2004) Mechanisms regulating the development of the corpus callosum and its agenesis in mouse and human. Clin Genet 66:276–289
45. Riise J, Pakkenberg B (2011) Stereological estimation of the total number of myelinated callosal fibers in human subjects. J Anat 218:277–284
46. Rippon G, Brock J, Brown C, Boucher J (2007) Disordered connectivity in the autistic brain: challenges for the "new psychophysiology". Int J Psychophysiol 63:164–172
47. Rushton WAH (1951) A theory on the effects of fibre size in medullated nerve. J Physiol 115:101–122
48. Seidl AH (2014) Regulation of conduction time along axons. Neuroscience 12:126–134
49. Singer W (1999) Neuronal synchrony: a versatile code for the definition of relations? Neuron:49–65
50. Stanfield AC, McIntosh AM, Spencer MD, Philip R, Gaur S, Lawrie SM (2008) Towards a neuroanatomy of autism: a systematic review and meta-analysis of structural magnetic resonance imaging studies. Eur Psychiatry 23:289–299
51. StataCorp. Stata: Release 15. Statistical Software. College Station, TX: StataCorp LLC. 2017
52. Uppal N, Wicinski B, Buxbaum JD, Heinsen H, Schmitz C, Hof PR (2014) Neuropathology of the anterior midcingulate cortex in young children with autism. J Neuropathol Exp Neurol 73:891–902
53. van Kooten IAJ, Palmen SJMC, von Cappeln P, Steinbusch HWM, Korr H, Heinsen H et al (2008) Neurons in the fusiform gyrus are fewer and smaller in autism. Brain 131:987–999
54. Vidal CN, Nicolson R, DeVito TJ, Hayashi KM, Geaga JA, Drost DJ et al (2006) Mapping corpus callosum deficits in autism: an index of aberrant cortical connectivity. Biol Psychiatry 60:218–225
55. Wahl M, Lauterbach-Soon B, Hattingen E, Jung P, Singer O, Volz S et al (2007) Human motor corpus callosum: topography, somatotopy, and link between microstructure and function. J Neurosci 27:12132–12138
56. Wegiel J, Flory M, Kaczmarski W, Brown WT, Chadman K, Wisniewski T et al (2017) Partial agenesis and hypoplasia of the corpus callosum in idiopathic autism. J Neuropathol Exp Neurol 76:225–237
57. Wegiel J, Kuchna I, Nowicki K, Imaki H, Wegiel J, Marchi E et al (2010) The neuropathology of autism: defects of neurogenesis and neuronal migration, and dysplastic changes. Acta Neuropathol 119:755–770
58. Westerhausen R, Grüner R, Specht K, Hugdahl K (2009) Functional relevance of interindividual differences in temporal lobe callosal pathways: a DTI tractography study. Cereb Cortex 19:1322–1329

59. Willsey AJ, Sanders SJ, Li M, Dong S, Tebbenkamp AT, Muhle RA et al (2013) Coexpression networks implicate human midfetal deep cortical projection neurons in the pathogenesis of autism. Cell 155:997–1007
60. Witelson SF (1989) Hand and sex differences in the isthmus and genu of the human corpus callosum. A postmortem morphological study. Brain 112:799–835
61. Zikopoulos B, Barbas H (2010) Changes in prefrontal axons may disrupt the network in autism. J Neurosci 30:14595–14609

Medulloblastoma therapy generates risk of a poorly-prognostic H3 wild-type subgroup of diffuse intrinsic pontine glioma: a report from the International DIPG Registry

Hunter C. Gits[2], Maia Anderson[2], Stefanie Stallard[2], Drew Pratt[1], Becky Zon[2], Christopher Howell[3], Chandan Kumar-Sinha[1,4], Pankaj Vats[1,4], Katayoon Kasaian[5], Daniel Polan[6], Martha Matuszak[6], Daniel E. Spratt[6], Marcia Leonard[2], Tingting Qin[7], Lili Zhao[8], James Leach[9], Brooklyn Chaney[10], Nancy Yanez Escorza[10], Jacob Hendershot[11], Blaise Jones[9], Christine Fuller[12], Sarah Leary[13], Ute Bartels[14], Eric Bouffet[14], Torunn I. Yock[15], Patricia Robertson[16], Rajen Mody[2], Sriram Venneti[1], Arul M. Chinnaiyan[1,4], Maryam Fouladi[17], Nicholas G. Gottardo[4,18,19] and Carl Koschmann[2*]

Abstract

With improved survivorship in medulloblastoma, there has been an increasing incidence of late complications. To date, no studies have specifically addressed the risk of radiation-associated diffuse intrinsic pontine glioma (DIPG) in medulloblastoma survivors. Query of the International DIPG Registry identified six cases of DIPG with a history of medulloblastoma treated with radiotherapy. All patients underwent central radiologic review that confirmed a diagnosis of DIPG. Six additional cases were identified in reports from recent cooperative group medulloblastoma trials (total $n = 12$; ages 7 to 21 years). From these cases, molecular subgrouping of primary medulloblastomas with available tissue ($n = 5$) revealed only non-WNT, non-SHH subgroups (group 3 or 4). The estimated cumulative incidence of DIPG after post-treatment medulloblastoma ranged from 0.3–3.9%. Posterior fossa radiation exposure (including brainstem) was greater than 53.0 Gy in all cases with available details. Tumor/germline exome sequencing of three radiation-associated DIPGs revealed an H3 wild-type status and mutational signature distinct from primary DIPG with evidence of radiation-induced DNA damage. Mutations identified in the radiation-associated DIPGs had significant molecular overlap with recurrent drivers of adult glioblastoma (e.g. *NRAS*, *EGFR*, and *PTEN*), as opposed to epigenetic dysregulation in H3-driven primary DIPGs. Patients with radiation-associated DIPG had a significantly worse median overall survival (median 8 months; range 4–17 months) compared to patients with primary DIPG. Here, it is demonstrated that DIPG occurs as a not infrequent complication of radiation therapy in survivors of pediatric medulloblastoma and that radiation-associated DIPGs may present as a poorly-prognostic distinct molecular subgroup of H3 wild-type DIPG. Given the abysmal survival of these cases, these findings provide a compelling argument for efforts to reduce exposure of the brainstem in the treatment of medulloblastoma. Additionally, patients with radiation-associated DIPG may benefit from future therapies targeted to the molecular features of adult glioblastoma rather than primary DIPG.

Keywords: Secondary malignant neoplasm, Diffuse intrinsic pontine glioma, Medulloblastoma, Cranial irradiation, Brainstem

* Correspondence: ckoschma@med.umich.edu
[2]Department of Pediatrics, Division of Pediatric Hematology-Oncology; Michigan Medicine, Ann Arbor, MI 48109, USA
Full list of author information is available at the end of the article

Introduction

Medulloblastoma is the most common malignant pediatric brain tumor, and standard treatment includes surgical resection followed by adjuvant external beam radiation therapy (EBRT) and systemic chemotherapy [14]. As the prognosis of medulloblastoma has improved, late complications such as secondary malignant neoplasms (SMNs) have increased in frequency [15, 39]. While 10-year survival rates of medulloblastoma are now near 80%, the 20-year cumulative incidence of SMNs is reported to be as high as 20%, comprising 11.8% of late mortality [15, 29, 31, 39].

The elevated risk of SMNs in medulloblastoma survivors may be due to high doses of EBRT. The risks of glioma, the most common SMN reported after primary medulloblastoma, increase linearly with radiation dose [9, 19, 31, 42]. Radiation dosing for medulloblastoma varies based on clinical and molecular risk stratification, and standard treatment involves craniospinal irradiation (CSI) with a posterior fossa boost. No previous studies have assessed the risk of the development of radiation-associated DIPG in medulloblastoma survivors, which could impact the future dose and modality of radiation therapy in future clinical trials.

Diffuse intrinsic pontine glioma (DIPG) is a rare infiltrative brainstem tumor, and patients rarely survive longer than 2 years after diagnosis. DIPG is diagnosed primarily by radiographic features showing an intrinsic lesion that encompasses at least 50% of the pons. When available, histology frequently shows features consistent with an infiltrating high-grade glioma (HGG). Approximately 80% of these tumors harbor a point mutation in the histone H3 (H3.3 and H3.1), which now defines the new histomolecular entity Diffuse Midline Glioma, H3K27M-mutant, and is associated with epigenetic dysregulation of neuro-developmental pathways and a worse prognosis than H3 wild-type DIPG [8]. Recent studies suggest that H3 mutants are distinct biological entities, and that H3.3 mutants alone may display a worse prognosis relative to H3 wild-type [10, 20].

There is a paucity of data specifically addressing the risk and molecular characteristics of radiation-associated DIPG among medulloblastoma survivors. A recent report performed genomic analysis of recurrent tumors of seventeen pediatric medulloblastoma patients [33]. The report revealed some of the tumors as secondary glioblastomas with known driver mutations and identified *PDGFRA* as a potential molecular target for therapy. Although this work addressed radiation-associated cancers following treatment for pediatric medulloblastoma, there were no pontine tumors and there remains no published incidence data for radiation-associated DIPG. Here, this report describes a poorly-prognostic molecular subgroup of H3 wild-type DIPG that occurs as a not infrequent complication of radiation therapy in survivors of pediatric medulloblastoma.

Materials and methods

Case acquisition

The International DIPG Registry (IDIPGR) was queried for cases of DIPG diagnosed after radiation treatment for primary medulloblastoma. Details of the registry's structure and recruitment are described elsewhere [12]. Diagnosis of DIPG was confirmed by central radiology review by two primary neuroradiologists (BJ, JL). A Medline/PubMed and Google Scholar search was performed to identify any additional published cases. Various combinations of keywords were used including: *medulloblastoma, diffuse intrinsic pontine glioma, brainstem glioma, pontine glioma, secondary malignant neoplasm.* Articles dated from 1999 to 2017 were obtained and demographic, treatment, and survival data extracted as available. All patients diagnosed with primary medulloblastoma from age 0–21 years who were subsequently diagnosed with brainstem glioma were included. While it was not possible to review imaging for all cases obtained from primary literature, care was taken to exclude patients with focal (non-diffuse) brainstem tumors.

Methylation analyses of primary medulloblastoma

The medulloblastoma methylation-derived subgrouping was performed using the Infinium Assay with the Illumina MethylationEPIC BeadChip platform. DNA was extracted and isolated according to standard protocols, and bisulfite conversion was performed using the Zymo EZ DNA methylation kit. A support vector machine was trained on a cohort of medulloblastoma samples to develop a methylation-derived sub-classification prediction algorithm. The MethylationEPIC BeadChip 46 CpG dinucleotide signature algorithm and R statistical program (version 3.0.0) were used to classify the medulloblastoma tumor into one of four subgroups: Sonic hedgehog pathway activated (SHH), Wnt-pathway activated (WNT), Group 3, or Group 4. Quality control parameters were assessed using Illumina Genome Studio V2011.1 (Methylation Module, version 1.9.1000).

Karyotyping and immunohistochemistry of DIPG

Fresh tumor was disaggregated mechanically and enzymatically using collagenase V (Sigma-Aldrich, St. Louis, MO). The suspension cultures were incubated overnight or 24 h before harvest, and in-situ cultures were harvested after 3–12 days in culture. Karyotype was interpreted according to the International System for Human Cytogenetic Nomenclature (ISCN 2013). Immunohistochemical studies were performed as previously published using the Discovery XT processor (Ventana Medical Systems) [40, 41]. In brief, immunostaining was performed using the rabbit polyclonal anti-H3K27me3 (07–449, Millipore, Billercia, MA; 1 μg/mL) or rabbit polyclonal anti-H3.3 K27 M (ABE419, Millipore, Billercia, MA; 0.5

μg/mL) antibodies. Streptavidin- HRP and DAB detection kit (Ventana Medical Systems) were used according to the manufacturer instructions.

Exome and transcriptome profiling of DIPG

For cases 1 and 3, frozen DIPG tumor and normal brain tissue from autopsy were submitted for whole exome (paired tumor and germline DNA) and transcriptome (tumor RNA) sequencing. Clinically-integrated sequencing was performed according to previously published methodology [26]. Nucleic acid preparation, high-throughput sequencing, and computational analysis were performed by the Michigan Center for Translational Pathology sequencing laboratory using standard protocols in adherence to the Clinical Laboratory Improvement Amendments.

For case 6, only frozen DIPG tissue (no germline DNA sample) was submitted for whole exome sequencing (WES). This sample was processed through the GATK 3.6 variant analysis pipeline as a germline sample. After variants were called, variant annotation using SnpEff was completed to assign mutation information. This annotated VCF file was filtered using bcftools on a known set of specific genes/histones of interest to examine the mutational landscape. These gene specific variants then were examined further based on mutation effect (high, moderate) as to elucidate the presence of any non-synonymous variants, processed to remove common variants with ≥5% allele frequencies in 1000 Genomes Project (2015) [18], NHLBI Exome Sequencing Project 6500 (ESP6500) [16], Exome Aggregation Consortium dataset (http://exac.broadinstitute.org), and Genome Aggregation Database (http://gnomad.broadinstitute.org), and removed of non-recurrent somatic variants using annotations in the COSMIC v70 database (http://cancer.sanger.ac.uk/cosmic).

Mutational signature

The somatic mutations in each of the DIPG samples from cases 1 and 3 processed through MiOncoseq sequencing platform [26] were categorized into one of the 96 possible categories: 6 classes of base substitution (C > A, C > G, C > T, T > A, T > C and T > G) × 16 combinations of bases immediately 5′ and 3′ to each mutation base (context information), and the frequency of each mutation category per sample was computed [2, 3]. The previously defined 30 mutational signatures were downloaded from COSMIC (http://cancer.sanger.ac.uk/cosmic/signatures). Assuming the mutational distribution of a single sample is a linear combination of the known 30 signatures, an iterative method was used that was implemented in a R package deconstructSigs [35] to decompose the mutational signatures (a 96 × 30 matrix) for the observed mutational distribution of each DIPG sample (a 96 × 1 vector). The contributions of each known mutational signature in cases 1 and 3, the radiation-associated DIPG samples, were compared to all other primary cases, which were taken from both diagnosis and autopsy (n = 9).

Statistical analysis

To estimate cumulative incidence of DIPG in survivors of pediatric medulloblastoma, the number of observed cases of DIPG during each observation period was divided by the total number of patients who underwent treatment of medulloblastoma [37]. For estimates of cumulative incidence from single institutions, data was obtained for the time period of January 2000 to December 2015. Survival data were extracted from the IDIPGR. Survival functions were estimated using Kaplan-Meier methods (GraphPad Prism version 7.00). For survival by histone status in primary DIPGs, analysis was limited to the subset of tumors for which the OS and sequencing information were available. Cox proportional hazards regression model was used to investigate the association between radiation exposure or histone status and survival after controlling for potential prognostic factors including age and sex (PROC PHREG in SAS 9.4). Mann Whitney test was performed using GraphPad Prism (version 7.00) to compare mutation and fusion frequency in DIPG in radiated and non-radiated setting.

Results

Twelve patients who developed DIPG after radiation treatment for primary pediatric medulloblastoma were identified. Six of these cases were acquired from the IDIPGR, and six were extracted from literature review, reported primarily in results from medulloblastoma cooperative group trials: COG A9961 (n = 2), HIT'91 (n = 1), HIT-SIOP-PNET4 (n = 1), and CCG 9892 (n = 1) [12, 30, 31, 36, 42, 46]. Within the limits of incomplete follow-up timing and records, the estimated cumulative incidence of DIPG after medulloblastoma ranged from 0.3–3.9% among the involved institutions and reported studies (Table 1). The cumulative incidence of radiation-associated DIPG survivors of the reported trials ranged from 0.3–1.5% with median follow-up of 4.7–10 years, while the estimated cumulative incidences at single institutions ranged from 0.7–3.9%.

Primary Medulloblastoma

Patient characteristics and treatments are described in Table 2. Of patients with known sex and age information (n = 7), six were male, and ages at diagnosis of primary medulloblastoma ranged from 2 to 9 years. For risk stratification of medulloblastoma based on clinical criteria, seven cases were average-risk, three were high-risk, and two were unreported. All cases with known histology (n = 6) showed classic histology (cases 1–6) with subgroup classification into either Group 3 (cases 3 and 4) or Group 4 (cases 1, 5, and 6). Cytogenetics was not performed on

Table 1 Observed cumulative incidence of radiation-associated malignancies in survivors of pediatric medulloblastoma

Source	Population	Size of cohort	Number of secondary malignant neoplasms	Number of gliomas	Number of DIPG (cumulative incidence %)	Median follow-up (years)
COG A9961 (Packer et al. 2013) [31]	December 1996–2000, age 3–21 years, average-risk only	397	15	7	2 (0.5)	9.7
HIT'91 (Von Hoff et al. 2009) [42]	August 1991–December 1997, age 3–18 years	280	12	4	1 (0.4)	10
HIT-SIOP-PNET4 (Sabel et al. 2016) [36]	2001–2006, age 4–21 years, average risk only	338	3	2	1 (0.3)	7.8
CCG 9892 (Packer et al. 1999) [30]	January 1990–December 1994, age 3–10 years, average risk only	65	1	1	1 (1.5)	4.7
Single institution (Michigan Medicine)	Pediatric patients diagnosed with medulloblastoma and treated between 2000 and 2015	77	*Not reported*	*Not reported*	3 (3.9)	*Not reported*
Single institution (Seattle Children's Hospital)	Pediatric patients diagnosed with medulloblastoma and treated between 2000 and 2015	91	*Not reported*	*Not reported*	1 (1.0)	*Not reported*
Single institution (Hospital for Sick Children)	Pediatric patients diagnosed with medulloblastoma and treated between 2000 and 2015	140	*Not reported*	*Not reported*	1 (0.7)	*Not reported*
Single institution (Princess Margaret Hospital for Children)	Pediatric patients diagnosed with medulloblastoma and treated between 2000 and 2015	41	*Not reported*	*Not reported*	1 (2.4)	*Not reported*

case 2, which was classified as Group 3/4 by immunohistochemistry (IHC).

Treatment details are described in Additional file 1: Table S1. All cases underwent surgical resection. Seven cases achieved gross total resection and three cases had partial resections; surgical details were not reported for two cases. All patients received CSI (dose range 18.0–36.0 Gy) with a posterior fossa boost (dose range 19.8–36.0 Gy) for total posterior fossa exposure 53.4–60.0 Gy. Ten cases received chemotherapy, which included a variety of established regimens including cytotoxic and/or targeted therapies, one case declined chemotherapy, and one case lacked chemotherapy details.

Radiation-associated DIPG

Diagnostic and outcome information on DIPG for all cases are described in Table 2. Median time to diagnosis of DIPG after completion of primary medulloblastoma therapy was 7 years (range 2–11 years). DIPG histologic diagnoses were reported for 7 patients and included HGG ($n = 3$), glioma, grade-unspecified ($n = 1$), anaplastic astrocytoma ($n = 2$), and pilocytic astrocytoma ($n = 1$; case was not reviewed centrally by trial). For cases with tissue (cases 1, 3 and 6), IHC revealed the tumors to be negative for H3K27 M staining and GFAP positive (Fig. 1a, positive and negative control tumor staining in Additional file 1: Figure S1). Additional staining in case 3 revealed retention of H3K27me3 (Fig. 1a), consistent with previous analyses of H3 wild-type DIPG, as tri-methyl is lost in H3K27M-mutant glioma [6]. Treatment for DIPG varied widely. Three patients received focal EBRT, two of whom received oral treatment with histone deacetylase inhibitors (panobinostat, vorinostat) and one of whom received everolimus. Three additional patients received chemotherapeutic agents, with temozolomide being the most common choice. One patient declined treatment, and treatment details were not available for five patients.

For cases with complete outcome data ($n = 8$), seven patients died at a median of 8 months after DIPG diagnosis (range 4–17 months), and one patient remains living 5 months after DIPG diagnosis. OS was shorter for radiation-associated DIPG as compared to radiographically-confirmed primary DIPG cases from IDIPGR ($n = 428$, $p = 0.046$; Fig. 1b). On multivariate analysis, radiation exposure (hazard ratio 2.87; $p = 0.014$) and age (hazard ratio 1.00; $p = 0.019$) were associated significantly with overall survival (Additional file 1: Table S2). Further, the radiation-associated DIPG cohort showed the shortest OS compared to two subgroups of primary DIPG with sequencing information ($n = 38$), separated by H3 status (vs. H3.3 K27 M mutant, $p = 0.038$; vs. H3.1 K27 M mutant, $p = 0.024$; Fig. 1c). On multivariate analysis, radiation exposure (hazard ratio 4.51; $p = 0.005$) and sex (hazard ratio 2.51; $p = 0.016$) were associated significantly with overall survival (Additional file 1: Table S2).

For illustration, case 1 presented with standard-risk medulloblastoma at age 8 (Fig. 2a), and underwent gross total resection (Fig. 3a), followed by 23.4 Gy CSI with a 32.4 Gy boost to primary tumor site (Fig. 2b) with concurrent vincristine. At 21 years of age (12 years after

Table 2 Characteristics of survivors of pediatric medulloblastoma who developed radiation-associated DIPG

	Primary Medulloblastoma				Radiation-associated DIPG				
Case number - location	Age at diagnosis (years)	Gender	Risk stratification	Histology, subgroup	Time in remission (years)	Age at diagnosis (years)	Treatment	Histology (sequencing)	Outcome (months after DIPG diagnosis)
1 - IDIPGR (Michigan Medicine)	8	F	Average	Classic, Group 4 (isochromosome 17q)	12	21	54 Gy focal radiation treatment, panobinostat	High-grade glioma (*TP53* loss, *PTEN* loss, *NRAS* mutation)	Died of disease (17)
2 - IDIPGR (Michigan Medicine)	9	M	Average	Classic, Group 3/4	7	16	Patient declined	Biopsy deferred	Died of disease (4)
3 - IDIPGR (Michigan Medicine)	4	M	High	Classic, Group 3	4	9	35 Gy focal radiation treatment, everolimus	High-grade glioma (*PIK3CA* and *EZH2* mutations)	Living (5)
4 - IDIPGR (Hospital for Sick Children)	2	M	High	Classic, Group 3	2	7	Etoposide, temozolomide, mechlorethamine, cyclophosphamide	*Not reported*	Died of disease (5)
5 - IDIPGR (Seattle Children's Hospital)	6	M	Average	Classic, Group 4	10	17	Temozolomide	Not biopsied; diagnosis made by imaging	Died of disease (10)
6 - IDIPGR (Princess Margaret Hospital for Children)	4	M	High	Classic, Group 4	11	15	Focal radiation treatment, vorinostat	High-grade glioma (*PIK3CA* mutation)	Died of disease (8)
7- Packer et al. 2013 [31] (COG A9961)	*3–21*	*Not reported*	Average	*Not reported*	6.5	*Not reported*	*Not reported*	Pilocytic astrocytoma[a]	Died of disease (10)
8 - Packer et al. 2013 [31] (COG A9961)	*3–21*	*Not reported*	Average	*Not reported*	9	*Not reported*	*Not reported*	Biopsy deferred	*Not reported*
9 - Von Hoff et al. 2009 [42] (HIT'91)	*3–18*	*Not reported*	*Not reported*	*Not reported*	*Not reported*	*Not reported*	*Not reported*	*Not reported*	*Not reported*
10 - Sabel et al. 2016 [36] (HIT-SIOP-PNET4)	*4–21*	*Not reported*	Average	*Not reported*	5.1	*Not reported*	*Not reported*	Anaplastic astrocytoma	Died of disease (Not reported)
11 - Packer et al. 1999 [30] (CCG 9892)	*3–10*	*Not reported*	Average	*Not reported*	4.8	*Not reported*	*Not reported*	Glioma, grade unspecified	Died of disease (Not reported)
12 - You et al. 2013 [46] (Yonsei Hospital)	8	M	*Not reported*	*Not reported*	7.8	*Not reported*	Temozolomide	Anaplastic astrocytoma	Died of disease (6)

[a]Pathology was not reviewed centrally by trial

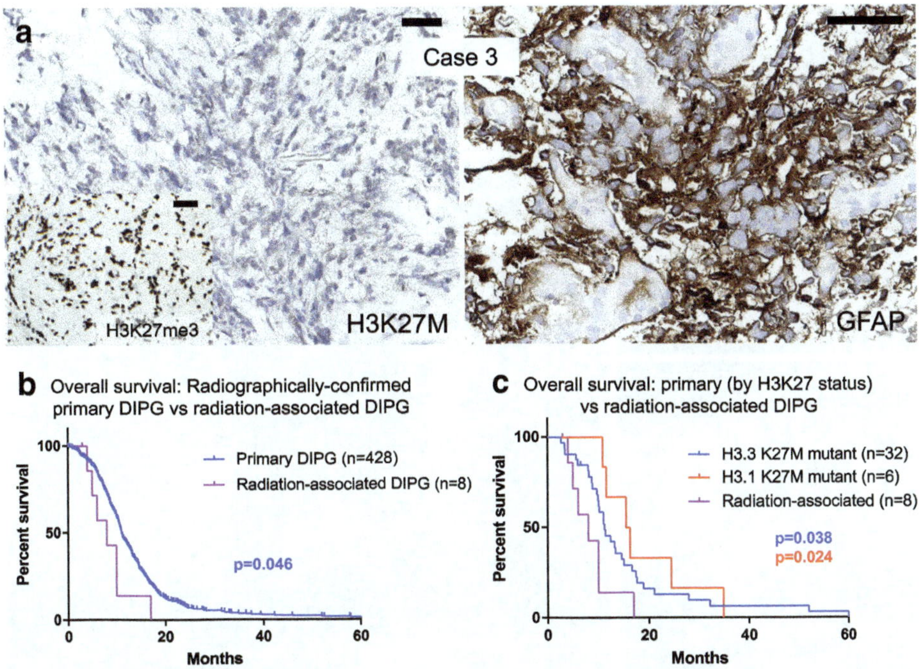

Fig. 1 Radiation-associated DIPGs define a distinct molecular subtype with poor prognosis. **a** Immunohistochemistry performed for case 3 showed wild-type status for histone H3 (H3K27M) with retention of H3K27me3, as well as diffuse GFAP expression, which was negative in primary medulloblastoma (not shown). Positive and negative controls are shown in Additional file 1: Figure S1. Insufficient tissue was available for such analysis in cases 1 and 6; however, H3 wild-type status was demonstrated by tumor sequencing in these cases. All scale bars are 50 μm. **b** OS data for primary DIPG in the IDIPGR ($n = 428$) was compared via Kaplan-Meier analysis to OS of radiation-associated DIPG cohort ($n = 8$). OS was significantly less for the radiation-associated DIPG group ($p = 0.046$). **c** OS data for primary DIPG patients with both genomic and OS information available ($n = 38$), as categorized by histone mutational status and compared via Kaplan-Meier analysis to OS of radiation-associated DIPG cohort ($n = 8$). The radiation-associated DIPG cohort showed the shortest OS in comparison to the two subgroups of primary DIPG with significantly shorter survival compared to H3.3 K27 M mutant ($p = 0.038$) and H3.1 K27 M mutant ($p = 0.024$) primary DIPGs

completion of therapy), she presented to her primary care physician with a one-month history of difficulty swallowing and clumsiness of her right hand. MRI brain revealed a new infiltrative mass with diffuse pontine T2 hyperintensity, consistent radiographically with a DIPG (Fig. 2c). MR spectroscopy revealed markedly elevated choline to creatinine peak with depressed NAA peak, consistent with malignancy (Fig. 2d). She underwent re-irradiation and ultimately died of disease 17 months after diagnosis. Histopathology at autopsy revealed a diffusively infiltrating glioma (Fig. 3b). Karyotyping of her initial medulloblastoma previously had revealed isochromosome for the long arm of chromosome 17, a hallmark feature of Group 4 medulloblastoma (Fig. 3c). In contrast, copy number analysis of her DIPG revealed copy number changes frequently seen in gliomas including homozygous loss of *RB1*, *SETDB2*, *CDKN2A* and *CDKN2B* with no abnormalities in chromosome 17, distinguishing it from the primary medulloblastoma (Fig. 3d-e). Similar imaging findings, radiation fields and MR spectroscopy images were obtained in cases 2 and 3 (Additional file 1: Figures S2 and S3).

DIPG sequencing

For cases 1, 3, and 6, DIPG frozen tissue from autopsy (cases 1 and 6) and diagnosis (case 3) were sequenced and mutations were analyzed (Additional file 1: Table S3). All three tumors were wild-type for *H3F3A* and *HIST1H3B*. Mutational signature analysis of DIPGs showed mutations consistent with radiation-induced DNA damage (e.g., insertional event in *TP53*), as well as mutations in other oncogenic drivers (e.g., *PTEN*, *EGFR*, and *NRAS*), suggestive of a distinct mutational process as compared with primary DIPGs. Mutations identified in radiation-associated DIPGs had molecular overlap with recurrent drivers of adult GBM, using previously published datasets of adult GBM and primary DIPG (Fig. 4a) [7, 44]. Within a cohort sequenced using the same sequencing methodology (UM MI-ONCOSEQ) ($n = 11$), COSMIC mutational signature analysis demonstrated that radiation-associated DIPGs had the highest predicted somatic mutation counts and were more likely to harbor Signature 24 than primary DIPGs, which has not previously been connected to previous malignancy or radiation exposure (Fig. 4b) [3]. Cases with previous radiation in the MI-ONCOSEQ cohort

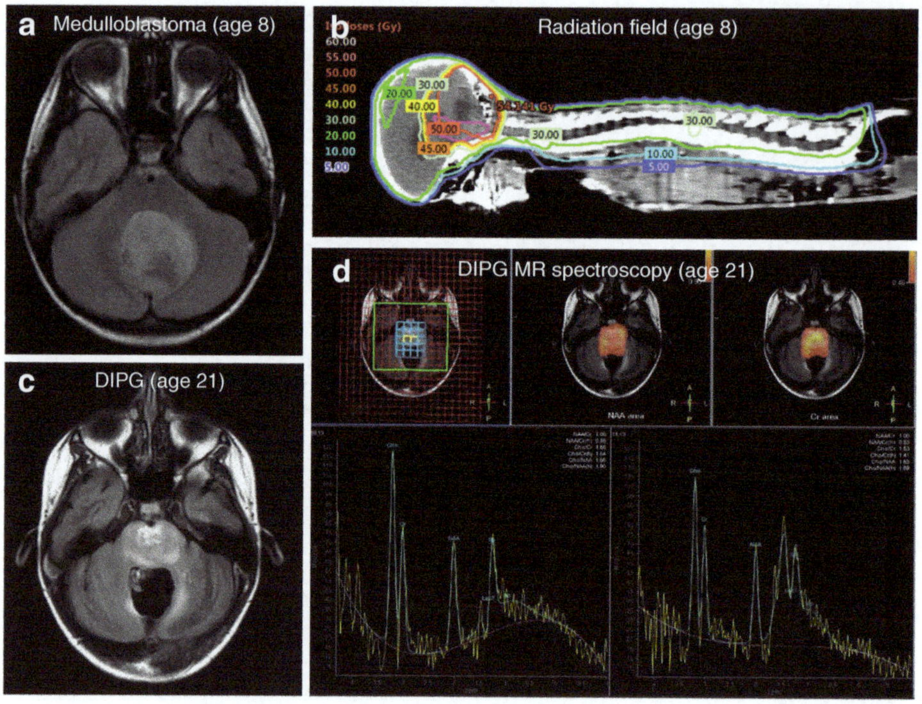

Fig. 2 Diagnosis and treatment with standard therapy of case 1, which included significant radiation to brainstem. **a** MR axial T2 FLAIR image of primary medulloblastoma diagnosed at age 8. **b** Radiation dose distribution showing craniospinal irradiation prescribed to 23.4 Gy and posterior fossa boost prescribed to 32.4 Gy. The brainstem is contoured in purple and received a mean dose of 50.1 Gy. **c** MR axial T2 FLAIR image of DIPG diagnosed at age 21, 13 years after treatment for primary medulloblastoma and in the area of the previously irradiated field. **d** MR spectroscopy with an elevated Chol/Cr ratio (1.66) that is consistent with malignancy (DIPG)

(including cases 1 and 3 form this series as well as primary DIPG cases at autopsy) harbored higher mutations (Fig. 4c; $p = 0.0043$) and fusions per exome (Fig. 4d; $p = 0.0135$), although overall mutational burden remained lower than what might be clinically significant in comparison to cancers with known mismatch repair deficiency [22]. Cases 1 and 3 underwent germline sequencing as well, which revealed no evidence of cancer predisposition syndromes.

Discussion

With cumulative incidences ranging from 4.2–12%, long-term survivors of medulloblastoma show an increased risk of central nervous system SMNs, particularly gliomas [15, 19, 29, 31, 39, 42]. The risk of glioma has been shown to increase linearly with radiation dose, with reported excess relative risk of 0.079–0.33 per Gy [4, 38]. Prior studies have commented on radiation-associated DIPG following individual cases of pediatric central nervous system cancers [1, 8, 9, 17], although no studies have commented specifically on the incidence or molecular characteristics of radiation-associated DIPGs following treatment for pediatric medulloblastoma. In standard medulloblastoma therapy, the brainstem receives high doses of EBRT due to its anatomic proximity to the posterior fossa boost. In this study, the estimated cumulative incidence of DIPG in children diagnosed with medulloblastoma and treated with EBRT ranged from 0.3–3.9%. The cumulative incidence reported in this study may have been impacted by incomplete or brief follow-up and may be underestimated as the cohort continues to age. While DIPG was diagnosed at a median of 7 years after completion of treatment for medulloblastoma, median follow-up was only 10 years or less for the cited studies. In large studies of pediatric survivors, median time to diagnosis of radiation-associated gliomas ranged from 6.6–17.4 years [9, 15, 28, 32, 38, 39]. Furthermore, in patients with treated primary medulloblastoma, posterior fossa tumors often are labeled as recurrent medulloblastomas based solely on radiographic evidence. It may be that some tumors that are not biopsied and assumed to be recurrent medulloblastomas may in fact be DIPGs.

Increasing attention has been given to the impact of radiation field and modality on efficacy and risk of SMNs. For medulloblastoma treatment, many centers now are shifting away from a posterior fossa boost and toward a primary site boost only [24, 43]. Preliminary results from a recently closed phase III COG trial (ACNS0331) of involved field radiotherapy with chemotherapy in average-risk medulloblastoma found no difference in 5-year event free survival or OS when boost volume was limited to the primary site vs. entire posterior fossa [25]. In patients treated with primary

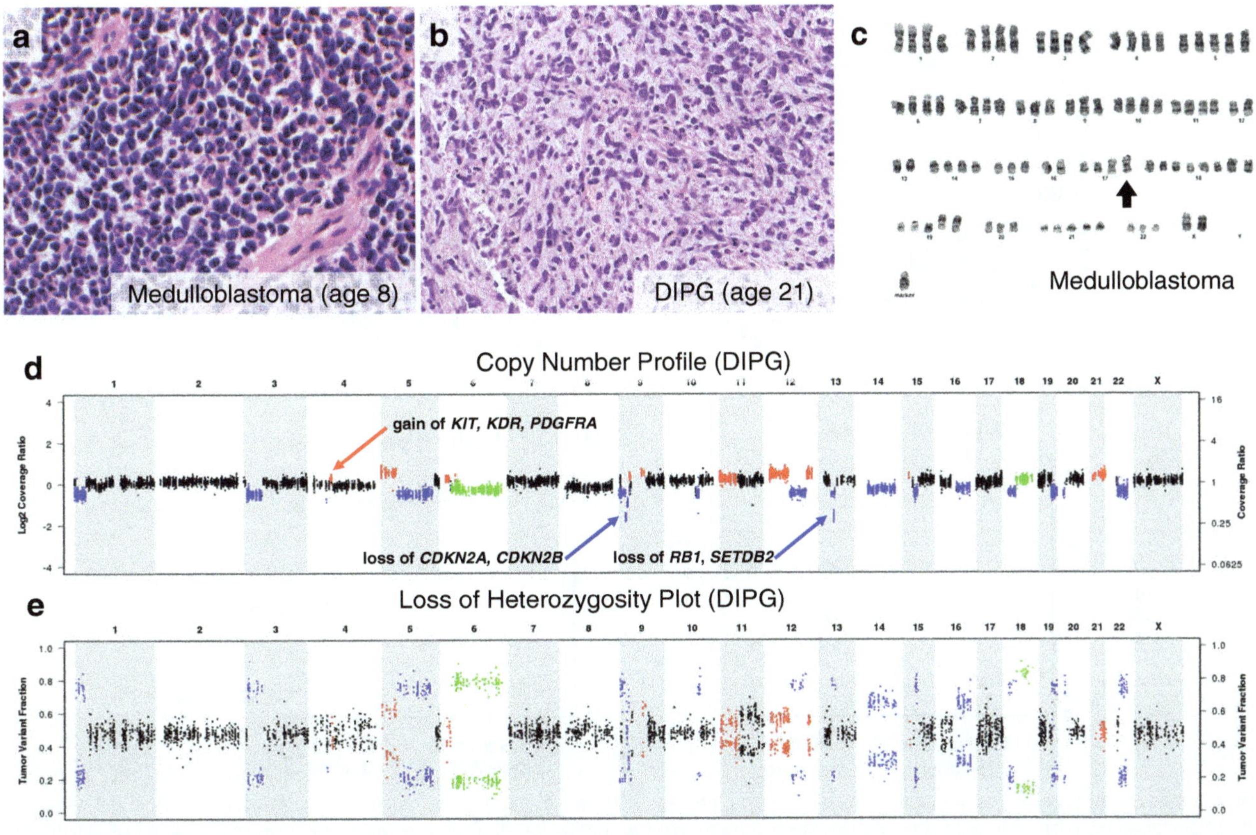

Fig. 3 Histology and molecular results distinguish primary medulloblastoma from radiation-associated DIPG. **a** Resected medulloblastoma from case 1 showing characteristic classic-type features including sheets of cells with primitive hyperchromatic nuclei and scant cytoplasm. **b** DIPG at autopsy showing an infiltrating glial tumor with small angulated nuclei and abundant amphophilic cytoplasm. **c** Karyotype analysis of medulloblastoma shows near-tetraploid clone with arrow indicating i(17q), most consistent with Group 4. **d** Copy number analysis of DIPG shows focal and structural changes distinct from primary tumor, including focal homozygous loss of *RB1*, *SETDB2*, *CDKN2A* and *CDKN2B*, focal 1 copy gain of *KIT*, *KDR* and *PDGFRA*, and activation mutations in *NRAS* and *TP53*. **e** Loss of heterozygosity plot showing regions on chromosomes 6 and 18 with copy-neutral loss of heterozygosity events

site boost only, proton radiotherapy may decrease brainstem radiation exposure even further relative to photon therapy [11, 34]. In a multi-institutional cohort study and phase II single center trial, there were no significant differences in recurrence-free survival or OS between patients treated with photon vs. proton radiotherapy, and in the phase II trial, no radiation-associated malignancies were reported within a median follow-up time of 7 years [45]. While longer follow-up is required to evaluate definitively its impact, smaller boost fields and proton radiotherapy show promise for reducing the risk of SMNs without sacrificing efficacy of treatment.

Radiation-associated gliomas are molecularly distinct from their primary counterparts. A previous report of non-brainstem radiation-associated pediatric GBM showed overexpression of a number of genes involved in tumorigenesis as compared to primary pediatric GBM [13]. Additionally, prior studies suggest that tumors can be differentiated based on these molecular signatures and that radiation-associated tumors may exhibit distinct patterns [3, 5]. It has been observed that radiation-associated tumors exhibit a significantly higher total number of mutations, as well as balanced inversions, with both small deletions and inversions generating driver mutations [5].

In this study, tumor exome sequencing of three radiation-associated DIPGs demonstrated tumors to be H3-wildtype. This finding is significant in the context of a recent large cohort of sequenced primary DIPGs, in which only 16.8% were found to be H3-wildtype [23]. Notably, sequencing confirmed that the tumors were indeed distinct from their primary malignancies and not local recurrences. Moreover, patients did not harbor germline mutations in known cancer predisposition genes. Although alterations in two of the most frequently mutated genes in primary DIPG (*H3F3A* and *ACVR1*) were not detected, the tumor mutations in the sequenced cases are established tumor drivers in adult GBM (e.g. *PTEN*, *NRAS*, and *EGFR*). Interestingly, case 3 was found to have an *EZH2* mutation in the radiation-associated DIPG, which has not been identified

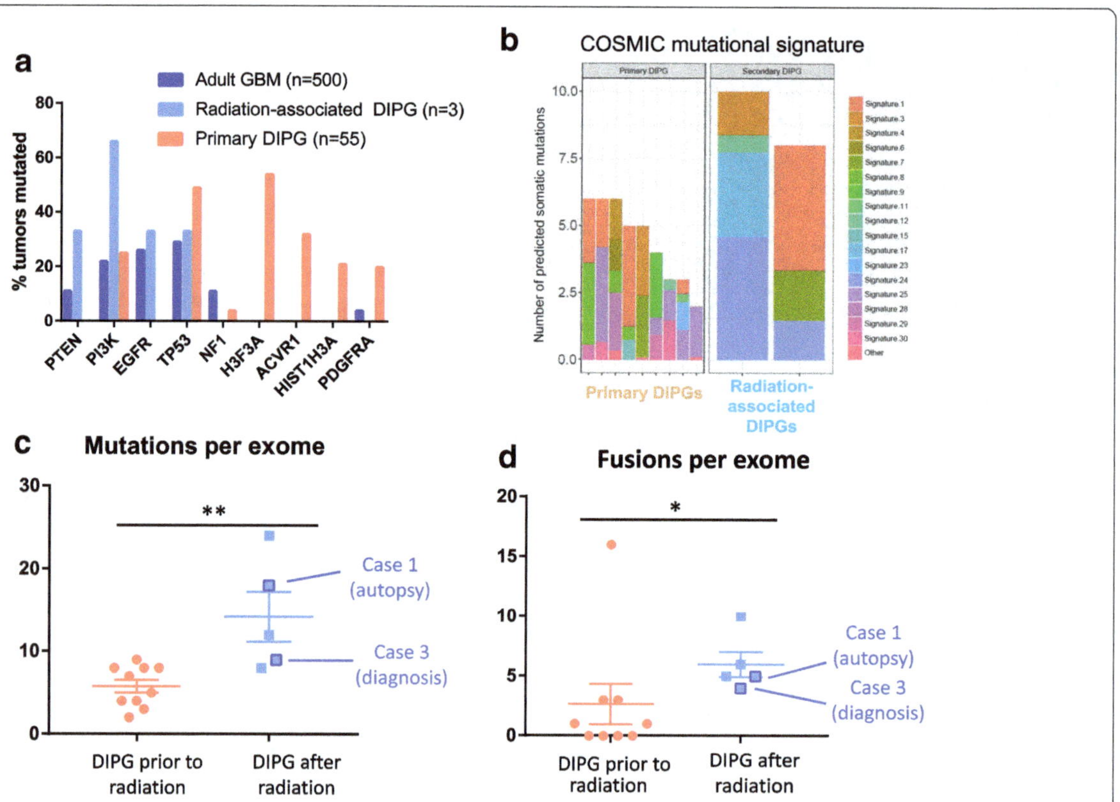

Fig. 4 Radiation-associated DIPGs are molecularly distinct from primary DIPGs. **a** Plot of recurrent mutations in previously published datasets (adult GBM [$n = 500$] [7]; primary DIPG [$n = 55$] [44]) demonstrates that the distribution of driving mutations in radiation-associated DIPG is more similar to recurrent alterations in adult GBM than primary DIPG. **b** Contributions of established COSMIC mutational signatures were determined for radiation-associated DIPG samples as compared to all other primary cases sequenced through same sequencing platform (MI-ONCOSEQ). **c-d** Cases with previous radiation in this cohort (including case 1 and 3 and primary DIPG at autopsy) show higher mutations and fusions per exome ($p = 0.0043$ and $p = 0.0135$, respectively using Mann Whitney test)

as a recurrent driver in DIPG, but has been established as a potential therapeutic target in pre-clinical DIPG models [27]. Mutational signature analysis of radiation-associated DIPGs showed mutations consistent with radiation-induced DNA damage (e.g., insertional event in *TP53*), at a rate similar to cases observed in a recent report of radiation-associated GBMs, in which two out of five sequenced GBMs had mutations in *TP53* [33]. Unlike radiation-associated GBMs, however, *PDGFRA* played a smaller role in the cohort of radiation-associated DIPG. The *PDGFRA* amplification in case 1 (Fig. 1d) was the isolated alteration in the three tumors. While non-silent mutations in *PDGFRA* were identified in all five radiation-associated GBMs, none of radiation-associated DIPGs from this study had mutations in *PDGFRA*.

Taken together, these results are suggestive of distinct mutational processes compared with primary DIPGs: primary DIPGs originate within a particular early developmental timespan that is amenable to transformation with *H3F3A* and *ACVR1* mutation [23], whereas radiation-associated DIPGs appear to arise as a result of radiation-induced DNA damage in established oncogenic drivers of primary adult GBM. Future sequencing of additional cases may elucidate patterns of distinct biology in radiation-associated DIPG, which may have implications in terms of clinical management. These data suggest that patients with radiation-associated DIPG may benefit from future therapies targeted to the molecular features of adult GBM rather than primary DIPG.

Significantly, the radiation-associated DIPG cohort demonstrated a shorter OS relative to patients with primary DIPG. The three molecularly sequenced cases of radiation-associated DIPG cohort additionally are distinguished as H3 wild-type designation, considered a positive prognostic variable in primary DIPG relative to H3.3 mutant DIPG [10, 20]. Prior studies report a similar decreased survival from radiation-induced HGG as compared to primary HGG [13, 21]. In conjunction with the molecular findings, these data suggest that radiation-associated DIPGs form a distinct molecular subgroup that has negative implications on survival. The findings in this work demonstrate the importance of tumor biopsy or resection at appearance of a second cancer. Most cases had no or limited histological and molecular diagnostic information, which was not clinically relevant until recently. Now, this information is critical for prognostic information, current clinical management, and potential future therapies [33].

Conclusions

In conclusion, patients treated for pediatric medulloblastoma are at increased risk for development of radiation-associated DIPG, which may represent a distinct molecular subtype with a worse prognosis relative to other DIPGs. This risk highlights the importance of radiation volume and modality in the treatment of children with medulloblastoma and provides a compelling argument for efforts to reduce exposure of the brainstem. Additionally, the presented molecular data suggest that patients with radiation-associated DIPG may benefit from future therapies targeted to the molecular features of adult GBM rather than primary DIPG.

Acknowledgements

The authors gratefully acknowledge all patients and their families who have supported and contributed to the International DIPG Registry. Supporting foundations include: The Cure Starts Now Foundation, Hope for Caroline Foundation, Julian Boivin Courage for Cures Foundation, Abbie's Army, Michael Mosier Defeat DIPG Foundation, Reflections of Grace Foundation, The Cure Starts Now Australia, Brooke Healey Foundation, Soar With Grace Foundation, Jeffrey Thomas Hayden Foundation, Cure Brain Cancer Foundation, The Jones Family Foundation, Musella Foundation, Pray, Hope Believe Foundation, Smiles for Sophie Foundation, Benny's World, Love Chloe Foundation, Aiden's Avengers, A Cure from Caleb Society, The Operation Grace White Foundation, Ryan's Hope, Wayland Villars DIPG Foundation, American Childhood Cancer Organization, Juliana Rose Donnelly Trust, Sheila Jones & Friends, The Ellie Kavalieros DIPG Research Fund, Voices Against Brain Cancer, The DIPG Collaborative.

Funding

CK is supported by NIH/NINDS Grant K08-NS099427-01, Michigan Medicine Department of Pediatrics Gorman Scholar Award, The University of Michigan Chad Carr Pediatric Brain Tumor Center, the Chad Tough Foundation and Hyundai Hope on Wheels. NGG is supported by a Cancer Council of Western Australia fellowship.

Authors' contributions

All authors have contributed meaningfully to the manuscript. All authors read and approved the final manuscript.

Competing interests

The authors declare that they have no competing interests.

Author details

[1]Department of Pathology, Michigan Medicine, Ann Arbor, MI 48109, USA. [2]Department of Pediatrics, Division of Pediatric Hematology-Oncology; Michigan Medicine, Ann Arbor, MI 48109, USA. [3]Department of Haematology and Oncology, Princess Margaret Hospital for Children, Perth, WA 6840, Australia. [4]Michigan Center for Translational Pathology, Michigan Medicine, Ann Arbor, MI 48109, USA. [5]Ontario Institute for Cancer Research, MG5 0A3, Toronto, ON, Canada. [6]Department of Radiation Oncology; Michigan Medicine, Ann Arbor, MI 48109, USA. [7]Department of Computational Medicine and Bioinformatics; Michigan Medicine, Ann Arbor, MI 48109, USA. [8]Department of Biostatistics, University of Michigan School of Public Health, Ann Arbor, MI 48109, USA. [9]Division of Radiology, Cincinnati Children's Hospital, Cincinnati, OH 45229, USA. [10]Division of Oncology, Cincinnati Children's Hospital, Cincinnati, OH 45229, USA. [11]Division of Biomedical Informatics, Cincinnati Children's Hospital, Cincinnati, OH 45229, USA. [12]Division of Pathology, Cincinnati Children's Hospital, Cincinnati, OH 45229, USA. [13]Department of Pediatrics, Division of Oncology, Seattle Children's Hospital, Seattle, WA 98105, USA. [14]Department of Pediatrics, Division of Haematology/Oncology, Hospital for Sick Children, Toronto, ON M5G 1X8, Canada. [15]Department of Radiation Oncology, Massachusetts General Hospital, Boston, MA 02114, USA. [16]Department of Pediatrics, Division of Neurology; Michigan Medicine, Ann Arbor, MI 48109, USA. [17]Department of Pediatrics, Division of Oncology, Cincinnati Children's Hospital, Cincinnati, OH 45229, USA. [18]Kids Cancer Centre, Telethon Kids Institute, Subiaco, WA 6008, Australia. [19]Division of Paediatrics, University of Western Australia, Crawley, WA 6009, Australia.

References

1. Abboud SE, Wolansky LJ, Manjila SV, Lo SS, Arafah BM, Selman WR, Couce ME, Rogers LR (2015) Histologically proven radiation-induced brainstem Glioma 93 months after external beam radiotherapy for pituitary macroadenoma: radiation treatment dose and volume correlation. J Neuroimaging 25:674–676. https://doi.org/10.1111/jon.12181
2. Alexandrov LB, Jones PH, Wedge DC, Sale JE, Campbell PJ, Nik-Zainal S, Stratton MR (2015) Clock-like mutational processes in human somatic cells. Nat Genet 47:1402–1407. https://doi.org/10.1038/ng.3441
3. Alexandrov LB, Nik-Zainal S, Wedge DC, Aparicio SA, Behjati S, Biankin AV, Bignell GR, Bolli N, Borg A, Borresen-Dale AL et al (2013) Signatures of mutational processes in human cancer. Nature 500:415–421. https://doi.org/10.1038/nature12477
4. Armstrong GT, Stovall M, Robison LL (2010) Long-term effects of radiation exposure among adult survivors of childhood cancer: results from the childhood cancer survivor study. Radiat Res 174:840–850. https://doi.org/10.1667/RR1903.1
5. Behjati S, Gundem G, Wedge DC, Roberts ND, Tarpey PS, Cooke SL, Van Loo P, Alexandrov LB, Ramakrishna M, Davies H, et al (2016) Mutational signatures of ionizing radiation in second malignancies. Nat Commun 7: 12605 https://doi.org/10.1038/ncomms12605
6. Bender S, Tang Y, Lindroth AM, Hovestadt V, Jones DT, Kool M, Zapatka M, Northcott PA, Sturm D, Wang W et al (2013) Reduced H3K27me3 and DNA hypomethylation are major drivers of Gene Expr in K27M mutant pediatric high-grade gliomas. Cancer Cell 24:660–672. https://doi.org/10.1016/j.ccr.2013.10.006
7. Brennan CW, Verhaak RG, McKenna A, Campos B, Noushmehr H, Salama SR, Zheng S, Chakravarty D, Sanborn JZ, Berman SH et al (2013) The somatic genomic landscape of glioblastoma. Cell 155:462–477. https://doi.org/10.1016/j.cell.2013.09.034
8. Broniscer A, Baker JN, Baker SJ, Chi SN, Geyer JR, Morris EB, Gajjar A (2010) Prospective collection of tissue samples at autopsy in children with diffuse intrinsic pontine glioma. Cancer 116:4632–4637. https://doi.org/10.1002/cncr.25405
9. Broniscer A, Ke W, Fuller CE, Wu J, Gajjar A, Kun LE (2004) Second neoplasms in pediatric patients with primary central nervous system tumors: the St. Jude Children's Research Hospital experience. Cancer 100: 2246–2252. https://doi.org/10.1002/cncr.20253
10. Castel D, Philippe C, Calmon R, Le Dret L, Truffaux N, Boddaert N, Pages M, Taylor KR, Saulnier P, Lacroix L et al (2015) Histone H3F3A and HIST1H3B K27M mutations define two subgroups of diffuse intrinsic pontine gliomas with different prognosis and phenotypes. Acta Neuropathol 130:815–827. https://doi.org/10.1007/s00401-015-1478-0
11. Chung CS, Yock TI, Nelson K, Xu Y, Keating NL, Tarbell NJ (2013) Incidence of second malignancies among patients treated with proton versus photon radiation. Int J Radiat Oncol Biol Phys 87:46–52. https://doi.org/10.1016/j.ijrobp.2013.04.030
12. Cooney T, Lane A, Bartels U, Bouffet E, Goldman S, Leary SES, Foreman NK, Packer RJ, Broniscer A, Minturn JE et al (2017) Contemporary survival endpoints: an international diffuse intrinsic Pontine Glioma registry study. Neuro Oncol 19:1279 1280. https://doi.org/10.1093/neuonc/nox107

13. Donson AM, Erwin NS, Kleinschmidt-DeMasters BK, Madden JR, Addo-Yobo SO, Foreman NK (2007) Unique molecular characteristics of radiation-induced glioblastoma. J Neuropathol Exp Neurol 66:740–749. https://doi.org/10.1097/nen.0b013e3181257190
14. Fossati P, Ricardi U, Orecchia R (2009) Pediatric medulloblastoma: toxicity of current treatment and potential role of protontherapy. Cancer Treat Rev 35: 79–96. https://doi.org/10.1016/j.ctrv.2008.09.002
15. Friedman DL, Whitton J, Leisenring W, Mertens AC, Hammond S, Stovall M, Donaldson SS, Meadows AT, Robison LL, Neglia JP (2010) Subsequent neoplasms in 5-year survivors of childhood cancer: the childhood Cancer survivor study. J Natl Cancer Inst 102:1083–1095. https://doi.org/10.1093/jnci/djq238
16. Fu W, O'Connor TD, Jun G, Kang HM, Abecasis G, Leal SM, Gabriel S, Rieder MJ, Altshuler D, Shendure J et al (2013) Analysis of 6,515 exomes reveals the recent origin of most human protein-coding variants. Nature 493:216–220. https://doi.org/10.1038/nature11690
17. Fuller GN, Kaba SE, Ginsberg LE, McCutcheon IE, Langford LA (1997) Late sequelae of treated pleomorphic xanthoastrocytoma: malignant brain stem astrocytoma occurring 15 years after radiation therapy. J Neuro-Oncol 32:57–61
18. Genomes Project C, Abecasis GR, Altshuler D, Auton A, Brooks LD, Durbin RM, Gibbs RA, Hurles ME, McVean GA (2010) A map of human genome variation from population-scale sequencing. Nature 467:1061–1073. https://doi.org/10.1038/nature09534
19. Goldstein AM, Yuen J, Tucker MA (1997) Second cancers after medulloblastoma: population-based results from the United States and Sweden. Cancer Causes Control 8:865–871
20. Hoffman LM, Veldhuijzen van Zanten SEM, Colditz N, Baugh J, Chaney B, Hoffmann M, Lane A, Fuller C, Miles L, Hawkins C et al (2018) Clinical, radiologic, pathologic, and molecular characteristics of long-term survivors of diffuse intrinsic Pontine Glioma (DIPG): a collaborative report from the international and European Society for Pediatric Oncology DIPG registries. J Clin Oncol 36:1963–1972. https://doi.org/10.1200/JCO.2017.75.9308
21. Kleinschmidt-Demasters BK, Kang JS, Lillehei KO (2006) The burden of radiation-induced central nervous system tumors: a single institution s experience. J Neuropathol Exp Neurol 65:204–216. https://doi.org/10.1097/01.jnen.0000205146.62081.29
22. Le DT, Durham JN, Smith KN, Wang H, Bartlett BR, Aulakh LK, Lu S, Kemberling H, Wilt C, Luber BS et al (2017) Mismatch repair deficiency predicts response of solid tumors to PD-1 blockade. Sci 357:409–413. https://doi.org/10.1126/science.aan6733
23. Mackay A, Burford A, Carvalho D, Izquierdo E, Fazal-Salom J, Taylor KR, Bjerke L, Clarke M, Vinci M, Nandhabalan M et al (2017) Integrated Molecular Meta-Analysis of 1,000 Pediatric High-Grade and Diffuse Intrinsic Pontine Glioma. Cancer Cell 32:520–537. https://doi.org/10.1016/j.ccell.2017.08.017 e525
24. Merchant TE, Kun LE, Krasin MJ, Wallace D, Chintagumpala MM, Woo SY, Ashley DM, Sexton M, Kellie SJ, Ahern V et al (2008) Multi-institution prospective trial of reduced-dose craniospinal irradiation (23.4 Gy) followed by conformal posterior fossa (36 Gy) and primary site irradiation (55.8 Gy) and dose-intensive chemotherapy for average-risk medulloblastoma. Int J Radiat Oncol Biol Phys 70:782–787. https://doi.org/10.1016/j.ijrobp.2007.07.2342
25. Michalski JM, Janss A, Vezina G, Gajjar A, Pollack I, Merchant TE, FitzGerald TJ, Booth T, Tarbell NJ, Li Y et al (2016) Results of COG ACNS0331: a phase III trial of involved-field radiotherapy (IFRT) and low dose Craniospinal irradiation (LD-CSI) with chemotherapy in average-risk Medulloblastoma: a report from the Children's oncology group. Int J Radiat Oncol Biol Phys 96:937–938
26. Mody RJ, Wu YM, Lonigro RJ, Cao X, Roychowdhury S, Vats P, Frank KM, Prensner JR, Asangani I, Palanisamy N et al (2015) Integrative Clinical Sequencing in the Management of Refractory or Relapsed Cancer in Youth. Jama 314:913–925. https://doi.org/10.1001/jama.2015.10080
27. Mohammad F, Weissmann S, Leblanc B, Pandey DP, Hojfeldt JW, Comet I, Zheng C, Johansen JV, Rapin N, Porse BT et al (2017) EZH2 is a potential therapeutic target for H3K27M-mutant pediatric gliomas. Nat Med 23:483–492. https://doi.org/10.1038/nm.4293
28. Neglia JP, Robison LL, Stovall M, Liu Y, Packer RJ, Hammond S, Yasui Y, Kasper CE, Mertens AC, Donaldson SS et al (2006) New primary neoplasms of the central nervous system in survivors of childhood cancer: a report from the childhood Cancer survivor study. J Natl Cancer Inst 98:1528–1537. https://doi.org/10.1093/jnci/djj411
29. Ning MS, Perkins SM, Dewees T, Shinohara ET (2015) Evidence of high mortality in long term survivors of childhood medulloblastoma. J Neuro-Oncol 122:321–327. https://doi.org/10.1007/s11060-014-1712-y
30. Packer RJ, Goldwein J, Nicholson HS, Vezina LG, Allen JC, Ris MD, Muraszko K, Rorke LB, Wara WM, Cohen BH et al (1999) Treatment of children with medulloblastomas with reduced-dose craniospinal radiation therapy and adjuvant chemotherapy: A Children's Cancer Group Study. J Clin Oncol 17: 2127–2136. https://doi.org/10.1200/JCO.1999.17.7.2127
31. Packer RJ, Zhou T, Holmes E, Vezina G, Gajjar A (2013) Survival and secondary tumors in children with medulloblastoma receiving radiotherapy and adjuvant chemotherapy: results of Children's oncology group trial A9961. Neuro-Oncology 15:97–103. https://doi.org/10.1093/neuonc/nos267
32. Pettorini BL, Park YS, Caldarelli M, Massimi L, Tamburrini G, Di Rocco C (2008) Radiation-induced brain tumours after central nervous system irradiation in childhood: a review. Childs Nerv Syst 24:793–805. https://doi.org/10.1007/s00381-008-0631-7
33. Phi JH, Park AK, Lee S, Choi SA, Baek IP, Kim P, Kim EH, Park HC, Kim BC, Bhak J et al (2018) Genomic analysis reveals secondary glioblastoma after radiotherapy in a subset of recurrent medulloblastomas. Acta Neuropathol 35(6):939–953. https://doi.org/10.1007/s00401-018-1845-8
34. Rechner LA, Howell RM, Zhang R, Newhauser WD (2012) Impact of margin size on the predicted risk of radiogenic second cancers following proton arc therapy and volumetric modulated arc therapy for prostate cancer. Phys Med Biol 57:N469–N479. https://doi.org/10.1088/0031-9155/57/23/N469
35. Rosenthal R, McGranahan N, Herrero J, Taylor BS, Swanton C (2016) DeconstructSigs: delineating mutational processes in single tumors distinguishes DNA repair deficiencies and patterns of carcinoma evolution. Genome Biol 17:31. https://doi.org/10.1186/s13059-016-0893-4
36. Sabel M, Fleischhack G, Tippelt S, Gustafsson G, Doz F, Kortmann R, Massimino M, Navajas A, von Hoff K, Rutkowski S et al (2016) Relapse patterns and outcome after relapse in standard risk medulloblastoma: a report from the HIT-SIOP-PNET4 study. J Neurooncol 129:515–524. https://doi.org/10.1007/s11060-016-2202-1
37. Stromberg U (1994) A distinction between cumulative incidence and risk as measures of disease occurrence. Int J Epidemiol 23:1104–1105
38. Taylor AJ, Little MP, Winter DL, Sugden E, Ellison DW, Stiller CA, Stovall M, Frobisher C, Lancashire ER, Reulen RC et al (2010) Population-based risks of CNS tumors in survivors of childhood cancer: the British Childhood Cancer Survivor Study. J Clin Oncol 28:5287–5293. https://doi.org/10.1200/JCO.2009.27.0090
39. Tsui K, Gajjar A, Li C, Srivastava D, Broniscer A, Wetmore C, Kun LE, Merchant TE, Ellison DW, Orr BA et al (2015) Subsequent neoplasms in survivors of childhood central nervous system tumors: risk after modern multimodal therapy. Neuro Oncol 17:448–456. https://doi.org/10.1093/neuonc/nou279
40. Venneti S, Garimella MT, Sullivan LM, Martinez D, Huse JT, Heguy A, Santi M, Thompson CB, Judkins AR (2013) Evaluation of histone 3 lysine 27 trimethylation (H3K27me3) and enhancer of zest 2 (EZH2) in pediatric glial and glioneuronal tumors shows decreased H3K27me3 in H3F3A K27M mutant glioblastomas. Brain Pathol 23:558–564. https://doi.org/10.1111/bpa.12042
41. Venneti S, Santi M, Felicella MM, Yarilin D, Phillips JJ, Sullivan LM, Martinez D, Perry A, Lewis PW, Thompson CB et al (2014) A sensitive and specific histopathologic prognostic marker for H3F3A K27M mutant pediatric glioblastomas. Acta Neuropathol 128:743–753. https://doi.org/10.1007/s00401-014-1338-3
42. von Hoff K, Hinkes B, Gerber NU, Deinlein F, Mittler U, Urban C, Benesch M, Warmuth-Metz M, Soerensen N, Zwiener I et al (2009) Long-term outcome and clinical prognostic factors in children with medulloblastoma treated in the prospective randomised multicentre trial HIT'91. Eur J Cancer 45:1209–1217. https://doi.org/10.1016/j.ejca.2009.01.015
43. Wolden SL, Dunkel IJ, Souweidane MM, Happersett L, Khakoo Y, Schupak K, Lyden D, Leibel SA (2003) Patterns of failure using a conformal radiation therapy tumor bed boost for medulloblastoma. J Clin Oncol 21:3079–3083. https://doi.org/10.1200/JCO.2003.11.140
44. Wu G, Diaz AK, Paugh BS, Rankin SL, Ju B, Li Y, Zhu X, Qu C, Chen X, Zhang J et al (2014) The genomic landscape of diffuse intrinsic pontine glioma and pediatric non-brainstem high-grade glioma. Nat Genet 46:444–450. https://doi.org/10.1038/ng.2938

45. Yock TI, Yeap BY, Ebb DH, Weyman E, Eaton BR, Sherry NA, Jones RM, MacDonald SM, Pulsifer MB, Lavally B et al (2016) Long-term toxic effects of proton radiotherapy for paediatric medulloblastoma: a phase 2 single-arm study. Lancet Oncol 17:287–298. https://doi.org/10.1016/S1470-2045(15)00167-9
46. You SH, Lyu CJ, Kim DS, Suh CO (2013) Second primary brain tumors following cranial irradiation for pediatric solid brain tumors. Childs Nerv Syst 29:1865–1870. https://doi.org/10.1007/s00381-013-2098-4

11

ABCA1 haplodeficiency affects the brain transcriptome following traumatic brain injury in mice expressing human APOE isoforms

Emilie L. Castranio, Cody M. Wolfe, Kyong Nyon Nam, Florent Letronne, Nicholas F. Fitz, Iliya Lefterov*
and Radosveta Koldamova*

Abstract

Expression of human Apolipoprotein E (*APOE*) modulates the inflammatory response in an isoform specific manner, with *APOE4* isoform eliciting a stronger pro-inflammatory response, suggesting a possible mechanism for worse outcome following traumatic brain injury (TBI). APOE lipidation and stability is modulated by ATP-binding cassette transporter A1 (ABCA1), a transmembrane protein that transports lipids and cholesterol onto APOE. We examined the impact *of Abca1* deficiency and APOE isoform expression on the response to TBI using 3-months-old, human $APOE3^{+/+}$ ($E3/Abca1^{+/+}$) and $APOE4^{+/+}$ ($E4/Abca1^{+/+}$) targeted replacement mice, and $APOE3^{+/+}$ and $APOE4^{+/+}$ mice with only one functional copy of the *Abca1* gene ($E3/Abca1^{+/-}$; $E4/Abca1^{+/-}$). TBI-treated mice received a craniotomy followed by a controlled cortical impact (CCI) brain injury in the left hemisphere; sham-treated mice received the same surgical procedure without the impact. We performed RNA-seq using samples from cortices and hippocampi followed by genome-wide differential gene expression analysis. We found that TBI significantly impacted unique transcripts within each group, however, the proportion of unique transcripts was highest in $E4/Abca1^{+/-}$ mice. Additionally, we found that *Abca1* haplodeficiency increased the expression of microglia sensome genes among only *APOE4* injured mice, a response not seen in injured *APOE3* mice, nor in either group of sham-treated mice. To identify gene networks, or modules, correlated to TBI, *APOE* isoform and *Abca1* haplodeficiency, we used weighted gene co-expression network analysis (WGCNA). The module that positively correlated to TBI groups was associated with immune response and featured hub genes that were microglia-specific, including *Trem2, Tyrobp, Cd68* and *Hexb*. The modules positively correlated with *APOE4* isoform and negatively to *Abca1* haplodeficient mice represented "protein translation" and "oxidation-reduction process", respectively. Our results reveal $E4/Abca1^{+/-}$ TBI mice have a distinct response to injury, and unique gene networks are associated with *APOE* isoform, *Abca1* insufficiency and injury.

Keywords: Traumatic brain injury, Apolipoprotein E, ABCA1, Transcriptome, WGCNA, Microglia sensome

Introduction

Traumatic brain injury (TBI) is a significant public health concern; it is a major cause of death and disability in the United States, and its occurrence is highest among multiple vulnerable populations, including the elderly, young adults, and military personnel [16]. No treatment currently exists for the approximately 2 million cases of TBI sustained each year in the United States, and the costs of medical care for 2010 were estimated at $76.5 billion annually [10].

TBI is caused by an initial external force, whether a physical object or inertia, contacting the head [27]. The impact and initial mechanical stress placed on the cells constitute the primary injury, whereas the secondary injury occurs after the inciting traumatic event, and involves multiple pathways and signaling cascades that can cause further damage [3, 13, 39]. Inflammation is a major component of the secondary injury. Inflammation is present in every case of TBI and may be a driving

* Correspondence: iliya@pitt.edu; radak@pitt.edu
Department of Environmental and Occupational Health, University of Pittsburgh, Pittsburgh, PA 15261, USA

force for secondary pathology [13]. Chronic neuroinflammation following TBI was closely associated with neuronal death and impaired cell proliferation in locations both immediately adjacent to, and more distant from, the site of injury [1]. Many studies have shown that levels of inflammation and inflammatory molecules are strongly correlated with multiple measures of outcome in patients, including neurobehavioral impairments and survival rates [29, 45]. Microglia are the brain's main form of immune response to infection, disease, and injury, as well as the source of inflammation. As such, inflammation and microglia have been recent concentrations of research as a means of developing therapies and improving outcomes of TBI.

Outcomes of TBI include possible changes in cognition, behavior, emotion, and sensory processing, all of which are influenced by injury severity and location [5, 18, 35]. Additionally, research has linked TBI to the risk of developing neurodegenerative diseases, including chronic traumatic encephalopathy and Alzheimer's disease (AD) [36]. The high level of heterogeneity in outcomes suggests a significant role for genetic influence on brain susceptibility and recovery [15, 44]. The apolipoprotein E (*APOE*) gene has been frequently studied to determine its role in TBI and its isoform-dependent impact on outcome. The *APOEε4* allele is the strongest genetic risk factor for late onset AD, and is thought to confer worse outcome after TBI [2]. *APOEε4* carriers have been found to have slower recovery, increased risk of posttraumatic seizures, and worse memory performance after TBI [2, 12, 14]. However, multiple studies also show that *APOEε4* carriers did not differ from non-carriers in cognitive performance, functional outcomes or recovery after TBI [11, 37]. The contradictory results so far emphasize the need for more research on APOE and TBI.

APOE is involved in several pathways after a TBI occurs, including inflammation [20]. Inheritance of the *APOEε4* allele is associated with increased inflammatory responses, including after TBI [28]. APOE4 may induce a more robust pro-inflammatory reaction from microglia and may suppress anti-inflammatory signaling [4, 23, 24, 26]. This may be a result of decreased stability and faster catabolic degradation of APOE4, compared to the other isoforms, which is possibly due to its lower lipidation levels [20]. APOE is secreted as nonlipidated apolipoprotein, cholesterol and phospholipid efflux to lipid-poor APOE is mediated by ATP Binding Cassette Transporter A1 (ABCA1) [7]. *Abca1* deficiency results in decreased APOE lipidation and APOE levels [17, 21]. ABCA1 may also play a role in modulating the inflammatory response in the brain. Mice lacking brain ABCA1 saw increased inflammatory gene expression, and the microglia cultured from these mice exhibited an increased pro-inflammatory response, as seen by higher levels of TNFα secretion and lower phagocytic activity, in response to lipopolysaccharide administration [19]. It is not known how *Abca1* haploinsufficiency may influence TBI.

We recently performed transcriptional profiling of APOE expressing mice after TBI using Next Generation Sequencing [9]. Using a network-based approach, we were able to identify distinct modules correlated to injury and APOE isoform, as well as a module driven by APOE isoform across TBI groups. The aim of this study was to examine the effect of *Abca1* haploinsufficiency on gene expression induced by TBI in APOE targeted replacement mice using transcriptional profiling and a network-based approach. We used 3-month-old mice expressing human *APOE3*$^{+/+}$ and *APOE4*$^{+/+}$ isoforms (*E3/Abca1*$^{+/+}$ and *E4/Abca1*$^{+/+}$, respectively), and compared them to their *Abca1* haploinsufficient counterparts (*E3/Abca1*$^{+/-}$ and *E4/Abca1*$^{+/-}$, respectively), after performing a controlled cortical impact. Transcriptional profiling of hippocampal and cortical tissue from the injury site was performed using RNA-sequencing (RNA-seq). *E4/Abca1*$^{+/-}$ mice had higher expression levels of the common up-regulated transcripts after TBI, which included genes related to the immune response and inflammatory response. We then examined how ABCA1 insufficiency impacted expression of the microglia sensome genes, and found that *E4/Abca1*$^{+/-}$ TBI mice expressed these genes higher than *E4/Abca1*$^{+/+}$ TBI mice, whereas no difference was found when comparing sham *Abca1*$^{+/-}$ to *Abca1*$^{+/+}$ mice of either isoform. There was no effect of *Abca1* haploinsufficiency on the expression of microglia genes in *APOE3* TBI mice. We were able to correlate the transcriptome to each phenotype using a network-based approach, Weighted Gene Co-expression Network Analysis (WGCNA). We found that the immune response module, although correlated positively to all TBI groups regardless of APOE isoform or *Abca1* copy number, consisted of genes expressed at higher levels in *E4/Abca1*$^{+/-}$TBI mice, and featured microglia-specific hub genes, including *Trem2, Tyrobp, Hexb,* and *Cd68.* Our results demonstrate an effect of ABCA1 deficiency on microglia gene expression after TBI in *APOE4* mice.

Materials and methods

Animals

All animal experiments were approved through the University of Pittsburgh Institutional Animal Care and Use Committee and carried out in accordance with PHS policies on the use of animals in research. Human *APOE3*$^{+/+}$ and *APOE4*$^{+/+}$ targeted replacement mice (referred to as *E3/Abca1*$^{+/+}$ and *E4/Abca1*$^{+/+}$) were bred to *Abca1*$^{+/-}$ mice to generate *APOE3*$^{+/+}$/*Abca1*$^{+/-}$ and *APOE4*$^{+/+}$/ *Abca1*$^{+/-}$ (referred to *E3/Abca1*$^{+/-}$ and *E4/Abca1*$^{+/-}$, respectively) [8, 17]. All mice were on the C57BL/6 genetic background and experimental groups consisted of both genders. Experimental mice were kept on a 12 h light-dark cycle with ad libitum access to food and water. At 3 months of age, these mice were randomly assigned to either sham or controlled cortical impact (CCI) experimental group. Mice were handled for 2 days (5 min per day) prior to surgical procedures. All materials were purchased through ThermoFisher Scientific, unless otherwise noted.

Traumatic brain injury

CCI model of brain injury was performed as previously described [9]. Anesthesia was induced using 5% isoflurane, after which it was maintained at 1.5% isoflurane. The head was secured using a stereotaxic frame, and core body temperature was held at 37 °C using a heating pad. After shaving the heads, two separate iodine - alcohol washes were performed to sterilize the surgical site. A 50% mixture of bupivacaine and lidocaine was applied to the area and ophthalmic ointment was applied to the eyes. The scalp was opened with a midline incision exposing the dorsal aspect of the skull and the skull leveled. A 4.5 mm diameter craniotomy was performed over the left parietal cortex using a dental drill. Once the bone flap was removed, mice in the CCI group received a single impact at 1.0 mm depth with a 3.0 mm diameter metal tip onto the cortex (3 m/s, 100 ms dwell time; Impact One, Leica). Sham mice received identical anesthesia and craniotomy, but did not receive impact and are considered negative controls. Following the impact, the surgical site was sutured, triple antibiotic cream applied, Buprenex (0.1 mg/kg, IP) provided for analgesia, and sterile saline administered for rehydration. Mice were allowed to recover on heating pad, until freely mobile, before returning to their home cage.

Tissue processing

Fourteen days post-injury, mice were anesthetized using Avertin (250 mg/kg of body weight, i.p.) and perfused transcardially with 20 mL of cold 0.1 M PBS pH 7.4 [9, 31]. Brains were rapidly removed and a 1.5 mm coronal section of the brain, including the injury site, taken by slicing the brain at − 2.5 mm and − 4.0 mm from bregma. Within the coronal slice, the hemispheres were separated, and the subcortical tissue was dissected out; hippocampal and cortical tissue were snap-frozen together for RNA isolation and RNA-seq.

RNA isolation and RNA sequencing

All procedures were performed as before [9, 32]. CCI and sham mice consisting of both genders for each genotype were used for RNA-seq. RNA was isolated from frozen cortices and hippocampi at the injury site and purified using RNeasy kit (Qiagen) according to the manufacturer recommendations. Quality control of all RNA samples was performed on a 2100 Bioanalyzer instrument (Agilent Technologies) and samples with RIN > 8 were further used for library construction using RNA Library Prep Reagent Set (Illumina). Libraries for Next Generation Sequencing were generated by PCR enrichment including incorporation of barcodes to enable multiplexing. Sequencing was performed by the Next Generation Sequencing Center (University of Pennsylvania, https://ngsc.med.upenn.edu/) on HiSeq 2500 machine. Following initial processing and quality control, the sequencing datasets were further analyzed for differential gene expression, which in all cases was calculated using Subread/featureCounts (v1.5.0; https://sourceforge.net/projects/subread/files/subread-1.5.0/) for read alignment and summarization and statistical package edgeR (v3.14.0; https://bioconductor.org/packages/release/bioc/html/edgeR.html). Lists of differentially expressed genes are further analyzed as described in the following section.

Weighted gene co-expression network analysis (WGCNA)

Network analysis was performed using WGCNA (v.1.51; https://cran.r-project.org/web/packages/WGCNA/index.html) [33, 48]. Libraries are clustered by gene expression enabling the detection of outliers and the power is determined by scale free topology model. Modules were generated automatically using a soft thresholding power, $\beta = 10$, a minimum module size of 18 genes and a minimum module merge cut height of 0.25. Modules were named by conventional color scheme and then correlated with trait data using Pearson's correlation (APOE isoform, Injury, *Abca1* copy number). Statistical significance was determined by student's *t*-test, $p < 0.05$. All modules were summarized by module eigengenes (ME), the first principle component of each module that was calculated as a synthetic gene representing the expression profile of all genes within a given module.

Representative networks were built using hub genes and the transcripts connected to them. Hub genes were identified using cutoffs of their interconnectivity within the module (module membership, > 0.8), and the correlation between expression level and trait (gene significance, > 0.2). Once the hub genes are selected, the connections to other transcripts are sorted by weight, with the first 150 connections used for visualization. Gene-association networks of interest were visualized using Cytoscape (v3.3.0). Unsupervised hierarchical clustering was performed on ME turquoise using pheatmap (v1.0.10; https://cran.r-project.org/web/packages/pheatmap/index.html) to identify 2 sub-modules. A representative network was built for each sub-module consisting of only genes within the sub-module.

Functional pathway analysis

Functional annotation clustering was performed using the Database for Annotation, Visualization and Integrated Discovery (DAVID v6.8, https://david.ncifcrf.gov) [25].

Results

TBI induces changes to the transcriptome that are common among both *Abca1*$^{+/+}$ and *Abca1*$^{+/-}$ mice expressing human APOE isoforms

To examine the effect of TBI on gene expression in the brains of *Abca1*$^{+/-}$ and *Abca1*$^{+/+}$ mice expressing human APOE isoforms (*E3/Abca1*$^{+/-}$, *E4/Abca1*$^{+/-}$, *E3/Abca1*$^{+/-}$, *E4/Abca1*$^{+/+}$), we collected hippocampal and cortical tissue from the injury site at 14 days post-injury. Total RNA was isolated from these tissues and used for RNA-sequencing. As shown in Fig. 1a-d, TBI significantly affected the transcriptome within each genotype. We highlighted several genes of interest on the scatterplots increased by TBI, and while they were differentially expressed within all the groups, the group with the highest CPM values was the *E4/Abca1*$^{+/-}$ TBI mice. To determine what similarities existed among the affected biological processes, we examined the differentially expressed genes that were significant and common among the 4 genotypes. The expression levels of the top 100 up- and down-regulated genes are shown in the heatmap (Fig. 1e). Although the genes are common, the *E4/Abca1*$^{+/-}$ mice show the highest expression levels of the upregulated genes. Figure 1f shows the biological processes associated with the common, upregulated genes. There were 1196 up-regulated genes common to all the groups and the top Gene Ontology (GO) terms derived from these genes were "immune system process", "innate immune response", and "inflammatory response". In comparison, Fig. 1g shows the biological processes associated with the common, downregulated genes. There were 579 downregulated genes common to the groups, and these genes were functionally associated with "regulation of ion transmembrane transport", and "potassium ion transport".

TBI significantly alters the expression of transcripts unique to each genotype, with the highest proportion of unique transcripts among *E4/Abca1*$^{+/-}$ mice

We were interested in whether the response to TBI was specifically influenced by each genotype, particularly by *Abca1* haplodeficiency in conjunction with *APOE4* isoform. To do this, we determined the proportion of genes that were differentially expressed, either in common among several of the groups or were uniquely expressed in only one group. These proportions are shown for each genotype in the donut plots in Fig. 2. As seen in Fig. 2a, *E4/Abca1*$^{+/-}$ mice have a higher proportion of unique transcripts (26%) that are up-regulated by TBI than the other genotypes (*E3/Abca1*$^{+/-}$: 5.5%; *E3/Abca1*$^{+/+}$: 10%; *E4/Abca1*$^{+/+}$: 5.0%). The biological processes derived from the unique genes of the *E4/Abca1*$^{+/-}$ mice, include "positive regulation of neuroblast differentiation" and "positive regulation of apoptotic process". The biological functions associated with the unique, upregulated genes within each group differ greatly; the other top terms include "determination of left/right symmetry", "negative regulation of cell proliferation", and "inner dynein arm assembly" for *E3/Abca1*$^{+/-}$, *E3/Abca1*$^{+/+}$, *E4/Abca1*$^{+/+}$ respectively. Expression plots show the distinct upregulation in *E4/Abca1*$^{+/-}$ TBI mice of several genes, including *Plekho1*, which has been shown to promote apoptosis (Fig. 2c) [47].

E4/Abca1$^{+/-}$ mice, again, have a higher proportion of unique, significant down-regulated transcripts (30%) than the other genotypes (*E3/Abca1*$^{+/-}$: 10%; *E3/Abca1*$^{+/+}$: 13%; *E4/Abca1*$^{+/+}$: 9.4%) (Fig. 2b). The top GO terms derived from the unique down-regulated genes for each group are "transcription, DNA-templated", "covalent chromatin modification", "GPI anchor biosynthetic process", and "Protein K63-linked ubiquitination" for *E3/Abca1*$^{+/-}$, *E4/Abca1*$^{+/-}$, *E3/Abca1*$^{+/+}$, *E4/Abca1*$^{+/+}$, respectively.

Abca1 haploinsufficiency upregulates microglia sensome genes in injured *APOE4* mice

To determine if there was any effect of *Abca1* haploinsufficiency on gene expression changes induced by injury, we examined the expression of microglial sensome genes. Although a clear effect of TBI is present in the differential expression of the microglia sensome by *Abca1* genotype, the heatmap also shows that *E4/Abca1*$^{+/-}$ TBI mice have higher expression levels of microglial sensome genes than the other groups (Fig. 3a). In contrast, there is no effect of *Abca1* copy number on synaptic transmission genes (Fig. 3b), although an injury effect on expression is still visible. We examined the expression levels of the microglia sensome genes within each *APOE* isoform, separated by injury status, for the effect of *Abca1* genotype. Sham mice in both *APOE* isoforms (Fig. 3c-d) and injured *APOE3* mice (Fig. 3e) have no significant changes in microglia sensome gene expression due to *Abca1* haploinsufficiency. In comparison, the injured *E4/Abca1*$^{+/-}$ mice demonstrate significant expression of the microglia sensome compared to *E4/Abca1*$^{+/+}$ TBI mice (Fig. 3f). In conclusion, these results demonstrate an effect of *Abca1* haploinsufficiency on the microglia sensome in *APOE4* mice after TBI.

WGCNA reveals interconnected gene clusters associated with each trait of interest- *APOE* isoform, *Abca1* copy number, and injury status

To identify interconnected gene clusters, or modules, associated within each trait of interest, we employed WGCNA. We were interested in the modules that were

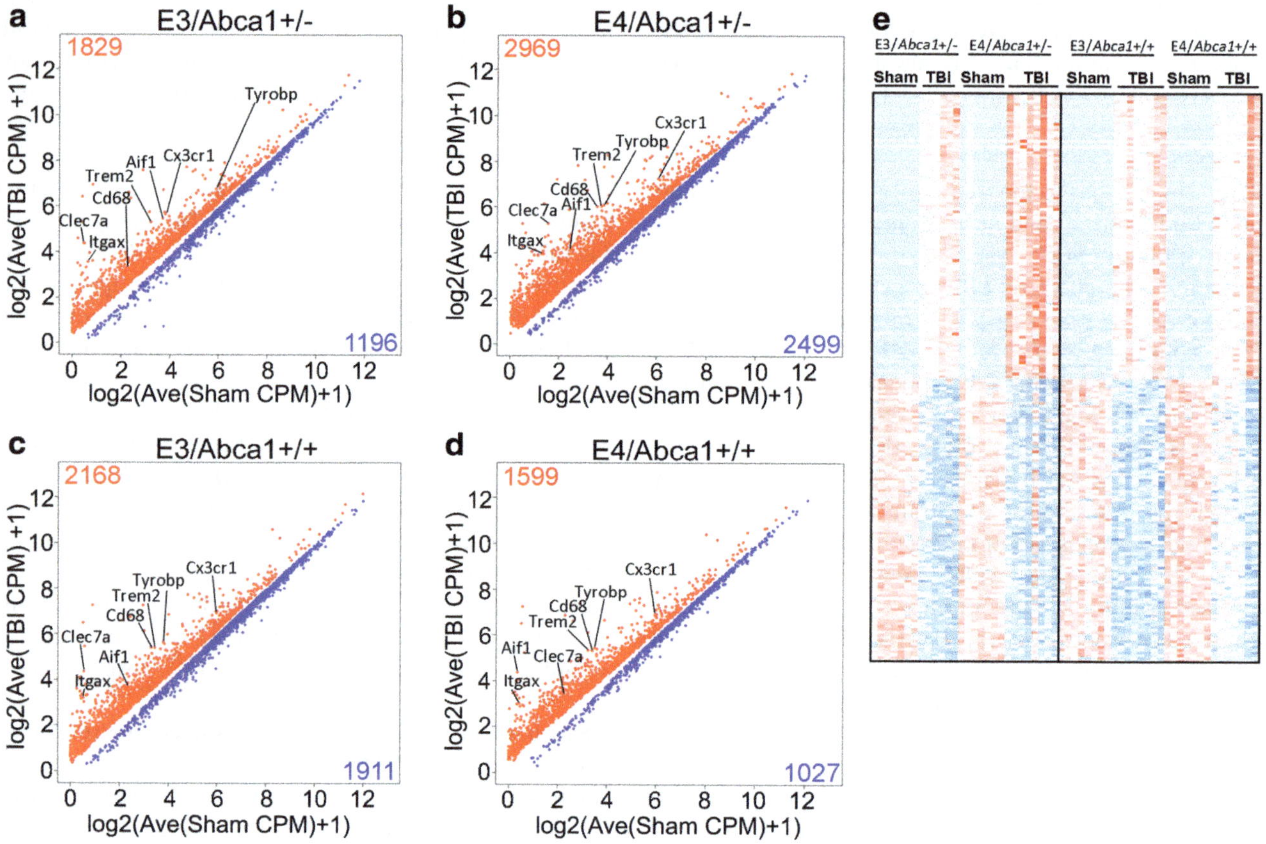

f

GO Term	Count	p-value
Immune system process	133	2.45E-66
Innate immune response	112	4.25E-45
Inflammatory response	91	1.90E-34
Defense response to virus	59	1.16E-29
Response to virus	33	2.41E-18
Chemotaxis	37	7.64E-17
Positive regulation of IκB kinase/NFκB signaling	39	8.53E-15
Positive regulation of inflammatory response	25	4.57E-14
Immune response	52	1.61E-13
Positive regulation of TNF production	23	8.76E-13

g

GO Term	Count	p-value
Regulation of ion transmembrane transport	29	8.50E-17
Potassium ion transport	27	1.73E-15
Potassium ion transmembrane transport	20	6.21E-12
Ion transport	49	3.57E-11
Exocytosis	20	1.07E-10
Regulation of membrane potential	18	1.73E-09
Ca2+-regulated exocytosis of neurotransmitter	11	2.52E-09
Neurotransmitter secretion	11	2.52E-09
Intracellular signal transduction	35	1.28E-08
Transport	93	2.43E-08

Fig. 1 TBI increases the expression of genes associated with immune response, and decreases the expression of genes connected to ion transmembrane transport. RNA was isolated from the hippocampal and cortical tissues collected 14 days after injury from *Abca1*$^{+/-}$ and *Abca1*$^{+/+}$ mice of both *APOE* isoforms and was then used to perform RNA-seq, $N = 6–8$ mice per group of both genders. **a-d** Scatter plots represent the RNA-seq results for differentially expressed genes. EdgeR analysis between sham and injured mice identified significant affected transcripts in (**a**) *APOE3/Abca1*$^{+/-}$, (**b**) *APOE4/Abca1*$^{+/-}$, (**c**) *APOE3/Abca1*$^{+/+}$ and (**d**) *APOE4/Abca1*$^{+/+}$ mice. Red denotes up-regulated, and blue denotes down-regulated genes, $p < 0.05$. **e** Heatmap of the top 100 upregulated and downregulated genes by TBI is shown. **f** A table shows the top annotated GO terms derived from the common, upregulated genes (total = 1215 genes). **g** A table shows the top annotated GO terms derived from the common, downregulated genes (total = 531 genes)

differentially expressed across our traits of interest - injury status, *APOE* isoform and *Abca1* copy number. The relationship table (Fig. 4a) shows the MEs of interest and the corresponding correlation coefficients per group.

ME tan (module size = 182 genes) correlated across the groups depending on *APOE* isoform, regardless of either injury status, or *Abca1* copy number. It positively correlated to *APOE4* groups and negatively with *APOE3* groups. As seen in the module heatmap (Fig. 4a; far right), the gene members are generally increased in APOE4 mice and decreased in APOE3 mice. The GO terms associated with the module genes were "tRNA aminoacylation for translation", "RNA processing", and "translation". We built a representative network using the hub genes associated with "tRNA aminoacylation for translation", such as *Yars, Gars,* and *Nars,* which are aminoacyl-tRNA synthetases.

ME pink correlated with injury status, however, it negatively correlated to TBI groups and positively correlated

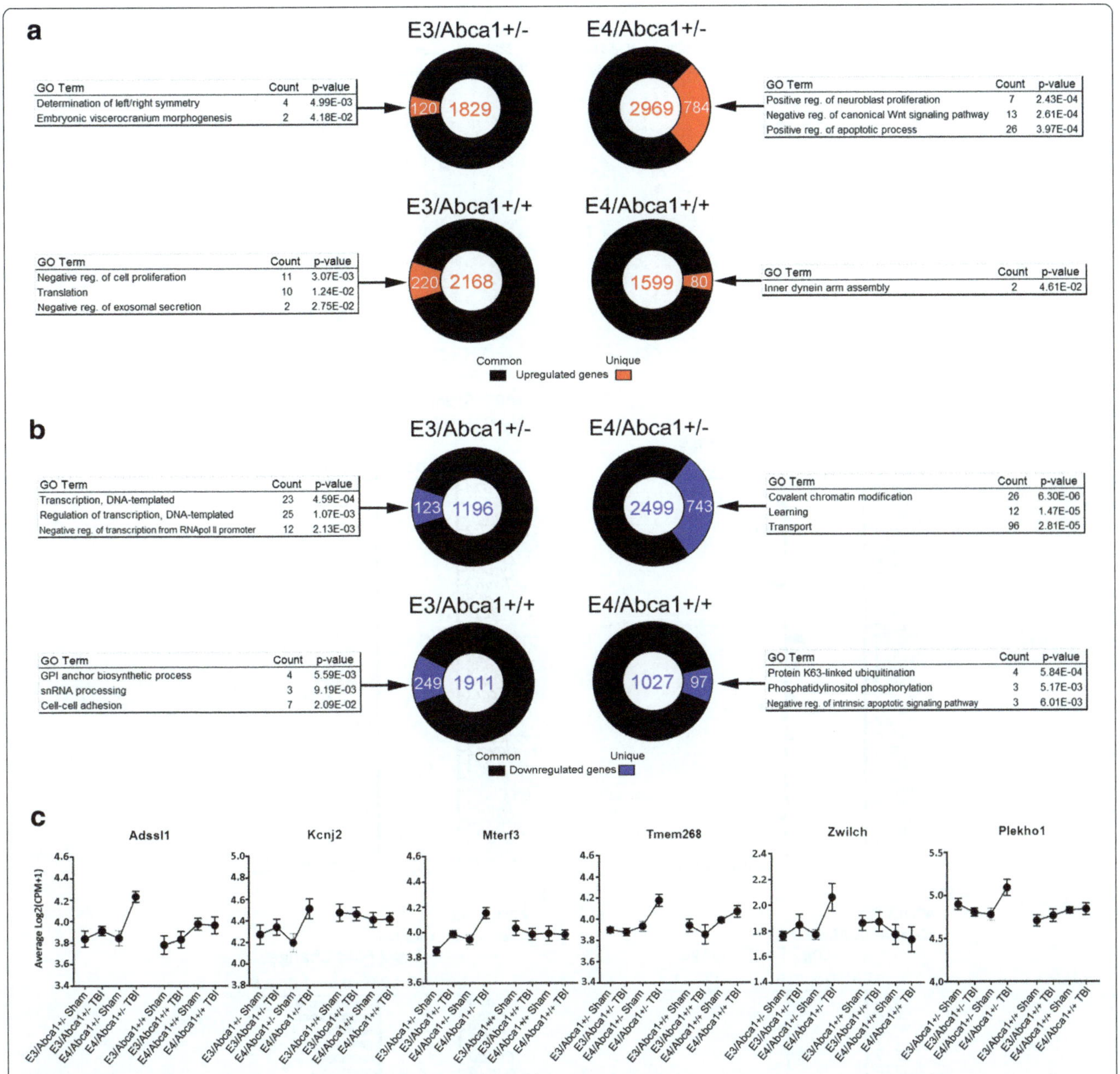

Fig. 2 TBI affects a greater proportion of unique genes in *E4/Abca1*$^{+/-}$ mice. **a**, **b** Donut plots for each genotype indicate the proportion of significantly (**a**) up- or (**b**) down-regulated genes that are either expressed in common among 2 or more groups (black), or unique to that group (**a**: red and **b**: blue). The total number of genes are shown within the center of the plot for each genotype. The top 3 GO terms for the unique genes in each genotype are shown to either side. **c** Expression plots for unique transcripts upregulated in *E4/Abca1*$^{+/-}$ mice are shown

with sham groups. The biological processes associated with ME pink (module size = 518 genes) included "synaptic vesicle docking", "long-term synaptic potentiation", and "chemical synaptic transmission", which suggests that injury decreases synaptic transmission. The representative network (Fig. 4c) was built around several hub genes related with synaptic transmission, including *Stx1a*, *Snap25*, and *Lamp5*, which are all associated with synaptic vesicle docking and neurotransmitter release. *Lamp5*, in particular, is associated with GABAergic synaptic transmission and short-term synaptic plasticity [40]. Another hub gene is *Prkcz*, which is necessary for long-term potentiation in hippocampal CA1 pyramidal cells [42].

ME grey60 correlated with *Abca1* copy number, specifically, it negatively correlated with *Abca1*$^{+/-}$ mice and positively correlated with *Abca1*$^{+/+}$ mice, regardless of injury or *APOE* isoform. As seen in Fig. 4d, the network was built around hub genes, which represented GO terms "oxidation-reduction process", "transport", and "aging". These hub genes included a number of the

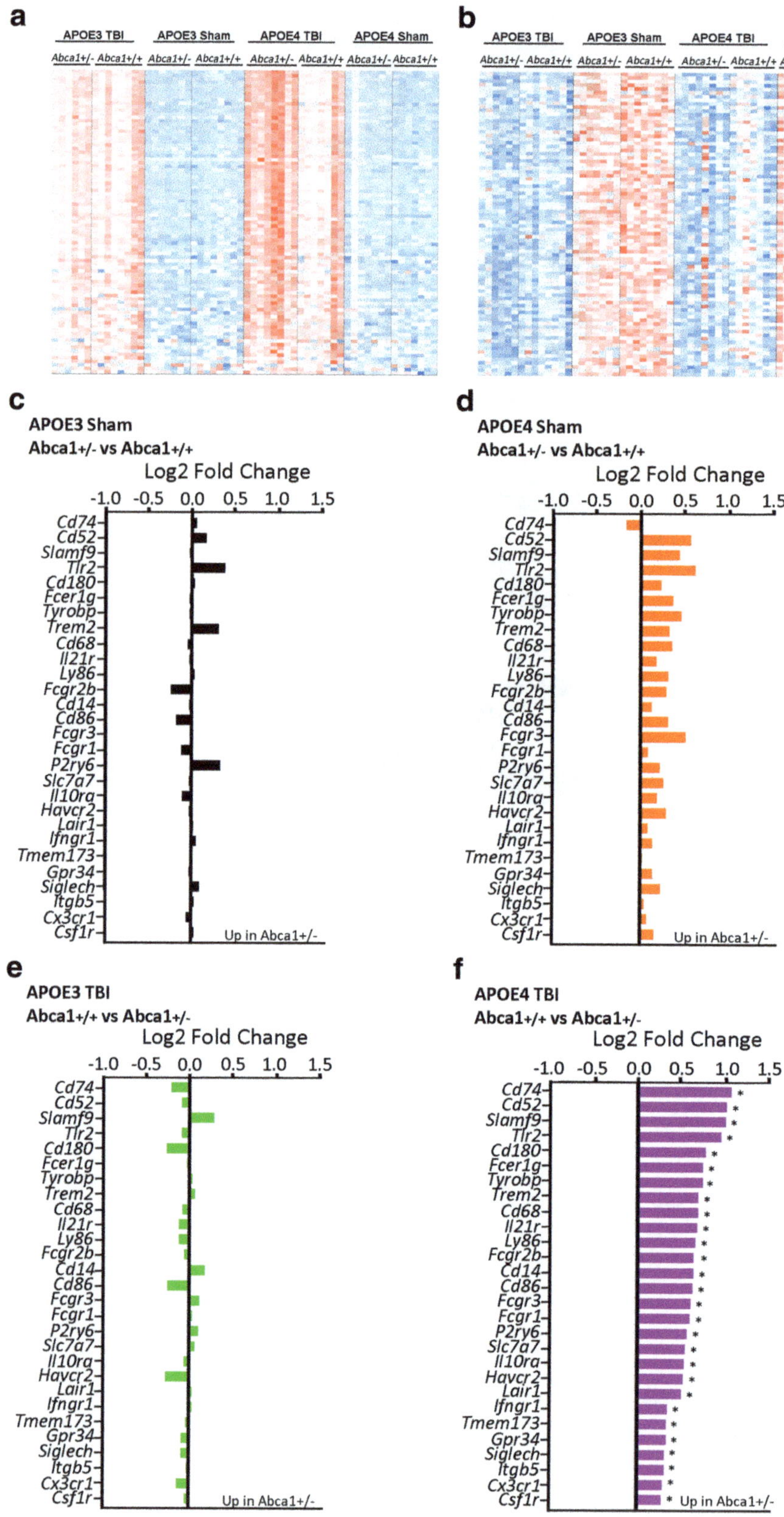

Fig. 3 (See legend on next page.)

(See figure on previous page.)
Fig. 3 *Abca1* deficiency affects the microglial response to TBI in an *APOE* isoform-dependent manner. **a-b** Heatmaps were generated using normalized *Abca1*$^{+/-}$ versus *Abca1*$^{+/+}$ CPM values for each group for (**a**) microglia sensome genes and (**b**) synaptic transmission genes. Red denotes higher expression values, and blue denotes lower expression values. $n = 6–8$ per group, including both males and females. **c**, **e** Selected genes from the microglia sensome of *APOE3/Abca1*$^{+/-}$ and *APOE3/Abca1*$^{+/+}$ mice are compared separately for (**c**) sham (black bars) and (**e**) TBI groups (green bars). Shown are the Log2-fold change values for each gene. **d**, **f** Selected genes from the microglia sensome of *APOE4/Abca1*$^{+/+}$ and *APOE4/Abca1*$^{+/-}$ mice are compared separately for (**d**) sham (orange bars) and TBI groups (purple bars). Shown are the Log2-fold change values. *: $p < 0.05$

NADH hydrogenase subunits, such as *ND1, ND2, ND4, ND5* and *ND6*. Other hub genes were *COX1, Atp5j2,* and *CYTB*. All of these hub genes are involved in the mitochondrial respiratory chain [46].

ME turquoise strongly correlated with injury status, however, unlike ME pink, it correlated positively to TBI groups and negatively to sham groups. The genes within ME turquoise are strongly connected and related to the module biological process, as seen in the module membership and gene significance scatterplot (Fig. 5a). Due to the size (module size = 3860 genes), we were interested in further separating the module. To do this, we ran a pheatmap function on the genes within the module, which aggregates the genes using hierarchical clustering. As shown in Fig. 5b, the pheatmap separated the module into 2 distinct clusters based on injury status and direction of expression. Additionally, the pheatmap shows the expression for all the genes in ME turquoise and the eigengene expression for each sample. The first cluster, Cluster 1, (size = 2605 genes) consisted of genes upregulated in TBI groups and downregulated in sham groups. The pheatmap suggests a stronger response of the cluster 1 genes within the *E4/Abca1*$^{+/-}$ mice, which is consistent with the correlation of ME turquoise to this group in the relationship table. The GO terms derived from Cluster 1 were "immune system response", "innate immune response", and "inflammatory response". Additionally, among the top 10 GO terms was "lipid metabolic process". The representative network (Fig. 5c) was built around hub genes associated with immune response, such as *Clec7a, C1qc,* and microglia-specific genes, *Trem2, Tyrobp, Hexb,* and *Cd68.*

The second cluster, Cluster 2, (size = 1111 genes) featured genes downregulated in the TBI mice, upregulated in the sham mice. Functionally, this cluster is enriched in genes connected to the GO term "transport", other transport-associated terms, such as "vesicle-mediated transport", but also GO terms "sterol biosynthetic process" and "cholesterol biosynthetic process". The network (Fig. 5d) built for Cluster 2 excluded any genes from Cluster 1, and functionally represents transport, however, while hub genes, *Gabrb2, Gabrg2,* and *Atp1b1* all relate directly to the transport of ions across the membrane, through this mechanism, these genes are also strongly associated with synaptic transmission. In conclusion, ME turquoise strongly correlated to injury status, but hierarchical clustering of the genes revealed two distinct clusters associated with the gene expression direction. Cluster 1 was larger and featured genes related to immune response and was more strongly upregulated in *E4/Abca1*$^{+/-}$ mice, while cluster 2 featured genes downregulated in TBI groups and represented transport, but functionally are also involved in synaptic transmission.

Discussion

We examined the impact of *Abca1* deficiency and *APOE* isoform expression on the response to traumatic brain injury. Our goal was to identify differences in the transcriptional response and trait-associated genome-wide correlated gene networks between *Abca1*$^{+/+}$ and *Abca1*$^{+/-}$ mice following a controlled cortical impact in human *APOE3*$^{+/+}$and *APOE4*$^{+/+}$ targeted replacement mice. We found that the four groups within our study -*E3/Abca1*$^{+/-}$, *E4/Abca1*$^{+/-}$, *E3/Abca1*$^{+/+}$, and *E4/Abca1*$^{+/+}$ – had common and distinct responses to TBI. *E4/Abca1*$^{+/-}$ mice had the highest proportion of unique transcripts affected by TBI, suggesting that *E4/Abca1*$^{+/-}$ mice are more disposed to changes in gene expression by TBI than the other groups, and demonstrate possible pathways that could be associated with worsened outcome, such as downregulated genes associated with learning. The common, up-regulated genes were associated with biological processes related to immune response, innate immune response and inflammatory response. While, these genes were common among the four groups, the *E4/Abca1*$^{+/-}$ mice had higher expression levels of the genes upregulated by TBI compared to the other groups, suggesting a role for *APOE* isoform and ABCA1 in the expression of inflammatory genes after TBI. Consequently, we examined the effect of *Abca1* insufficiency on microglia sensome genes by injury status and APOE isoform. When comparing injured *Abca1*$^{+/-}$ to *Abca1*$^{+/+}$ mice, we found *E4/Abca1*$^{+/-}$ TBI mice had increased expression of the microglia sensome genes. In contrast, there was no effect of *Abca1* copy number in *APOE3* mice, sham or TBI. These results suggest that *Abca1* haploinsufficiency may influence the inflammatory response following TBI, particularly through an impact on microglia and their gene expression. This effect is seen only among *APOE4* mice, not *APOE3* mice; this response may be related to the isoform-specific effect on inflammation. Additionally, the *APOE4* isoform may be more vulnerable to the

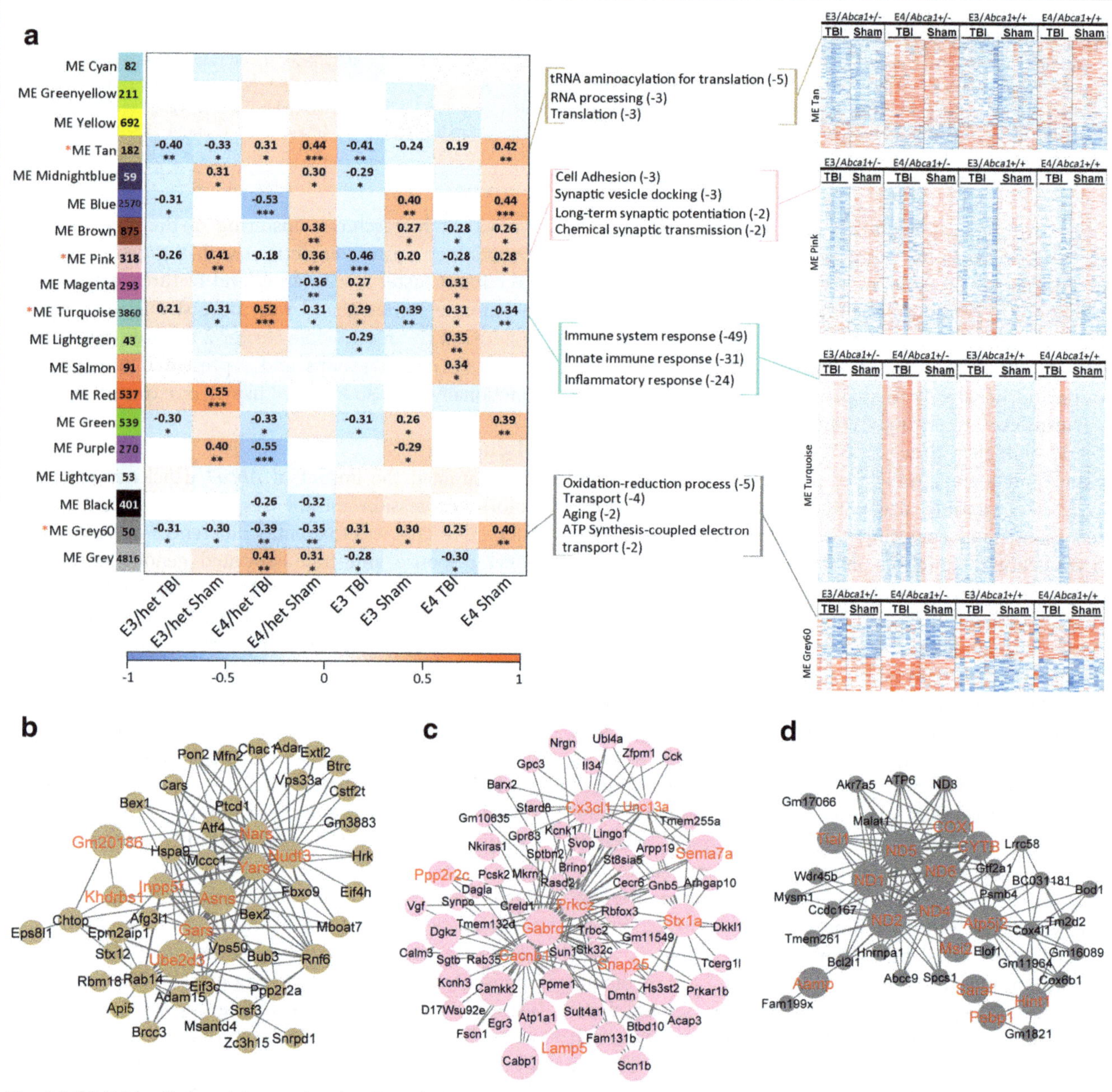

Fig. 4 WGCNA identified modules correlated to TBI and *Abca1* haploinsufficiency. WGCNA was used to determine the correlation of module eigengenes to Injury and *Abca1* genotype. **a** The relationship tables shows the correlation between the module eigengene (rows) and group (columns) with *p*-value. Red denotes a positive, and blue is a negative correlation. Modules of interest are differentially expressed between trait conditions. MEs turquoise and pink correlated with TBI in opposite directions, ME tan correlated with *APOE* isoform, and ME grey60 with *Abca1* genotype. Top assigned GO terms and their log10 of the *p*-values are shown to the right of the table, aside heatmaps of the genes within each module, for each animal. Red denotes higher expression values, and blue denotes lower expression values. **b-d** Representative networks for (**b**) ME tan, (**c**) ME pink, and (**d**) ME grey60 were built using module hub genes. Hub genes are identified in red font. Size of the nodes represents the module membership value and the width of the edge, the interaction between genes shown as connecting lines, represents the weight of the connection

consequences of *Abca1* haploinsufficiency due to a gene-gene interaction, a result also demonstrated by data from AD-model mice [17]. These results suggest a possible mechanism for worse outcome after TBI associated with *APOE4* isoform.

Using WGCNA, we identified modules associated with each trait – injury, *APOE* isoform and *Abca1* copy number. ME tan was associated with *APOE* isoform; the module positively correlated with *APOE4* mice and negatively correlated with *APOE3* mice, regardless of *Abca1* copy

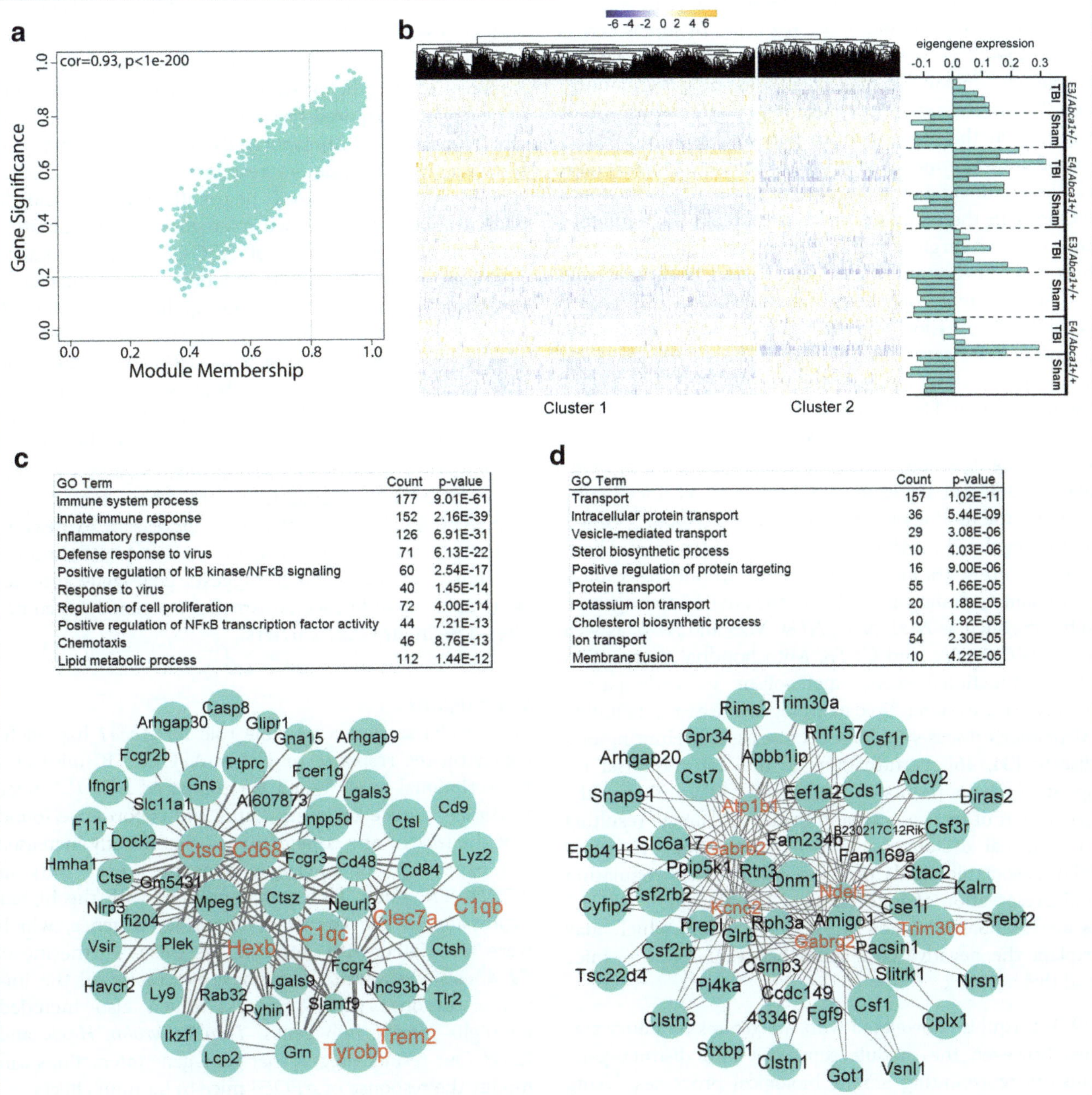

GO Term	Count	p-value
Immune system process	177	9.01E-61
Innate immune response	152	2.16E-39
Inflammatory response	126	6.91E-31
Defense response to virus	71	6.13E-22
Positive regulation of IκB kinase/NFκB signaling	60	2.54E-17
Response to virus	40	1.45E-14
Regulation of cell proliferation	72	4.00E-14
Positive regulation of NFκB transcription factor activity	44	7.21E-13
Chemotaxis	46	8.76E-13
Lipid metabolic process	112	1.44E-12

GO Term	Count	p-value
Transport	157	1.02E-11
Intracellular protein transport	36	5.44E-09
Vesicle-mediated transport	29	3.08E-06
Sterol biosynthetic process	10	4.03E-06
Positive regulation of protein targeting	16	9.00E-06
Protein transport	55	1.66E-05
Potassium ion transport	20	1.88E-05
Cholesterol biosynthetic process	10	1.92E-05
Ion transport	54	2.30E-05
Membrane fusion	10	4.22E-05

Fig. 5 ME turquoise consists of two sub-modules correlated to injury status. **a** Scatter plot for ME turquoise module showing the correlation between module membership and gene significance. The degree of association between MM and GS was evaluated by Pearson correlation. Correlation value and p value are indicated in the plot. **b** The pheatmap shows normalized gene expression values beside module eigengene expression values for each sample for ME turquoise. The pheatmap function aggregated the module into 2 sub-modules (Cluster 1 and Cluster 2) by hierarchical clustering. **c-d** Tables of top assigned GO terms are shown above representative networks for (**c**) cluster 1 and (**d**) cluster 2. Hub genes are identified in red font. Size of the nodes represents the module membership value and width of the edge represents the weight of the connection

number or injury status. The representative network was associated with the GO term "tRNA aminoacylation for translation", and included hub genes *Yars, Gars,* and *Nars,* which are aminoacyl-tRNA synthetases. Mutations in these genes are associated with Charcot-Marie-Tooth disease, one of the most commonly inherited neurological disorders [6]. Additionally, a metabolomics study on AD patient CSF and plasma found that a pathway significantly affected in plasma by AD severity was aminoacyl-tRNA biosynthesis, however, the mechanisms associated with altered aminoacyl-tRNA synthetases and AD remain unknown [41].

The "synaptic transmission" module, ME pink was significantly correlated and down-regulated by injury across the groups. The network represented GO terms "synaptic vesicle docking", "long-term synaptic potentiation", and "chemical synaptic transmission". The hub genes featured in the representative network, included *Stx1a, Snap25,* and *Lamp5,* which are all associated with synaptic vesicle docking and neurotransmitter release. *Lamp5* localizes in the synapse, where it may play a regulatory role in GABAergic synaptic transmission [40]. Another hub gene in this network, *Prkcz,* has an important role in hippocampal long term potentiation and learning [42]. Its expression mediates the storage of specific forms of long term memory [38]. The negative association between this network and injury is consistent with the impact that TBI is known to have on memory.

The network representing ME grey60 was associated with "oxidation-reduction process" and "aging". This module was differentially expressed dependent on *Abca1* copy number; the module was downregulated in *Abca1*$^{+/-}$ mice, and upregulated in *Abca1*$^{+/+}$ mice. The network was built around hub genes involved in the mitochondrial respiratory chain, including a number of the NADH hydrogenase subunits, such as *ND1, ND2, ND4, ND5* and *ND6*, as well as, *COX1, Atp5j2,* and *CYTB.* Mitochondrial dysfunction and dysfunctional energy metabolism are early pathological features of multiple neurological diseases, including Alzheimer's disease, Parkinson's disease and Huntington's disease [34, 46]. Perturbations in the mitochondrial respiratory chain results in decreased ATP synthesis, the generation of free radicals and oxidative damage resulting in neuronal dysfunction and apoptosis [30]. HDL and HDL-associated lipids play key roles in the regulation and preservation of mitochondrial function [43]. ABCA1 is an essential mediator of HDL formation, which may explain the negative correlation between *Abca1*$^{+/-}$ mice and this network.

ME turquoise correlated with the groups by injury status, however, the module separated into distinct gene clusters representing unique biological processes. Using the pheatmap function, we were able to separate ME turquoise into 2 sub-modules by hierarchical clustering. The clusters were separated based on injury status and the direction of gene expression. The first cluster was larger and consisted of genes upregulated by injury. This cluster represented the "immune response" and the network was built from several microglia-specific genes including *Trem2, Tyrobp, Hexb,* and *Cd68.* Although there was no specific modulatory effect of *APOE* isoform or *Abca1* copy number on the module, the expression of the module genes was much higher in *E4/Abca1*$^{+/-}$ injured mice, which is consistent with our other results.

ABCA1 is a major regulator of cholesterol transport and an essential mediator of high density lipoprotein generation [22]. ABCA1 may have a crucial role in the response to TBI by providing essential cholesterol and phospholipids required for repair. However, ABCA1 may also influence the TBI response through its modulatory effects on the inflammatory response. Mice lacking brain ABCA1 exhibit increased neuroinflammation, and in particular have an increased microglial pro-inflammatory response [19]. The effect of ABCA1 on inflammation could also occur through its functional role in mediating cholesterol efflux onto lipid-poor apolipoprotein, including APOE. It was previously shown that the loss of ABCA1 function results in a reduction of APOE, and data from experimental animals show that *Abca1* deficiency abolishes the lipidation of APOE [21]. The isoform-dependent effect of APOE is possibly driven by lipidation status, which has been shown to affect its stability and degradation rate. Our study shows that ABCA1 haploinsufficiency increased expression of the microglia sensome genes in an APOE isoform dependent manner, which suggests gene-gene interactions as a possible mechanism for worsened outcomes after TBI in *APOEε4* carriers.

Conclusions

Our results suggest a possible role for *Abca1* haplodeficiency on the response to TBI in APOE4 TBI mice at a transcriptional level. When we compared *Abca1*$^{+/+}$ mice to *Abca1*$^{+/-}$ mice by injury status and isoform, we found that the lack of one copy of *Abca1* significantly increased the expression of microglia sensome genes only in *APOE4* TBI mice. This was consistent with the higher expression of the common, upregulated genes, which were associated with immune response. Furthermore, *E4/Abca1*$^{+/-}$ showed the highest expression of the immune response gene network, which also included microglia-specific hub genes, *Trem2, Tyrobp, Hexb,* and *Cd68.* Our results suggest that gene-gene interactions can modify the response of *APOE4* mice to harmful effects.

Abbreviations

ABCA1: ATP Binding Cassette Transporter A1; AD: Alzheimer's disease; APOE: Apolipoprotein E; CCI: Controlled Cortical Impact; GO: Gene Ontology; ME: Module Eigengene; RNA-seq: RNA-sequencing; TBI: Traumatic brain injury; WGCNA: Weighted Gene Co-expression Network Analysis

Acknowledgements

The authors would like to acknowledge the excellent assistance of Dr. Anais Mounier for her assistance in library generation. We would also like to acknowledge the assistance of the lab technicians Brittany Playso, and Heather Wells, who were responsible for the management of the animal colony during the course of the project.

Funding
This study was supported by the National Institutes of Health (ES024233, AG056371, AG057565 and K01AG044490) and the U.S. Department of Defense (W81XWH-13-1-0384).

Authors' contributions
ELC was a major contributor to all the experimental work, data analysis, and manuscript writing. FL generated the libraries for RNA-seq. CMW, KNN, NFF, IL and RK were involved in data analysis. NFF, RK and IL were involved in project administration. RK and IL conceptualized the project, acquired funding and assisted with manuscript development. All authors read and approved the final manuscript.

Competing interests
The authors declare that they have no competing interests.

References

1. Acosta SA, Tajiri N, Shinozuka K, Ishikawa H, Grimmig B, Diamond DM et al (2013) Long-term upregulation of inflammation and suppression of cell proliferation in the brain of adult rats exposed to traumatic brain injury using the controlled cortical impact model. PLoS One 8:e53376. https://doi.org/10.1371/journal.pone.0053376
2. Alexander S, Kerr ME, Kim Y, Kamboh MI (2007) Apolipoprotein E4 allele presence and functional outcome after severe traumatic brain injury. J Neurotrauma 24:790–797
3. Anthonymuthu TS, Kenny EM, Bayir H (2016) Therapies targeting lipid peroxidation in traumatic brain injury. Brain Res 1640:57–76. https://doi.org/10.1016/j.brainres.2016.02.006
4. Barger SW, Harmon AD (1997) Microglial activation by Alzheimer amyloid precursor protein and modulation by apolipoprotein E. Nature 388:878–881. https://doi.org/10.1038/42257
5. Bazarian JJ, Cernak I, Noble-Haeusslein L, Potolicchio S, Temkin N (2009) Long-term neurologic outcomes after traumatic brain injury. J Head Trauma Rehabil 24:439–451. https://doi.org/10.1097/HTR.0b013e3181c15600
6. Blocquel D, Li S, Wei N, Daub H, Sajish M, Erfurth ML et al (2017) Alternative stable conformation capable of protein misinteraction links tRNA synthetase to peripheral neuropathy. Nucleic Acids Res 45:8091–8104. https://doi.org/10.1093/nar/gkx455
7. Brooks-Wilson A, Marcil M, Clee SM, Zhang LH, Roomp K, van Dam M et al (1999) Mutations in ABC1 in tangier disease and familial high-density lipoprotein deficiency. Nature Genet 22:336–345
8. Carter AY, Letronne F, Fitz NF, Mounier A, Wolfe CM, Nam KN et al (2017) Liver X receptor agonist treatment significantly affects phenotype and transcriptome of APOE3 and APOE4 Abca1 haplo-deficient mice. PLoS One 12:e0172161. https://doi.org/10.1371/journal.pone.0172161
9. Castranio EL, Mounier A, Wolfe CM, Nam KN, Fitz NF, Letronne F et al (2017) Gene co-expression networks identify Trem2 and Tyrobp as major hubs in human APOE expressing mice following traumatic brain injury. Neurobiol Dis 105:1–14. https://doi.org/10.1016/j.nbd.2017.05.006
10. Centers for Disease C, Prevention (2013) CDC grand rounds: reducing severe traumatic brain injury in the United States. MMWR Morb Mortal Wkly Rep 62: 549–552
11. Chamelian L, Reis M, Feinstein A (2004) Six-month recovery from mild to moderate traumatic brain injury: the role of APOE-epsilon4 allele. Brain 127:2621–2628. https://doi.org/10.1093/brain/awh296
12. Crawford FC, Vanderploeg RD, Freeman MJ, Singh S, Waisman M, Michaels L et al (2002) APOE genotype influences acquisition and recall following traumatic brain injury. Neurology 58:1115–1118
13. Das M, Mohapatra S, Mohapatra SS (2012) New perspectives on central and peripheral immune responses to acute traumatic brain injury. J Neuroinflammation 9:236. https://doi.org/10.1186/1742-2094-9-236
14. Diaz-Arrastia R, Gong Y, Fair S, Scott KD, Garcia MC, Carlile MC et al (2003) Increased risk of late posttraumatic seizures associated with inheritance of APOE epsilon4 allele. Arch Neurol 60:818–822. https://doi.org/10.1001/archneur.60.6.818
15. Draper K, Ponsford J (2008) Cognitive functioning ten years following traumatic brain injury and rehabilitation. Neuropsychology 22:618–625. https://doi.org/10.1037/0894-4105.22.5.618
16. Faul M, Coronado V (2015) Epidemiology of traumatic brain injury. Handb Clin Neurol 127:3–13. https://doi.org/10.1016/B978-0-444-52892-6.00001-5
17. Fitz NF, Cronican AA, Saleem M, Fauq AH, Chapman R, Lefterov I et al (2012) Abca1 deficiency affects Alzheimer's disease-like phenotype in human ApoE4 but not in ApoE3-targeted replacement mice. J Neurosci 32:13125–13136. https://doi.org/10.1523/JNEUROSCI.1937-12.2012
18. Ghroubi S, Feki I, Chelly H, Elleuch MH (2016) Neuropsychological and behavioral disorders and their correlations with the severity of the traumatic brain injury. Ann Phys Rehabil Med 59S:e134–e135. https://doi.org/10.1016/j.rehab.2016.07.302
19. Karasinska JM, de Haan W, Franciosi S, Ruddle P, Fan J, Kruit JK et al (2013) ABCA1 influences neuroinflammation and neuronal death. Neurobiol Dis 54:445–455. https://doi.org/10.1016/j.nbd.2013.01.018
20. Kim J, Basak JM, Holtzman DM (2009) The role of apolipoprotein E in Alzheimer's disease. Neuron 63:287–303. https://doi.org/10.1016/j.neuron.2009.06.026
21. Koldamova R, Staufenbiel M, Lefterov I (2005) Lack of ABCA1 considerably decreases brain ApoE level and increases amyloid deposition in APP23 mice. J Biol Chem 280:43224–43235. https://doi.org/10.1074/jbc.M504513200
22. Koldamova R, Fitz NF, Lefterov I (2014) ATP-binding cassette transporter A1: from metabolism to neurodegeneration. Neurobiol Dis 72 Pt A:13–21. https://doi.org/10.1016/j.nbd.2014.05.007
23. Laskowitz DT, Fillit H, Yeung N, Toku K, Vitek MP (2006) Apolipoprotein E-derived peptides reduce CNS inflammation: implications for therapy of neurological disease. Acta Neurol Scand (Suppl 185):15–20. https://doi.org/10.1111/j.1600-0404.2006.00680.x
24. Li X, Montine KS, Keene CD, Montine TJ (2015) Different mechanisms of apolipoprotein E isoform-dependent modulation of prostaglandin E2 production and triggering receptor expressed on myeloid cells 2 (TREM2) expression after innate immune activation of microglia. FASEB J 29:1754–1762. https://doi.org/10.1096/fj.14-262683
25. Liliang PC, Liang CL, Weng HC, Lu K, Wang KW, Chen HJ *et al.* (2010) Tau proteins in serum predict outcome after severe traumatic brain injury. J Surg Res 160:302–307. https://doi.org/10.1016/j.jss.2008.12.022
26. Lynch JR, Morgan D, Mance J, Matthew WD, Laskowitz DT (2001) Apolipoprotein E modulates glial activation and the endogenous central nervous system inflammatory response. J Neuroimmunol 114:107–113
27. Maas AI, Stocchetti N, Bullock R (2008) Moderate and severe traumatic brain injury in adults. Lancet Neurol 7:728–741. https://doi.org/10.1016/s1474-4422(08)70164-9
28. Mannix RC, Zhang J, Park J, Zhang X, Bilal K, Walker K et al (2011) Age-dependent effect of apolipoprotein E4 on functional outcome after controlled cortical impact in mice. J Cereb Blood Flow Metab 31:351–361. https://doi.org/10.1038/jcbfm.2010.99
29. Morganti JM, Jopson TD, Liu S, Riparip L-K, Guandique CK, Gupta N et al (2015) CCR2 antagonism alters brain macrophage polarization and ameliorates cognitive dysfunction induced by traumatic brain injury. J Neurosci 35:748–760
30. Muller WE, Eckert A, Kurz C, Eckert GP, Leuner K (2010) Mitochondrial dysfunction: common final pathway in brain aging and Alzheimer's disease--therapeutic aspects. Mol Neurobiol 41:159–171. https://doi.org/10.1007/s12035-010-8141-5
31. Nam KN, Mounier A, Fitz NF, Wolfe C, Schug J, Lefterov I et al (2016) RXR controlled regulatory networks identified in mouse brain counteract deleterious effects of Abeta oligomers. Sci Rep 6:24048. https://doi.org/10.1038/srep24048
32. Nam KN, Mounier A, Wolfe CM, Fitz NF, Carter AY, Castranio EL et al (2017) Effect of high fat diet on phenotype, brain transcriptome and lipidome in Alzheimer's model mice. Sci Rep 7:4307. https://doi.org/10.1038/s41598-017-04412-2
33. Nam KN, Wolfe CM, Fitz NF, Letronne F, Castranio EL, Mounier A et al (2018) Integrated approach reveals diet, APOE genotype and sex affect immune response in APP mice. Biochim Biophys Acta 1864:152–161. https://doi.org/10.1016/j.bbadis.2017.10.018
34. Pathania D, Millard M, Neamati N (2009) Opportunities in discovery and delivery of anticancer drugs targeting mitochondria and cancer cell metabolism. Adv Drug Deliv Rev 61:1250–1275. https://doi.org/10.1016/j.addr.2009.05.010
35. Perez-Garcia G, Gama Sosa MA, De Gasperi R, Lashof-Sullivan M, Maudlin-Jeronimo E, Stone JR et al (2016) Chronic post-traumatic stress disorder-related traits in a rat model of low-level blast exposure. Behav brain res. In: https://doi.org/10.1016/j.bbr.2016.09.061

36. Plassman BL, Grafman J (2015) Traumatic brain injury and late-life dementia. Handb Clin Neurol 128:711–722. https://doi.org/10.1016/B978-0-444-63521-1.00044-3
37. Ponsford J, Rudzki D, Bailey K, Ng KT (2007) Impact of apolipoprotein gene on cognitive impairment and recovery after traumatic brain injury. Neurology 68:619–620. https://doi.org/10.1212/01.wnl.0000254609.04330.9d
38. Shema R, Sacktor TC, Dudai Y (2007) Rapid erasure of long-term memory associations in the cortex by an inhibitor of PKM zeta. Science 317:951–953. https://doi.org/10.1126/science.1144334
39. Simon DW, McGeachy MJ, Bayir H, Clark RS, Loane DJ, Kochanek PM (2017) The far-reaching scope of neuroinflammation after traumatic brain injury. Nat Rev Neurol 13:171–191. https://doi.org/10.1038/nrneurol.2017.13
40. Tiveron MC, Beurrier C, Ceni C, Andriambao N, Combes A, Koehl M *et al.* (2016) LAMP5 fine-tunes GABAergic synaptic transmission in defined circuits of the mouse brain. PLoS One 11:e0157052. https://doi.org/10.1371/journal.pone.0157052
41. Trushina E, Dutta T, Persson XM, Mielke MM, Petersen RC (2013) Identification of altered metabolic pathways in plasma and CSF in mild cognitive impairment and Alzheimer's disease using metabolomics. PLoS One 8:e63644. https://doi.org/10.1371/journal.pone.0063644
42. Wang S, Sheng T, Ren S, Tian T, Lu W (2016) Distinct roles of PKCiota/lambda and PKMzeta in the initiation and maintenance of hippocampal long-term potentiation and memory. Cell Rep 16:1954–1961. https://doi.org/10.1016/j.celrep.2016.07.030
43. White CR, Datta G, Giordano S (2017) High-density lipoprotein regulation of mitochondrial function. Adv Exp Med Biol 982:407–429. https://doi.org/10.1007/978-3-319-55330-6_22
44. Whitnall L, McMillan TM, Murray GD, Teasdale GM (2006) Disability in young people and adults after head injury: 5-7 year follow up of a prospective cohort study. J Neurol Neurosurg Psychiatry 77:640–645. https://doi.org/10.1136/jnnp.2005.078246
45. Yang SH, Gangidine M, Pritts TA, Goodman MD, Lentsch AB (2013) Interleukin 6 mediates neuroinflammation and motor coordination deficits after mild traumatic brain injury and brief hypoxia in mice. Shock (Augusta, Ga) 40:471–475
46. Yu H, Lin X, Wang D, Zhang Z, Guo Y, Ren X et al (2018) Mitochondrial molecular abnormalities revealed by proteomic analysis of hippocampal organelles of mice triple transgenic for Alzheimer disease. Front Mol Neurosci 11:74. https://doi.org/10.3389/fnmol.2018.00074
47. Zhang L, Xing G, Tie Y, Tang Y, Tian C, Li L et al (2005) Role for the pleckstrin homology domain-containing protein CKIP-1 in AP-1 regulation and apoptosis. EMBO J 24:766–778. https://doi.org/10.1038/sj.emboj.7600532
48. Zhao W, Langfelder P, Fuller T, Dong J, Li A, Hovarth S (2010) Weighted gene coexpression network analysis: state of the art. J Biopharm Stat 20: 281–300. https://doi.org/10.1080/10543400903572753

12

Characterisation of tau in the human and rodent enteric nervous system under physiological conditions and in tauopathy

Arthur Lionnet[1,2,3], Matthew A. Wade[4], Anne-Gaëlle Corbillé[1,2,3], Alice Prigent[1,3], Sébastien Paillusson[4], Maddalena Tasselli[4], Jacques Gonzales[1,3], Emilie Durieu[1], Malvyne Rolli-Derkinderen[1,3], Emmanuel Coron[1,3], Emilie Duchalais[1,3], Michel Neunlist[1,3], Michael S. Perkinton[5], Diane P. Hanger[4], Wendy Noble[4*] and Pascal Derkinderen[1,2,3*]

Abstract

Tau is normally a highly soluble phosphoprotein found predominantly in neurons. Six different isoforms of tau are expressed in the adult human CNS. Under pathological conditions, phosphorylated tau aggregates are a defining feature of neurodegenerative disorders called tauopathies. Recent findings have suggested a potential role of the gut-brain axis in CNS homeostasis, and therefore we set out to examine the isoform profile and phosphorylation state of tau in the enteric nervous system (ENS) under physiological conditions and in tauopathies. Surgical specimens of human colon from controls, Parkinson's disease (PD) and progressive supranuclear palsy (PSP) patients were analyzed by Western Blot and immunohistochemistry using a panel of anti-tau antibodies. We found that adult human ENS primarily expresses two tau isoforms, localized in the cell bodies and neuronal processes. We did not observe any difference in the enteric tau isoform profile and phosphorylation state between PSP, PD and control subjects. The htau mouse model of tauopathy also expressed two main isoforms of human tau in the ENS, and there were no apparent differences in ENS tau localization or phosphorylation between wild-type and htau mice. Tau in both human and mouse ENS was found to be phosphorylated but poorly susceptible to dephosphorylation with lambda phosphatase. To investigate ENS tau phosphorylation further, primary cultures from rat enteric neurons, which express four isoforms of tau, were pharmacologically manipulated to show that ENS tau phosphorylation state can be regulated, at least in vitro. Our study is the first to characterize tau in the rodent and human ENS. As a whole, our findings provide a basis to unravel the functions of tau in the ENS and to further investigate the possibility of pathological changes in enteric neuropathies and tauopathies.

Keywords: Tau, Tau isoform, Tau phosphorylation, Enteric nervous system, Progressive supranuclear palsy, Parkinson's disease, Gut, Biopsy, Htau mouse

* Correspondence: Wendy.Noble@kcl.ac.uk; pascal.derkinderen@chu-nantes.fr; derkinderenp@yahoo.fr
Arthur Lionnet, Matthew A. Wade, Anne-Gaëlle Corbillé and Alice Prigent contributed equally to this work.
Wendy Noble and Pascal Derkinderen Co-senior authorship
[4]King's College London, Institute of Psychiatry, Psychology and Neuroscience, Department of Basic and Clinical Neuroscience, Maurice Wohl Clinical Neuroscience Institute, Rm 1.23, 5 Cutcombe Road, Camberwell, London SE5 9RX, UK
[1]Inserm, U1235, 1 rue Gaston Veil, F-44035 Nantes, France
Full list of author information is available at the end of the article

Introduction

The microtubule-associated protein tau is found predominantly in neurons, where it exists as a highly soluble protein that interacts with the cytoskeleton [25, 28]. Six different isoforms of tau are expressed in the adult human CNS via alternative splicing of the *MAPT* gene, which comprises 16 exons. Regulated inclusion of exons 2 and 3 yields tau isoforms with 0, 1, or 2 N-terminal inserts (0 N, 1 N, 2 N, respectively), whereas exclusion or inclusion of exon 10 leads to expression of tau isoforms with three (3R) or four (4R) microtubule-binding repeats [28]. The various splice combinations of tau are thus abbreviated 0N3R, 0N4R, 1N3R, 1N4R, 2N3R, 2N4R, encoding six protein isoforms ranging from 352 to 441 amino acids in length [25]. The function of tau is strongly affected by its phosphorylation status, which influences its ability to interact with microtubules and various signaling proteins [20, 57], as well as its localization and association with membranes [56, 63]. Under pathological conditions, aberrant assembly of highly phosphorylated tau into insoluble aggregates is observed in a range of neurodegenerative disorders, collectively referred to as tauopathies. Tauopathies encompass more than 20 clinicopathological entities, including Alzheimer's disease (AD), progressive supranuclear palsy (PSP), Pick's disease, all of which can be biochemically subclassified according to the predominance of tau isoforms found in the intracellular aggregates [43]. Tau aggregates found in tauopathies generally contain tau in an elevated state of phosphorylation [7, 29, 34] that is often aberrantly cleaved [31, 51]. Highly phosphorylated forms of tau are also found in other neurodegenerative diseases, including Parkinson's disease (PD), where it often colocalises with abnormal alpha-synuclein [39, 66].

The enteric nervous system (ENS) is an integrated neuronal network distributed from the lower esophagus to the rectum. Compared to other components of the peripheral nervous system, the ENS shows some unique features that closely resemble the CNS and is sometimes referred to as 'the brain-in-the-gut' or the 'second brain'. This close homology between the CNS and ENS suggests that a disease process affecting the CNS could also involve its enteric counterpart, as has already been described in variant Creutzfeldt-Jakob disease [33, 41] and PD [6, 21, 65]. Whether such a scenario can be extended to other neurodegenerative disorders such as tauopathies remains to be demonstrated, and this was one focus of the current study.

A few studies have shown that tau is expressed in rodent [30] and human [8, 17, 61] gastrointestinal (GI) tract, but no data are available about the distribution and phosphorylation pattern of tau isoforms in the ENS. Here, we examined the expression levels of tau isoforms, their phosphorylation profile and truncation in sigmoid colon biopsy specimens from PSP patients and compared them to samples from PD patients and controls. We examined the same tau characteristics in a mouse model of tauopathy in comparison to wild-type mice. Our results show the expression of two main human tau isoforms in the ENS. ENS tau is phosphorylated but is remarkably resistant to dephosphorylation with lambda phosphatase. We then examined the isoform profile and phosphorylation state of tau under physiological conditions in rat primary enteric neuron cultures, which showed that ENS tau phosphorylation can be modified, at least in vitro. These data provide the first detailed characterization of ENS tau in humans and rodents in health and tauopathies. Further investigation of tau modifications in the ENS in disease may provide valuable information about tau modifications that promote or prevent tau abnormalities spreading between the gut and brain in neurodegenerative diseases.

Material and methods

Human tissues

Samples of frozen temporal cortex from one post-mortem human brain devoid of neurodegeneration were obtained from the Neuropathology Department of Angers (Dr Franck Letournel) to serve as a control for the following experiments. Specimens of human colon were obtained from three neurologically unimpaired subjects who underwent colon resection for colorectal cancer. For all three tissues specimens, sampling was performed in macroscopically normal segments of uninvolved resection margins. Colonic sections were separated into muscle and submucosal/mucosal layer [36], which contain the myenteric and submucosal plexus respectively. Two out of three samples were frozen and kept at – 80 °C until further analysis by Western blot. The remaining sample was analyzed by immunohistochemistry.

Routine sigmoid colon biopsies were obtained during sigmoidoscopy/colonoscopy from 24 subjects, 10 with PD, 5 with PSP and 9 controls. All patients were recruited from the movement disorder clinic at Nantes University Hospital, France. Diagnosis of PD was made according to criteria provided by the United Kingdom Parkinson's Disease Survey Brain Bank. PSP patients fulfilled the diagnostic criteria for possible or probable PSP. Control subjects were healthy subjects who had a routine colonoscopy performed for colorectal cancer screening. All controls subjects underwent a detailed neurological examination to rule out PD symptoms and cognitive deficiency. Except for control subjects 183 and 208 (Table 1) who had 6 biopsies, 4 biopsies per patient were taken during the endoscopic procedure. Biopsies were stored at – 80 °C until required.

The sampling of human brain and colon was approved by the *Fédération des biothèques* of the University Hospital of Nantes, according to the guidelines of the French Ethics Committee for Research on Humans and registered under the no. DC-2008-402. Regarding sigmoid biopsies sampling,

Table 1 Demographics and characteristics of controls subjects and patients

Patient #	Age/sex	Diagnosis	DD
183	49/F	Control	–
188	67/F	Control	–
189	63/F	Control	–
190	45/M	Control	–
191	19/F	Control	–
208	76/M	Control	–
210	63/F	Control	–
214	69/F	Control	–
227	56/F	Control	–
162	56/F	PD	12
166	64/F	PD	11
167	67/M	PD	10
168	55/F	PD	4
171	71/M	PD	3
173	67/M	PD	11
175	70/M	PD	12
177	70/F	PD	8
178	53/F	PD	1
179	52/F	PD	4
170	63/F	PSP	4
176	72/M	PSP	4
185	72/F	PSP	11
187	75/M	PSP	5
228	76/F	PSP	1

Patient ID, age, sex, diagnosis of PD or PSP (including probable PSP) are shown in addition to disease duration (DD) in years

the study protocol was approved by the local Committee on Ethics and Human Research (*Comité de Protection des Personnes Ouest VI*), and registered on ClinicalTrials.gov (EnteroLark and ColoBioParker, identifier NCT01618383 and NCT01353183, respectively). Written informed consent was obtained from each patient and from each normal volunteer.

Mouse tissues

Htau mice (B6.Cg-Mapt$^{tm1(EGFP)Klt}$Tg(MAPT)8cPdav/J) were originally purchased from the Jackson laboratory (Bar Harbor, ME, USA) and maintained at King's College London. Wild-type and tau knockout offspring of an identical background strain (C57Bl/6 J) were obtained via breeding. All housing and experimental procedures were carried out in compliance with the local ethical review panel of King's College London under a UK Home Office project license held in accordance with the Animals (Scientific Procedures) Act 1986 and the European Directive 2010/63/EU. Two-month old male and female mice were used in this study. Animals were housed at 19–22 °C, humidity 55%, 12 h:12 h light: dark cycle with lights on at 07:30. Animals were culled using Schedule 1 methods, brains removed and snap-frozen on dry-ice. Sections of colon tissue were removed, with tissue from the distal portion of each part being cleaned and snap-frozen on dry-ice, prior to storage at – 80 °C for RNA extraction or biochemical analysis. The proximal portion from each part of colon along with the duodenum, jejunum and ileum were dissected with fine forceps to reveal the myenteric plexus as described previously [62].

Rat tissues

Sciatic nerve sections were taken from two pregnant Sprague-Dawley rats (used for the generation of primary culture of rat ENS, see below) to serve as a positive control for big tau experiments [60].

Primary cultures of rat ENS

Primary culture of rat ENS were generated using pregnant Sprague–Dawley rats (Janvier Laboratories SA, Le Genest-St-Isle, France) as previously described [11]. All housing and experimental procedures were carried out in compliance with the local ethical review panel of INSERM (agreement E. 44,011; INSERM, Nantes, France). Pregnant rats were killed by an overdose of CO_2 followed by severing the carotid arteries. The small intestines of rat embryos were removed, diced in Hank's Buffered Salt Solution (Sigma, Saint-Quentin Fallavier, France) and collected in 5 mL of Dulbecco's modified Eagle's medium (DMEM)-F12 (Gibco®, Life Technologies, Villebon sur Yvette, France) (1:1) for digestion at 37 °C for 15 min in 0.1% (*v*/v) trypsin (Sigma). The trypsin reaction was stopped by adding medium containing 10% fetal calf serum and then treatment with DNase I 0.01% (*v*/v) (Sigma) for 10 min at 37 °C. After triturating with a 10 mL pipette, cells were centrifuged at 750 rpm for 10 min. Cells were counted and then seeded at a density of 2.4×10^5 cells/cm^2 on 24-well plates previously coated with a solution of 0.5% (v/v) gelatin in sterile phosphate buffered saline. After 24 h, the medium was replaced with a serum-free medium DMEM-F12 (1:1) containing 1% (v/v) of N-2 supplement (Life Technologies). Cultures were maintained for 14 days.

Treatment of rat ENS primary cultures with serine/threonine phosphatases inhibitors

After 14 days in vitro (DIV), cells were treated with a cocktail of three phosphatase inhibitors including 1 μM okadaic acid, 1 μM ciclosporine A and 6.75 μM sanguinarine (Sigma) for broad-spectrum inhibition of serine/threonine phosphatases, or with vehicle (DMSO, Sigma) for one hour.

Dephosphorylation of tissues and cell lysates

For dephosphorylation experiments, cells or tissues were homogenised in a buffer containing 100 mM NaCl and 50 mM Tris-Cl at pH 7.4 with 1% (v/v) IGEPAL® CA-630 and a protease inhibitors cocktail without EDTA (Roche, Neuilly sur Seine, France) using either a "Precellys 24" (Bertin technologies, St Quentin-en-Yvelines, France) or a Tissue Master 125 (Omni International, Kennesaw, GA, USA) tissue homogenizer and followed by sonication with "vibracell 75 186" device (Sonics, Newton CT, USA). Homogenates were centrifuged at 16,300 g for 20 min at 4 °C with an Eppendorf 5415R centrifuge (Eppendorf, Hamburg, Germany), sonicated for 10 s and protein amounts normalized following a BCA protein assay (ThermoFisher, Waltham, MA, USA). Samples were diluted to 1.0 mg/mL protein using homogenisation buffer and incubated with 20 U/μL lambda phosphatase in $MnCl_2$ and enzyme buffer as supplied with the lambda protein phosphatase kit (New England Biolabs, Ipswich, MA, USA) for 3 h at 30 °C. The reaction was stopped by the addition of sample buffer (National Diagnostics, Hull, UK or Life Technologies, Courtaboeuf, France) and heating to 95 °C for 5 min. Control samples were treated identically without the addition of lambda phosphatase.

SDS-PAGE and western blot

For dephosphorylation experiments, cells or tissues were processed as described above. For experiments that did not require dephosphorylation, cells or tissues were lysed in RIPA lysis buffer (Merck Millipore, Fontenay sous Bois, France). Western blots were performed as we previously described [10] using NuPAGE™ 10% Bis-Tris Protein Gels (Life Technologies, Courtaboeuf, France). The primary anti-tau antibodies used are listed in Table 2. Phospho-ERK (Cell signaling, Ozyme, France 1:2000 dilution) and PGP 9.5 antibodies (Abcam, France, 1:1000 dilution) were used for the evaluation of phosphatase treatment and as loading control, respectively.

Immunohistochemistry

For mouse GI tract tissues, following the excision of myenteric plexus from mouse colon, tissue segments were incubated in combined blocking (50 mM tris-buffered saline [TBS] pH 7.4 containing 5% bovine serum albumen [BSA] and 0.05% tween-20) and permeabilisation (50 mM TBS pH 7.4, 0.1% triton X-100) solutions overnight at 4 °C. Primary antibodies (Table 2) in blocking solution were incubated with gut tissues overnight at 4 °C. Following washing in 50 mM TBS the appropriate fluorescently-tagged secondary antibody was added for 3 h at ambient temperature, the antibodies removed by washing and Hoechst 33258 added for 3 min. Images were acquired using a CTR5000 digital camera (Leica Microsystems, Cambridge, UK) attached to a Leica DM5000B fluorescence microscope with Leica AIF lite software.

For human tissues, fixed human tissues were embedded in paraffin using an embedding station (LEICA EG1150C) and sections (3 μm) were cut using a microtome (LEICA RM2255). The sections were deparaffinised by bathing twice in xylene (for 5 min each) and taken through graded concentrations of ethanol (100, 95, 70, 70%, respectively for 3 min each). After a rinse in distilled water, slides were washed in PBS and antigen retrieval was performed using a sodium citrate solution (2.94 g Sodium Citrate Tribase; 1 L ultrapure water; 500 μL Tween 20; pH 6) at 95 °C for 20 min. Slides were incubated in NH_4Cl (100 mM) for 15 min before incubation in PBS-0.5% triton X-100 for 1 h and blocking for

Table 2 Tau antibodies used in this study

Name	Specificity	Epitope (a.a)	Source and dilution
A0024 Tau	All tau isoforms	243–441 (2N4R)	Dako, rp (WB 1:1000; IHC 1:500)
TAU-5	All tau isoforms	210–241 (2N4R)	ThermoFisher, mm (WB 1:1000)
Tau-1	All tau isoforms	189–207 (2N4R)	Merck, mm, clone PC1C6 (WB 1:2000)
TP70	All tau isoforms	428–441 (2N4R)	IOP, KCL, rp (WB 1:500)
Anti tau RD3	3R tau Isoforms	267–282 (2N3R)	Merck, mm, clone 8E6 (WB 1:1000; IHC 1:500)
Anti tau RD4	4R tau isoforms	275–291 (2N4R)	Merck, mm, clone 1E1/A6 (WB 1:1000)
Anti 4R-tau	4R tau isoforms	NS	Cosmo bio co., rp (WB 1:2000; IHC 1:1000)
Anti 0 N-tau	0 N Tau isoforms	39–50 (0N3R)	BioLegend, mm (WB 1:500)
AT8	Tau Ⓟ S202/T205	Tau Ⓟ S202/T205	Innogenetics, mm (WB 1:1000)
PHF-1	Tau Ⓟ S396/S404	Tau Ⓟ S396/S404	Gift from Peter Davies, mm (WB 1:500)
PHF13	Tau Ⓟ S396	Tau Ⓟ S396	Cell Signaling, mm (WB 1:1000)
CP13	Tau Ⓟ S202	Tau Ⓟ S202	Gift from Peter Davies, mm (IHC 1:200)

The name, specificity, epitope, source and dilution of the antibodies used in this study are shown.
Abbreviations: *a.a.* amino-acids, *IHC* immunohistochemistry, *IOP, KCL* Institute of Psychiatry, King's college London, *mm* mouse monoclonal, *NS* not specified, *rp* rabbit polyclonal, *WB* western blot

2 h in 10% horse serum in PBS-0.5% triton X-100. Primary antibodies (Table 2) were incubated overnight at 4 °C, and following washing, secondary antibodies were added for 2 h at room temperature. Images were acquired with an Olympus IX 50 fluorescence microscope coupled to a digital camera (model DP71, Olympus).

RNA extraction and RT-PCR

Frozen proximal colon and cortex from htau, wild-type and tau knockout mice was homogenised in approximately 100 mg/mL Quiazol® supplied with the Quiagen RNA LipidEasy kit (Qiagen, Hilden, Germany), and RNA was extracted following the manufacturer's protocol. The RNA obtained was eluted in ultrapure H_2O and its concentration and purity determined using a NanoDrop spectrophotometer (Thermo Scientific, Waltham, MA, USA). Samples were diluted to 1 µg RNA/15 µL RNAse-free H_2O, heat-shocked for 3 min at 72 °C to break down double-stranded structures and returned immediately to ice. One µg RNA per sample was reverse transcribed using a Superscript III reverse transcriptase assay kit (Life Technologies, Paisley, UK) according to the manufacturer's instructions. The resulting cDNA was stored at − 20 °C until use. To examine the alternate splicing of the microtubule binding domain repeat region encoded by exon 10, primers were used that specifically recognize mouse or human exons 9 and 11 as described by Duff et al. [16]. Primer sequences were: mouse exon 9F 5'-CCCCCTAAGTCACCATCAGCTAGT, mouse exon 11R 5'-CACTTTGCTCAGGTCCACCGGC, human exon 9F 5'-CTCCAAAATCAGGGGATCGC, human exon 11R 5'-CCTTGCTCAGGTCAACTGGT. Splicing around the N terminal insert domain encoded by exons 2 and 3 was detected using primers that recognize exons 1 and 5. Primer sequences used were: mouse exon 1F 5'-TCCG CTGTCCTCTTCTGTC, mouse exon 5R 5′- TTCTCG TCATTTCCTGTCC, human exon 1F 5′- TGAACCA GGATGGCTGAGC, human exon 5R 5'-TTGTCATCG CTTCCAGTCC. Annealing temperatures were 64 °C (all *MAPT* primers), 62 °C (M1F/M5R) and 68 °C (M9F/M11R). 35 reaction cycles were used for all. Mouse and human-specific RT-PCR products were analysed by agarose gel electrophoresis. Products corresponding to exon 10+ tau mRNA (4R) are 390 base pairs (bp), while products corresponding to exon 10- mRNA (3R) are 297 bp. RT-PCR products containing tau mRNA with exons 2 and 3 (2 N) are 428 bp, 2 + 3- mRNA products (1 N) are 341 bp, and 2–3- mRNA products (0 N) are 253 bp.

Statistics

All data shown are mean ± SEM. Statistical analyses was conducted using GraphPad software version 5.00 (San Diego California, USA). For comparisons of means between groups, Kruskal-Wallis tests were performed. Differences were deemed statistically significant when $p < 0.05$.

Results

The expression pattern of tau isoforms is different in adult human brain and gut

In adult human brain, the six tau isoforms are phosphorylated resulting in reduced electrophoretic mobility on SDS-PAGE compared to recombinant tau [25]. In order to identify the tau isoforms expressed in the human ENS, colonic samples from healthy subjects treated or not with lambda phosphatase [35] were analyzed by western blot using the A0024 tau antibody that recognizes all six tau isoforms. ENS samples were compared to dephosphorylated and non-dephosphorylated brain samples as well as to a recombinant tau ladder. The banding pattern was markedly different between brain and colonic samples (Fig. 1a). The A0024 Tau antibody detected one major band migrating at 53–54 kDa in both the submucosal and muscle layers (which contain the submucosal and myenteric plexus, respectively and therefore are referred to as SMP and MP) (Fig. 1a). This band migrated only slightly faster after dephosphorylation of SMP and MP samples despite the efficiency of the dephosphorylation treatment being validated by phospho-ERK immunoblot, Fig. 1a). The major band detected in ENS samples comigrated with 0N4R-1N3R detected in human brain samples and the recombinant tau ladder (red line in Fig. 1a) and was also observed when a pan-tau (TAU-5) antibody was used (Fig. 1b). In addition, a fainter band around 57–58 kDa in SMP and a strong immunoreactive band at 62 kDa (white arrow) in both SMP and MP were also observed when the A0024 tau antibody was used (Fig. 1a). These two bands are most likely non-specific as they were not observed with TAU-5 (Fig. 1b) or with other specific antibodies subsequently used in this study (Figs. 1b, c and 2).

To further refine this analysis, we used 3 commercially available isoform-specific tau antibodies. Two of these antibodies directed against 3R and 0 N-tau have been shown to be highly specific in a recent comprehensive study that tested the specificity of tau antibodies using immunoblotting [19]. In addition, we used a 4R-tau antibody that only detects 4R tau isoforms in human brain lysates and in tau ladder (Additional file 1: Figure S1). All of these antibodies detected a single 53–54 kDa-band that comigrates with the major band detected by TAU-5 and with 0N4R-1N3R in the recombinant tau ladder (Fig. 1b).

Until recently, analysis of the ENS in humans was mainly performed using full thickness specimens of the gut obtained during surgery or autopsy. However, several recent studies have shown that the ENS is accessible and analyzable through routine GI biopsies, which can be processed to measure quantitative differences in neuronal

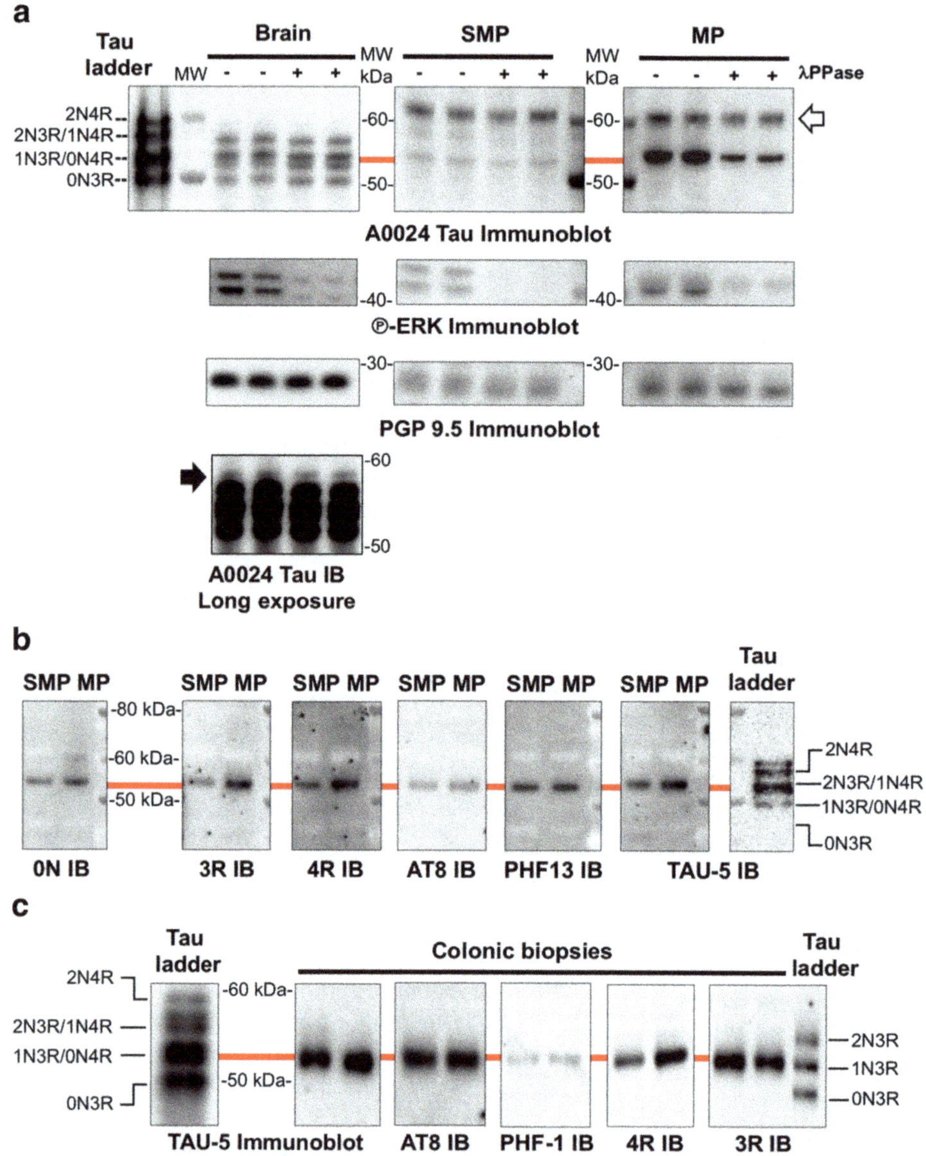

Fig. 1 Tau isoforms and phosphorylation in adult human ENS. **a** Human brain and colon tissue lysates (submucosal and muscle layers, which contains the submucosal (SMP) and myenteric plexus (MP), respectively) were subjected to immunoblot analysis using the pan-Tau antibody A0024. Lysates were treated (+) or not (–) with lambda phosphatase before immunoblotting. The effectiveness of dephosphorylation was confirmed by phospho-ERK immunoblot (P-ERK immunoblot). Tau antibody A0024 detected all six tau isoforms in the recombinant human tau ladder and brain samples (the 2N4R was only visible on long exposure immunoblots, black arrow). The non-specific band detected by Tau antibody A0024 in the ENS is marked by a white arrow. An antibody against protein gene product (PGP) 9.5 was used as a loading control. **b** Colon tissue lysates (SMP and MP) were subjected to immunoblot analysis using antibodies specific to 0 N, 3R, 4R tau, the pan-tau TAU-5 antibody, and the phospho-specific tau antibodies AT8 (phos-Ser202/Thr205) and PHF13 (phos-Ser396). **c** Sigmoid colon biopsies lysates from 2 control subjects (#183 and 208, Table 2) were subjected to immunoblot analysis using the TAU-5 antibody, antibodies specific to the 3R and 4R tau isoforms and the phospho-specific tau antibodies AT8 and PHF-1. In all experiments, the banding pattern was compared to that of tau ladder which contains all six recombinant tau isoforms. The red line shows the comigration of the observed bands with 1 N3/0N4R. The results shown in (**a**), (**b**) and (**c**) are representative of 3, 2 and 5 independent experiments, respectively

and/or glial markers [5, 23, 44]. We therefore analyzed the expression levels of tau in routine sigmoid biopsies from 2 control subjects (#183 and 208, Table 1) with the pan-tau antibody TAU-5 and with the 3R and 4R isoform-specific antibodies. The immunoblotting pattern observed with these 3 antibodies in biopsies was similar to those observed in colonic SMP and MP samples (Fig. 1c).

"Big" or peripheral tau is a tau isoform specifically expressed in the peripheral nervous system, including trigeminal, dorsal root and sympathetic ganglia as well as

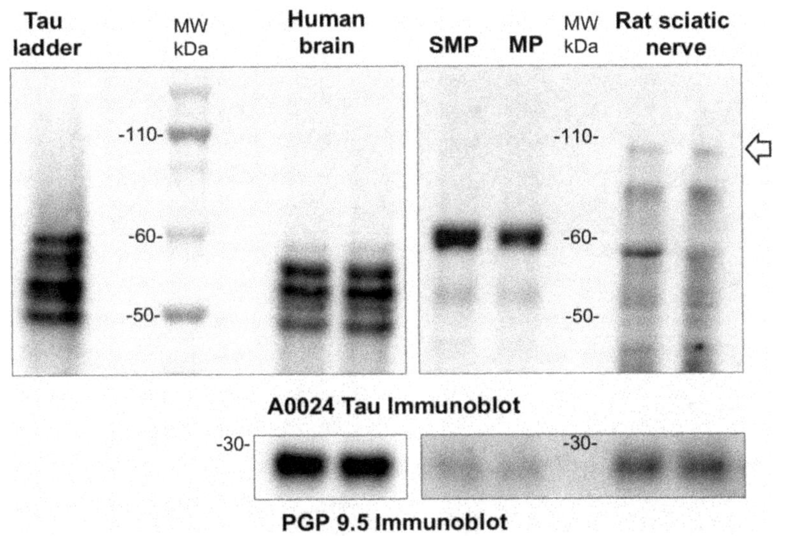

Fig. 2 Big tau is not detected in adult human ENS. Human brain and colon tissue lysates (SMP and MP) were subjected to immunoblot analysis using the pan-Tau antibody A0024. Rat sciatic nerve lysates were used as positive control to detect big tau (white arrow). PGP 9.5 was used as a loading control. Images are representative of five independent experiments

sciatic nerve. It differs from the 2N4R tau isoform by a 254 amino-acid insert located in the amino-terminal half and migrates at 110 kDa on SDS/PAGE [27]. To determine whether big tau is expressed in the ENS, human colon tissue lysates were analyzed by Western blot using Tau A0024 antibody. Rat sciatic nerve lysates were used as positive controls [60]. Tau A0024 detected the expected low molecular weight tau isoforms between 45 and 60 kDa in human colon and rat sciatic nerve, however a 110 kDa migrating band was only observed with rat sciatic nerve lysates (Fig. 2).

When taken together, these results show that 1N3R and 0N4R are the two main tau isoforms that are expressed in human adult colon and these two isoforms can be detected in routine GI biopsies. In addition, our work indicates that big tau is not expressed in the adult human ENS.

Tau isoforms are differentially expressed in the gut and brain of tauopathy mice

To determine if tau is also differentially expressed in the ENS and CNS of mice, we used the transgenic htau mouse model which expresses exclusively the six wild-type human isoforms of tau under the control of the *MAPT* promoter [3]. The enteric expression profile of tau isoforms in these transgenic mice was compared to that observed in wild-type mice. Tau knockout mice were examined as an additional control. To this end, RNA from 2-month-old, wild-type, tau knockout and htau mouse proximal colon was reverse transcribed to cDNA and amplified with PCR. Brain tissue from the same mice was used for comparison. Primers were designed, based on those previously described by Duff et al. [16], to detect splicing of human tau exons 2 and 3. This allowed amplification of products corresponding to 0 N, 1 N and 2 N human tau that were detected in htau, but not wild-type or tau knockout, brain and proximal colon (Fig. 3). Transcripts of 3R and 4R *MAPT* were also observed in htau brain and proximal colon when inclusion of exon 10 was assessed using primers specific to human tau exons 9 and 11 (Fig. 3). Thus, 0 N, 1 N, 2 N, 3R and 4R human tau transcripts are expressed in htau proximal colon.

Primers against mouse tau were also used to allow detection of 0 N, 1 N and 2 N transcripts in wild-type mouse brain and proximal colon (Fig. 3a). A weak non--specific PCR product corresponding to the predicted size of 0 N tau was also amplified in htau and tau knockout samples with these primers. In addition, 4R, but not 3R *Mapt* was detected in WT mouse brain, and a weak signal was also apparent in proximal colon (Fig. 3a). These transcripts were not amplified in htau or tau knockout tissues. Thus, wild-type mice express 0 N, 1 N, 2 N and mainly 4R tau in gut and brain.

The detection of multiple products in a single lane, each corresponding to a different tau isoform, allows each product to act as an internal control for the other transcripts. This allowed us to make comparisons between the relative abundance of tau isoforms in different tissues. Htau brain showed 0 $N > 1$ $N > 2$ N relative abundance of tau isoforms, in keeping with previous observations in adult mice [59]. Htau mouse brain also showed greater exclusion of tau exon 10 (3R > 4R), as previously reported [3] and in contrast to adult wild-type mice where mainly 4R tau is expressed (Fig. 3; [59]). Thus, htau brain mirrors human brain in that both 3R and 4R tau are expressed,

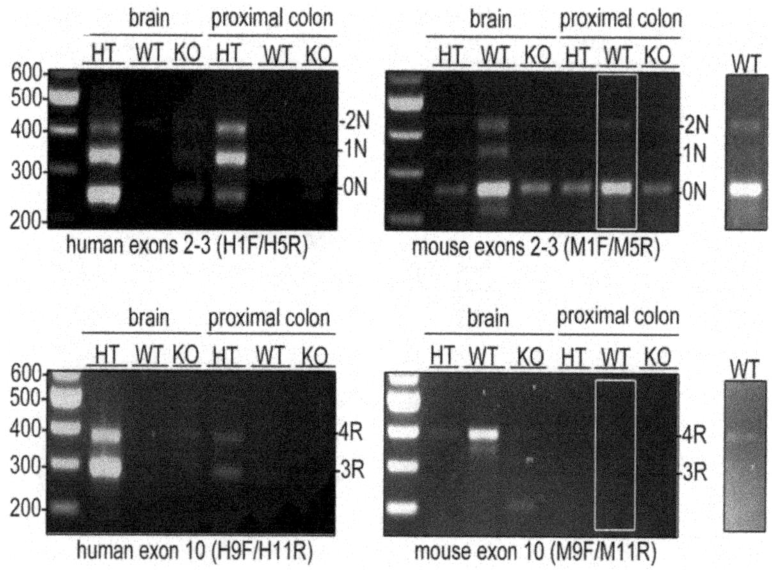

Fig. 3 Detection of tau in htau and wild-type mouse brain and proximal colon. Htau (HT), wild-type (WT) and tau knock-out (KO) brain and proximal colon cDNA was amplified using PCR with human- and mouse-tau specific primers to detect the expression of exon 2–3- (0 N), 2 + 3- (1 N), 2 + 3+ (2 N), 10- (3R) and 10+ (4R) tau isoforms. Gel images show detection of human and mouse 0 N, 1 N and 2 N tau. An insert shows a higher intensity portion of image to illustrate tau products in proximal colon. Numbers correspond to base pairs of a DNA ladder. The expected position of PCR product is indicated to the right of each panel. $N = 3$

albeit that under physiological conditions these isoforms are expressed in approximately equal proportions in the human CNS [4]. In contrast, in htau proximal colon, PCR transcripts showed an altered relative abundance of 1 N > 2 N > 0 N (Fig. 3), and there appeared to be approximately equal inclusion and exclusion of exon 10 (3R ≈ 4R). These data suggest differential expression of tau isoforms in the ENS of htau mice in comparison to those in brain.

Tau protein is expressed throughout the human and mouse myenteric plexus

Immunohistochemistry was used to examine the localization of tau proteins in the human and rodent ENS. Human colonic myenteric plexus showed intense tau immunoreactivity in both neuronal cell bodies and processes when pan-tau A0024, 3R and 4R antibodies were used, which nearly completely overlapped with beta-tubulin immunostaining (Fig. 4a).

In order to examine the localization of tau proteins in the ENS of htau and wild-type mice, sections of small intestine (duodenum, jejunum, ileum) and large intestine (proximal colon and distal colon) were dissected from 2-month-old htau, wild-type and tau knockout mice to isolate the myenteric plexus. Tissue was immunolabelled with an antibody against total tau (A0024). eGFP fluorescence was also imaged as it is inserted in tau exon 1 to disrupt tau expression in tau knockout and htau mice [3]. Tau proteins were found to be abundant throughout the GI tract of htau and wild-type mice, including in the duodenum, jejunum, ileum, proximal colon and distal colon (Fig. 4b). There were no apparent differences in neuronal tau localisation between these regions. Htau proximal colon exhibited dense ganglia and axons and a robust tau signal, whereas the axons and ganglia in htau ileum and WT jejunum were less dense and the resulting tau signal was comparatively less intense. Tau KO express GFP which is observed, and show no tau immunoreactivity. Thus, tau protein is expressed in the myenteric plexus throughout the GI tract of wild-type and htau mice.

Tau isoforms are phosphorylated in mature ENS but are not susceptible to dephosphorylation with lambda phosphatase

The phosphorylation of tau at multiple serine and threonine sites has been described in both developing and adult brain and is the predominant mechanism by which tau functions are regulated [32]. This logically led us to analyze tau phosphorylation in mature human ENS. Two antibodies specific for tau phosphorylated at Ser202/Thr205 (AT8) [26] and Ser396 (PHF13) [19] detected one single band at 53–54 kDa in colon surgical specimen and biopsies (Fig. 1b and c), thereby demonstrating that the enteric 1N3R and 0N4R tau isoforms are phosphorylated on serine residues under physiological conditions.

We were nevertheless struck by the fact that, in contrast to the brain, lambda phosphatase treatment did not appear to influence the charge/mobility of tau bands in human colon samples when the pan-Tau antibody A0024 was used (Fig. 1a). To further investigate if tau can be dephosphorylated in adult human ENS, colonic biopsy

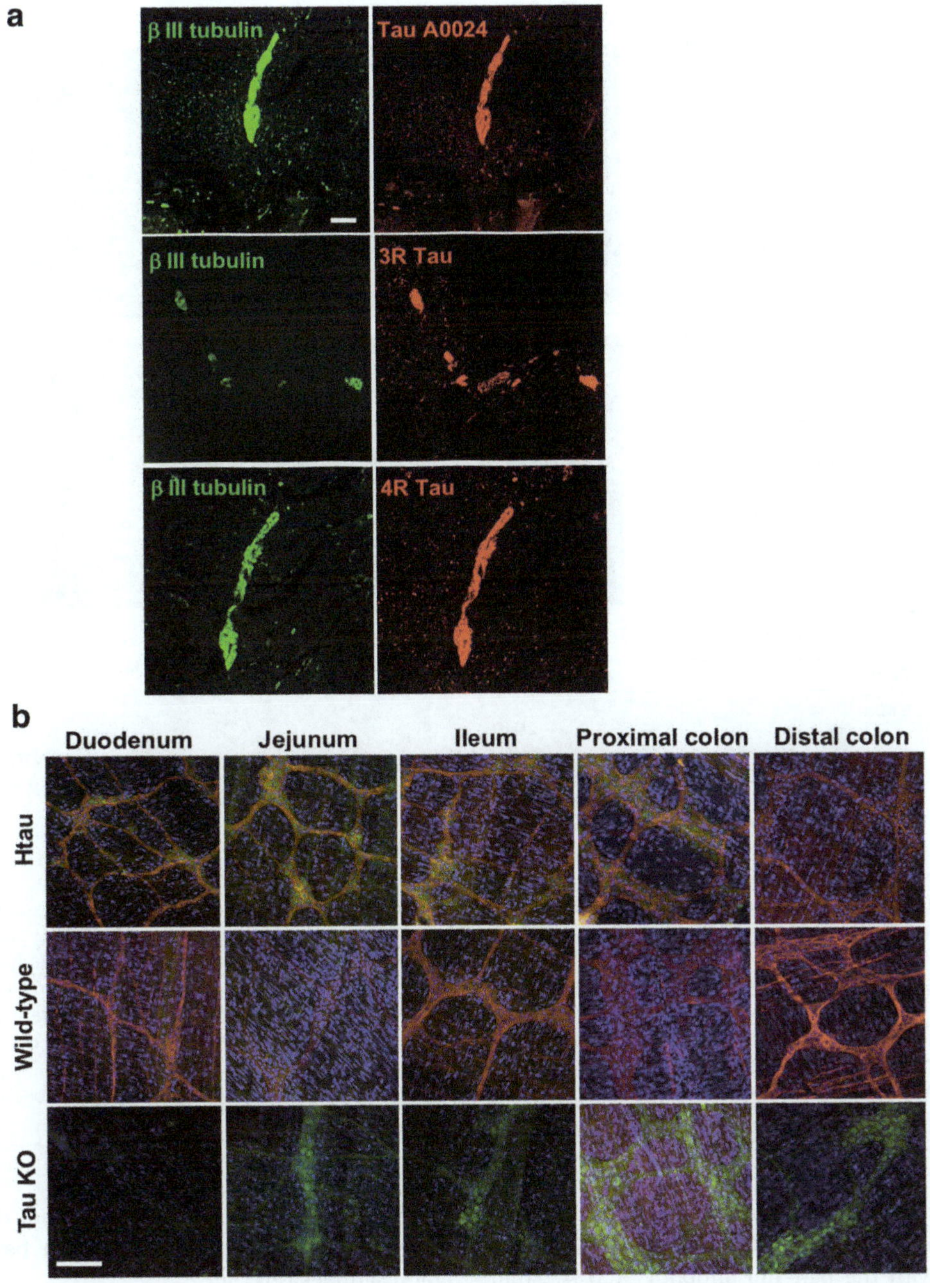

Fig. 4 Distribution and localization of tau in human, htau and wild-type mouse myenteric plexus. **a** Total tau antibody A0024 and the isoform-specific antibodies against 3R and 4R-tau were used to detect tau in the myenteric plexus of a human colonic sample. An antibody specific to betaIII-tubulin was used to specifically label neurons. Scale bar is 200 μm (**b**) Total tau antibody A0024 was used to detect tau in the myenteric plexus of the duodenum, jejunum, ileum, proximal colon and distal colon of 2-month-old htau, wild-type (WT) and tau knockout (KO) mice. Merged images show tau (red), EGFP (green) and Hoechst 33342 labelling of nucleic acids (blue). Scale bar is 100 μm. Images are representative of three independent experiments

lysates were treated with lambda phosphatase and western blots of these samples were probed with the phospho tau-specific antibodies AT8 (phospho-Ser202/Thr205), PHF1 (phospho-Ser396/404) and Tau-1 (dephospho--Ser199/202/Thr205). Antibodies against ERK were used to check the efficiency of treatment. Although lambda phosphatase efficiently dephosphoylated ERK, it did not modify the phosphorylation state of tau, suggesting that tau is relatively resistant to dephosphorylation in the human adult ENS (Fig. 5a). To further examine tau phosphorylation in mature ENS, we analyzed enteric tau phosphorylation and dephosphorylation in htau mice. Tau is phosphorylated in htau mouse ENS, as phospho-Ser202-positive tau was detected in their

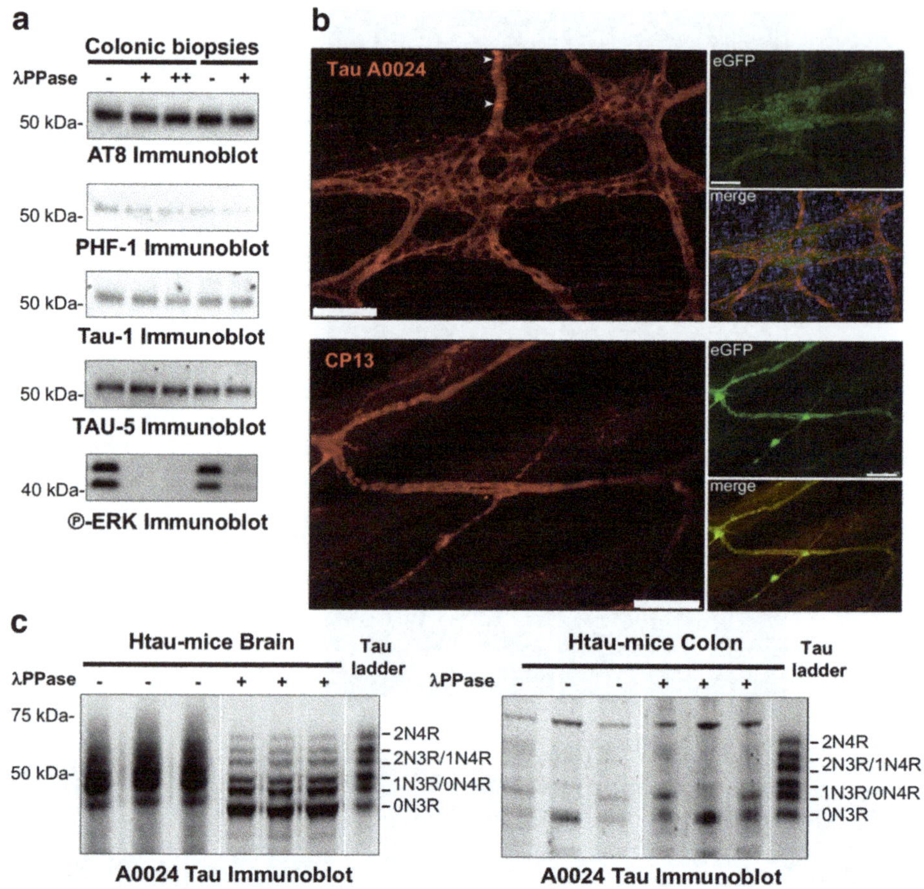

Fig. 5 Poor susceptibility of tau to dephosphorylation in mature ENS. **a** Sigmoid colon biopsies lysates from 2 control subjects (#183, 3 first lanes and #208, 2 last lanes) were subjected to immunoblot analysis using TAU-5, AT8, PHF-1 and Tau-1 and antibodies. Lysates were treated with (+ for 1 h and ++ for 3 h) or without (–) lambda phosphatase before immunoblotting. The effectiveness of dephosphorylation was confirmed by phospho-ERK immunoblot (P-ERK immunoblot). **b** Antibodies against total tau (Tau antibody A0024) and tau phosphorylated at serine 202 (CP13) (both shown in red) were used to detect tau in the myenteric plexus of the proximal colon from htau mice. eGFP expression is shown in green, together with merged images including Hoechst 33,342 nuclear labelling (blue). Scale bars are 100 μm. *N* = 3. **c.** Brain and proximal colon homogenates from htau mice were treated with or without lambda phosphatase (+ or -) and immunoblotted with pan-Tau antibody A0024. A recombinant human tau ladder was included on each blot. White lines indicate rearrangement of lanes within the same blot for clarity. Data and images are representative of three independent experiments

colonic myenteric plexus (Fig. 5b). Samples of brain and proximal colon from 2-month-old htau mice were immunoblotted with the pan-Tau antibody A0024. Htau brain showed prominent tau bands ranging from 45 to 70 kDa, in agreement with previous reports [3, 52]. Treatment of these samples with lambda phosphatase to dephosphorylate tau showed that all six major isoforms of tau are expressed in htau brain; these showed good alignment with a recombinant human tau ladder (Fig. 5c). Multiple tau immunoreactive bands ranging from approximately 25–70 kDa were detected in samples from htau proximal colon (Fig. 5c). Two doublets of bands at approximately 45-50 kDa and 55–60 kDa were apparent, which likely corresponds to full-length tau with 0 or 1 N terminal inserts, likely the 1N3R and ON4R isoforms. Moreover, and in contrast to results with htau brain samples, lambda phosphatase treatment did not appear to influence the charge/mobility of tau bands in proximal colon samples (Fig. 5c).

Tau expression levels are unaltered in the ENS in PSP

An increase of the 4R tau to 3R tau isoform ratio has been described in some brain regions in PSP [38]. We thus analyzed the expression levels of tau and the relative abundance of 3R and 4R isoforms in the ENS in colonic biopsies from 5 PSP patients in comparison to colonic samples from 10 PD patients and 9 controls devoid of neurodegenerative disorders. Clinical features of the study population are shown in Table 1. The expression levels of total tau as assessed by immunoblots using the Tau-5 antibody, and the 3R/4R ratio was found not to differ between PSP samples and those from PD and controls (Fig. 6a).

Tau phosphorylation and truncation in the ENS are similar in PSP, PD and control subjects

Abnormal phosphorylation of tau is a characteristic feature of PSP brain [24, 49] and we therefore analyzed the phosphorylation state of tau in colonic biopsies from PSP patients using the AT8 and PHF-1 antibodies. There were no apparent alterations in tau phosphorylation at these sites in PSP samples in comparison to those from PD and controls, or between PD and controls (Fig. 6b).

Besides abnormal phosphorylation, tau is also truncated in the pathological deposits observed in tauopathies, and especially in PSP [31, 51]. C-terminal tau truncation by caspase-3 was evaluated using a Tau Asp421 antibody, which is specific for tau cleaved at Asp421, along with an antibody against the extreme C-terminus of tau (TP70) [59]. Quantification of the immunoreactive bands detected by Tau Asp421 and TP70 showed no difference in tau truncation at Asp421 and the presence of an intact C-terminus, between PD, PSP and control subjects (Fig. 6c).

Four tau isoforms are expressed and phosphorylated in primary culture of rat ENS

Primary neuronal cultures of rat CNS neurons, which primarily express the shortest tau isoforms 0N3R and 0N4R, have been widely used for studying tau expression, aggregation and secretion [13, 52, 56]. The brain is not the only source from which neurons can be cultured and there are now established protocols for the isolation of enteric neurons from rodents and especially rats. These have already been shown to be useful for studying the expression of neuronal proteins involved in neurodegeneration such as alpha-synuclein [54], however the expression pattern of tau isoforms in rat primary ENS culture remains to be determined. As a first approach to identify tau isoforms in cultured rat enteric neurons, we compared the banding pattern on western blots of total tau as evaluated with the A0024 pan-Tau antibody between primary culture of ENS and cortical neurons. In keeping with previous observations [13, 52], this antibody detected a tau doublet with one major band at 50 kDa and a fainter one around 53 kDa in CNS neurons, which likely correspond to 0N3R and 0N4R isoforms, respectively (Fig. 7a). In ENS neurons, the observed banding pattern was markedly different with a triplet of 50, 53 and 58 kDa bands observed, the latter showing the most intense labelling (Fig. 7a). Further blotting with the 3R and 4R specific antibodies identified 0N3R, 1N3R/0N4R and 2N3R as the main component of the tau triplet observed in primary culture of ENS, while 0N3R and 0N4R were the two primary tau isoforms expressed by primary culture of CNS (Fig. 7a).

Phosphorylation of tau at multiple serine and threonine sites can be modulated in primary culture of CNS [13]. To determine whether tau phosphorylation can be also regulated in primary culture of rat ENS, we treated the cells with either lambda phosphatase or a combination of serine/threonine phosphatase inhibitors. Treatment with lambda phosphatase caused tau dephosphorylation, as evidenced by a significant downward shift in mobility of the tau triplet detected with either the pan-Tau A0024 or 3R antibodies (Fig. 7b). Conversely, treatment with phosphatase inhibitors induced tau phosphorylation as shown by upward shift in mobility of the protein on Western blots probed with the pan-Tau A0024 antibody, and the disappearance of all immunoreactive bands when the Tau-1 antibody against dephosphorylated tau was used (Fig. 7c). When the AT8 antibody was used, no signal was observed under basal conditions, while 3 immunoreactive bands were detected in the presence of phosphatase inhibitors (Fig. 7c). The PHF-1 antibody also detected three immunoreactive bands in untreated cells. An increase in signal intensity along with a mobility shift of all 3 bands was observed following treatment of primary ENS cultures with phosphatases inhibitors (Fig. 7c). Thus, the phosphorylation of ENS tau can be modified, at least in an in vitro setting.

3R and 4R tau are differentially expressed in rat primary enteric neuron cultures

Lastly, the distribution of tau in rat enteric neurons in culture was examined by immunohistochemistry using pan-Tau A0024, 3R and 4R-tau antibodies at 14 days in vitro. Total tau immunoreactivity was observed in both soma and neuronal processes and the staining patterns produced by pan-Tau A0024, 3R-tau and beta III tubulin antibodies were virtually superimposable (Fig. 8). The 4R-tau staining pattern was markedly different from that observed with 3R-tau and was primarily limited to the cell bodies (Fig. 8). These data indicate that 3R and 4R tau species have different localization in rat primary ENS neurons.

Discussion

Here, we have used samples of brain and gut from humans, htau transgenic mice and rat primary cultures to show that the isoform profile of tau differs between the ENS and the CNS. We identified 1N3R and 0N4R as the two main tau isoforms expressed in adult human ENS and observed an apparent difference in the relative abundance of different tau isoforms in htau gut and brain, with 1 N and 2 N tau isoforms being over-represented at mRNA levels in htau gut tissues, although 0 N and 1 N isoforms were the predominant protein species detected. We also found that primary culture of rat ENS express four isoforms of tau contrasting with the predominant

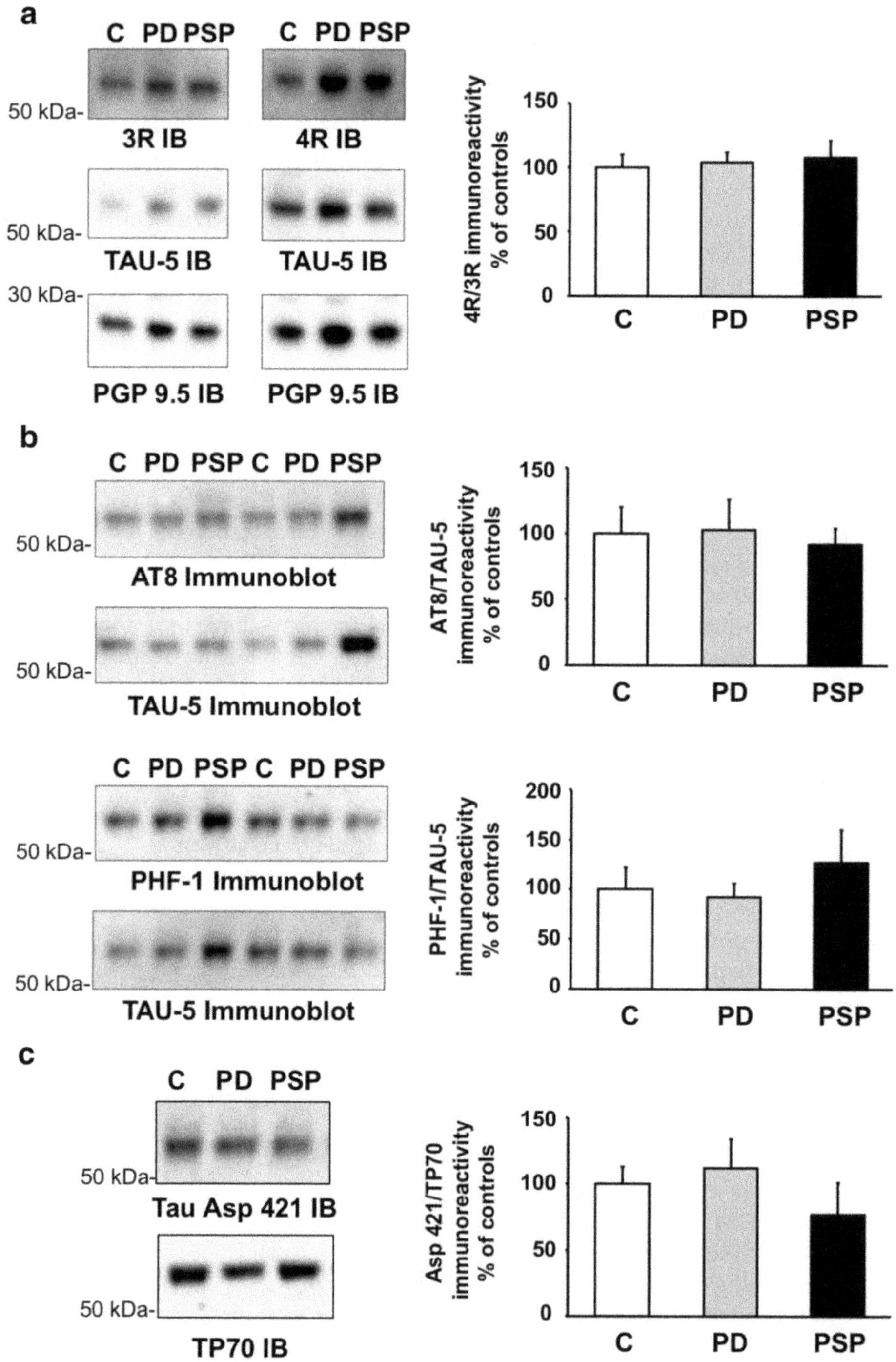

Fig. 6 Tau expression and post-translational modifications in colonic biopsies from patients progressive supranuclear palsy, Parkinson's disease and control subjects. **a** Biopsies lysates were subjected to immunoblot analysis using antibodies against total tau (TAU-5) and against 3R and 4R isoforms. An antibody against protein gene product (PGP) 9.5 was used as a loading control. For quantification, the optical densities of tau-immunoreactive bands were measured, normalized to the optical densities of PGP9.5 immunoreactive bands in the same samples and expressed as percentages of controls. **b** Biopsies lysates were subjected to immunoblot analysis using AT8, PHF-1 and TAU-5 antibodies. The optical densities of phospho-tau-immunoreactive bands were measured, normalized to the optical densities of TAU-5 immunoreactive bands in the same samples, expressed as percentages of controls. **c** Biopsies lysates were subjected to immunoblot analysis using antibodies Tau Asp421 and TP70. The optical densities of immunoreactive bands were measured, normalized to the optical densities of TP70 immunoreactive bands in the same samples, expressed as percentages of controls. Data correspond to mean ± SEM for 9 control samples (C), 10 from Parkinson's disease (PD) patients and 5 from progressive supranuclear palsy (PSP). Immunoblot (IB)

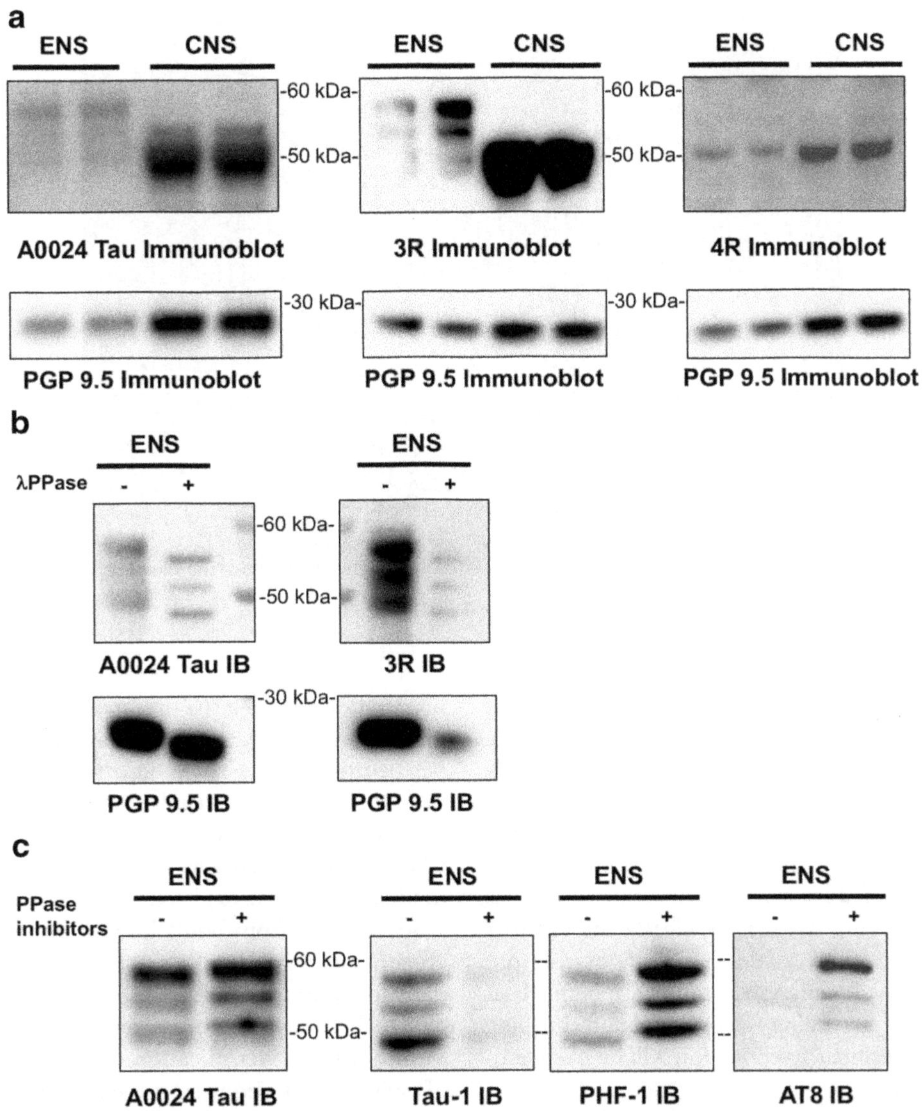

Fig. 7 Tau isoform profile and phosphorylation state in rat primary culture of ENS. **a** Lysates of rat primary ENS and CNS cultures were subjected to immunoblot analysis using the pan-Tau antibody A0024 and the isoform specific antibodies 3R and 4R. **b** Primary culture lysates were treated with (+) or without (–) lambda phosphatase before immunoblot analysis with the pan-Tau antibody A0024 and the isoform specific antibody against 3R-tau. PGP9.5 was used as a loading control. **c** Primary culture of rat ENS were treated (+) or not (–) with a cocktail of 3 phosphatase inhibitors including 1 μM okadaic acid, 1 μM ciclosporine A and 6.75 μM sanguinarine (Ppase inhibitors) for 1 h. Fifteen μg of cell lysates were subjected to immunoblot analysis using Tau-1, AT8 and PHF-1 antibodies. IB is for immunoblot. The results shown in (**a**), (**b**) and (**c**) are representative of 2, 4 and 3 independent experiments, respectively

expression of the single fetal tau 0N3R isoform in rat primary cortical neurons. The ENS and the CNS both contain integrated nervous networks and the similarities between them, including between neurons and glia at a morphological level, have led to the ENS being described as the 'brain in the gut' or the 'second brain' [22]. Our current and previous results suggest that this anatomical resemblance does not extend to the molecular level as the ENS expresses only a limited number of isoforms of neuronal and glial markers as compared to the brain [10], although the functional consequences of these differences still remain to be determined.

Tau was found to be expressed in both the myenteric and submucosal plexus of human colon and throughout the ENS of wild-type mice and in the htau mouse model. In both the human and rodent ENS, tau protein had a mainly axonal and somatic distribution, which might be expected since in physiological conditions, tau is described as being a predominantly axonal protein [32]. The presence of nuclear tau has been documented in a wide variety

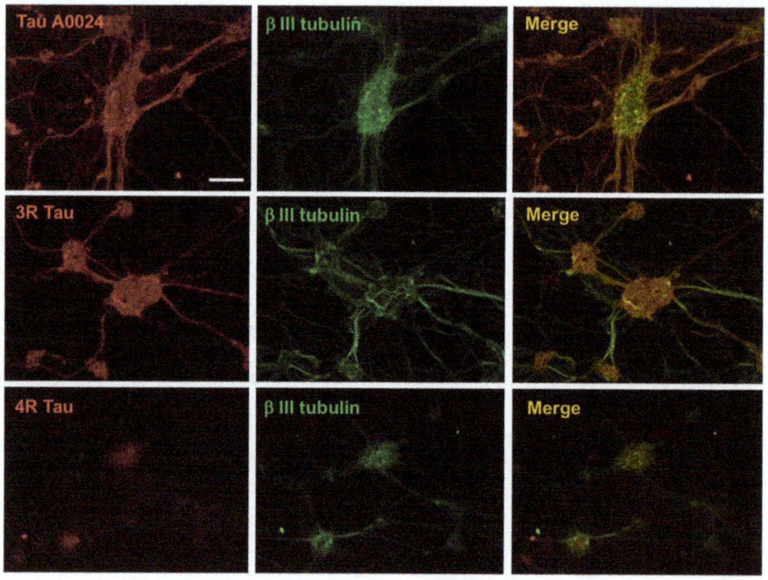

Fig. 8 Distribution and localization of tau in primary culture of rat ENS. After 14 days in culture, primary culture of rat ENS were immunostained with the pan-Tau antibody A0024 and the isoforms specific antibodies against 3R and 4R-tau. Scale bar is 100 μM

of cell and animal systems, including human and rodent brains and neuronal cell lines (reviewed in [32]). So far, the transcript encoding nuclear tau has not been formally identified but findings obtained in mouse brain suggest that the 1N4R isoform is preferentially localised in the nucleus [46]. Although our immunofluorescence experiments performed in mature human and mouse ENS, as well as in primary culture of rat ENS, clearly showed that tau was mainly axonal and somatic, we cannot rule out that a small proportion of enteric tau could also be nuclear. Further experiments including high resolution imaging and biochemical subcellular fractionation will be needed to answer this question.

A panel of well-characterised phospho-specific tau antibodies were used to show that tau is phosphorylated in the ENS of healthy subjects at Ser202/Thr205 and Ser396/Ser404. Tau is known to be phosphorylated at these sites under physiological conditions, with elevated phosphorylation at these epitopes chacteristic of pathological conditions in the CNS (reviewed in [53]). There is mounting evidence to suggest that tau phosphorylation plays a key role in neuronal physiology. The function of tau is strongly affected by its phosphorylation status, influencing its ability to interact with signaling proteins and kinases [57], its association with microtubules and membranes and its ability to regulate axonal transport [58]. Phosphorylation of Ser202/Thr205 and Ser396/Ser404 is commonly found in primary cortical neurons under basal conditions [2] as well as in snap-frozen brain biopsies from subject devoid of neurodegenerative conditions [47], suggesting that these sites are involved in the normal physiology of the CNS. This is further reinforced by the recent observation showing the presence of endogenous tau phosphorylated at these sites at postsynaptic sites in hippocampal neurons where tau interacts with the PSD95-NMDA receptor complex to regulate synaptic activity [48]. These results obtained in the CNS could be extended to the ENS where neuronal plasticity has also been described following modulation of neuronal activity [9, 37].

Soluble tau from adult human brain consist of a heterogeneous mixture of tau isoforms in multiple states of phosphorylation [25, 28]. Because normal electrophoresis techniques do not separate the individual tau isoforms, correct identification of the isoform composition of soluble tau requires an efficient dephosphorylation reaction with lambda phosphatase before immunoblotting [35]. Dephosphorylation of tau from normal adult human brain classically produces a downwards shift enabling a more precise separation and identification of the six tau isoforms [25, 35]. We therefore used the same approach in mature human ENS and the gut of htau mice. In sharp contrast to results with human and htau brain samples, lambda phosphatase treatment did not change the charge/mobility of tau bands in colon samples, suggesting that gut tau may have not been efficiently dephosphorylated. Since western blot and immunohistochemical findings showed that tau in gut is in fact phosphorylated, at least at Ser202, Thr205 and Ser396/ Ser404 this raises the possibility that ENS tau is modified in such a way that it is not susceptible to dephosphorylation. This relative resistance to dephosphorylation, which might be due to conformational changes occurring in case of phosphorylation at some specific sites [18], is specific to adult ENS tau as lambda phosphatase efficiently dephosphorylated tau in rat primary ENS cultures prepared from fetal rats.

In 1978 the first study on cultured myenteric neurons was published [40] and since then there has been a growing interest in this method with several different culture preparations being developed. Using primary cultures of rat ENS [9, 11], we have shown that fetal rat enteric neurons express four isoforms of tau, including the three 3R isoforms. This again stands in sharp contrast to the CNS as rat primary cortical neurons primarily express the 0N3R isoform (our study and [13]). We also show that tau isoforms present in primary ENS culture are phosphorylated under basal conditions and their levels of phosphorylation can be down or upregulated. This suggests that cultured ENS might be helpful to study the regulation of tau expression, phosphorylation and secretion not only in physiological conditions but also in the context of enteric neuropathies [55].

We did not observe any pathological tau changes in the ENS of PSP patients. This stands in sharp contrast with the fact that PSP is considered a prototypical tauopathy of the CNS characterized by tau hyperphosphorylation and truncation [31, 67] and an imbalance in 4R/3R ratio [38]. We have recently proposed that the ENS may be a mirror on to the PD pathology of the CNS since it recapitulates several of the neuronal and glial changes observed in the brain [10, 15, 45]. Our results suggest that, unlike PD, the pathological process in PSP is limited to the CNS and does not involve the ENS. This is supported by the paucity of studies reporting that the peripheral nervous systems are affected in PSP (reviewed in [64]) and by our observation of a lack of glial reaction in the gut in PSP patients [10]. One obvious limitation of this work is that our analysis of PSP samples was restricted to the analysis of the submucosal plexus. We can therefore not rule out that the absence of overt pathological changes in tau in colonic samples from our PSP patients may be due to this limited regional analysis and perhaps different findings would have been obtained had we examined the myenteric plexus. The refinement of new endoscopic procedures, such as full thickness biopsies [50], which provide access to both myenteric and submucosal plexi, may help to answer these critical questions. A second limitation in our study is the lack of neuropathological confirmation of PD and PSP, as the clinical diagnosis of both disorders may have a relatively poor accuracy [1, 42], especially for PD patients for whom signs and symptoms have been present for less than 5 years [1]. In addition, we can not rule out that some of our control subjects may have asymptomatic tauopathy [12].

Conclusions

We have characterised tau in the human and rodent ENS under physiological conditions and tauopathies. We show differences in tau isoform expression at mRNA and protein level, and in the susceptibility of tau to be dephosphorylated in the CNS and ENS. The data we have acquired on tau in the ENS strongly supports additional future studies aimed at expanding our knowledge of peripheral pathology in neurodegenrative disorders of the CNS and in enteric neuropathies [14].

Abbreviations

AD: Alzheimer's disease; BSA: Bovine serum albumen; CNS: Central nervous system; ENS: Enteric nervous system; ERK: Extracellular signal-regulated kinases; GI: Gastrointestinal; KO: Knockout; MP: Myenteric plexus; PCR: Polymerase chain reaction; PD: Parkinson's disease; PSP: Progressive supranuclear palsy; Ser: Serine; SMP: Submucosal plexus; TBS: Tris-buffered saline; Thr: Threonine

Acknowledgements

We are grateful to Professor Peter Davies (Feinstein Institute for Medical Research, NY, USA) for his generous gift of tau antibodies.

Funding

This work was supported by BBSRC/AstraZeneca (BB/L502601/1 to WN), the National Centre for the Replacement, Refinement and Reduction of Animals in Research (NC3Rs, NC/K500343/1 to WN), CECAP, FFGP and PSP France (to AP and PD).

Authors' contributions

AL, MAW, AGC, AP, SP, MT and JG performed the experiments and analyzed the data. ED and MRD managed the biobanking and dissected the colonic samples. ED and EC performed the endoscopy. MAW, PD, MN, DPH, MSP and WN designed the research and MAW, WN and PD wrote the manuscript. All authors read and approved the final manuscript.

Competing interests

Michael S. Perkinton is an employee of MedImmune.

Author details

[1]Inserm, U1235, 1 rue Gaston Veil, F-44035 Nantes, France. [2]Department of Neurology, CHU Nantes, F-44093 Nantes, France. [3]University Nantes, F-44000 Nantes, France. [4]King's College London, Institute of Psychiatry, Psychology and Neuroscience, Department of Basic and Clinical Neuroscience, Maurice Wohl Clinical Neuroscience Institute, Rm 1.23, 5 Cutcombe Road, Camberwell, London SE5 9RX, UK. [5]Neuroscience, IMED Biotech Unit, AstraZeneca, Cambridge CB21 6GH, UK.

References

1. Adler CH, Beach TG, Hentz JG, Shill HA, Caviness JN, Driver-Dunckley E et al (2014) Low clinical diagnostic accuracy of early vs advanced Parkinson disease: clinicopathologic study. Neurology 83:406–412. https://doi.org/10.1212/WNL.0000000000000641
2. Anderton BH, Brion JP, Couck AM, Davis DR, Gallo JM, Hanger DP et al (1995) Modulation of PHF-like tau phosphorylation in cultured neurones and transfected cells. Neurobiol Aging 16:389–397 discussion 398-402
3. Andorfer C, Kress Y, Espinoza M, de Silva R, Tucker KL, Barde Y-A et al (2003) Hyperphosphorylation and aggregation of tau in mice expressing normal human tau isoforms. J Neurochem 86:582–590
4. Andreadis A (2005) Tau gene alternative splicing: expression patterns, regulation and modulation of function in normal brain and neurodegenerative diseases. Biochim Biophys Acta 1739:91–103. https://doi.org/10.1016/j.bbadis.2004.08.010
5. Barrenschee M, Zorenkov D, Böttner M, Lange C, Cossais F, Scharf AB et al (2017) Distinct pattern of enteric phospho-alpha-synuclein aggregates and gene expression profiles in patients with Parkinson's disease. Acta Neuropathol Commun 5:1. https://doi.org/10.1186/s40478-016-0408-2

6. Beach TG, Adler CH, Sue LI, Vedders L, Lue L, White Iii CL et al (2010) Multi-organ distribution of phosphorylated alpha-synuclein histopathology in subjects with Lewy body disorders. Acta Neuropathol 119:689–702. https://doi.org/10.1007/s00401-010-0664-3
7. Bussière T, Hof PR, Mailliot C, Brown CD, Caillet-Boudin ML, Perl DP et al (1999) Phosphorylated serine422 on tau proteins is a pathological epitope found in several diseases with neurofibrillary degeneration. Acta Neuropathol 97:221–230
8. Chambonnière ML, Mosnier-Damet M, Mosnier JF (2001) Expression of microtubule-associated protein tau by gastrointestinal stromal tumors. Hum Pathol 32:1166–1173
9. Chevalier J, Derkinderen P, Gomes P, Thinard R, Naveilhan P, Vanden Berghe P, Neunlist M (2008) Activity-dependent regulation of tyrosine hydroxylase expression in the enteric nervous system. J Physiol Lond 586:1963–1975. https://doi.org/10.1113/jphysiol.2007.149815
10. Clairembault T, Kamphuis W, Leclair-Visonneau L, Rolli-Derkinderen M, Coron E, Neunlist M et al (2014) Enteric GFAP expression and phosphorylation in Parkinson's disease. J Neurochem 130:805–815. https://doi.org/10.1111/jnc.12742
11. Coquenlorge S, Duchalais E, Chevalier J, Cossais F, Rolli-Derkinderen M, Neunlist M (2014) Modulation of lipopolysaccharide-induced neuronal response by activation of the enteric nervous system. J Neuroinflammation 11:202. https://doi.org/10.1186/s12974-014-0202-7
12. Crary JF, Trojanowski JQ, Schneider JA, Abisambra JF, Abner EL, Alafuzoff I et al (2014) Primary age-related tauopathy (PART): a common pathology associated with human aging. Acta Neuropathol 128:755–766. https://doi.org/10.1007/s00401-014-1349-0
13. Davis DR, Brion JP, Couck AM, Gallo JM, Hanger DP, Ladhani K et al (1995) The phosphorylation state of the microtubule-associated protein tau as affected by glutamate, colchicine and beta-amyloid in primary rat cortical neuronal cultures. Biochem J 309(Pt 3):941–949
14. De Giorgio R, Bianco F, Latorre R, Caio G, Clavenzani P, Bonora E (2016) Enteric neuropathies: yesterday, today and tomorrow. Adv Exp Med Biol 891:123–133. https://doi.org/10.1007/978-3-319-27592-5_12
15. Devos D, Lebouvier T, Lardeux B, Biraud M, Rouaud T, Pouclet H et al (2013) Colonic inflammation in Parkinson's disease. Neurobiol Dis 50:42–48. https://doi.org/10.1016/j.nbd.2012.09.007
16. Duff K, Knight H, Refolo LM, Sanders S, Yu X, Picciano M et al (2000) Characterization of pathology in transgenic mice over-expressing human genomic and cDNA tau transgenes. Neurobiol Dis 7:87–98. https://doi.org/10.1006/nbdi.1999.0279
17. Dugger BN, Whiteside CM, Maarouf CL, Walker DG, Beach TG, Sue LI et al (2016) The presence of select tau species in human peripheral tissues and their relation to Alzheimer's disease. J Alzheimers Dis 51:345–356. https://doi.org/10.3233/JAD-150859
18. Dupont-Wallois L, Sautière PE, Cocquerelle C, Bailleul B, Delacourte A, Caillet-Boudin ML (1995) Shift from fetal-type to Alzheimer-type phosphorylated tau proteins in SKNSH-SY 5Y cells treated with okadaic acid. FEBS Lett 357:197–201
19. Ercan E, Eid S, Weber C, Kowalski A, Bichmann M, Behrendt A et al (2017) A validated antibody panel for the characterization of tau post-translational modifications. Mol Neurodegener 12:87. https://doi.org/10.1186/s13024-017-0229-1
20. Fischer D, Mukrasch MD, Biernat J, Bibow S, Blackledge M, Griesinger C et al (2009) Conformational changes specific for pseudophosphorylation at serine 262 selectively impair binding of tau to microtubules. Biochemistry 48:10047–10055. https://doi.org/10.1021/bi901090m
21. Gelpi E, Navarro-Otano J, Tolosa E, Gaig C, Compta Y, Rey MJ et al (2014) Multiple organ involvement by alpha-synuclein pathology in Lewy body disorders. Mov Disord 29:1010–1018. https://doi.org/10.1002/mds.25776
22. Gershon MD (1999) The enteric nervous system: a second brain. Hosp Pract (1995) 34:31–32 35–38, 41–42 passim
23. Giancola F, Torresan F, Repossi R, Bianco F, Latorre R, Ioannou A et al (2017) Downregulation of neuronal vasoactive intestinal polypeptide in Parkinson's disease and chronic constipation. Neurogastroenterol Motil 29. https://doi.org/10.1111/nmo.12995
24. Gibb GM, de Silva R, Revesz T, Lees AJ, Anderton BH, Hanger DP (2004) Differential involvement and heterogeneous phosphorylation of tau isoforms in progressive supranuclear palsy. Brain Res Mol Brain Res 121:95–101. https://doi.org/10.1016/j.molbrainres.2003.11.007
25. Goedert M, Jakes R (1990) Expression of separate isoforms of human tau protein: correlation with the tau pattern in brain and effects on tubulin polymerization. EMBO J 9:4225–4230
26. Goedert M, Jakes R, Vanmechelen E (1995) Monoclonal antibody AT8 recognises tau protein phosphorylated at both serine 202 and threonine 205. Neurosci Lett 189:167–169
27. Goedert M, Spillantini MG, Crowther RA (1992) Cloning of a big tau microtubule-associated protein characteristic of the peripheral nervous system. Proc Natl Acad Sci U S A 89:1983–1987
28. Goedert M, Spillantini MG, Jakes R, Rutherford D, Crowther RA (1989) Multiple isoforms of human microtubule-associated protein tau: sequences and localization in neurofibrillary tangles of Alzheimer's disease. Neuron 3:519–526
29. Grundke-Iqbal I, Iqbal K, Tung YC, Quinlan M, Wisniewski HM, Binder LI (1986) Abnormal phosphorylation of the microtubule-associated protein tau (tau) in Alzheimer cytoskeletal pathology. Proc Natl Acad Sci U S A 83:4913–4917
30. Gu Y, Oyama F, Ihara Y (1996) Tau is widely expressed in rat tissues. J Neurochem 67:1235–1244
31. Guillozet-Bongaarts AL, Glajch KE, Libson EG, Cahill ME, Bigio E, Berry RW, Binder LI (2007) Phosphorylation and cleavage of tau in non-AD tauopathies. Acta Neuropathol 113:513–520. https://doi.org/10.1007/s00401-007-0209-6
32. Guo T, Noble W, Hanger DP (2017) Roles of tau protein in health and disease. Acta Neuropathol 133:665–704. https://doi.org/10.1007/s00401-017-1707-9
33. Haïk S, Faucheux BA, Sazdovitch V, Privat N, Kemeny J-L, Perret-Liaudet A, Hauw J-J (2003) The sympathetic nervous system is involved in variant Creutzfeldt-Jakob disease. Nat Med 9:1121–1123. https://doi.org/10.1038/nm922
34. Hanger DP, Byers HL, Wray S, Leung K-Y, Saxton MJ, Seereeram A, Reynolds CH et al (2007) Novel phosphorylation sites in tau from Alzheimer brain support a role for casein kinase 1 in disease pathogenesis. J Biol Chem 282:23645–23654. https://doi.org/10.1074/jbc.M703269200
35. Hanger DP, Gibb GM, de Silva R, Boutajangout A, Brion J-P, Revesz T et al (2002) The complex relationship between soluble and insoluble tau in tauopathies revealed by efficient dephosphorylation and specific antibodies. FEBS Lett 531:538–542
36. Heumüller-Klug S, Sticht C, Kaiser K, Wink E, Hagl C, Wessel L, Schäfer K-H (2015) Degradation of intestinal mRNA: a matter of treatment. World J Gastroenterol 21:3499–3508. https://doi.org/10.3748/wjg.v21.i12.3499
37. Hons IM, Storr MA, Mackie K, Lutz B, Pittman QJ, Mawe GM, Sharkey KA (2012) Plasticity of mouse enteric synapses mediated through endocannabinoid and purinergic signaling. Neurogastroenterol Motil 24:e113–e124. https://doi.org/10.1111/j.1365-2982.2011.01860.x
38. Ingelsson M, Ramasamy K, Russ C, Freeman SH, Orne J, Raju S et al (2007) Increase in the relative expression of tau with four microtubule binding repeat regions in frontotemporal lobar degeneration and progressive supranuclear palsy brains. Acta Neuropathol 114:471–479. https://doi.org/10.1007/s00401-007-0280-z
39. Ishizawa T, Mattila P, Davies P, Wang D, Dickson DW (2003) Colocalization of tau and alpha-synuclein epitopes in Lewy bodies. J Neuropathol Exp Neurol 62:389–397
40. Jessen KR, McConnell JD, Purves RD, Burnstock G, Chamley-Campbell J (1978) Tissue culture of mammalian enteric neurons. Brain Res 152:573–579
41. Joiner S, Linehan JM, Brandner S, Wadsworth JDF, Collinge J (2005) High levels of disease related prion protein in the ileum in variant Creutzfeldt-Jakob disease. Gut 54:1506–1508. https://doi.org/10.1136/gut.2005.072447
42. Joutsa J, Gardberg M, Röyttä M, Kaasinen V (2014) Diagnostic accuracy of parkinsonism syndromes by general neurologists. Parkinsonism Relat Disord 20:840–844. https://doi.org/10.1016/j.parkreldis.2014.04.019
43. Kovacs GG (2017) Tauopathies. Handb Clin Neurol 145:355–368. https://doi.org/10.1016/B978-0-12-802395-2.00025-0
44. Lebouvier T, Coron E, Chaumette T, Paillusson S, Bruley d, Varannes S, Neunlist M, Derkinderen P (2010) Routine colonic biopsies as a new tool to study the enteric nervous system in living patients. Neurogastroenterol Motil 22:e11–e14. https://doi.org/10.1111/j.1365-2982.2009.01368.x
45. Lebouvier T, Neunlist M, Bruley d, Varannes S, Coron E, Drouard A, N'Guyen J-M et al (2010) Colonic biopsies to assess the neuropathology of Parkinson's disease and its relationship with symptoms. PLoS One 5:e12728. https://doi.org/10.1371/journal.pone.0012728

46. Liu C, Götz J (2013) Profiling murine tau with 0N, 1N and 2N isoform-specific antibodies in brain and peripheral organs reveals distinct subcellular localization, with the 1N isoform being enriched in the nucleus. PLoS One 8:e84849. https://doi.org/10.1371/journal.pone.0084849
47. Matsuo ES, Shin RW, Billingsley ML, Van deVoorde A, O'Connor M, Trojanowski JQ, Lee VM (1994) Biopsy-derived adult human brain tau is phosphorylated at many of the same sites as Alzheimer's disease paired helical filament tau. Neuron 13:989–1002
48. Mondragón-Rodríguez S, Trillaud-Doppia E, Dudilot A, Bourgeois C, Lauzon M, Leclerc N, Boehm J (2012) Interaction of endogenous tau protein with synaptic proteins is regulated by N-methyl-D-aspartate receptor-dependent tau phosphorylation. J Biol Chem 287:32040–32053. https://doi.org/10.1074/jbc.M112.401240
49. Morris HR, Gibb G, Katzenschlager R, Wood NW, Hanger DP, Strand C et al (2002) Pathological, clinical and genetic heterogeneity in progressive supranuclear palsy. Brain 125:969–975. https://doi.org/10.1093/brain/awf109
50. Neunlist M, Coquenlorge S, Aubert P, Duchalais-Dassonneville E, des Varannes SB, Meurette G, Coron E (2011) Colonic endoscopic full-thickness biopsies: from the neuropathological analysis of the myenteric plexus to the functional study of neuromuscular transmission. Gastrointest Endosc 73: 1029–1034. https://doi.org/10.1016/j.gie.2011.01.041
51. Newman J, Rissman RA, Sarsoza F, Kim RC, Dick M, Bennett DA et al (2005) Caspase-cleaved tau accumulation in neurodegenerative diseases associated with tau and alpha-synuclein pathology. Acta Neuropathol 110:135–144. https://doi.org/10.1007/s00401-005-1027-3
52. Noble W, Garwood C, Stephenson J, Kinsey AM, Hanger DP, Anderton BH (2009) Minocycline reduces the development of abnormal tau species in models of Alzheimer's disease. FASEB J 23:739–750. https://doi.org/10.1096/fj.08-113795
53. Noble W, Hanger DP, Miller CCJ, Lovestone S (2013) The importance of tau phosphorylation for neurodegenerative diseases. Front Neurol 4:83. https://doi.org/10.3389/fneur.2013.00083
54. Paillusson S, Tasselli M, Lebouvier T, Mahé MM, Chevalier J, Biraud M et al (2010) α-Synuclein expression is induced by depolarization and cyclic AMP in enteric neurons. J Neurochem 115:694–706. https://doi.org/10.1111/j.1471-4159.2010.06962.x
55. Phillips RJ, Walter GC, Ringer BE, Higgs KM, Powley TL (2009) Alpha-synuclein immunopositive aggregates in the myenteric plexus of the aging Fischer 344 rat. Exp Neurol 220:109–119. https://doi.org/10.1016/j.expneurol.2009.07.025
56. Pooler AM, Usardi A, Evans CJ, Philpott KL, Noble W, Hanger DP (2012) Dynamic association of tau with neuronal membranes is regulated by phosphorylation. Neurobiol Aging 33:431.e27–431.e38. https://doi.org/10.1016/j.neurobiolaging.2011.01.005
57. Reynolds CH, Garwood CJ, Wray S, Price C, Kellie S, Perera T et al (2008) Phosphorylation regulates tau interactions with Src homology 3 domains of phosphatidylinositol 3-kinase, phospholipase Cgamma1, Grb2, and Src family kinases. J Biol Chem 283:18177–18186. https://doi.org/10.1074/jbc.M709715200
58. Rodríguez-Martín T, Cuchillo-Ibáñez I, Noble W, Nyenya F, Anderton BH, Hanger DP (2013) Tau phosphorylation affects its axonal transport and degradation. Neurobiol Aging 34:2146–2157. https://doi.org/10.1016/j.neurobiolaging.2013.03.015
59. Takuma H, Arawaka S, Mori H (2003) Isoforms changes of tau protein during development in various species. Brain Res Dev Brain Res 142:121–127
60. Taleghany N, Oblinger MM (1992) Regional distribution and biochemical characteristics of high molecular weight tau in the nervous system. J Neurosci Res 33:257–265. https://doi.org/10.1002/jnr.490330209
61. Tam PK (1990) An immunohistological study of the human enteric nervous system with microtubule-associated proteins. Gastroenterology 99:1841–1844
62. Tasselli M, Chaumette T, Paillusson S, Monnet Y, Lafoux A, Huchet-Cadiou C et al (2013) Effects of oral administration of rotenone on gastrointestinal functions in mice. Neurogastroenterol Motil 25:e183–e193. https://doi.org/10.1111/nmo.12070
63. Usardi A, Pooler AM, Seereeram A, Reynolds CH, Derkinderen P, Anderton B et al (2011) Tyrosine phosphorylation of tau regulates its interactions with Fyn SH2 domains, but not SH3 domains, altering the cellular localization of tau. FEBS J 278:2927–2937. https://doi.org/10.1111/j.1742-4658.2011.08218.x
64. Wakabayashi K, Mori F, Tanji K, Orimo S, Takahashi H (2010) Involvement of the peripheral nervous system in synucleinopathies, tauopathies and other neurodegenerative proteinopathies of the brain. Acta Neuropathol 120:1–12. https://doi.org/10.1007/s00401-010-0706-x
65. Wakabayashi K, Takahashi H, Takeda S, Ohama E, Ikuta F (1988) Parkinson's disease: the presence of Lewy bodies in Auerbach's and Meissner's plexuses. Acta Neuropathol 76:217–221
66. Wills J, Credle J, Haggerty T, Lee J-H, Oaks AW, Sidhu A (2011) Tauopathic changes in the striatum of A53T α-synuclein mutant mouse model of Parkinson's disease. PLoS One 6:e17953. https://doi.org/10.1371/journal.pone.0017953
67. Wray S, Saxton M, Anderton BH, Hanger DP (2008) Direct analysis of tau from PSP brain identifies new phosphorylation sites and a major fragment of N-terminally cleaved tau containing four microtubule-binding repeats. J Neurochem 105:2343–2352. https://doi.org/10.1111/j.1471-4159.2008.05321.x

13

Rac1 activation links tau hyperphosphorylation and Aβ dysmetabolism in Alzheimer's disease

Mirta Borin[1†], Claudia Saraceno[2†], Marcella Catania[3], Erika Lorenzetto[1], Valeria Pontelli[1], Anna Paterlini[2], Silvia Fostinelli[2], Anna Avesani[1], Giuseppe Di Fede[3], Gianluigi Zanusso[1], Luisa Benussi[2], Giuliano Binetti[4], Simone Zorzan[5], Roberta Ghidoni[2], Mario Buffelli[1] and Silvia Bolognin[2,6*]

Abstract

One of the earliest pathological features characterizing Alzheimer's disease (AD) is the loss of dendritic spines. Among the many factors potentially mediating this loss of neuronal connectivity, the contribution of Rho-GTPases is of particular interest. This family of proteins has been known for years as a key regulator of actin cytoskeleton remodeling. More recent insights have indicated how its complex signaling might be triggered also in pathological conditions. Here, we showed that the Rho-GTPase family member Rac1 levels decreased in the frontal cortex of AD patients compared to non-demented controls. Also, Rac1 increased in plasma samples of AD patients with Mini-Mental State Examination < 18 compared to age-matched non demented controls. The use of different constitutively active peptides allowed us to investigate in vitro Rac1 specific signaling. Its activation increased the processing of amyloid precursor protein and induced the translocation of SET from the nucleus to the cytoplasm, resulting in tau hyperphosphorylation at residue pT181. Notably, Rac1 was abnormally activated in the hippocampus of 6-week-old 3xTg-AD mice. However, the total protein levels decreased at 7-months. A rescue strategy based on the intranasal administration of Rac1 active peptide at 6.5 months prevented dendritic spine loss. This data suggests the intriguing possibility of a dual role of Rac1 according to the different stages of the pathology. In an initial stage, Rac1 deregulation might represent a triggering co-factor due to the direct effect on Aβ and tau. However, at a later stage of the pathology, it might represent a potential therapeutic target due to the beneficial effect on spine dynamics.

Introduction

Alzheimer's disease (AD) is an age-associated disorder, characterized by the abnormal depositions of hyperphosphorylated tau protein in the form of neurofibrillary tangles (NFT) and of β-amyloid (Aβ) peptide in the form of senile plaques (SP). These thoroughly studied hallmarks are certainly key contributors to the development of the pathology. However, it is the synaptic and dendritic loss, which seems to be the best predictor of the clinical symptoms. This dysfunction has been extensively described in early studies [45, 57, 64] and confirmed in more recent years [17, 42, 59]. Reduction in spine number correlates well with the degree of memory loss and cognitive impairment of the patients [16].

Despite the plethora of evidence showing that the loss of spine stability is associated with AD, the identification of pathways responsible for this abnormal disruption is still elusive. In physiological conditions, spine morpho-dynamics rely on changes of F-actin-rich cytoskeleton, which are highly regulated by Rho-GTPases (see reviews [10, 27]). Rho-GTPases maintain the equilibrium between actin monomer (G-actin) and filament (F-actin) pools [18]. This family of proteins comprises of three main members: Ras-related C3 botulinum toxin substrate 1 (Rac1), cell division protein 42 (Cdc42), and Ras homologous member A (RhoA). By switching

* Correspondence: silvia.bolognin@gmail.com
[†]Mirta Borin and Claudia Saraceno contributed equally to this work.
[2]Molecular Markers Laboratory, IRCCS Istituto Centro San Giovanni di Dio Fatebenefratelli, Via Pilastroni, 4, 25125 Brescia, Italy
[6]Luxembourg Centre for Systems Biomedicine, University of Luxembourg, 6 avenue du Swing, L-4367 Belvaux, Luxembourg
Full list of author information is available at the end of the article

between an active GTP and inactive GDP-bound form, they convert signals of the postsynaptic receptors into changes of actin binding proteins, ultimately resulting in the remodeling of spine shape and density [15, 47]. The signaling pathway is quite complex: the timely activation of the 3 proteins is strictly regulated by the association to several additional regulatory proteins, which are in turn activated during synaptic transmission or neurotrophic factor release [47, 58, 65].

As impaired actin cytoskeleton stability [6, 48], including the formation of actin rod-like inclusions [33], has been shown in AD brains, Rho-GTPase signaling deregulation might contribute to the synaptic degeneration observed in the disease [1, 10]. Among the different members, Rac1 has shown to be connected to amyloid precursor protein (APP) processing. Studies on hippocampal primary neurons showed that Rac1-specific inhibitor decreased APP protein levels in a concentration-dependent manner by modulating its transcriptional activity [67]. However, studies examining the direct connection between Aβ and Rac1 are contradictory, leaving a rather unclear scenario regarding the potential contribution of the protein to disease-relevant mechanisms [34, 44, 52].

More puzzling is the connection with the other key pathological hallmark of the pathology, tau hyperphosphorylation. In the context of cancer and cell migration studies, Rac1 was shown to directly bind the oncoprotein SET [61, 63]. Interestingly for the AD field, SET is the inhibitor of the protein phosphatase 2A (PP2A), the major regulator of tau phosphorylation. The laboratory of Dr. Khalid Iqbal showed extensively how SET abnormally translocated from the nucleus to the cytoplasm in the brain of AD patients compared to controls [62]. In the cytosol, SET can directly bind PP2A and decrease its activity [4]. The pathological relevance of this pathway was demonstrated by the fact that the overexpression of SET by adeno-associated viral vectors generated a rat model of sporadic AD [9, 68, 69].

Here, we provide evidence showing that Rac1 was altered in fronto-cortical brain lysate and plasma of AD patients compared to healthy age-matched controls. Importantly, the degree of the alteration in the circulating Rac1 pool reflected the severity of the cognitive impairment, suggesting a potential role of Rac1 as a biomarker for AD. In vitro studies on mouse primary cortical neurons and SH-SY5Y showed that the triggering of selective Rac1 signaling induced the generation of pathogenic Aβ fragments and the translocation of SET from the nucleus to the cytoplasm. This resulted in an increase of tau phosphorylation (at pT181). Active Rac1 increased in 6-week-old 3xTgAD mouse hippocampus while the total level decreased at 7 months compared to controls. Intranasal treatment with a constitutively active form of the peptide at 6.5 months resulted in a rescue of the number of dendritic spines compared to vehicle-treated animals.

Materials and methods

Human subjects

Brain samples were provided by the Biobank of the IRCCS Foundation – Carlo Besta Neurological Institute and from the Brain Bank of the Department of Pathology at Indiana University School of Medicine. We included 24 brains from AD patients and 12 from age-matched non-demented controls (Table 1). The neuropathological diagnosis was performed according to international guidelines for the assessment of AD [25].

For the plasma samples, the patients considered for this study underwent clinical and neurological examination at the MAC Memory Center of the IRCCS Centro San Giovanni di Dio-Fatebenefratelli, Brescia. Clinical diagnosis of AD, MCI was made according to international guidelines [37, 38, 51]. We included AD (n = 114) and MCI (n = 47) patients, and age and sex-matched cognitively healthy controls (CTRL, n = 102). Biological samples were

Table 1 Braak stage and known co-pathologies of the brain samples in the AD study group

ID	Braak stage	Co-pathology
AD 1	VI	Cerebrovascular disease
AD 2	VI	Hemorrage
AD 3	VI	–
AD 4	VI	Lewy body
AD 5	V	Cerebrovascular disease
AD 6	V-VI	–
AD 7	VI	–
AD 8	III-IV	–
AD 9	VI	–
AD 10	VI	Lewy body
AD 11	VI	–
AD 12	III-IV	Lewy body
AD 13	IV	–
AD 14	III [31]	–
AD 15	V-VI	Hemorrhage
AD 16	VI	–
AD 17	VI	Vascular formation
AD 18	VI	Cerebrovascular disease
AD 19	VI	Cerebrovascular disease
AD 20	V-VI	Cerebrovascular disease
AD 21	V-VI	Cerebrovascular disease
AD 22	VI	Diffuse Lewy body
AD 23	V-VI	–
AD 24	IV	–

collected and stored in the Biobank of the IRCCS Centro San Giovanni di Dio-Fatebenefratelli, Brescia, Italy, after obtaining informed consent, as approved by the local ethics committee (approval No. 26/2014). The study was approved by the local ethics committee (approval No. 03/2015). Plasma was isolated according to standard procedures. The demographic characteristics of the patients in the study are shown in Tables 2, 3.

Cell culture and drug/peptide treatments

Primary neuronal cultures (from E18 C57BL/6 J mouse embryos) and SH-SY5Y neuroblastoma cells were obtained and cultured as previously described [11]. For the okadaic acid (OA) treatment, the powder (Sigma-Aldrich) was dissolved in dimethylsulfoxide (DMSO, Sigma-Aldrich) at 50 μM concentration as stock solution. In working solutions, DMSO never exceeded the concentration of 0.02% and the final OA concentration was 10 nM. The TAT fusion proteins were prepared as previously described [12]. For a scheme of the recombinant mutant proteins with all the mutations and a complete amino acid sequence, see Lorenzetto et al. [30]

For the Leptomycin B (LMB, #9676 Cell Signaling Technology) treatment, SH-SY5Y cells were differentiated to a mature neuron-like phenotype by retinoic acid (RA, R2625 Sigma-Aldrich). Cells were seeded on a glass slide (12 mm Ø), at a density of 10^3 cells per well, and RA was added 3 h after plating at a final concentration of 10 μM in DMEM with 2% FBS and maintained for 10 days. LMB was added to the medium at a final concentration of 10 pM for 48 h, alone or together with Tat-Rac1L61F37A or Tat-Rac1WT (1 μM).

The MTT assays were performed 3 or 4 times, in triplicate as previously described [11]. The various treatments were compared to the control (untreated cells), which represented 100% viability.

Animal housing and intranasal delivery

Animal breeding and handling were performed following a protocol approved by the Animal Care and Use Committee of the University of Verona (CIRSAL), and authorized by the Italian Ministry of Health, in strict adherence to the European Communities Council directives (86/609/EEC). Mice were housed with water and food ad libitum and with 12 h/12 h light/dark cycle, under standard environmental conditions (temperature, humidity). For these studies, female 3xTg-AD mice harboring APP_{swe}, $PS1_{M146V}$, tau_{P301L} transgenes [43], and age-matched control (C57BL/6) were purchased from the Jackson laboratory (New Harbor, ME, USA), and used at 4 different ages: 6 week-, 3, 7, and 16 month-old.

Intranasal administration is a non-invasive method for delivering therapeutic agents to the central nervous system. Animals were randomly assigned to the two treatment groups (PBS solution, Veh; Rac1-L61F37A mutant peptide solution (100 μM)). Mice were anesthetized with isofluorane, and a total volume of 16 μl solution was administered (alternating smaller injections of 4 μl each to the left and to the right nares with 10 min between each administration). Animals were treated 3 times a week for two weeks, starting from 6.5 months.

ELISA analyses

Brain samples from frontal cortex of AD patients and age-matched non-demented controls were homogenized in 9 volumes of 1X PBS using a manual Dounce homogenizer and centrifuged at 1500 xg for 15 min. Supernatants were collected and the total protein amount was measured by BCA Protein Assay kit (Pierce).

Blood samples from AD and MCI patients and CTRL were kept at 4 °C for at least 20 min and then centrifuged for 5 min (4 °C, 1,000×g). Plasma was collected and centrifuged 5 min (4 °C, 1,000×g) after the addition of the protease inhibitors.

Rac1 and RhoA levels were measured in plasma, in duplicate, using commercially available ELISA kits (Human Ras-Related C3 Botulinum Toxin Substrate 1, RAC1, and Human Transforming Protein RhoA, MyBioSource).

Rac1-GTP level was assayed using Rac1 Activation Biochem Kit™ (#BK035, Cytoskeleton). Proteins were separated by SDS-PAGE and, after blotting, membranes were probed with anti-Rac1 antibody (mouse anti-Rac1, 1:1000, #05–389, Upstate). GAPDH antibody (rabbit anti-GAPDH, 1:20000, #G9545, Sigma-Aldrich) was used as a loading control. All the kits were used according to the manufacturer's instructions.

Table 2 Demographic characteristic and plasma Rac1 levels in the four groups

	Controls	MCI	AD MMSE≥18	AD MMSE< 18
No. of subject	102	47	72	42
Gender (% female)	49	64	60	48
Mean age (SD), years	70 (5)	75 (6)	72 (5)	73 (6)
Rac1 (ng/ml) range, median and mean	0.10–1.82 0.38, 0.45	0.16–8.21 0.42, 0.77	0.14–6.81 0.37, 0.62	0.16–6.95 0.76, 1.00
MMSE (SD)	28 (1)	26 (2)	22 (2)	9 (6)

Table 3 Demographic characteristic and plasma RhoA levels in the four groups

	Controls	MCI	AD MMSE≥18	AD MMSE< 18
No. of subjects	83	45	47	27
Gender (% female)	52	64	58	41
Mean age (SD), years	70 (5)	75 (6)	72 (5)	73 (5)
RhoA (pg/ml) range, median and mean	3.26–1634 27.82, 153.7	1.78–849 28.77, 98.17	1.38–812.3 56.35, 132.5	2.96–1395 38.8, 144.1
MMSE (SD)	28 (1)	26 (2)	22 (2)	8 (7)

Immunoprecipitation and immunoproteomic analysis

The mouse brain dissection was performed in a plastic petri dish on ice, after collecting the whole brain from the mouse skull. The two cortexes and the hippocampus were collected, flash frozen in liquid nitrogen, and stored at − 80 °C until analysis. The whole procedure did not exceed 5 min to preserve brain integrity.

Brain homogenates (10% weight/volume) were obtained using a micro-pestle on ice in cold lysis buffer containing: 50 mM Tris-HCl (pH 7.5), 2% Igepal, 10 mM $MgCl_2$, 0.5 M NaCl, 2 mM ethylenediaminetetraacetic acid (EDTA), 2 mM ethylene glycol tetraacetic acid (EGTA), 5 mM benzamidine, 0.5 mM phenylmethylsulfonyl fluoride (PMSF), 8 mg/mL pepstatin A and 20 mg/mL leupeptin, 50 mM b-glycerolphosphate, 100 mM sodium fluoride, 1 mM sodium vanadate, 20 mM sodium pyrophosphate, and 100 nM OA. Homogenates were clarified by a centrifugation at 4 °C (10000xG 1 min). After assessment of protein concentration by Precision red protein quantification assay (Cytoskeleton #ADV02), lysates were processed for either Rac1 activation assay (Cytoskeleton # BK035) or Western Blot. For the Rac1 activation assay, lysates were diluted to a concentration of 0.5 mg/ml.

The following primary antibodies were used: GluR1 (Anti-GluR, Recombinant rabbit monoclonal antibody, 1:250, 05-855R, Millipore/Merk); Lamin B1 (mouse monoclonal antibody, 1:1000, B-10:sc-374,015, Santa Cruz Biotechnology); PSD95 (anti-PSD 95 mouse monoclonal antibody, 1:1000, #124 01, Synaptic Systems); pT181 (Phospho-Tau (Thr181) mouse monoclonal antibody (AT270), 1:1000, MN1050 Thermo Fisher Scienific); Tau-5 (Mouse (monoclonal) Anti-tau (Neurofibrillary Tangles Marker), clone Tau-5, 1:1000, Invitrogen/Thermo Fisher Scienific, USA); TuJ1 (rabbit anti-ß-tubulin III antibody, 1:2500, T2200 Sigma-Aldrich).

Primary cortical neurons were seeded into 6-well plates, at a density of 9,5 × 10^5 cells per well. Following treatments, cells were washed 1× in Tris-buffered saline (TBS), then lysed and scraped with 50 μl of pre-warmed Laemmli buffer and boiled for 10 min. The cell lysate was assayed for protein using the Bradford method (Sigma-Aldrich). The lysates were separated using a 4–12% Bis-Tris gels (Novex pre-cast gel, Invitrogen) and transferred to 0.45 μm nitrocellulose membrane (Invitrogen) for probing with antibodies. Blots were blocked for 1 h at room temperature in 1X Odyssey blocking buffer (TBS) and incubated with primary antibodies overnight in Odyssey blocking buffer (TBS) plus 0.1% Tween-20 at 4 °C. Then the membranes were washed 3 × 10 min in TBST (Tween-20 TBS) at room temperature, followed by incubation with secondary antibody conjugated to IRDye diluted in Odyssey blocking buffer (TBS) plus 0.1% Tween-20 for 1 h at room temperature. Blots were washed 2 × 10 min in TBST, 1 × 10 min in TBS and visualized with Odyssey Infrared Imaging System.

Levels of total tau and tau phosphorylation at each specific site were determined by using phosphorylation-dependent and site-specific tau antibodies from Invitrogen (rabbit anti-tau (pS262) phosphospecific antibody, 1:1000 for WB and 1:100 IF, #44-750G; mouse anti-tau, 1:1000, #AHB0042; rabbit anti-tau (pS202) phosphospecific antibody 1:1000 for WB and 1:100 IF, #44779G). The pan-Actin antibody (mouse anti-actin, 1:2500, #MAB1501, Millipore or rabbit anti-actin, 1:2500, #A2066, Sigma-Aldrich) was used as a loading control. Primary antibodies were detected using anti-mouse IRDye 800 (1:2500; Li-Cor) or anti-rabbit Alexa Flour 680 (1:5000; Invitrogen). Blots were scanned and subsequently quantified using the Odyssey Imaging System (Li-Cor) by quantifying fluorescent signals as Integrated Intensities (I.I. K Counts) using the Odyssey Infrared Imaging System, Application Version 1.2 software. After background subtraction, ratios were calculated for each antibody against the pan-actin loading control using I.I. K Counts. The respective antibody to pan-actin ratio was then used to calculate phosphorylated protein to total protein ratio.

The subcellular fractionation on primary cortical neurons was performed as previously described [74]. For these experiments, 5-6 × 10^6 neurons were seeded. The membrane fraction was not quantified but directly suspended in 15 μl of sample buffer and loaded on the gel.

Surface-enhanced laser desorption/ionization time-of-flight mass spectrometry (SELDI TOF MS)

Immunoproteomic analyses of Aβ isoforms were performed as previously described with minor modifications

[2]. Briefly, 3 μl of the specific monoclonal human antibodies (mAbs) 4G8 (anti-Aβ_{17-24}) + 6E10 (anti-Aβ_{1-16}) (BioLegend, San Diego, CA, USA) at the total mAbs concentration of 0.125 mg/ml (concentration of each mAbs was 0.0625 mg/ml), were incubated in a humidity chamber for 3 h at room temperature (RT) to allow covalent binding to the PS20 ProteinChip Array (Bio-Rad, Laboratories, Inc.). Unreacted sites were blocked with Tris-HCl 0.5 M, pH 8 in a humid chamber at RT for 1 h. Each spot was washed first 3 times with PBS containing 0.5% (*v*/v) Triton X-100 and then twice with PBS. The spots were coated with 5 μl of cell lysate and incubated in a humid chamber at 4 °C overnight. Each spot was washed first 3 times with PBS containing 0.1% (v/v) Triton X-100, twice with PBS, and finally with deionized water. One microliter of α-ciano-4-hydroxy cinnamic acid (CHCA, Bio-Rad Laboratories, Inc.) was added to each spot. Mass identification was made using the ProteinChip SELDI System, Enterprise Edition (Bio-Rad Laboratories, Inc.). The analysis was performed with mass focus FM5500 and laser energy E1500.

Golgi staining and spine count

Animals were sacrificed by terminal anaesthesia with 2,2,2-Tribromoethanol (ip dose of 0,8 g/kg body weight; T48402, Sigma-Aldrich, Germany), and intracardially perfused with 0.9% saline solution added with 0.5% heparin (H3393, Sigma-Aldrich) to remove blood, followed by 4% paraformaldehyde buffered pH 7, to fix brain tissue. Brains were impregnated in 50 fold volume of staining solution, containing 1% $HgCl_2$ (Sigma-Aldrich, n°7,487,947, Germany), 1% $K_2Cr_2O_7$ (Sigma-Aldrich) and 1% K_2CrO_4 (Sigma-Aldrich), and stored at room temperature for 2 weeks in the dark.

Brain sections (80 to 100 μm thickness) were obtained using a vibratome (Leica VT1200, Leica Biosystems, Germany). Sections were placed in a mixture consisting of 1 part Developer Replenisher solution (GBX n° 3,101,508, Carestream Dental) and 2 parts Milli-Q water for 5 min, rinsed in Milli-Q water for 5 min, placed in a mixture consisting of 1 part Fixer Replenisher solution (GBX n° 3,101,557, Carestream Dental) and two parts Milli-Q water for 15 min, and then rinsed again in Milli-Q water for 5 min. Sections were then dehydrated in 60, 80 and 100% ethanol 2 min each, cleared in 100% xylene for 2 min and then mounted in Eukitt (03989 Sigma-Aldrich). Dendrites and spines of 4 sub regions of the cortex (primary motor cortex, secondary motor cortex, posterior parietal area and visual cortex) were imaged by a 100X oil objectives using Olympus BX63 microscope (Olympus Corporation, Japan) and acquired by Neurolucida 64-Bit software (MBF Bioscience, USA). Dendritic spines were manually counted scrolling along the z-stack (0.35 μm) of acquired images, and tagging spines using the multi-point selection tool of ImageJ 1.47v software (NIH, USA). The dendritic length was determined using the ImageJ software segmented line tool.

Immunostaining and confocal analysis

Immunocytochemistry and image acquisition was performed as previously described [11]. The following primary antibodies were used: anti-active Rac1 (1:1000, #26903, NewEast), pan-axonal neurofilament marker (1:1000, #SMI-312R, Covance), rhodamine phalloidin (1:40, #R415, Thermo Fisher Scientific), Aβ_{17-24} (1:200, 4G8; #SIG-39200, Covance), anti-map2 (1:500, #M9942, Sigma-Aldrich), anti-SET (I2PP2A; 1:50, #sc-25,564, Santa Cruz Biotechnology), site-specific tau antibodies were purchased from Invitrogen (1:100, #44779G).

Aβ_{1-42} oligomer preparation and dot-blot

The preparation of Aβ1$_{-42}$ synthetic oligomers was performed according to a previously described protocol [32]. The supernatant with Aβ_{1-42} oligomers was assayed for protein content using the Bradford kit (Sigma-Aldrich). The oligomerization of Aβ_{1-42} was checked by dot blotting using two different antibodies: 6E10 (beta amyloid antibody; #SIG-39320, Covance) and A11 (anti-oligomer antibody; #AHB0052, Invitrogen). 0.1 to 1 μg of each oligomeric preparation were applied on a nitrocellulose membrane and allowed to air dry. The membrane was then washed with TBS for 5 min and blocked with Odyssey Blocking Buffer (Li-Cor, #FE3092750000) for 1 h at room temperature. The membranes were then incubated overnight at 4 °C with 6E10 (1:2000) or the conformation dependent antibody A11 (1:500) in Odyssey Blocking Buffer with 0.1% Tween-20. Following 3 10-min washes, the blot was incubated with secondary antibody (anti-mouse IRDye 800, 1:2500 (Li-Cor) or anti-rabbit Alexa Flour 680, 1:5000 (Invitrogen)) for 1 h at room temperature, washed again and scanned on Odyssey Imaging System (Li-Cor).

Regulatory context of Rac1 and AD by bioinformatics tools

The role of Rac1 in the AD was investigated starting from the genes related to the disease through GWAS (Genome Wide Association Studies). The GWAS Catalog [70] allowed collecting 720 genes statistically linked to the pathology. In order to reconstruct a network connecting the selected genes, including others likely involved in the process, we started from ANAT [3]. ANAT is a bioinformatics tool to chart molecular pathways including direct high confidence interactors to connect all the input genes. SET and PP2A were added to the list. Of the GWAS list, ANAT did not recognize 269 genes. The resulting network was enriched by ANAT with 182 high confidence interactors connecting GWAS nodes, including Rac1.

Statistical analysis

Data were analysed with Prism 5 (GraphPad Software). Statistical significant differences are reported as $^{*}p < 0.05$, $^{**}p < 0.01$, and $^{***}p < 0.001$. The correlation of plasma Rac1 with MMSE was performed using the Spearman's correlation procedure with SPSS 20.0 software for Windows (IBM). The sample size and the used statistical tests are indicated in Table 4.

Results

Rac1 protein levels are altered in human AD fronto-cortical brain and plasma samples

To investigate the role of Rac1 in the pathogenesis of AD, fronto-cortical brain homogenates from 24 neuropathologically confirmed AD patients and 12 age-matched non-demented controls were analysed. Rac1 levels decreased in AD brains as compared to controls (Fig. 1a). We also evaluated Rac1 protein levels in the plasma of 114 patients affected by AD, 47 subjects with mild cognitive impairment (MCI), and 102 sex and age-matched non-demented controls. To investigate the link between Rac1 and cognitive decline, a correlation analysis was performed between Rac1 levels and the Mini-Mental State Examination (MMSE) in AD: Rac1 plasma levels were negatively correlated with MMSE score ($r = -0.208$; $p = 0.026$). We stratified AD patients based on their MMSE score (AD patients with MMSE< 18, $n = 42$; AD patients with MMSE≥18, $n = 72$). Rac1 levels significantly increased in the plasma of the AD patients with MMSE< 18 compared to controls ($p = 0.0002$), MCI ($p = 0.045$), and the AD group with MMSE≥18 ($p = 0.0051$) (Kruskal-Wallis followed by Dunn's multiple comparison test) (Fig. 1b). No alterations were detected in RhoA plasma levels in AD patients and MCI subjects (AD MMSE≥18 $n = 47$; AD MMSE< 18 $n = 27$; MCI $n = 45$; CTRL 83; $p = 0.104$ Kruskal-Wallis test) (Fig. 1c).

Rac1 perturbation affects APP metabolism

To modulate Rac1 activity, we generated TAT-Rac1 mutant proteins. These mutants contained a sequence coding for TAT, derived from the 86-amino acid transactivation protein involved in HIV replication, which allows the internalization of the protein into the cell. The produced proteins were: (i) Rac1-WT, which contained the wild type sequence of the protein; (ii) Rac1-L61F37A and Rac1-L61Y40C, two double mutants with a point mutation, which tonically activated the protein (Q61L,

Table 4 Sample size and performed statistical analysis

Figure Number	Experiment	Test	Sample size (n)	*P* value
Fig. 1a	Rac1 AD human brain	Mann Whitney	12, 24	0.028
Fig. 1b	Rac1 human AD plasma	Kruskal-Wallis Dunn's test	102, 47, 72, 42	0.0005 CTRL vs AD MMSE< 18 p = 0.0002; MCI vs AD MMSE< 18 p = 0.045; AD MMSE≥18 vs AD MMSE< 18p = 0.0051;
Fig. 3b	SET/GluR1	Two-tailed One sample t test	10,10,10,6,6	CTRL vs Rac1-WT *p* = 0.039 CTRL vs Rac1-L61F37A, *p* = 0.037
Fig. 3d	pT181/GAPDH	Two-tailed One sample t test	9, 7, 7,3,3	CTRL vs Rac1-WT *p* = 0.023; CTRL vs Rac1-L61F37A, *p* = 0.014
Fig. 3d	pT181/Tau5	Two-tailed One sample t test	9, 7, 7,3,3	CTRL vs Rac1-L61F37A, p = 0.045;
Fig. 5b	Hippocampus 6 weeks Rac1GTP/Rac1	Student t test	4, 4	0.044
Fig. 5c	Cortex 7 months Rac1/GAPDH	Student t test	6, 6	0.005
Fig. 6a	PSD95/Tuj1	One-way Anova Turkey's MC	8, 9, 8	0.0064 C57 + Veh vs 3xTgAD + Veh * 3xTgAD + Veh vs 3xTgAD + Rac1 *
Fig. 6d	Spine density	One-way Anova Turkey's MC	4, 4, 4	0.0061 C57 + Veh vs 3xTgAD + Veh * 3xTgAD + Veh vs 3xTgAD + Rac1 **
Additional file 1: Figure S2A	Aβ toxicity	Two-tailed One sample t test	4, 4, 4	Aβ0.1 μM 0.0044 Aβ0.5 μM 0.0088 Aβ1μM 0.0414
Additional file 1: Figure S4C	3 h OA	Two-Tailed paired t test	6, 6	0.003
Additional file 1: Figure S4C	6 h OA	Two-Tailed paired t test	4, 4	0.038

$^{*}p < 0.05$; $^{**}p < 0.01$

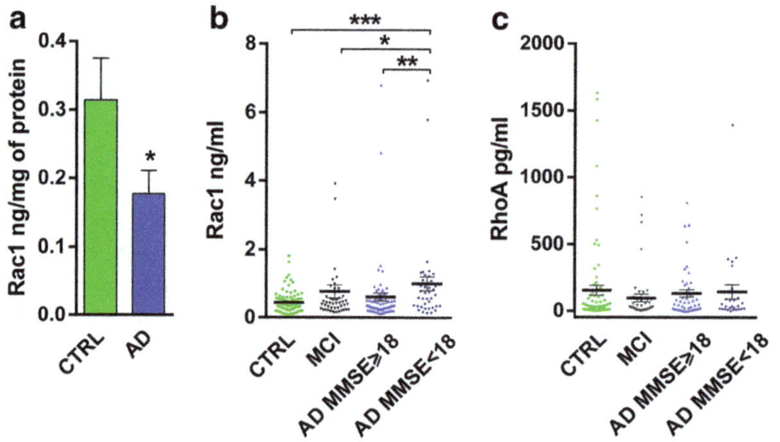

Fig. 1 Rac1 is altered in AD brain and plasma samples. **a** Rac1 (ng/mg of protein) was measured in brain homogenates from CTRL subjects and AD patients. **b** Rac1 (ng/ml of protein) was measured in plasma samples from CTRL subjects, MCI, and AD patients (MMSE≥18 and MMSE< 18). **c** RhoA (pg/ml of protein) was measured in plasma samples from CTRL subjects, MCI, and AD patients. The data represented are mean ± SEM

abbreviated in L61), and a second point mutation (F37A or Y40C), which conferred selectivity to the downstream signalling delivery [28]; (iii) Rac1-N17, a dominant negative (DN) mutant with a single point mutation (T17 N, abbreviated as N17).

The TAT trojan sequence efficiently allowed the internalization of the proteins in primary cortical neurons. Confocal pictures showed that TAT-GFP was internalized within 1 h after the treatment (Additional file 1: Figure S1A, B). Optical sectioning showed GFP-rich endosome-like structures in the cytoplasm. TAT-GFP signal was also evident in live cells imaged 1 h after the treatment (Additional file 1: Figure S1C). GFP fluorescence was found both in the somas as well as in neurites. We also checked, with MTT assay, whether the mutant peptides were toxic in primary cortical neurons (Additional file 1: Figure S1D). After 24 h treatment with 2 μM concentration, no toxic effect was observed. Mature cortical neurons were then treated for 24 h with 1 μM constitutively active (CA) double mutants (Rac1-L61F37A or Rac1-L61Y40C), Rac1-WT, or Rac1-DN and stained for F-actin to verify that the peptides were active (Fig. 2a). Both CA mutants increased F-actin reactivity compared to controls, Rac1-WT, and Rac1 DN. Rac1-DN reduced F-actin levels as expected. Staining against 4G8 antibody showed that both CA mutants enhanced the immunoreactivity of Aβ and/or its precursor compared to controls, Rac1-WT, and Rac1-DN (Fig. 2a). The second mutation of the CA proteins, F37A or Y40C, did not exert any differential effect, indicating that the observed effect on Aβ metabolism is dependent on the Q61L mutation, which the double mutants had in common. To determine which Aβ isoform was increasingly generated after the Rac1 peptide treatments, we performed immunoproteomic analysis, using surface-enhanced laser desorption/ionization time of flight mass spectrometry (SELDI-TOF MS). Human SH-SY5Y neuroblastoma cells were treated with TAT-Rac1 mutant proteins or vehicle: the lysate analysis revealed the presence of two main Aβ peptides, $A\beta_{1-42}$ and $A\beta_{11-42}$ pyr, after Rac1-L61F37A treatment (Fig. 2b).

To evaluate the directionality of the signalling, we also checked whether Aβ administration was able to modulate Rac1. We used synthetic $A\beta_{1-42}$, which we allowed to aggregate at 4 °C for 12 h. To test $A\beta_{1-42}$ toxicity, an MTT assay was performed (Additional file 1: Figure S2A). Mature cortical neurons, cultured for 10 days in vitro (DIV), were incubated for 24 h with different $A\beta_{1-42}$ concentrations, ranging from 0.1 μM to 1 μM. The preparation induced a significant toxicity at 0.1, 0.5, and 1 μM. A dot-blot assay verified the presence of a detectable oligomeric population, using two different antibodies: 6E10 and A11 (Additional file 1: Figure S2B). 6E10 is reactive to the amino acid residues 1–16 of Aβ. It is sequence-specific and recognizes both fibrillar and oligomeric forms as well as the precursor form. A11 antibody selectively recognizes amino acid sequence-independent oligomers but does not recognize monomers or mature fibers. All $A\beta_{1-42}$ samples were 6E10 and A11 positive, confirming that the preparations contained oligomers (representative image of the dot-blot assay).

Primary cortical neurons were treated with 0.1 μM $A\beta_{1-42}$ at different time points: 1 h, 3 h, 6 h or 24 h. After treatments, cells were fixed and immunostained against the active form of Rac1 protein (Rac1-GTP), neurofilaments, and F-actin. After 0.1 μM $A\beta_{1-42}$ administration, no differences were observed in Rac1 activation or localization in all the time points analysed.

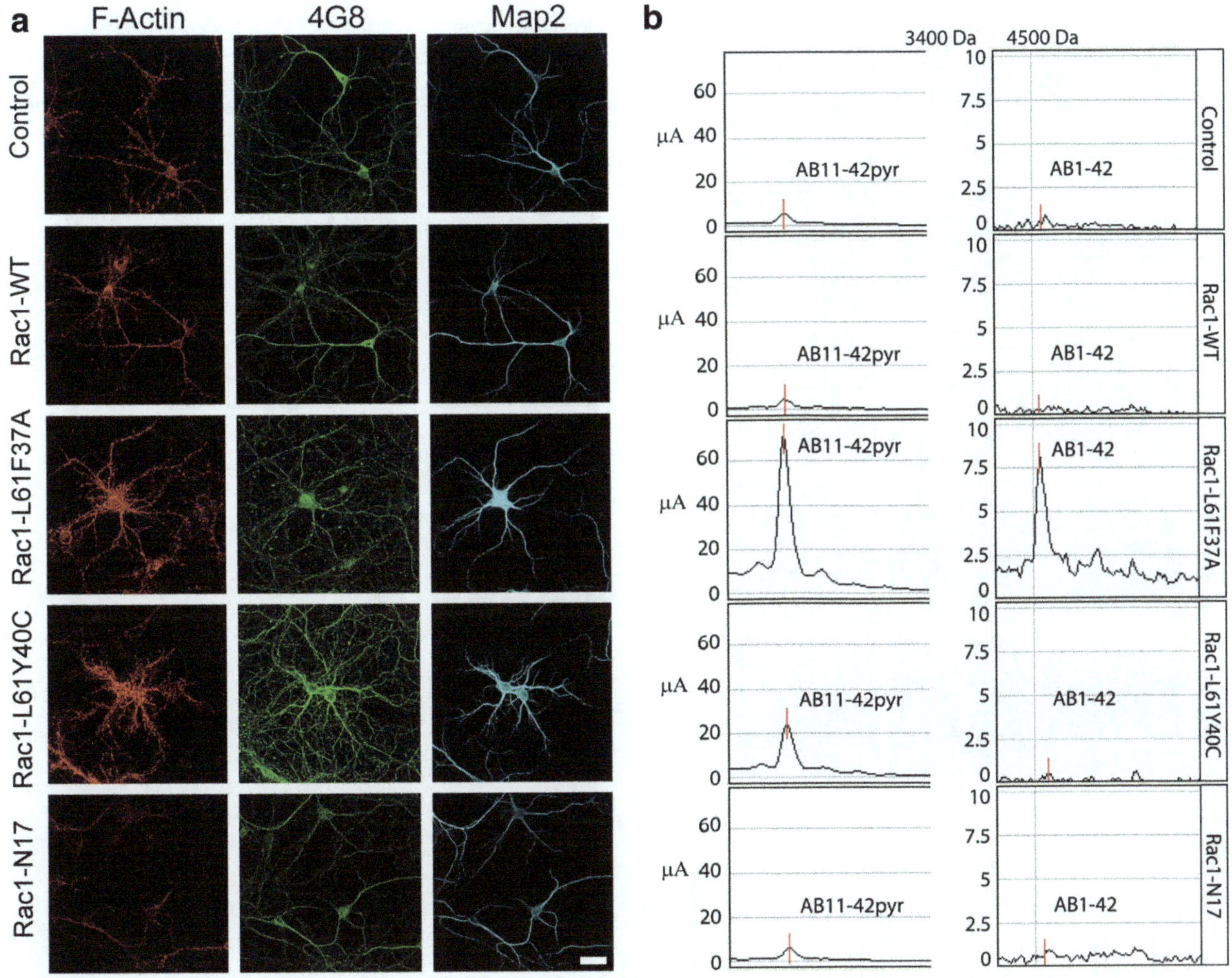

Fig. 2 Rac1 mutant peptides interfere with APP metabolism. **a** Primary cortical neurons treated between DIV11 and DIV14 with Rac1 mutant peptides. After 24 h, cells were stained to visualize of F-actin, APP and Aβ (4G8), and dendrites (MAP2). The increase in F-actin fluorescence was considered as a sign of Rac1 activation. These representative images were obtained from one of four independent experiments. Scale bar 30 μm. **b** Representative spectra of Aβ isoforms from SH-SY5Y cell lysates treated with TAT-Rac1 mutant proteins or vehicle ($n = 2$ experiments in triplicate). Peak intensity is expressed in μA (μAmpere). 3400 and 4500 are m/z reference values expressed in Dalton in the spectrum scale (m/z for Aβ 11–42pyr is 3325 Da, m/z for Aβ 1–42 is 4520 Da)

From the images acquired, F-actin seemed also not affected by the treatment (Additional file 1: Figure S2C). A few clots were observed along the neurites in the neurofilament staining after $A\beta_{1-42}$ administration, at both concentrations, confirming the toxicity measured with the MTT assay. These findings indicate that the interference with Rac1 signalling promotes an increased APP processing.

Rac1 perturbation affects SET translocation and results in tau hyperphosphorylation

Next, we evaluated whether Rac1 mutant peptides were able to interfere with tau phosphorylation. Neurons treated for 24 h with TAT-Rac1 mutants were stained against SET. SET, which is normally localized in the nucleus, translocated to the cytoplasm in AD brains. Its translocation from the nucleus to the cytoplasm was shown in AD temporal cortex and hippocampus, compared to age-matched controls [62]. Interestingly, the administration of both Rac1-WT and CA mutants was able to elicit the translocation of SET from the nucleus to the neurites (Fig. 3a). These results indicate that the protein itself is sufficient to induce SET translocation. To quantitative confirm this observation, we performed subcellular fractionation to purify the membranous and nuclear fractions. SET concentration in the membrane fraction significantly increased after Rac1-L61F37A and Rac1-WT treatments, while no change was observed in the nuclear fraction (Fig. 3b, c). Controls experiments to ensure the successful enrichment of the 2 fractions were

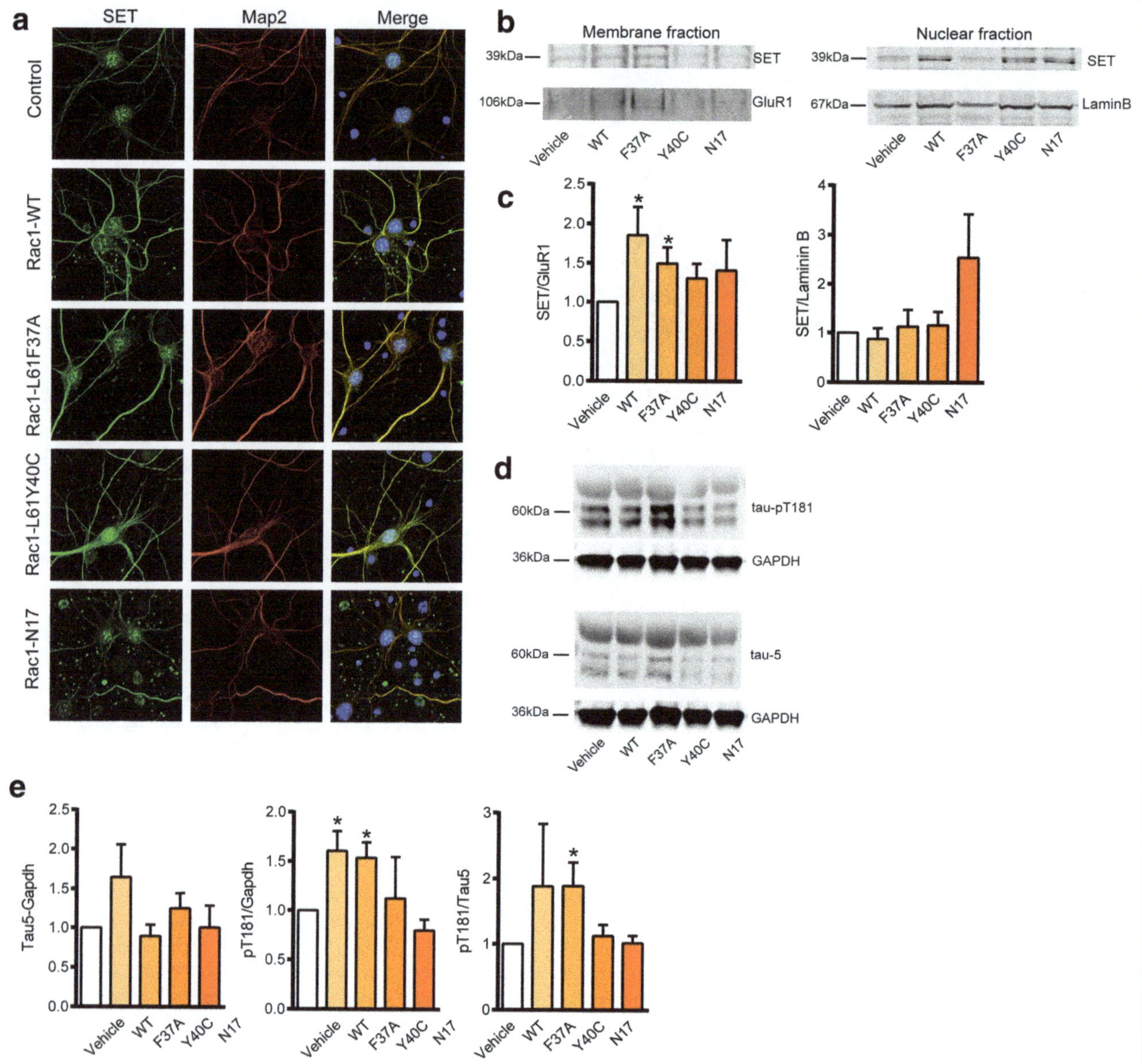

Fig. 3 Rac1 mutant peptides induce tau hyperphosphorylation mediated by SET translocation. **a** Primary cortical neurons treated between DIV11 and DIV14 with Rac1 mutant peptides. After 24 h, cells were fixed and stained to visualize SET, dendrites (MAP2), and nuclei (DAPI). These representative images were obtained from one of three independent experiments. Scale bar 10 µm. **b-c** Representative blots and densitometry of subcellular fractionation indicating the levels of SET in the membrane (SET/GluR1) and nuclear fractions (SET/LaminB) in the same conditions as in A. 15 µg of protein lysate were loaded for the nuclear fraction. Due to the low yield, the membrane fraction was not quantified. **d-e** Representative blots and corresponding quantification of tau pT181 phosphorylation and Tau-5 normalized against GAPDH levels, and pT181/Tau5. ~ 7 µg of protein lysate were loaded

performed in SH-SH5Y cells (Additional file 1: Figure S3). The abundancy of Lamin B was checked in the membrane fraction and the levels of GluR1 were assessed in the nuclear fraction. To check whether SET translocation resulted in an increased tau phosphorylation, cortical neurons were treated for 48 h with the peptides. We chose pT181 phospho-site as this is one of the major AD abnormally hyperphosphorylated sites regulated by PPA [66]. pT181/Tau5 was significantly increased after treatment with Rac1-L61F37A compared to vehicle treated cells (Fig. 3d, e). The use of a nuclear transporter inhibitor (LMB) reduced SET translocation from the nucleus to the membrane in SH-SY5Y when Rac1 peptides where administered. This impeded the increase in tau phosphorylation (Fig. 4).

As for Aβ, we checked whether tau-induced hyperphosphorylation altered Rac1 activation. We used okadaic acid (OA), a synthetic inhibitor of PP2A and PP1, which is a well-known tool to study AD pathology in vitro [46]. The evaluation of tau phosphorylation was performed for two of the main phosphorylated epitopes, pS262 and pS202 (Additional file 1: Figure S4A). Immunostaining after 6 h from OA treatment showed an enhanced tau immunoreactivity against both sites compared to the vehicle treated cells, with a pronounced accumulation in the somatodendritic compartment. The increased ratio pS262/tau was also detected via Western Blot at both time points (Additional file 1: Figure S4B, C). Neurons were also treated with 10 nM OA for 3 h or 6 h, and then analysed for Rac1-GTP by pull down assay. Rac1-GTP pull down assay showed no difference in the levels of activated proteins between OA treatment and control (Additional file 1: Figure S4D-E). In addition, the total expression of the protein was unchanged between the conditions. Overall, this data establishes a new direct pathway in which Rac1 induces SET translocation and, consequently, increases tau phosphorylation.

Rac1 is biphasically altered in 3xTg-AD mice

We investigated whether the reduced Rac1 expression observed in post-mortem AD brains was also recapitulated in a mouse model of familial AD. The 3xTg-AD model was selected. Pull-down assay for Rac1 and Rac1-GTP was performed to evaluate Rac1 levels and activation in the cortex and hippocampus of control (C57BL/6 J) and 3xTg-AD mice. We first checked in young animals, at 6 weeks, and found increased ratio Rac1-GTP/Rac1 in the hippocampus of 3xTg-AD mice compared to age-matched controls (Fig. 5a, b). We next evaluated how the levels of the protein changed over time at 3, 7, and 16 months. The analysis revealed a statistically significant decrease in total Rac1 in the cortex

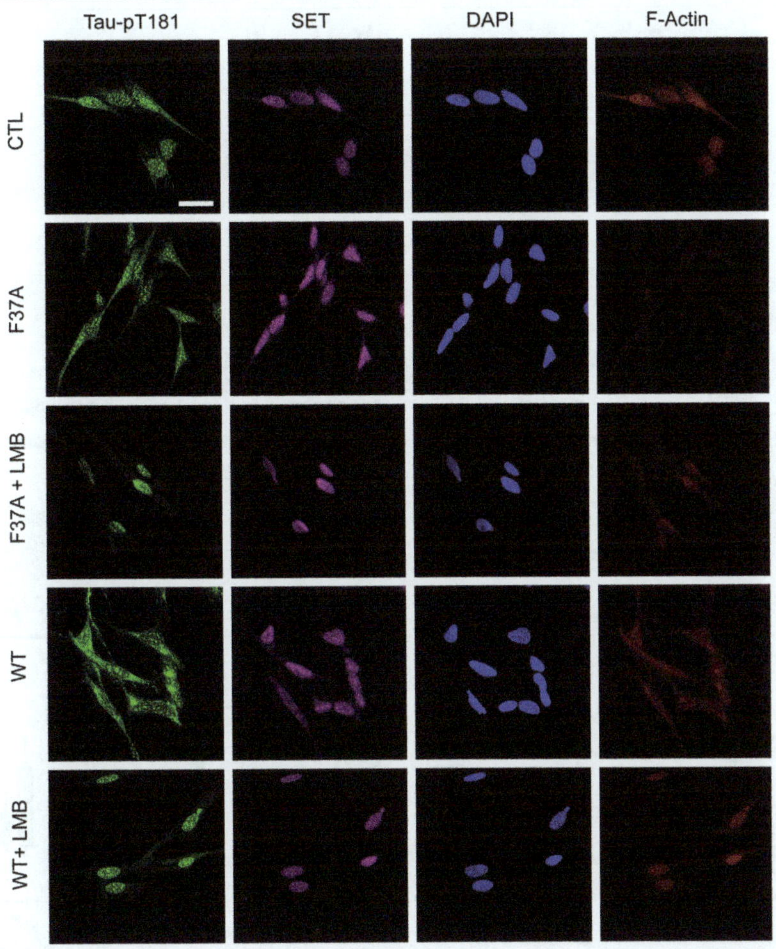

Fig. 4 The nuclear transporter inhibitor LMB blocks Rac1-induced translocation of SET. SH-SY5Y cells at 10 days of RA differentiation were treated with Rac1-WT and Rac1-L61F37A, with or without LMB in order to block Rac1-peptide mediated SET translocation. After 48 h, cells were fixed and stained against pT181 tau epitope (green), SET (magenta), F-actin (red), and DAPI (blue) was used to visualize nuclei. Representative images show that, in control condition SET expression is restricted to the nucleus. After Rac1-WT and Rac1-L61F37A treatments, SET presence is observed also outside cell nuclei, whereas SET translocation doesn't occur when LMB is added together with Rac1 mutant treatments. In the same way, tau phosphorylation at the epitope pT181 is decreased in presence of LMB. Scale bar 25 μm

of 3xTg-AD mice at 7-month-old compared to the controls. These findings suggest an abnormal activation of Rac1 at a very early stage of the pathology. This is followed by a decrease of the total level of the protein at a later stage, 7 months, when the cognitive impairment starts to become apparent according to published behavioral studies [60].

Rac1-L61F37A mutant peptide rescues spine loss in 3xTg-AD mice

Since Rac1 decreased in 7-month-old 3xTg-AD mice, we administered Rac1-L61F37A to evaluate its potential effect in ameliorating the known synaptic impairments [5]. We evaluated Rac1-L61F37A effect on the expression levels of PSD95 by Western Blot (Fig. 6a, b). Three experimental groups of animals were tested: C57BL/6 J mice treated with vehicle, 3xTg-AD mice treated with vehicle, and 3xTg-AD mice treated with Rac1-L61F37A ($n = 7$–9 animals per group). We observed a significant increase of the post-synaptic marker PSD95 in cortical homogenate of 7-month-old 3xTg-AD mice compared to controls. Importantly, after Rac1-L61F37A intranasal treatment, PSD95 levels normalized back to the control levels.

In order to analyze in more details the effect of Rac1-L61F37A mutant treatment, we evaluated the spine density in the 3 different groups. A total mean number of 5056 ± 1158.04 spines per animal were counted on mean dendrite length of 3903.49 ± 888.59 μm per animal, from 12 animals. Neuronal dendrites were acquired from four different cortical areas (primary motor cortex, premotor motor cortex, posterior parietal area and visual cortex) in each animal: C57BL/6 J treated with vehicle ($N = 4$), 3xTg-AD treated with vehicle (N = 4) and 3xTg-AD treated with Rac1-L61F37A (N = 4). 3xTg-AD mice showed a significant decrease in cortical spine

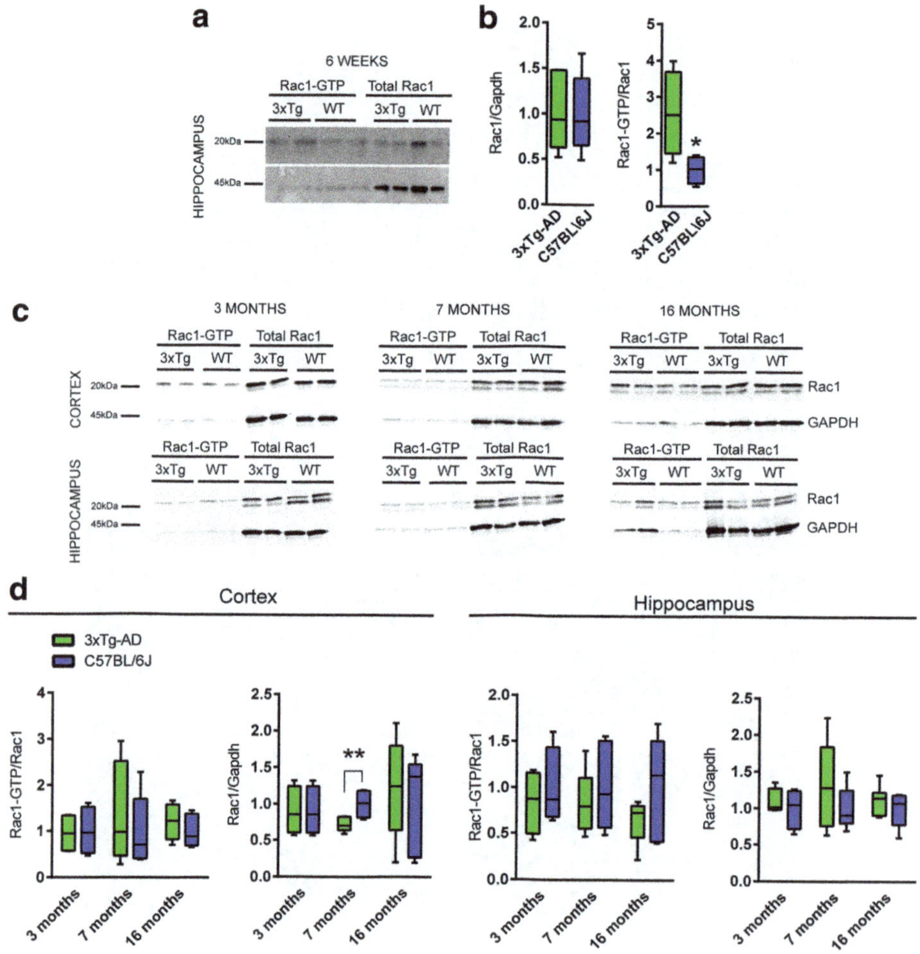

Fig. 5 Rac1 is altered in 3xTg-AD mice. **a**, **c** Representative blots of active and total Rac1 protein levels in homogenates of 3xTg-AD mice and age-matched control mice. **b**, **d** Quantification of Rac1 protein activation (ratio between Rac1-GTP level and total Rac1), and total Rac1 level (total Rac1 on GAPDH ratio), at the 4 time points (6 weeks: C57BL/6 J $n = 6$, 3xTG-AD n = 6; 3 months: C57BL/6 J $n = 4$, 3xTG-AD n = 4; 7 months: C57BL/6 J $N = 7$, 3xTG-AD N = 7; 16 months: C57BL/6 J $n = 5$, 3xTG-AD n = 5). Box and whiskers graphs represent boxes as min to max values, whiskers as standard errors and means as black lines

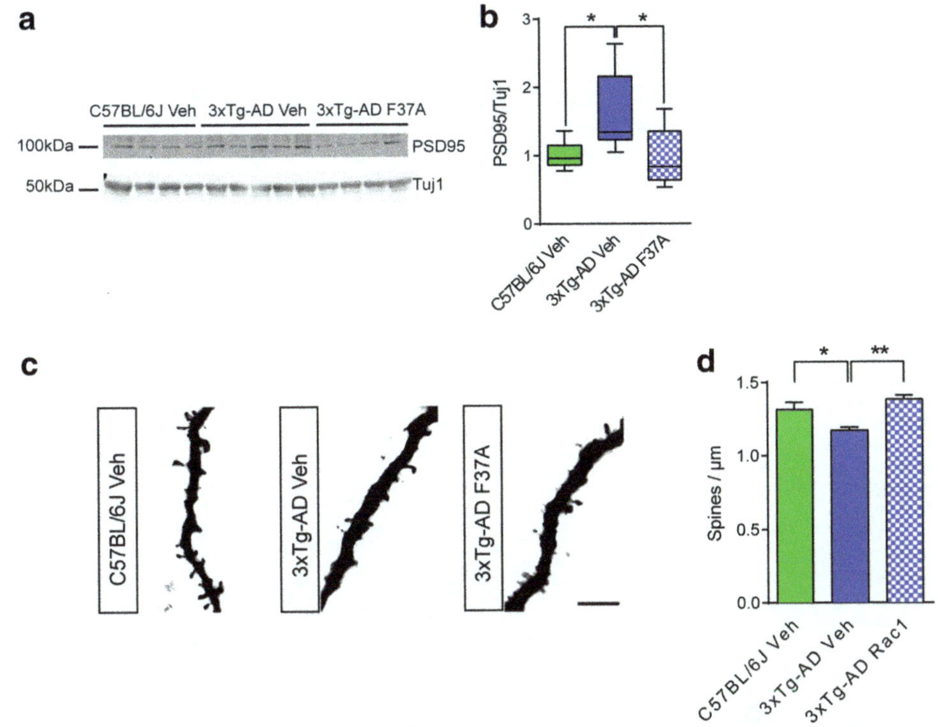

Fig. 6 Rac1 administration rescued dendritic impairment. **a** Representative immunoblots and (**b**) densitometry of PSD95 were normalized on Tuj1 in cortical homogenates of control mice (C57BL/6 J treated with vehicle) and 3xTg-AD mice treated with vehicle or with Rac1-L61F37A mutant peptide (100 μM) at 7 months old. Box and whiskers graphs represent boxes as min to max values, whiskers as standard errors and means as black lines. 10 μg of protein lysate were loaded. **c** Representative Golgi-cox stained dendrite portions with spines and (**d**) quantification of spine density (numbers of spine per μm of dendrite length) in the three condition analyzed (C57BL/6 J treated with vehicle: 1.32 ± 0.05 spines/μm; 3xTg-AD treated with vehicle: 1.17 ± 0.02 spines/μm; and 3xTg-AD treated with vehicle: 1.39 ± 0.03 spines/μm). The data represented are mean ± SEM

density at 7-month-old respect to control mice (Fig. 6c-d). Rac1-L61F37A intranasal treatment in 3xTg-AD increased significantly spine density respect to age-match vehicle-treated mice, restoring spine density at the same level of control mice. This data indicates the beneficial effect of timely activating Rac1 signaling to reverse spine and synaptic abnormalities in a disease-relevant animal model.

Pathways analysis indicates the interactome connecting Rac1 to tau and APP

To further characterize functional interactions between Rac1 and AD relevant proteins we performed ANAT analysis (Fig. 7a). In the resulting network, Rac1 is connected to the GWAS identified genes PAK2 [23], CHN2 [19], and IQGAP2 [76]. PAK2 is a well know Rac1 activator, CHN2 is an inhibitor upon EGF receptor stimulation in fibroblast-like cell lines cells [14], and IQGAP2 modulates Rac1 activity [13]. These proteins are all involved in cytoskeleton reorganization [21].

To deepen the analysis and to expand interactomic consistency to the network, the list of 720 GWAS identified genes, plus SET, PP2A and the 182 high confidence interactors were submitted to STRING. Of the 269 GWAS genes ANAT did not recognize, STRING recognized only 20, confirming the set of GWAS genes of interest. The interactions of Rac1 with PAK2, IQGAP2, and CHN2 were confirmed.

The neighborhood of Rac1, PP2A, and SET was selected from the STRING interactome. As recommended by STRING for higher confidence, we kept edges with a "STRING combined score" greater or equal to 0.7. The resulting network of 113 genes included also APP and TAU. The 113 set of genes were analyzed with ClueGO [7] for a Gene Ontology biological process (BP) evaluation. The first 10 BP classes sorted by number of interested genes resulting from the analysis contain between 81 and 31 genes and 4 of them are cytoskeleton-related: "adherens junction assembly" (47 genes), positive regulation of protein complex assembly (43 genes), dendritic spine development (34 genes) and stress fiber assembly (31 genes).

To increase the resolution of the analysis, we calculated the shortest paths connecting Rac1, SET, PP2A, and APP or tau. According to STRING results, these proteins are connected through CDK5 and HSP90AA1

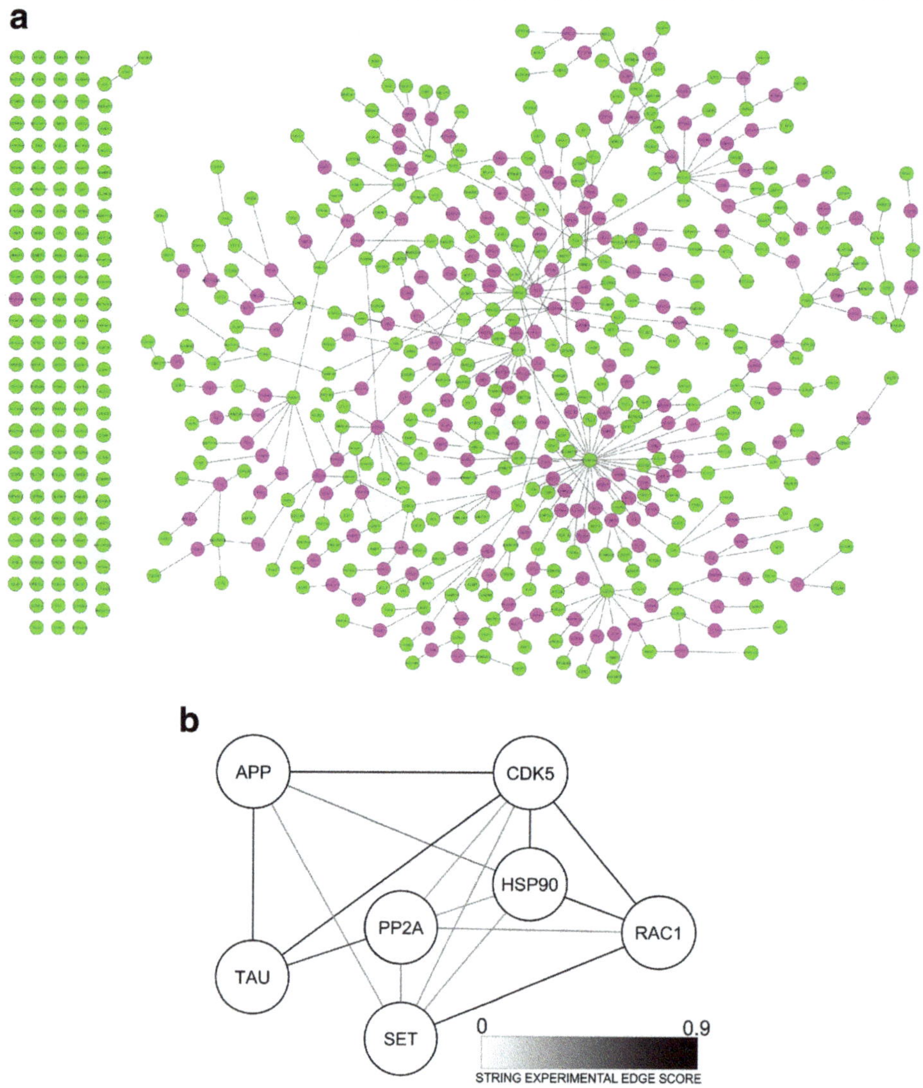

Fig. 7 Rac1 is linked to PP2A in protein network analysis. **a** ANAT network starting from the AD GWAS genes plus SET and PP2A (in green) leading to the addition of high-confidence connecting genes (in purple). **b** Rac1, SET, PP2A shortest paths to either APP or Tau according to the STRING network based on GWAS AD plus ANAT high confidence interactors genes. Edges score opacity mapping STRING experimental score

(HSP90A), both added by ANAT to the genes set to build a possible high confidence interactome (Fig. 7b). CDK5 is related to several of the aforementioned proteins. HSP90A was recently identified as a promising AD treatment target [8] and a HSP90A knockout has been associated with Rac1 down regulation [20]. It also interferes with PP2A mediated AKT phosphorylation with implications for AD [35].

CDK5 silencing has a Rac1-mediated neuroprotective effect [54]. SET is required for the stimulatory effect on the CDK5 region p35(nck5a) and CDK5 phosphorylates APP and TAU [26, 29], and it has been suggested that PP2A might act in functional association with GSK3 in tau hyperphosphorylation [53].

Discussion

Members of the Rho-GTPase family have been previously connected to pathogenic events contributing to synaptic deficits in AD. Our data describes a putative pathway in which Rac1 is up-stream and timely elicits the alterations of AD relevant proteins. The selective activation of Rac1 signalling enhanced Aβ levels and promoted SET translocation from the nucleus to the plasma membrane. The directionality of this alteration was confirmed as Aβ administration and tau-induced hyperphosphorylation did not perturb Rac1 cellular distribution or activation. At the same time, Rac1 increased in young 3xTg-AD mice and later decreased at 7 months. At this latter time point, the intranasal administration of

Rac1 active peptide restored spine loss. This indicates that the involvement of Rac1 in AD pathological cascade of events is rather complex. The contribution of Rac1 to these apparently contradicting pathways might be explained with the tight spatiotemporal regulation of the Rho-GTPases. Previous studies showing contrasting data on the levels of Rac1 in AD versus age-matched controls might be reconsidered in this light. Rac1 was reported to be increased [49], or decreased [36] in AD autoptic brain samples. A closer look into the pathological diagnoses of the mentioned studies shows that Rac1 increased when patients with mild AD were selected [49], meanwhile it decreased in samples with NTF stage V or VI [36] because of the extensive neuronal death. In the present study, we found a reduction of Rac1 protein levels in human AD brain. This decrease was accompanied by an increased protein plasma levels in AD patients with the most severe cognitive decline (MMSE < 18). In addition, Rac1 plasma levels weakly correlated with the cognitive decline in AD, thus suggesting that this protein might represent a marker of AD disease progression: further investigation are mandatories to confirm these preliminary results. At this stage, therapeutic intervention boosting Rac1 signalling to support spine maintenance might represent an interesting option. 3xTg-AD mice treated for 2 weeks at 6.5 months with Rac1-L61F37A showed a rescue of spine deficits. Both male and female 3xTg-AD mice showed a subtle deficit in spatial learning and memory exactly at 6.5 months of age [60], this underlying the spine impairment. Rac1-L61F37A peptide was previously shown to boost cell survival and regeneration after optic nerve crush by the activation of the Pak\MEK\Erk pathway [30]. The protective effect might also be ascribed to the release of neurotrophic factors as activation of Erk1/2 resulted in the secretion of endogenous CNTF [40]. Importantly, intranasal treatment with Rac1-L61F37A did not significantly interfere with tau phosphorylation and APP processing when administered in 3xTg-AD (data not shown). Rac1- L61F37A also normalized the levels of PSD95 proteins in 3xTg-AD compared to 3xTg-AD treated with vehicle. It was previously reported that PSD95 decreased in 3xTg-AD 7 month-old animals [55]. One of the reasons for this discrepancy might lie in the different loading controls used (Tuj1 in this study versus actin in Revilla et al.). Since we administered an actin modulating protein, Tuj1 seemed a better choice. Moreover, many papers have described how AD impairs actin stability [48] and its levels might change over the course of the pathology.

The pathway analysis offered a high-level view of the pathways connecting Rac1 to AD relevant proteins and highlighted the strong interaction between Rac1 and tau through SET and PP2A. The use of mutant peptides allowed us to better dissect Rac1 signaling, which is executed by several effectors. In these mutants, the L61 mutation, which tonically activated the protein, was coupled to a second mutation (F37A or Y40C) that gave signal specificity [30]. The Y40C blocked the binding to PAK and JNK mediated pathways meanwhile, F37A activated them. The specific effect of Rac1-L61F37A on tau hyperphosphorylation might be mediated by the effector protein PAK. Rac1-induced PAK activation has been shown to activate p38MAPK [75], which phosphorylates tau [72].

The reduction of PP2A activity via SET has been shown to affect APP regulation [24]. Coherently, we observed that Rac1-L61F37A was also the most effective mutant in determining an increase of Aβ fragments 11–42pyr and 1–42. Overexpression of both C and N terminals of SET in rats determined Aβ accumulation starting from 4-month old rats [9]. When we tested whether Rac1 could be altered following Aβ administration, we could not observe any impairment. Other studies used synthetic Aβ peptide and showed a consequent Rac1 activation. However, in these studies, the used Aβ concentration was in the μM range (above 1 μM) [34, 39]. Concentrations higher than 1 μM have been defined by many as "supraphysiological" [22] and the data obtained thus require careful consideration.

Despite the use of Rac1 mutant peptides, which allowed different signaling cascades to be triggered, they do not provide insights into the kinetic of the activation. As already proposed [50], Rho-GTPases alteration needs to be studied with tools allowing to follow their spatiotemporal dynamics. The different functions of Rac1 suggest a highly controlled regulation, which is also dependent on the cellular compartment. In this regard, new imaging tools based on sophisticated fluorescent biosensors can help to resolve the dynamics of these proteins, which are activated on a micrometer length and sub-minute time scales [41, 71, 73]. These tools might highlight even more subtle defects in either their compartmentalization or crosstalk between the family members.

Additional studies are clearly necessary to further unravel this intricate signaling. Elucidating the molecular mechanisms underlying the loss of spines is certainly essential for the development of disease-modifying therapeutics. Moreover, the possibility of further investigating Rho-GTPase members as potential indicator of disease progression for AD in plasma represents an interesting option. We showed here that Rac1 increased in AD patients with MMSE< 18 and, in a recent work, that Cdc42 decreased in fronto-temporal dementia patients [56].

Conclusion

Rac1 might have a role in AD as a triggering co-factor, participating both to Aβ and tau alteration. However, at a later stage of the pathology, it might represent a potential therapeutic target due to its beneficial effect on dendritic spine dynamics.

Additional file

Additional file 1: Figure S1. Rac1 mutant peptides have high penetration due to the TAT sequence. (**A-C**) Representative confocal images of cortical neurons treated at DIV3 with different concentrations of TAT-GFP: 5 μM (**A**), 10 μM (**B, C**). After treatment, cells were fixed and stained for visualization of dendrites (MAP2) and nuclei (DAPI). Confocal analysis showed that TAT-GFP was internalized (single plane), also in live cells directly imaged 1h after treatment. Scale bars 10 μm. (**D**) MTT assay on primary cortical neurons after 24h from the administration of 2 μM Rac1 mutant peptides. The cell viability is expressed as % as compared to control. The data represented are mean ±SEM of four independent experiments, each done in triplicate. **Figure S2.** $A\beta_{1-42}$ administration does not interfere with Rac1 localization or activation. (**A**) MTT assay on primary cortical neurons after 24h $A\beta_{1-42}$ treatment at the indicated concentrations The Aβ peptide suspension was incubated 12h at 4°C prior treatment. The cell viability is expressed as % as compared to control. The data represented are mean ±SEM of four independent experiments, each done in triplicate. One-sample *t* test to a hypothetical mean of 100% was performed. (**B**) Representative dot-blot analysis of $A\beta_{1-42}$ preparations with 6E10 and A11 antibodies. The protein concentration was 0.12 μg for 6E10 and 0.72 μg for A11 (**C**) Representative confocal images of primary cortical neurons treated with 0.1 μM $A\beta_{1-42}$ between DIV11 and DIV14. Cells were stained against Rac1-GTP, F-actin, and neurofilament. Scale bars 30 μm. **Figure S3.** Efficacy of the subcellular fractionation. Representative blots of the subcellular fractionation experiments showing the levels of GluR1, LaminB, and SET in the membrane and nuclear fractions of SH-SY5Y cells. Four independent samples were assessed for the 2 fractions. **Figure S4.** Tau induced hyperphosphorylation does not alter Rac1 levels or activation. (**A**) Representative confocal pictures of mature cortical neurons treated with 10nM OA for 6h and immunostained against pS262 tau. Scale bar 30 μm. (**B-C**) Tau pS262 phosphorylation was analysed by western blot after 3 and 6h from OA administration. The data represented are mean with SEM of four or six independent experiments (3h treatment n=6, 6h treatment n=4). (**D-E**) Rac1-GTP pull done assay was performed after 3 and 6h from OA administration. The data represented are mean with SEM of three independent experiments. ns, not significant. Asterisks indicate unspecific bands.

Acknowledgements
We thank prof. C. Laudanna and members of his laboratory for invaluable help in the preparations of Rac1 mutant peptides.

Funding
This work was supported by a grant from the Italian Ministry of Health, Convenzione n.173/GR-2011-02348526 and Ricerca Corrente. SZ is partially supported by FEDER, project "Digital Analytics Platform".

Authors' contributions
RG, LB, AP designed the experiments on plasma samples. SF and GB recruited and clinically evaluated the patients. MC and GDF designed the experiments on human brains. CS carried out the human plasma assays and the analyses of the Aβ peptides. VP performed the in vitro experiments with Aβ and OA treatments. MiBo performed the animal analysis and the in vitro treatments with mutant Rac1 peptides. EL produced the Rac1 mutant peptides and performed the internalization experiments with TAT-GFP. MiBo, CS, MC, and PV wrote materials, methods, and results of their corresponding part. MaBu, GZ, LB, and RG critically revised the manuscript. SZ performed the bioinformatics analysis. SB conceived the project, coordinated the study, and finalized the manuscript. All authors reviewed and approved the final manuscript.

Competing interests
SB is co-founder of the spin-off company from the University of Luxembourg Braingineering Technologies s.a.r.l.

Author details

[1]Department of Neurosciences, Biomedicine and Movement Sciences, University of Verona, Strada Le Grazie, 8, 37134 Verona, Italy. [2]Molecular Markers Laboratory, IRCCS Istituto Centro San Giovanni di Dio Fatebenefratelli, Via Pilastroni, 4, 25125 Brescia, Italy. [3]Division of Neurology 5 and Neuropathology, IRCCS Foundation - Carlo Besta Neurological Institute, Via Celoria 11, 20133 Milan, Italy. [4]MAC Memory Center, IRCCS Istituto Centro San Giovanni di Dio Fatebenefratelli, Brescia, Italy. [5]Environmental Research and Innovation (ERIN) Department, Luxembourg Institute of Science and Technology (LIST), Belvaux L-4422, Luxembourg. [6]Luxembourg Centre for Systems Biomedicine, University of Luxembourg, 6 avenue du Swing, L-4367 Belvaux, Luxembourg.

References

1. Aguilar BJ, Zhu Y, Lu Q (2017) Rho GTPases as therapeutic targets in Alzheimer's disease. Alzheimers Res Ther 9:97. https://doi.org/10.1186/s13195-017-0320-4
2. Albertini V, Bruno A, Paterlini A, Lista S, Benussi L, Cereda C, Binetti G, Ghidoni R (2010) Optimization protocol for amyloid-beta peptides detection in human cerebrospinal fluid using SELDI TOF MS. Proteomics Clin Appl 4:352–357
3. Almozlino Y, Atias N, Silverbush D, Sharan R (2017) ANAT 2.0: reconstructing functional protein subnetworks. BMC Bioinformatics 18:495. https://doi.org/10.1186/s12859-017-1932-1
4. Arnaud L, Chen S, Liu F, Li B, Khatoon S, Grundke-Iqbal I, Iqbal K (2011) Mechanism of inhibition of PP2A activity and abnormal hyperphosphorylation of tau by I2(PP2A)/SET. FEBS Lett 585:2653–2659. https://doi.org/10.1016/j.febslet.2011.07.020
5. Baazaoui N, Iqbal K (2017) Prevention of dendritic and synaptic deficits and cognitive impairment with a neurotrophic compound. Alzheimers Res Ther 9:45. https://doi.org/10.1186/s13195-017-0273-7
6. Bamburg JR, Bloom GS (2009) Cytoskeletal pathologies of Alzheimer disease. Cell Motil Cytoskeleton 66:635–649. https://doi.org/10.1002/cm.20388
7. Bindea G, Mlecnik B, Hackl H, Charoentong P, Tosolini M, Kirilovsky A, Fridman WH, Pages F, Trajanoski Z, Galon J (2009) ClueGO: a Cytoscape plug-in to decipher functionally grouped gene ontology and pathway annotation networks. Bioinformatics 25:1091–1093. https://doi.org/10.1093/bioinformatics/btp101
8. Blair LJ, Sabbagh JJ, Dickey CA (2014) Targeting Hsp90 and its co-chaperones to treat Alzheimer's disease. Expert Opin Ther Targets 18:1219–1232. https://doi.org/10.1517/14728222.2014.943185
9. Bolognin S, Blanchard J, Wang X, Basurto-Islas G, Tung YC, Kohlbrenner E, Grundke-Iqbal I, Iqbal K (2012) An experimental rat model of sporadic Alzheimer's disease and rescue of cognitive impairment with a neurotrophic peptide. Acta Neuropathol 123:133–151. https://doi.org/10.1007/s00401-011-0908-x
10. Bolognin S, Lorenzetto E, Diana G, Buffelli M (2014) The potential role of rho GTPases in Alzheimer's disease pathogenesis. Mol Neurobiol 50:406–422. https://doi.org/10.1007/s12035-014-8637-5
11. Bolognin S, Zatta P, Lorenzetto E, Valenti MT, Buffelli M (2013) Beta-amyloid-aluminum complex alters cytoskeletal stability and increases ROS production in cortical neurons. Neurochem Int 62:566–574. https://doi.org/10.1016/j.neuint.2013.02.008
12. Bolomini-Vittori M, Montresor A, Giagulli C, Staunton D, Rossi B, Martinello M, Constantin G, Laudanna C (2009) Regulation of conformer-specific activation of the integrin LFA-1 by a chemokine-triggered rho signaling module. Nat Immunol 10:185–194. https://doi.org/10.1038/ni.1691
13. Brill S, Li S, Lyman CW, Church DM, Wasmuth JJ, Weissbach L, Bernards A, Snijders AJ (1996) The Ras GTPase-activating-protein-related human protein IQGAP2 harbors a potential actin binding domain and interacts with calmodulin and rho family GTPases. Mol Cell Biol 16:4869–4878
14. Caloca MJ, Wang H, Delemos A, Wang S, Kazanietz MG (2001) Phorbol esters and related analogs regulate the subcellular localization of beta 2-chimaerin, a non-protein kinase C phorbol ester receptor. J Biol Chem 276:18303–18312. https://doi.org/10.1074/jbc.M011368200
15. Cerri C, Fabbri A, Vannini E, Spolidoro M, Costa M, Maffei L, Fiorentini C, Caleo M (2011) Activation of rho GTPases triggers structural remodeling and functional plasticity in the adult rat visual cortex. J Neurosci 31:15163–15172. https://doi.org/10.1523/JNEUROSCI.2617-11.2011
16. Falke E, Nissanov J, Mitchell TW, Bennett DA, Trojanowski JQ, Arnold SE (2003) Subicular dendritic arborization in Alzheimer's disease correlates with neurofibrillary tangle density. Am J Pathol 163:1615–1621. https://doi.org/10.1016/S0002-9440(10)63518-3
17. Forner S, Baglietto-Vargas D, Martini AC, Trujillo-Estrada L, LaFerla FM (2017) Synaptic impairment in Alzheimer's disease: a dysregulated symphony. Trends Neurosci 40:347–357. https://doi.org/10.1016/j.tins.2017.04.002

18. Fu AK, Ip NY (2017) Regulation of postsynaptic signaling in structural synaptic plasticity. Curr Opin Neurobiol 45:148–155. https://doi.org/10.1016/j.conb.2017.05.016
19. Furney SJ, Simmons A, Breen G, Pedroso I, Lunnon K, Proitsi P, Hodges A, Powell J, Wahlund LO, Kloszewska I et al (2011) Genome-wide association with MRI atrophy measures as a quantitative trait locus for Alzheimer's disease. Mol Psychiatry 16:1130–1138. https://doi.org/10.1038/mp.2010.123
20. Ghosh S, Shinogle HE, Garg G, Vielhauer GA, Holzbeierlein JM, Dobrowsky RT, Blagg BS (2015) Hsp90 C-terminal inhibitors exhibit antimigratory activity by disrupting the Hsp90alpha/Aha1 complex in PC3-MM2 cells. ACS Chem Biol 10:577–590. https://doi.org/10.1021/cb5008713
21. Gutierrez-Uzquiza A, Colon-Gonzalez F, Leonard TA, Canagarajah BJ, Wang H, Mayer BJ, Hurley JH, Kazanietz MG (2013) Coordinated activation of the Rac-GAP beta2-chimaerin by an atypical proline-rich domain and diacylglycerol. Nat Commun 4:1849. https://doi.org/10.1038/ncomms2834
22. Haass C, Selkoe DJ (2007) Soluble protein oligomers in neurodegeneration: lessons from the Alzheimer's amyloid beta-peptide. Nat Rev Mol Cell Biol 8:101–112. https://doi.org/10.1038/nrm2101
23. Harrison TM, Mahmood Z, Lau EP, Karacozoff AM, Burggren AC, Small GW, Bookheimer SY (2016) An Alzheimer's disease genetic risk score predicts longitudinal thinning of hippocampal complex subregions in healthy older adults. eNeuro 3. https://doi.org/10.1523/ENEURO.0098-16.2016
24. Holzer M, Bruckner MK, Beck M, Bigl V, Arendt T (2000) Modulation of APP processing and secretion by okadaic acid in primary guinea pig neurons. J Neural Transm 107:451–461. https://doi.org/10.1007/s007020070087
25. Hyman BT, Phelps CH, Beach TG, Bigio EH, Cairns NJ, Carrillo MC, Dickson DW, Duyckaerts C, Frosch MP, Masliah E et al (2012) National Institute on Aging-Alzheimer's Association guidelines for the neuropathologic assessment of Alzheimer's disease. Alzheimers Dement 8:1–13. https://doi.org/10.1016/j.jalz.2011.10.007
26. Kimura T, Ishiguro K, Hisanaga S (2014) Physiological and pathological phosphorylation of tau by Cdk5. Front Mol Neurosci 7:65. https://doi.org/10.3389/fnmol.2014.00065
27. Koleske AJ (2013) Molecular mechanisms of dendrite stability. Nat Rev Neurosci 14:536–550. https://doi.org/10.1038/nrn3486
28. Lamarche N, Tapon N, Stowers L, Burbelo PD, Aspenstrom P, Bridges T, Chant J, Hall A (1996) Rac and Cdc42 induce actin polymerization and G1 cell cycle progression independently of p65PAK and the JNK/SAPK MAP kinase cascade. Cell 87:519–529
29. Liu F, Su Y, Li B, Zhou Y, Ryder J, Gonzalez-DeWhitt P, May PC, Ni B (2003) Regulation of amyloid precursor protein (APP) phosphorylation and processing by p35/Cdk5 and p25/Cdk5. FEBS Lett 547:193–196
30. Lorenzetto E, Ettorre M, Pontelli V, Bolomini-Vittori M, Bolognin S, Zorzan S, Laudanna C, Buffelli M (2013) Rac1 selective activation improves retina ganglion cell survival and regeneration. PLoS One 8:e64350. https://doi.org/10.1371/journal.pone.0064350
31. Maderna E, Cattaneo C, Cacciatore F, Catania M, Di Fede G, Tagliavini F, Giaccone G (2015) Divergent cognitive status with the same Braak stage of neurofibrillary pathology: does the pattern of amyloid-beta deposits make the difference? J Alzheimers Dis 43:375–379. https://doi.org/10.3233/JAD-140540
32. Mairet-Coello G, Courchet J, Pieraut S, Courchet V, Maximov A, Polleux F (2013) The CAMKK2-AMPK kinase pathway mediates the synaptotoxic effects of Abeta oligomers through tau phosphorylation. Neuron 78:94–108. https://doi.org/10.1016/j.neuron.2013.02.003
33. Maloney MT, Bamburg JR (2007) Cofilin-mediated neurodegeneration in Alzheimer's disease and other amyloidopathies. Mol Neurobiol 35:21–44
34. Manterola L, Hernando-Rodriguez M, Ruiz A, Apraiz A, Arrizabalaga O, Vellon L, Alberdi E, Cavaliere F, Lacerda HM, Jimenez S et al (2013) 1–42 beta-amyloid peptide requires PDK1/nPKC/Rac 1 pathway to induce neuronal death. Transl Psychiatry 3:e219. https://doi.org/10.1038/tp.2012.147
35. Martin L, Latypova X, Wilson CM, Magnaudeix A, Perrin ML, Terro F (2013) Tau protein phosphatases in Alzheimer's disease: the leading role of PP2A. Ageing Res Rev 12:39–49. https://doi.org/10.1016/j.arr.2012.06.008
36. Matsui C, Inoue E, Kakita A, Arita K, Deguchi-Tawarada M, Togawa A, Yamada A, Takai Y, Takahashi H (2012) Involvement of the gamma-secretase-mediated EphA4 signaling pathway in synaptic pathogenesis of Alzheimer's disease. Brain Pathol 22:776–787. https://doi.org/10.1111/j.1750-3639.2012.00587.x
37. McKhann G, Drachman D, Folstein M, Katzman R, Price D, Stadlan EM (1984) Clinical diagnosis of Alzheimer's disease: report of the NINCDS-ADRDA work group under the auspices of Department of Health and Human Services Task Force on Alzheimer's disease. Neurology 34:939–944
38. McKhann GM, Knopman DS, Chertkow H, Hyman BT, Jack CR Jr, Kawas CH, Klunk WE, Koroshetz WJ, Manly JJ, Mayeux R et al (2011) The diagnosis of dementia due to Alzheimer's disease: recommendations from the National Institute on Aging-Alzheimer's Association workgroups on diagnostic guidelines for Alzheimer's disease. Alzheimers Dement 7:263–269. https://doi.org/10.1016/j.jalz.2011.03.005
39. Mendoza-Naranjo A, Gonzalez-Billault C, Maccioni RB (2007) Abeta1-42 stimulates actin polymerization in hippocampal neurons through Rac1 and Cdc42 rho GTPases. J Cell Sci 120:279–288. https://doi.org/10.1242/jcs.03323
40. Muller A, Hauk TG, Leibinger M, Marienfeld R, Fischer D (2009) Exogenous CNTF stimulates axon regeneration of retinal ganglion cells partially via endogenous CNTF. Mol Cell Neurosci 41:233–246. https://doi.org/10.1016/j.mcn.2009.03.002
41. Murakoshi H, Wang H, Yasuda R (2011) Local, persistent activation of rho GTPases during plasticity of single dendritic spines. Nature 472:100–104. https://doi.org/10.1038/nature09823
42. Neuman KM, Molina-Campos E, Musial TF, Price AL, Oh KJ, Wolke ML, Buss EW, Scheff SW, Mufson EJ, Nicholson DA (2015) Evidence for Alzheimer's disease-linked synapse loss and compensation in mouse and human hippocampal CA1 pyramidal neurons. Brain Struct Funct 220:3143–3165. https://doi.org/10.1007/s00429-014-0848-z
43. Oddo S, Caccamo A, Shepherd JD, Murphy MP, Golde TE, Kayed R, Metherate R, Mattson MP, Akbari Y, LaFerla FM (2003) Triple-transgenic model of Alzheimer's disease with plaques and tangles: intracellular Abeta and synaptic dysfunction. Neuron 39:409–421
44. Otth C, Mendoza-Naranjo A, Mujica L, Zambrano A, Concha II, Maccioni RB (2003) Modulation of the JNK and p38 pathways by cdk5 protein kinase in a transgenic mouse model of Alzheimer's disease. Neuroreport 14:2403–2409. https://doi.org/10.1097/01.wnr.0000099988.54721.3c
45. Palop JJ, Chin J, Mucke L (2006) A network dysfunction perspective on neurodegenerative diseases. Nature 443:768–773. https://doi.org/10.1038/nature05289
46. Pei JJ, Gong CX, An WL, Winblad B, Cowburn RF, Grundke-Iqbal I, Iqbal K (2003) Okadaic-acid-induced inhibition of protein phosphatase 2A produces activation of mitogen-activated protein kinases ERK1/2, MEK1/2, and p70 S6, similar to that in Alzheimer's disease. Am J Pathol 163:845–858. https://doi.org/10.1016/S0002-9440(10)63445-1
47. Penzes P, Rafalovich I (2012) Regulation of the actin cytoskeleton in dendritic spines. Adv Exp Med Biol 970:81–95. https://doi.org/10.1007/978-3-7091-0932-8_4
48. Penzes P, Vanleeuwen JE (2011) Impaired regulation of synaptic actin cytoskeleton in Alzheimer's disease. Brain Res Rev 67:184–192. https://doi.org/10.1016/j.brainresrev.2011.01.003
49. Perez SE, Getova DP, He B, Counts SE, Geula C, Desire L, Coutadeur S, Peillon H, Ginsberg SD, Mufson EJ (2012) Rac1b increases with progressive tau pathology within cholinergic nucleus basalis neurons in Alzheimer's disease. Am J Pathol 180:526–540. https://doi.org/10.1016/j.ajpath.2011.10.027
50. Pertz O (2010) Spatio-temporal rho GTPase signaling - where are we now? J Cell Sci 123:1841–1850. https://doi.org/10.1242/jcs.064345
51. Petersen RC, Smith GE, Waring SC, Ivnik RJ, Tangalos EG, Kokmen E (1999) Mild cognitive impairment: clinical characterization and outcome. Arch Neurol 56:303–308
52. Petratos S, Li QX, George AJ, Hou X, Kerr ML, Unabia SE, Hatzinisiriou I, Maksel D, Aguilar MI, Small DH (2008) The beta-amyloid protein of Alzheimer's disease increases neuronal CRMP-2 phosphorylation by a rho-GTP mechanism. Brain 131:90–108. https://doi.org/10.1093/brain/awm260
53. Plattner F, Angelo M, Giese KP (2006) The roles of cyclin-dependent kinase 5 and glycogen synthase kinase 3 in tau hyperphosphorylation. J Biol Chem 281:25457–25465. https://doi.org/10.1074/jbc.M603469200
54. Posada-Duque RA, Lopez-Tobon A, Piedrahita D, Gonzalez-Billault C, Cardona-Gomez GP (2015) p35 and Rac1 underlie the neuroprotection and cognitive improvement induced by CDK5 silencing. J Neurochem 134:354–370. https://doi.org/10.1111/jnc.13127
55. Revilla S, Sunol C, Garcia-Mesa Y, Gimenez-Llort L, Sanfeliu C, Cristofol R (2014) Physical exercise improves synaptic dysfunction and recovers the loss of survival factors in 3xTg-AD mouse brain. Neuropharmacology 81:55–63. https://doi.org/10.1016/j.neuropharm.2014.01.037
56. Saraceno C, Catania M, Paterlini A, Fostinelli S, Ciani M, Zanardini R, Binetti G, Di Fede G, Caroppo P, Benussi L et al (2018) Altered expression of circulating Cdc42 in frontotemporal lobar degeneration. J Alzheimers Dis 61:1477–1483. https://doi.org/10.3233/JAD-170722

57. Scheff SW, Price DA, Schmitt FA, DeKosky ST, Mufson EJ (2007) Synaptic alterations in CA1 in mild Alzheimer disease and mild cognitive impairment. Neurology 68:1501–1508. https://doi.org/10.1212/01.wnl.0000260698.46517.8f
58. Sekino Y, Kojima N, Shirao T (2007) Role of actin cytoskeleton in dendritic spine morphogenesis. Neurochem Int 51:92–104. https://doi.org/10.1016/j.neuint.2007.04.029
59. Serrano-Pozo A, Frosch MP, Masliah E, Hyman BT (2011) Neuropathological alterations in Alzheimer disease. Cold Spring Harbor Perspect Med 1: a006189. https://doi.org/10.1101/cshperspect.a006189
60. Stover KR, Campbell MA, Van Winssen CM, Brown RE (2015) Early detection of cognitive deficits in the 3xTg-AD mouse model of Alzheimer's disease. Behav Brain Res 289:29–38. https://doi.org/10.1016/j.bbr.2015.04.012
61. Switzer CH, Cheng RY, Vitek TM, Christensen DJ, Wink DA, Vitek MP (2011) Targeting SET/I(2)PP2A oncoprotein functions as a multi-pathway strategy for cancer therapy. Oncogene 30:2504–2513. https://doi.org/10.1038/onc.2010.622
62. Tanimukai H, Grundke-Iqbal I, Iqbal K (2005) Up-regulation of inhibitors of protein phosphatase-2A in Alzheimer's disease. Am J Pathol 166:1761–1771. https://doi.org/10.1016/S0002-9440(10)62486-8
63. ten Klooster JP, Leeuwen I, Scheres N, Anthony EC, Hordijk PL (2007) Rac1-induced cell migration requires membrane recruitment of the nuclear oncogene SET. EMBO J 26:336–345. https://doi.org/10.1038/sj.emboj.7601518
64. Terry RD, Masliah E, Salmon DP, Butters N, DeTeresa R, Hill R, Hansen LA, Katzman R (1991) Physical basis of cognitive alterations in Alzheimer's disease: synapse loss is the major correlate of cognitive impairment. Ann Neurol 30:572–580. https://doi.org/10.1002/ana.410300410
65. Tolias KF, Duman JG, Um K (2011) Control of synapse development and plasticity by rho GTPase regulatory proteins. Prog Neurobiol 94:133–148. https://doi.org/10.1016/j.pneurobio.2011.04.011
66. Wang JZ, Grundke-Iqbal I, Iqbal K (2007) Kinases and phosphatases and tau sites involved in Alzheimer neurofibrillary degeneration. Eur J Neurosci 25: 59–68. https://doi.org/10.1111/j.1460-9568.2006.05226.x
67. Wang PL, Niidome T, Akaike A, Kihara T, Sugimoto H (2009) Rac1 inhibition negatively regulates transcriptional activity of the amyloid precursor protein gene. J Neurosci Res 87:2105–2114. https://doi.org/10.1002/jnr.22039
68. Wang X, Blanchard J, Kohlbrenner E, Clement N, Linden RM, Radu A, Grundke-Iqbal I, Iqbal K (2010) The carboxy-terminal fragment of inhibitor-2 of protein phosphatase-2A induces Alzheimer disease pathology and cognitive impairment. FASEB J 24:4420–4432. https://doi.org/10.1096/fj.10-158477
69. Wang X, Blanchard J, Tung YC, Grundke-Iqbal I, Iqbal K (2015) Inhibition of protein phosphatase-2A (PP2A) by I1PP2A leads to hyperphosphorylation of tau, neurodegeneration, and Cognitive Impairment in Rats. J Alzheimers Dis 45:423–435. https://doi.org/10.3233/JAD-142403
70. Welter D, MacArthur J, Morales J, Burdett T, Hall P, Junkins H, Klemm A, Flicek P, Manolio T, Hindorff L et al (2014) The NHGRI GWAS catalog, a curated resource of SNP-trait associations. Nucleic Acids Res 42:D1001–D1006. https://doi.org/10.1093/nar/gkt1229
71. Yasuda R, Murakoshi H (2011) The mechanisms underlying the spatial spreading of signaling activity. Curr Opin Neurobiol 21:313–321. https://doi.org/10.1016/j.conb.2011.02.008
72. Yoshida H, Goedert M (2006) Sequential phosphorylation of tau protein by cAMP-dependent protein kinase and SAPK4/p38delta or JNK2 in the presence of heparin generates the AT100 epitope. J Neurochem 99:154–164. https://doi.org/10.1111/j.1471-4159.2006.04052.x
73. Yoshizaki H, Ohba Y, Kurokawa K, Itoh RE, Nakamura T, Mochizuki N, Nagashima K, Matsuda M (2003) Activity of rho-family GTPases during cell division as visualized with FRET-based probes. J Cell Biol 162:223–232. https://doi.org/10.1083/jcb.200212049
74. Yu Z, Huang Z, Lung ML (2013) Subcellular fractionation of cultured human cell lines. Bio-protocol 3:e754. https://doi.org/10.21769/BioProtoc.754
75. Zhang S, Han J, Sells MA, Chernoff J, Knaus UG, Ulevitch RJ, Bokoch GM (1995) Rho family GTPases regulate p38 mitogen-activated protein kinase through the downstream mediator Pak1. J Biol Chem 270:23934–23936
76. Zhu JB, Tan CC, Tan L, Yu JT (2017) State of play in Alzheimer's disease genetics. J Alzheimers Dis 58:631–659. https://doi.org/10.3233/JAD-170062

14

Phosphorylation of different tau sites during progression of Alzheimer's disease

Joerg Neddens[1], Magdalena Temmel[1], Stefanie Flunkert[1], Bianca Kerschbaumer[1,2], Christina Hoeller[1,3], Tina Loeffler[1], Vera Niederkofler[1], Guenther Daum[2], Johannes Attems[4] and Birgit Hutter-Paier[1*]

Abstract

Alzheimer's disease is characterized by accumulation of amyloid plaques and tau aggregates in several cortical brain regions. Tau phosphorylation causes formation of neurofibrillary tangles and neuropil threads. Phosphorylation at tau Ser202/Thr205 is well characterized since labeling of this site is used to assign Braak stage based on occurrence of neurofibrillary tangles. Only little is known about the spatial and temporal phosphorylation profile of other phosphorylated tau (ptau) sites. Here, we investigate total tau and ptau at residues Tyr18, Ser199, Ser202/Thr205, Thr231, Ser262, Ser396, Ser422 as well as amyloid-β plaques in human brain tissue of AD patients and controls. Allo- and isocortical brain regions were evaluated applying rater-independent automated quantification based on digital image analysis. We found that the level of ptau at several residues, like Ser199, Ser202/Thr205, and Ser422 was similar in healthy controls and Braak stages I to IV but was increased in Braak stage V/VI throughout the entire isocortex and transentorhinal cortex. Quantification of ThioS-stained plaques showed a similar pattern. Only tau phosphorylation at Tyr18 and Thr231 was already significantly increased in the transentorhinal region at Braak stage III/IV and hence showed a progressive increase with increasing Braak stages. Additionally, the increase in phosphorylation relative to controls was highest at Tyr18, Thr231 and Ser199. By contrast, Ser396 tau and Ser262 tau showed only a weak phosphorylation in all analyzed brain regions and only minor progression. Our results suggest that the ptau burden in the isocortex is comparable between all analyzed ptau sites when using a quantitative approach while levels of ptau at Tyr18 or Thr231 in the transentorhinal region are different between all Braak stages. Hence these sites could be crucial in the pathogenesis of AD already at early stages and therefore represent putative novel therapeutic targets.

Keywords: Microtubule-associated protein tau, Phosphorylation, Cingulate, Frontal, Occipital and temporal cortex, Transentorhinal region, Immunofluorescent labeling

Introduction

Alzheimer's disease (AD) is neuropathologically characterized by two hallmark lesions, which are extracellular amyloid-β (Aβ) plaques and intracellular accumulations of abnormally phosphorylated tau. Aβ plaques initially develop in neocortical regions and then progress to the limbic system, subcortical nuclei and reach the cerebellum at late stages of the disease [41]. Tau pathology manifests as neurofibrillary tangles (NFTs) and neuropil threads (NTs) and primarily accumulates in the entorhinal region and subsequently progresses to the limbic system and neocortical regions as reflected by NFT Braak stages [8]. Tau aggregation depends on several posttranslational modifications, including but not limited to, truncation, acetylation, ubiquitination, sumoylation and phosphorylation [13, 29, 34]. The best analyzed posttranslational modification in AD is abnormal phosphorylation of tau which in AD is referred to as hyperphosphorylation and that is characterized by an at least 3-fold increase of tau phosphorylation relative to controls. Over 70 potential tau phosphorylation (ptau) sites spanning almost the entire protein structure and including some phosphorylation sites are assumed to be pathologically relevant [40]. Some of these ptau sites are known to be abnormally phosphorylated in paired helical filaments (PHFs), NFTs or NTs during progression of AD but are not

* Correspondence: birgit.hutter-paier@qps.com
[1]QPS Austria GmbH, Neuropharmacology, Parkring 12, 8074 Grambach, Austria
Full list of author information is available at the end of the article

phosphorylated in healthy brains [10, 15, 22, 26, 28]. Several of these ptau sites are also phosphorylated in the fetal brain and are thus associated with embryonic brain developmental processes [10, 15, 22]. In the diseased brain, ptau sites are commonly associated with tau aggregation processes such as incomplete binding and destabilization of microtubules, causing the transition from pre-tangles to NFT formation [12, 16, 21, 28, 32].

The temporal and spatial phosphorylation pattern of tau residue Ser202/Thr205 has already been well characterized since staging of AD cases is based on labeling of ptau Ser202/Thr205 using the specific antibody AT8 [1, 9, 20, 46]. However, the phosphorylation profile of other ptau sites known to be involved in AD pathology is so far strongly neglected and analyzed only by ELISA and dot blots of AD brain tissue [20, 46].

Phosphorylation at several residues during disease progression has been repeatedly investigated in tau animal models but barely in AD brain tissue [4, 7, 14, 20, 46]. Hence, the present study aims to evaluate the spatial and temporal phosphorylation profile of different tau sites by applying rater-independent automated quantification based on digital image analysis. We therefore quantified the expression of total tau, amyloid-β plaques and the phosphorylation of seven different tau sites (Tyr18, Ser199, Ser202/Thr205, Thr231, Ser262, Ser396, Ser422) in iso- and allocortical brain regions of AD cases and controls.

Materials and methods

Human brain samples

Paraffin sections of 6 μm thickness and frozen samples from AD patients of three different disease stages [Braak stages I/II, III/IV and V/VI; [9]] and control individuals with Braak stage 0 were provided by the Newcastle Brain Tissue Resource (NBTR), Newcastle University, UK in accordance with the approval of the joint Ethics Committee of Newcastle and North Tyneside Health Authority and following NBTR brain banking procedures. The neuropathological diagnosis was performed according to

Table 1 Individual case information. Demographic patient information

	Case	Age	Sex	Braakstage	Post mortem delay to fixation in hours	Fixation time in weeks	Clinical information	# on Western blot
Controls	1	68	M	0	54	7	Cognitively normal	1
	2	55	M	0	41	11	Cognitively normal	2
	3	70	M	0	72	6	Cognitively normal	–
	4	78	F	0	34	8	Cognitively normal	4
	5	73	M	0	25	9	Cognitively normal	–
	mean	68.8		0	45.2	8.2		
Braak I/II Low	6	96	F	2	114	49	Mild dementia	6
	7	77	M	2	83	15	Cognitively normal	7
	8	94	F	2	15	9	Cognitively normal	–
	9	70	M	2	39	7	Multiple psychiatric and physical problems	9
	10	74	F	1	49	10	Cognitively normal	10
	mean	82.2		1.8	60	18		
Braak III/IV Mod.	11	75	M	4	82	23	Cognitively normal	11
	12	79	M	3	13	15	Cognitively normal	12
	13	81	M	3	82	8	Unspecified dementia	13
	14	98	F	3	59	8	Cognitively normal	14
	15	91	M	3	48	9	Moderate cognitive impairment and vascular disease	15
	mean	84.8		3.2	56.8	12.6		
Braak V/VI High	16	84	F	6	47	16	Severe dementia, anxiety and depression	–
	17	77	F	6	63	5	Dementia	17
	18	80	F	6	32	16	Dementia	18
	19	86	F	6	5	6	Dementia	19
	20	89	F	6	85	8	Dementia	–
	mean	83.2		6	46.4	10.2		

internationally accepted criteria [31] (Table 1). Tissue of one patient of Braak stage V/VI yielded clearly stronger pathology compared to the other cases, representing an overshoot outlier. Data of this patient were excluded for pThr231 calculations. For histological analyses, five cases per group were investigated using sections of 5 brain regions per individual: temporal cortex (TeCtx), frontal cortex (FrCtx), cingulate cortex (CiCtx), occipital cortex (OcCtx), and transentorhinal region (TEntR). All histological assessments were done blinded to clinical and neuropathological data.

Immunofluorescent labeling

Sections were deparaffinized for 10 min in Tissue Clear (Sakura, 1466, Netherlands) and 5 min in Tissue Clear/ 100% ethanol, washed for 5 min in 100% ethanol and then subsequently rehydrated with decreasing alcohol concentrations (96, 70 and 50% ethanol for 2 min each). Thereafter, sections were washed twice for 5 min in PBS. Sections for labeling of human total tau, ThioflavinS (ThioS), pThr231 tau, pSer199 tau, pTyr18 tau and pSer396 tau were treated for 15 min with citrate buffer (Thermo Scientific, AP-9003) at 95 °C in a steamer and cooled down to room temperature (RT) for another 15 min. For ThioS staining, sections were then labeled with 0.5% ThioS (Sigma-Aldrich, T1892) for 7 min under light protection. Afterwards, sections were pretreated with ice-cold sodium borohydride/PBS solution (1 mg/ml; Sigma-Aldrich, 213,462) for 4 min and tissue was then permeabilized with 1% Triton® X-100/PBS (AppliChem, A1388) for 10 min. Non-specific labeling was blocked by incubating sections for 30–60 min either with 10% donkey serum/PBS or 10% horse serum/PBS. For antigen detection, sections were incubated with primary antibodies (Table 2) in a damp chamber. Primary antibody binding was visualized by incubating sections with secondary fluorophore conjugated antibodies (Table 3) for 60 min. Furthermore, cell nuclei of tissue labeled for pSer199 tau, pSer396 tau and pSer422 tau were visualized by counterstaining with 4′,6-Diamidin-2-phenylindol-working solution (DAPI, AppliChem, A1001) for 15 min, differentiated with 70% ethanol and washed for 5 min in PBS. Finally, sections were washed in ddH_2O and covered with Moviol and coverslips.

In order to use the human sections efficiently, double labelings were performed: pSer202/Thr205 combined with pSer262 (Additional file 1: Online Source 9); ThioS combined with HT7; pSer199 combined with pSer396. Labelings against pTyr18, pThr231 and pSer422 were performed without co-labeling of another ptau residue.

Imaging and image analysis

Of each labeled brain section two mosaic images with a size of approximately 3 mm^2 including white and grey matter were captured at different z-levels and finally projected to 2D.

Imaging of immunofluorescent labelings of pSer202/ Thr205 tau, pSer199 tau, pSer396 tau, pSer262 tau and pSer422 tau was performed using a Nikon Eclipse E400 microscope with a high aperture lens (20× lens, numerical aperture 0.5, 1× optocoupler), equipped with a Nikon DS-Qi1MC camera and controlled by the Nikon NIS-Elements AR software. For quantification of total human tau, pThr231 tau, pTyr18 tau and ThioS images were recorded using a Zeiss AxioImager.Z1 microscope with a high aperture lens and an AxioVision 4.8 software-driven AxioCam MRm digital camera (20× lens, numeric aperture 0.8, 1× optocoupler).

Quantitative image analysis was performed using Image-Pro Plus (version 6.2, Media Cybernetics, Inc., Rockville, USA). Grey-scale single channel images were corrected for background intensities using lowpass filtering, and signal from autofluorescent objects (mostly lipofuscin and erythrocytes) was subtracted from the channel used for immunofluorescent labeling. Objects were identified by a combination of Edge+ filter, adequate thresholding, and size and shape restrictions (Additional file 1: Online Source 10). After defining the parameters for detecting the targeted objects, macro-driven quantitative image analysis ran automatically using the same parameters on all images. The

Table 2 List of primary antibodies for histology

Antigen	Clone	Species	Source	Order #	Dilution	Incubation time
Total human tau	HT7	mouse	Pierce Biotechnology, USA	MN1000	1:500	overnight at 4 °C
pThr231	AT180	mouse	Pierce Biotechnology, USA	MN1040	1:200	overnight at 4 °C
pSer202 + pThr205	AT8	mouse	Thermo Scientific, USA	MN1020	1:200	overnight at 4 °C
pSer199	polyclonal	rabbit	Invitrogen Corporation, USA	44-734G	1:500	1 h at RT
pTyr18	9G3	mouse	MédiMabs, Canada	MM-0194-P	1:1000	1 h at RT
pSer396	PHF13	mouse	Cell Signaling Technology, USA	9632	1:500	1 h at RT
pSer262	polyclonal	rabbit	MBL International, USA	AT-5020	1:250	overnight at 4 °C
pSer422	5.6.11	mouse	Roche Diagnostic, Switzerland	N/A	1:800	overnight at 4 °C

Table 3 List of secondary antibodies

Antibody	Conjugation	Source	Order #	To visualize
Donkey Anti-Mouse IgG H&L	AlexaFluor 555	abcam, UK	ab150110	pSer422 tau
				pSer396 tau
				pSer202/ pThr205 tau
Donkey Anti-Rabbit IgG H&L	DyLight 650	abcam, UK	ab96922	pSer199 tau
				pSer262 tau
Donkey Anti-Mouse IgG H&L	DyLight 550	abcam, UK	ab98795	pThr231 tau
Donkey Anti-Mouse IgG (H + L)	Cyanine Cy3	Jackson ImmunoResearch, USA	715–165-151	Total human tau
Donkey Anti-Mouse IgG H&L	DyLight 650	abcam, UK	ab98797	pTyr18 tau

Dilution of all secondary antibodies was 1:500

quantitative results are therefore unbiased, rater-independent and fully reproducible.

Preparation of sarcosyl insoluble fraction from frozen samples

Frozen transentorhinal cortex tissue was homogenized in 9 μl per mg of cold extraction buffer 1 (25 mM Tris HCl pH = 7.4, 150 mM NaCl, 1 mM EDTA, 1 mM EGTA, 10 mM ß-glycerophosphate, 30 mM NaF, 2 mM Na3VO4, protease and phosphatase inhibitor cocktails) and centrifuged at 80,000 g for 15 min at 4 °C. The pellet was resuspended in extraction buffer 2 (10 mM Tris HCl pH = 7.4, 800 mM NaCl, 300 mM sucrose, 1 mM EGTA, protease and phosphatase inhibitor cocktails) and centrifuged at 4000 g for 10 min at 4 °C. The Supernatant was transferred to a fresh tube and sarcosyl (30% aqueous solution) was added to a final concentration of 1% and incubated for 1.5 h at room temperature. After centrifugation at 80,000 g for 30 min at 4 °C, the supernatant was discarded and the pellet resuspended in buffer 3 (50 mM Tris-HCl, pH = 7.4). The suspension was aliquoted and frozen until used for Western blotting.

Western blot analyses

Equal amounts of proteins of each brain extract were separated by molecular weight on a SDS-PAGE polyacrylamide gel. A protein marker visualized correct separation of the proteins and confirmed the correct protein band size. Subsequently, proteins were transferred onto a 0.45 μm nitrocellulose membrane using a wet blot chamber (Bio-Rad, Hercules, USA), blocked with 5% non-fat dry milk in 1 × TBS and incubated with primary antibodies overnight. To blot for total tau, pTyr18, and pSer199 the same primary antibodies as used for histological labeling were used in a concentration of 1:1000 (Table 2). Antibodies AT180 and AT8 to label pThr231 and pSer202/Thr205, respectively, did not give any signal in the Western blot and were thus substituted by a rabbit polyclonal antibody against pThr231 (1:1000; Signalway, SAB1110–2, College Park, USA) and a rabbit monoclonal antibody, clone EPR2402, against pSer202/Thr205 (1:3000; abcam, ab108387, Cambridge, UK). All primary antibodies were incubated for 2 h at room temperature. Afterwards, membranes were washed in TBS and incubated in horseradish peroxidase-coupled secondary antibodies for 1 h at room temperature (1:5000; donkey-anti-rabbit IgG: NA934/ GENA934; sheep-anti-mouse IgG: NXA931; GE-Healthcare, Little Chalfont, UK). Proteins were detected using Wester-Bright ECL spray (Advansta, Menlo Park, USA) and placed on a X-ray film. As loading control a rabbit polyclonal GAPDH antibody (Sigma-Aldrich, G9545, St. Louis, USA) was co-labeled on each blot. To blot with 6 different antibodies, three blots were used to prevent false signals due to stripping, reblotting or blocking of antibody binding by previously used antibody. For each of the three blots, GAPDH as loading control is shown. Densitometry of all bands was performed using Image J software. Cases 3, 5, 8, 16 and 20 according to Table 1 are missing in this analysis since no frozen tissue was available from the corresponding brain region.

Statistical analyses

All statistical analyses and preparation of graphs were conducted using Graph-Pad Prism (version 4.03, San Diego, CA, USA). Descriptive statistical analyses were performed on all evaluated parameters including the evaluation of normal distribution using the Kolmogorov-Smirnov test.

Group variances were calculated either by one-way or two-way ANOVA. If a significant interaction among groups was detected, Newman-Keuls or Bonferroni's *post-hoc* analysis was followed. A detailed description of performed statistical analyses is given in the appropriate figure legend. Data were averaged and represented as mean + standard error of mean (SEM). An α-error level of $p < 0.05$ was considered significant.

Results

In order to evaluate the overall human tau expression levels, brain tissues were labeled with the anti-human

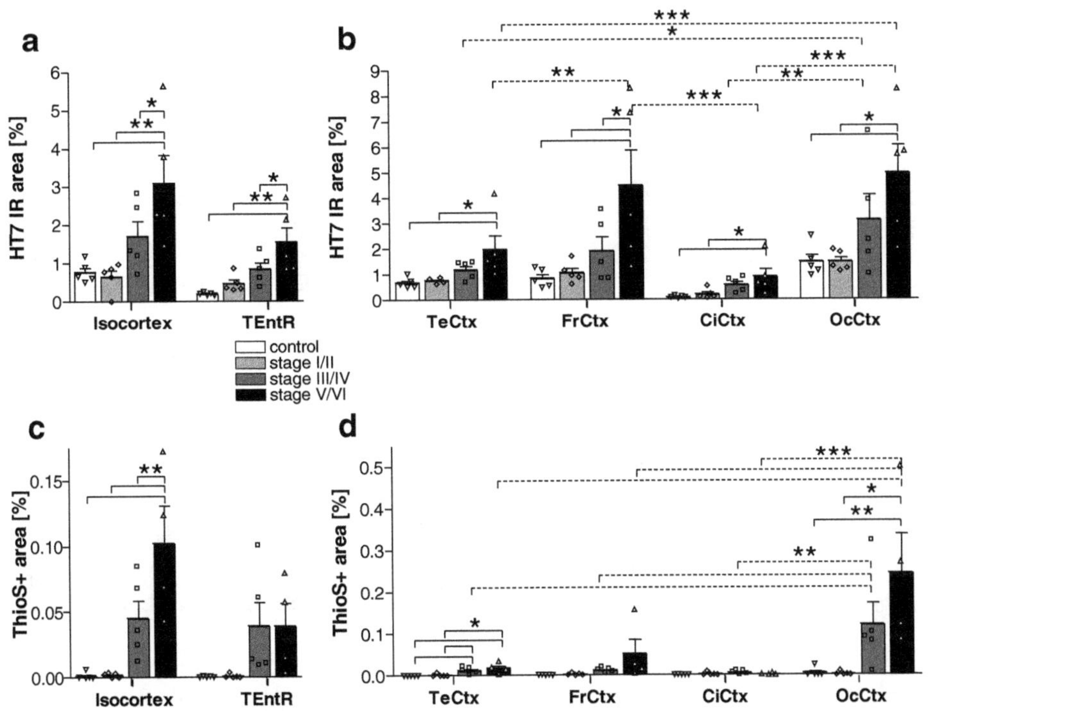

Fig. 1 Quantification of total human tau and plaque load in the cortex of AD cases. **a** Total human tau (HT7 antibody) immunoreactive area in percent in the isocortex and TEntR (allocortex) of different AD stages. **c** β sheet (ThioflavinS staining) positive area in percent in the isocortex and TEntR of different AD stages. Isocortical data in **a** and **c** present mean of different cortical regions that are shown separately in **b** and **d**, respectively. Two way ANOVA followed by Bonferroni's *posthoc* test. Mean + SEM; $N = 5$.*$p < 0.05$; **$p < 0.01$; ***$p < 0.001$. **c**, **d** One outlier of Braak stage V/VI excluded for all ThioS labelings. Solid lines: Comparison of AD stages within a brain region; dotted lines: comparison of brain regions of the same AD Braak stage. CiCtx: cingulate cortex; FrCtx: frontal cortex; OcCtx: occipital cortex; TeCtx: temporal cortex; TEntR: transentorhinal region

total tau HT7 antibody. Quantification of the HT7 immunoreactive area in the isocortex and transentorhinal region (TEntR) revealed a progressive increase of total tau with advancing Braak stages. This increase in total tau signal was significant at Braak stage V/VI compared to healthy controls but also compared to both lower Braak stages I/II and III/IV (Fig. 1a). Data of the isocortex in Fig. 1a are the mean of four distinct isocortical regions, namely temporal, frontal, cingulate and occipital cortex. Separate presentation of these brain regions shows that total tau expression levels were very low in the cingulate cortex but relatively high in the occipital cortex of healthy controls (Fig. 1b). In all isocortical regions total tau levels started to increase at Braak stage III/IV and were significantly higher at Braak stage V/VI compared to controls or Braak stage I/II. At Braak stage V/VI total tau levels were highest in the frontal and occipital cortex (Fig. 1b). Since amyloid-β plaques are, next to tau, a major pathological hallmark of AD, the ThioflavinS (ThioS) stained area was quantified in the isocortex and TEntR of AD patients and healthy controls (Fig. 1c). Threshold-based identification of ThioS staining combined with a size restriction filter enabled automatic quantification of plaque cores while excluding NFTs (Additional file 1: Online Source 11). We found that ThioS signal was significantly increased in the isocortex of Braak stage V/VI patients compared to controls and all other Braak stages; a first non-significant increase in the isocortical ThioS signal could be observed at Braak stage III/IV (Fig. 1c). In the TEntR, ThioS signal slightly increased at Braak stage III/IV but did not further increase at Braak V/VI. Almost no ThioS signal could be observed in the cingulate cortex. In the temporal and frontal cortex the ThioS signal was very low while only in the occipital cortex the ThioS signal was well measurable and progressively increasing starting at Braak stage III/IV (Fig. 1d).

After evaluating the overall total tau and plaque load, we measured the phosphorylation levels of ptau at seven different residues. First, we show phosphorylation of pSer202/Thr205 tau since this phosphorylation site is already well characterized and thus can be utilized to validate the results.

pSer202/Thr205 tau

Phosphorylation of tau at Ser202/Thr205 was low in all isocortical brain regions up to Braak stage III/IV. Only at

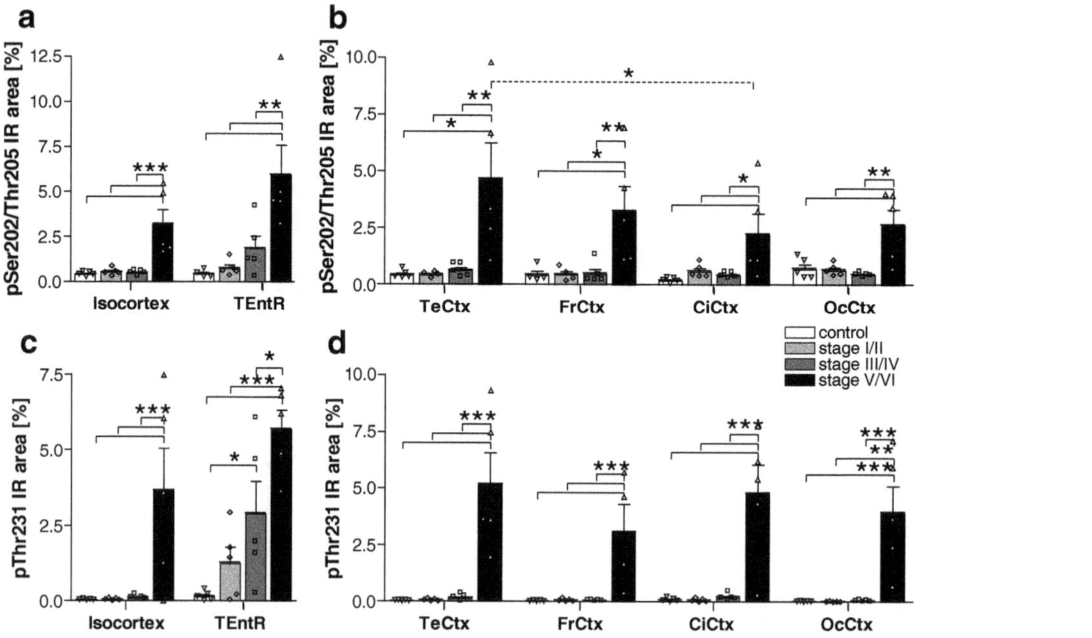

Fig. 2 Quantification of tau phosphorylation at pSer202/Thr205 and pThr231 in the cortex of AD cases. **a** Human pSer202/Thr205 tau immunoreactive area in percent in the isocortex and TEntR (allocortex) of different AD stages. **c** Human pThr231 tau immunoreactive area in percent in the isocortex and TEntR of different AD stages. Isocortical data in **a** and **c** present mean of different cortical regions that are shown separately in **b** and **d**, respectively. Two way ANOVA followed by Bonferroni's *posthoc* test. Mean + SEM; $N = 5$.$^*p < 0.05$; $^{**}p < 0.01$; $^{***}p < 0.001$. Solid lines: Comparison of AD stages within a brain region; dotted lines: comparison of brain regions of the same AD Braak stage. CiCtx: cingulate cortex; FrCtx: frontal cortex; OcCtx: occipital cortex; TeCtx: temporal cortex; TEntR: transentorhinal region

Braak stage V/VI Ser202/Thr205 tau phosphorylation significantly increased about 4- to 13-fold above control levels (Fig. 2a, b). At Braak stage V/VI the highest Ser202/Thr205 tau phosphorylation could be observed in the temporal cortex while the lowest phosphorylation was in the cingulate cortex (Fig. 2b). Analysis of the allocortical TEntR showed a first but not significant pSer202/Thr205 tau increase in Braak stage III/IV that became significant at Braak stage V/VI (Fig. 2a).

pThr231 tau

Phosphorylation of tau at Thr231 was also low in all isocortical brain regions up to Braak stage III/IV. Only at Braak stage V/VI Thr231 tau phosphorylation significantly increased about 30- to 160-fold above control levels (Fig. 2c, d). At Braak stage V/VI the highest Thr231 tau phosphorylation could be observed in the temporal cortex while the lowest phosphorylation was in the frontal cortex (Fig. 2d). Analysis of the allocortical TEntR showed a first but not significant pThr231 tau increase at Braak stage I/II that became significant compared to healthy controls at Braak stage III/IV and further significantly increased at Braak stage V/VI compared to all other groups (Fig. 2c).

pSer199 tau

Overall, phosphorylation of tau at Ser199 was low compared to other phosphorylation sites. In all isocortical regions Ser199 tau phosphorylation started late, namely at Braak stage V/VI (Fig. 3a, b). At Braak stage V/VI Ser199 tau phosphorylation was at comparable levels in all isocortical regions and about 50- (occipital cortex) to 1300-fold (cingulate cortex) increased compared to healthy controls (Fig. 3b). Analysis of the allocortical TEntR showed a progressive increase in Ser199 tau phosphorylation starting at Braak stage I/II but differences between Braak stages became significant only at stage V/IV and were then about 3.5-fold compared to Braak stage III/IV and 160-fold compared to healthy controls (Fig. 3a). Phosphorylation of Ser199 tau in the TEntR of late Braak stages was higher compared to all isocortical regions.

pTyr18 tau

The phosphorylation profile of pTyr18 tau in the isocortex and TEntR was very similar to pSer199 tau although the final phosphorylation levels in Braak stage V/VI were slightly higher and increased about 20- to 140-fold above control levels (Fig. 3c, d). The progression in the TEntR was already significant at Braak stage III/IV compared to

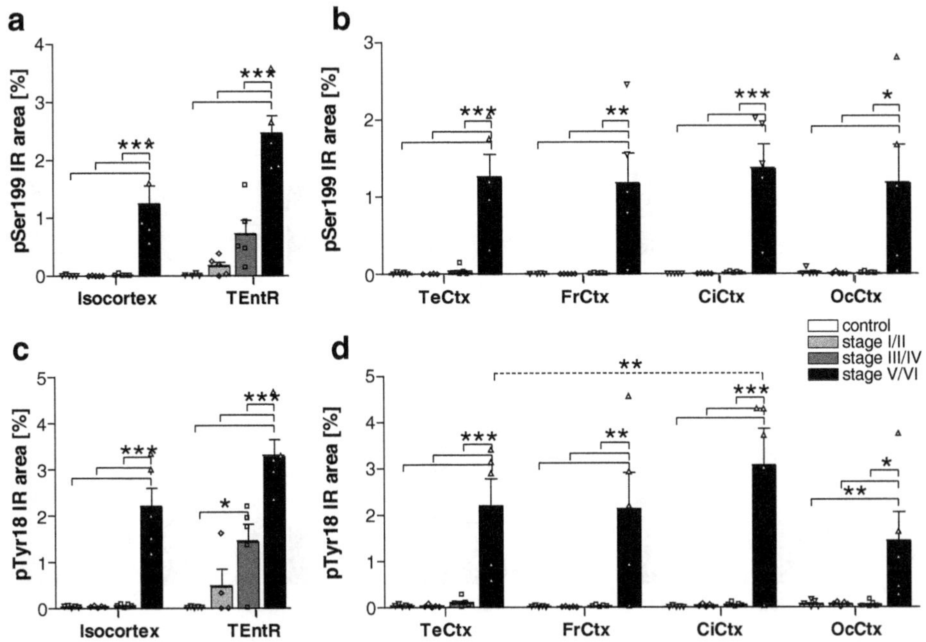

Fig. 3 Quantification of tau phosphorylation at pSer199 and pTyr18 in the cortex of AD cases. **a** Human pSer199 tau immunoreactive area in percent in the isocortex and TENtR (allocortex) of different AD stages. **c** Human pTyr18 tau immunoreactive area in percent in the isocortex and TEntR of different AD stages. Isocortical data in **a** and **c** present mean of different cortical regions that are shown separately in **b** and **d**, respectively. Two way ANOVA followed by Bonferroni's *posthoc* test. Mean + SEM; $N = 5$.*$p < 0.05$; **$p < 0.01$; ***$p < 0.001$. Solid lines: Comparison of AD stages within a brain region; dotted lines: comparison of brain regions of the same AD Braak stage. CiCtx: cingulate cortex; FrCtx: frontal cortex; OcCtx: occipital cortex; TeCtx: temporal cortex; TEntR: transentorhinal region

stage I/II and controls. At Braak stage V/VI the highest pTyr18 tau could be observed in the TEntR and cingulate cortex while the lowest phosphorylation was measured in the occipital cortex (Fig. 3c, d).

pSer396 tau

Phosphorylation of tau at Ser396 shows higher variability than other ptau sites. In the isocortex overall Ser396 tau phosphorylation starts very late in Braak stage V/VI (Fig. 4b). When analyzing different isocortical regions separately, it seemed that pSer396 tau was slightly increased in the temporal cortex of healthy controls and in the cingulate cortex at early Braak stage I/II but these variations were not significant (Fig. 4b). Otherwise, Ser396 tau phosphorylation in all investigated regions only increased about 2- to 5-fold at Braak stage V/VI compared to control levels. Due to the high inter-individual variation this difference was only significant in the temporal cortex compared to Braak stage I/II and III/IV (Fig. 4b). In the allocortical TEntR pSer396 tau increased significantly at Braak stage V/VI compared to all other groups (Fig. 4a).

pSer262 tau

Phosphorylation of tau at Ser262 was very low and showed almost no progression in higher Braak stages (Fig. 4c). Only in the TEntR a significant 8-fold increase could be noted between Braak stage I/II and V/VI (Fig. 4d).

pSer422 tau

The phosphorylation profile of tau at Ser422 was comparable in all isocortical regions and also the allocortical TEntR. In all analyzed regions phosphorylation significantly increased at Braak stage V/VI (Fig. 4e, f) and was about 3- to 13-fold higher compared to the control group.

Relative signal increase of ThioS and ptau

When evaluating the signal increase relative to healthy control brain tissue, it became evident that ThioS labeling in the TEntR increased already at Braak stage III/IV and reached a plateau at Braak stages III/IV and V/VI. The phosphorylation at tau Tyr18 and tau Ser199 severely increased at the same time at Braak stage V/VI with already a slight increase at Braak stage III/IV (Fig. 5a). Compared to this tau phosphorylation the changes at other sites were only minor. Analysis of the phosphorylation relative to healthy control tissue in isocortical brain regions (Fig. 5b-e) showed a progressive increase of ThioS labeling in all regions with highest values in the temporal cortex (Fig. 5b). Additionally, tau

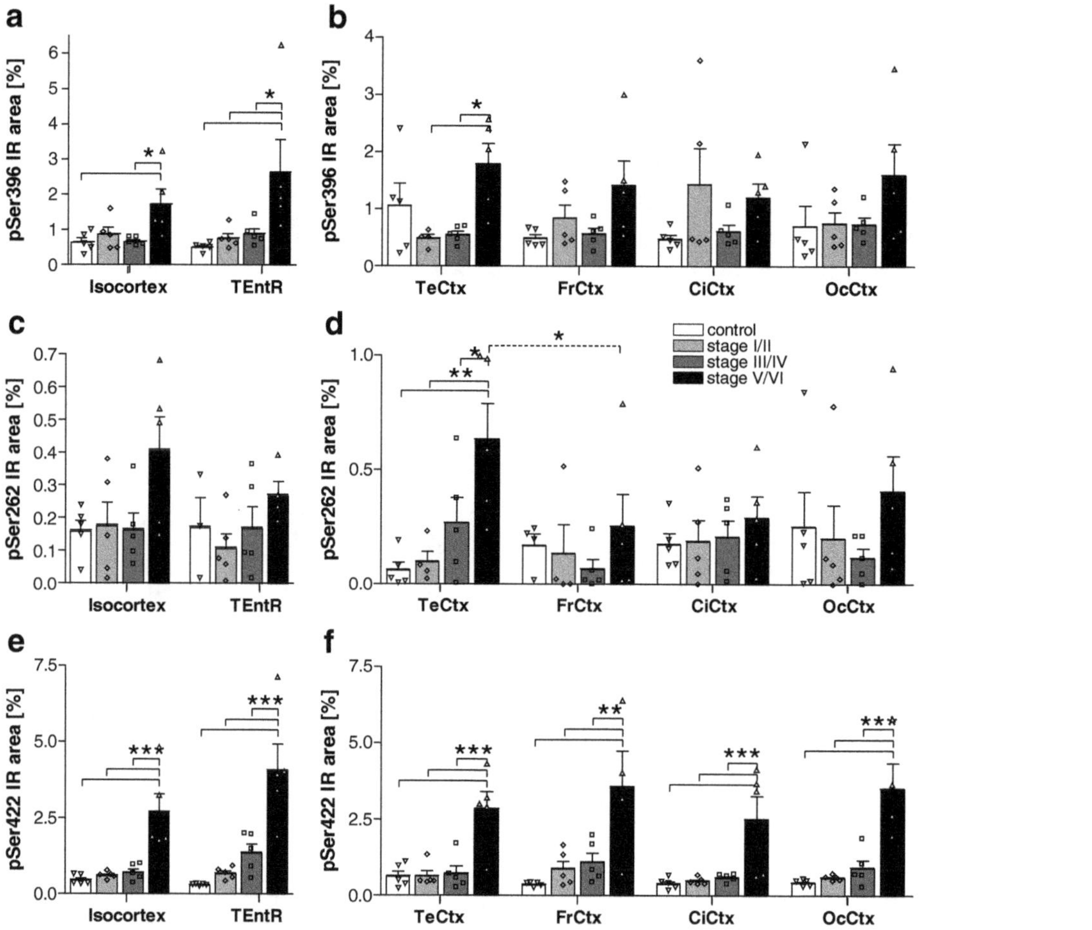

Fig. 4 Quantification of tau phosphorylation at pSer396, pSer262 and pSer422 in the cortex of AD cases. **a** Human pSer396 tau immunoreactive area in percent in the isocortex and TEntR (allocortex) of different AD stages. **c** Human pSer422 tau immunoreactive area in percent in the isocortex and TEntR of different AD stages. **e** Human pSer262 tau immunoreactive area in percent in the isocortex and TEntR of different AD stages. Isocortical data in **a**, **c** and **e** present mean of different cortical regions that are shown separately in **b**, **d** and **f**, respectively. Two way ANOVA followed by Bonferroni's *posthoc* test. Mean + SEM; $N = 5$.$^{*}p < 0.05$; $^{**}p < 0.01$; $^{***}p < 0.001$. Solid lines: Comparison of AD stages within a brain region; dotted lines: comparison of brain regions of the same AD Braak stage. CiCtx: cingulate cortex; FrCtx: frontal cortex; OcCtx: occipital cortex; TeCtx: temporal cortex; TEntR: transentorhinal region

phosphorylation was severely increased at Thr231, Ser199 and Tyr18. Tau phosphorylation at Thr231 was highest in the temporal cortex (Fig. 5b) while phosphorylation at Ser199 and Tyr18 was highest in the frontal cortex (Fig. 5c). Also in the isocortex tau phosphorylation changes at other sites are only minor compared to Thr231, Ser199 and Tyr18.

To validate results obtained by immunofluorescent labeling, Western blots of AD and control cases were performed (Fig. 6a). Cases showing an increase in total tau also showed a signal increase in all analyzed ptau residues. Furthermore, the majority of human cases with strong ptau signals were clinically diagnosed with mild dementia (case 6 and 12) or dementia (case 17, 18, 19, Table 1), suggesting a correlation between phosphorylation and cognitive status. Even that tissue of some cases was missing the quantification of Western blot results shows an overall conformity compared to immunofluorescent labelings. Total tau by HT7 labeling (Fig. 6b), pSer202/205 (Fig. 6c), and pSer199 (Fig. 6e) also showed a late signal increase in Braak stage V/VI with only weak signals in earlier Braak stages. Western blotting of pThr231 (Fig. 6d) and pTyr18 (Fig. 6f) also showed a late signal increase in Braak stage V/VI while signals of immunofluorescent labelings increased already in earlier Braak stages (Fig. 2c and Fig. 3c).

Representative images of total tau (Fig. 7) and pSer202/Thr205, pThr231, pSer199 and pTyr18 (Fig. 8) immunofluorescent labelings in the transentorhinal cortex of one case per Braak stage show the progressive increase of these ptau sites.

In summary, our quantitative analysis of AD brain tissue shows that a progressive increase of total tau and ptau expression can be observed in the

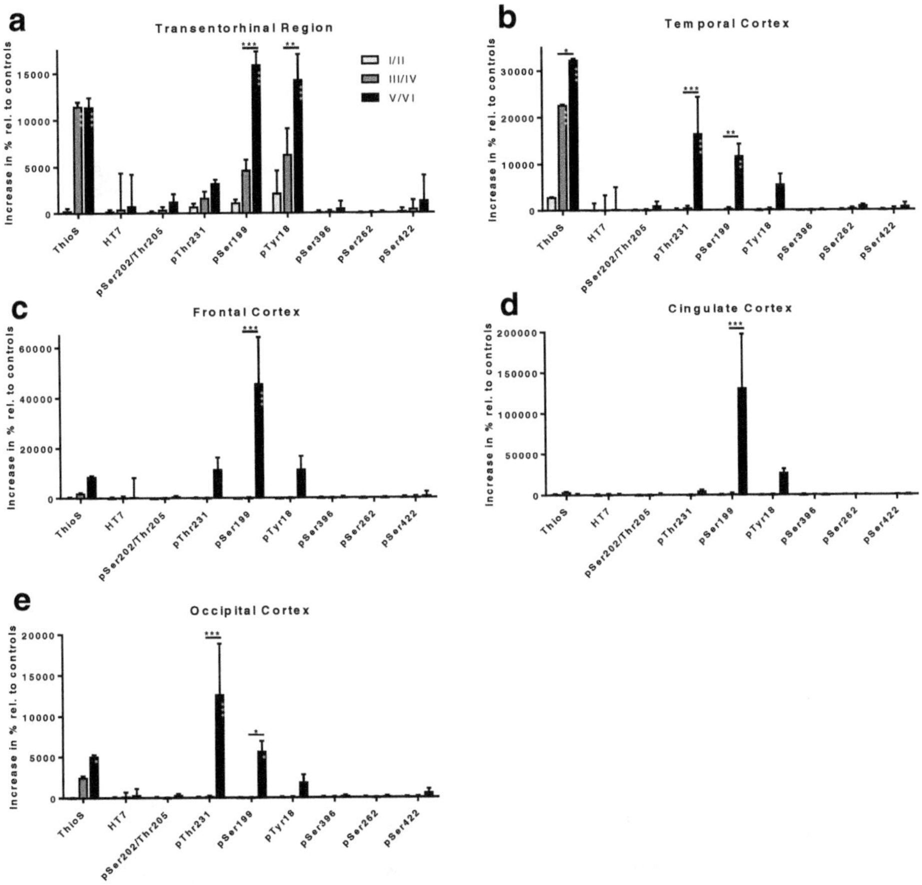

Fig. 5 Signal increase relative to control brain tissue in the cortex of AD cases. Values are calculated from means of absolute IR area from Figs. 1, 2, 3 and 4 relative to control brain samples. Calculation is shown for ThioS, HT7, pSer202/Thr205, pThr231, pSer199, pTyr18, pSer396, pSer262 and pSer422 in the transentorhinal (**a**), temporal (**b**), frontal (**c**), cingulate (**d**) and occipital cortex (**e**) at Braak stages I/II; III/IV and V/VI. Two way ANOVA followed by Bonferroni's *posthoc* test. Mean + SEM. White asterisks within bars mark significance compared to Braak stage I/II. $*p < 0.05$; $**p < 0.01$; $***p < 0.001$

transentorhinal region and in most of the analyzed isocortical regions. Staining of tissue with ThioS resulted only in a weak signal in the transentorhinal region and the occipital cortex. Labeling with antibodies against different tau phosphorylation sites showed that most ptau sites, namely pSer202/Thr205, pThr231, pSer199, pTyr18 and pSer422 were highly increased throughout the cortex at Braak stage V/VI. Only tau phosphorylation at Thr231 and Thy18 was already significantly increased at Braak stage III/IV. By contrast, pSer396 tau and pSer262 tau are only weakly expressed in all analyzed brain regions and only minor progression was observed. When comparing tau phosphorylation in percent relative to healthy controls, phosphorylation is specifically increased at tau Thr231, Ser199 and Tyr18. These data indicate that tau phosphorylation is a complex feature of AD progression, involving many but not all potential phosphorylation sites.

Discussion

The current study was designed to analyze spatial patterns of tau phosphorylation at multiple residues in discrete anatomical regions during AD progression. Our data revealed a very similar phosphorylation profile of most of the analyzed ptau sites in the allo- and isocortex while expression levels of phosphorylated tau at Tyr18 and Thr231 was distinguishable between Braak stages. When normalized to controls, phosphorylation of tau at Tyr18, Ser231 and also Ser199 was much more increased at Braak stage V/VI compared to other residues, suggesting a relevance of these sites for AD progression and a crucial role in pathogenesis.

Since tau phosphorylation is a main characteristic of AD progression, several groups have already analyzed the temporal phosphorylation pattern of different tau sites by histological methods. According to Luna-Munoz and colleagues the phosphorylation of Thr231 tau is an early event in the neuronal pathology of AD [27]. The

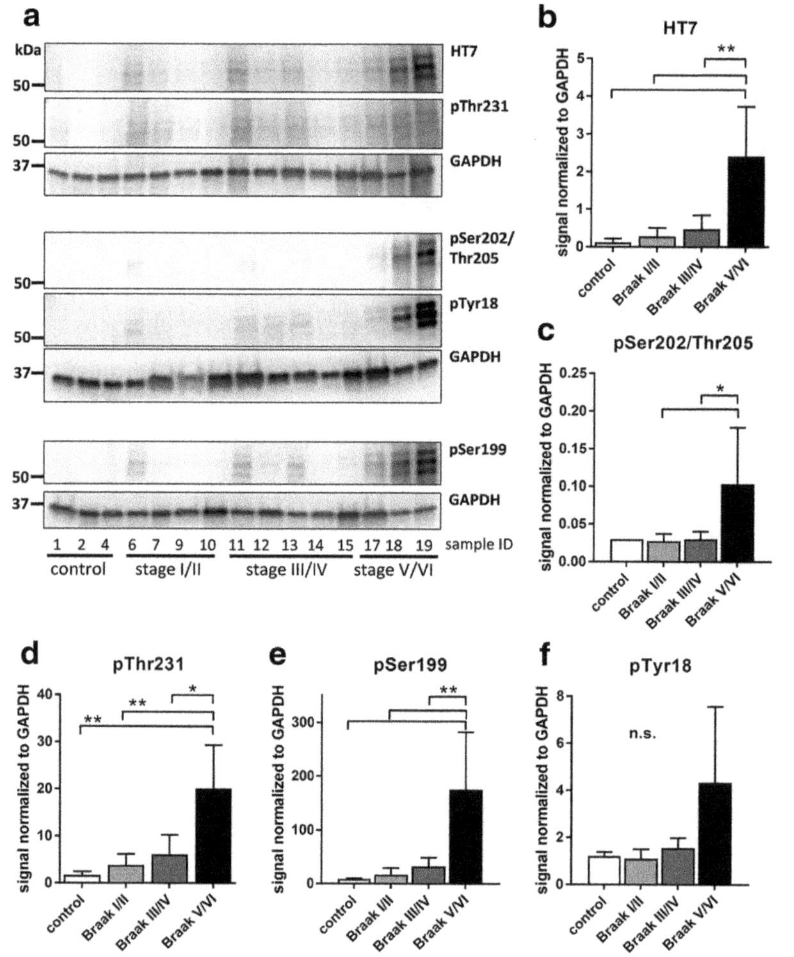

Fig. 6 Quantification of total tau and tau phosphorylation in the transentorhinal cortex of AD cases by Western blotting. **a**: Western blots of total tau, pThr231, pSer202/Thr205, pTyr18, and pSer199. GAPDH was used as loading control. Quantification of Western blot for (**b**) total tau by HT7 antibody, (**c**) pSer202/Thr205, (**d**) pThr231, (**e**) pSer199, and (**f**) pTyr18. All samples shown in (**a**) were used for quantification of (**b**-**f**). One way ANOVA followed by Tukey's multiple comparisons test. Mean + SEM. $^{*}p < 0.05$; $^{**}p < 0.01$; n.s.: not significant

temporal analysis of pSer202/Thr205 tau and pSer396 tau is very controversial. Simic and co-workers found higher phosphorylation of Ser396 and Ser202/Thr205 tau in mild cognitive impairment (MCI) cases [38] suggesting a parallel phosphorylation of both residues. Temporal phosphorylation analyses by two other groups contradict these results, while one group found an earlier phosphorylation of pSer202/Thr205 tau [39], the other reported an earlier phosphorylation of pSer396 tau [30]. Zhou and colleagues performed dot blots and ELISA analyses of a whole series of ptau sites of AD medial temporal cortex samples and found mostly a similar temporal phosphorylation pattern as shown here for the entorhinal cortex. Tau phosphorylation of Ser396 for example, was a late event and only measurable in Braak stage V and VI [46]. Dot blot analyses of AD lateral temporal lobe samples by another group showed that tau phosphorylation at Ser202/205 and Ser396 simultaneously increases with increasing Braak staging, but also that they observe the increase already in Braak stage III/IV compared to results shown here or by Zhou and colleagues [20, 46]. By quantitatively analyzing ptau sites in different brain regions of AD cases we found that most analyzed ptau sites, pSer202/Thr205, pThr231, pSer199, pTyr18 and pSer422, have a very similar allo- and isocortical phosphorylation profile, suggesting that pSer202/Thr205 tau analysis by AT8 antibody could be replaced by any of these ptau sites when using this quantitative histological approach. For neuropathological assessment of Braak stages ptau Ser202/Thr205 immunopositivity is scored semi-quantitatively and is required to be mild (+) in the transentorhinal cortex to assign Braak stage I, while for stages higher than Braak stage I, severity of ptau Ser202/Thr205 immunopositivity in the transentorhinal cortex is not decisive as other areas, e.g. entorhinal cortex, occipitotemporal gyrus,

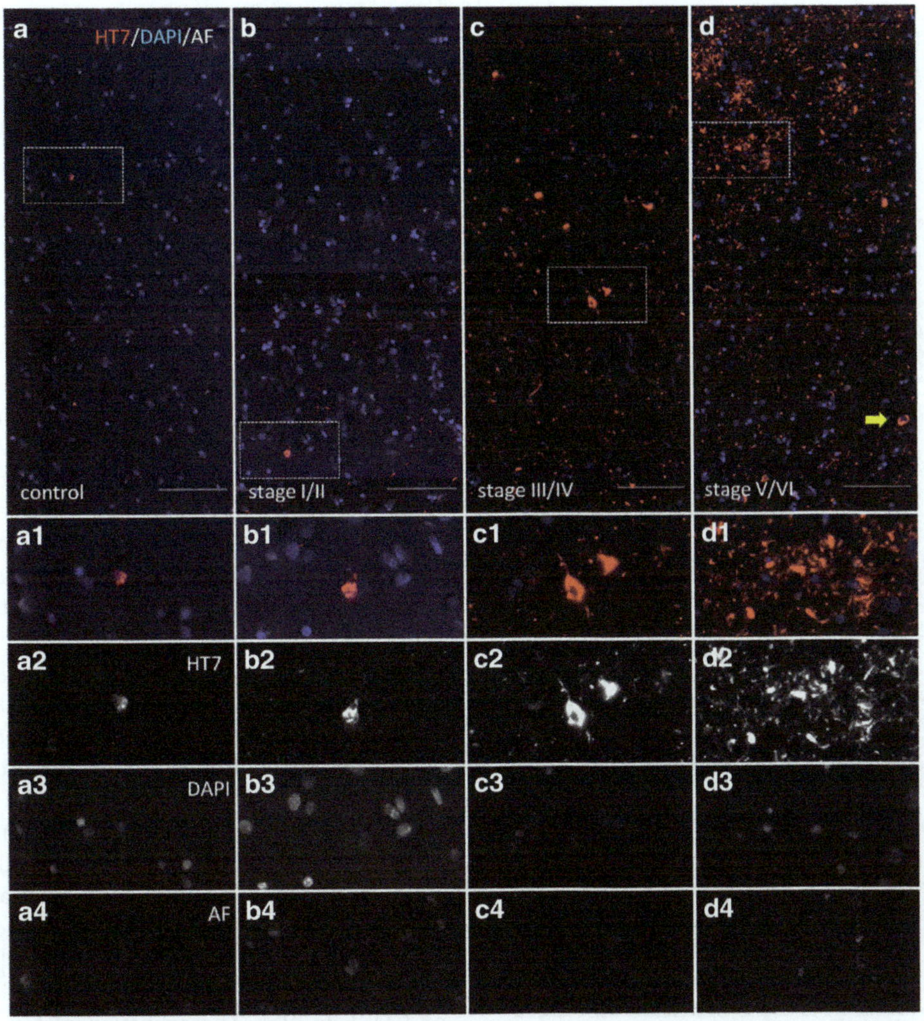

Fig. 7 Representative images of total tau labeling in the transentorhinal cortex grey matter of AD cases. Labeling of total tau using HT7 antibody (a-d, a1-d1, a2-d2) and nuclei by DAPI staining (**a-d**, a1-d1,a3-d3). Autofluorescence is shown in white (**a-d**, a1-d1, a4-d4). Samples of healthy control tissue (a-a4, case 2), Braak stage I/II (b-b4, case 8), Braak stage III/IV (c-c4, case 11) and Braak stage V/VI (d-d4, case 16) are shown. Scale bar: 100 µm

middle temporal gyrus, occipital cortex need to show at least moderate (++) severity of immunopositivity [1]. Therefore our findings of a lack of a significant increase of ptau Ser202/Thr205 immunopositivity in the transentorhinal cortex between Braak stages 0 to IV is not at variance with neuropathological Braak staging. However, pSer202/Thr205 immunopositivity does indeed increase continuously in the transentorhinal cortex with increasing Braak stages. In our study the difference is only statistically significant between Braak stages 0-IV and V/VI. The phosphorylation of the analyzed tau sites starts in the isocortical regions late at Braak stage V/VI and consequently the analysis of the allocortical transentorhinal region is preferable. Indeed, the quantitative analysis of pTyr18 tau and pThr231 tau in the transentorhinal region shows a distinct progression between Braak stages that might be caused by highly increased phosphorylation relative to control tissue that could also be observed for pSer199 tau. Since phosphorylation of tau at residues Tyr18 and Thr231 occurs earlier compared to other sites, these phosphorylation sites might trigger disease progression and thus would represent an early therapeutic target.

Western blot analyses of total tau and the most prominent phosphorylation residues pSer202/Thr205, pThr231, pSer199, and pTyr18 overall validated the results obtained by immunofluorescent labeling. It has to be mentioned that tissue of not all human cases was available for blotting, so quantification and statistical analyses should be understood as preliminary. Blots for pSer202//Thr205 and pThr231 are not directly comparable, since different antibodies as used

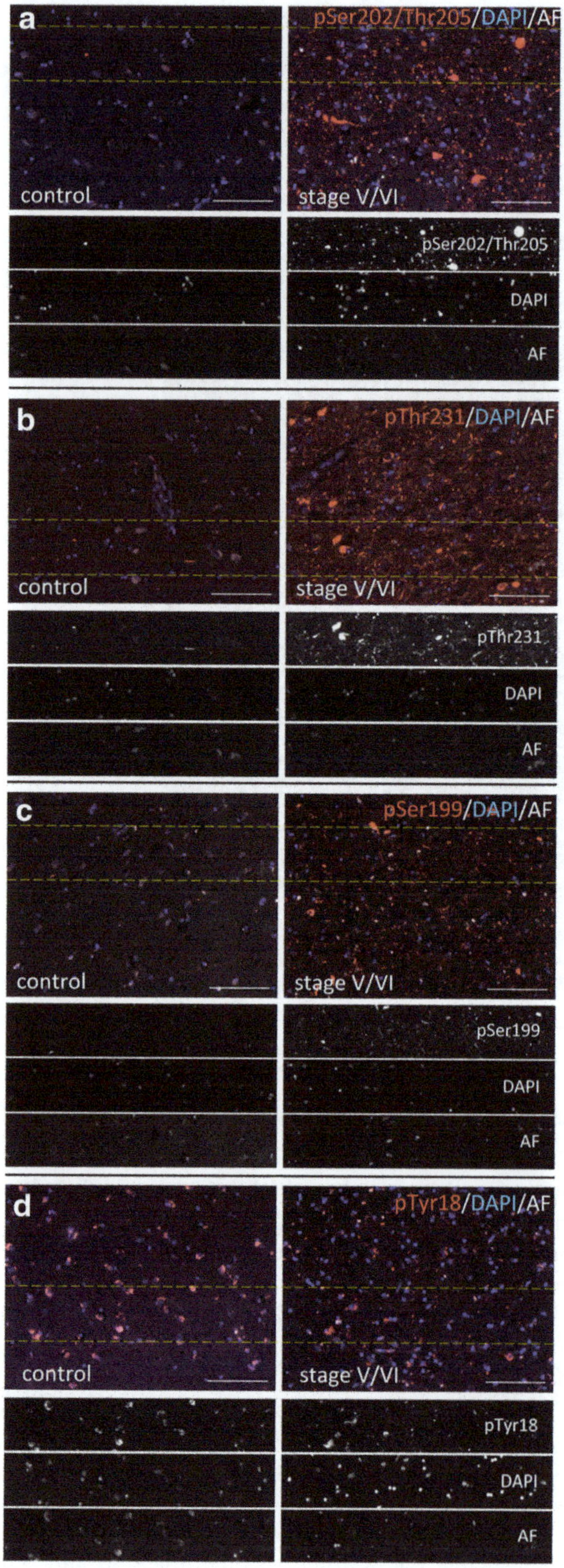

Fig. 8 Representative images of ptau labelings in the transentorhinal cortex of control and AD cases. Labeling of pSer202/Thr205 (**a**), pThr231 (**b**), pSer199 (**c**), and pTyr18 (**d**) are shown. Tissues were additionally stained with DAPI to visualize nuclei. Autofluorescence is shown in white. Dotted lines indicate areas shown in grey scale images. Samples of healthy control tissue (case 2) and Braak stage V/VI (case 16) are shown. Scale bar: 100 μm

for immunofluorescent labelings had to be used for methodical reasons.

AD is known to affect women with a higher prevalence, which is among other reasons caused by hormonal differences. The higher frequency of AD cases in women is reflected in this study by the fact that only tissue of women was provided for Braak stage V/VI, while tissue of earlier Braak stages was predominantly from men. Although several in vitro and in vivo studies in rodents suggest an impact of sex hormones (e.g. estrogens, progesterone and prolactin) and therefore gender on tau phosphorylation, there are so far no studies in AD patients validating these results in humans (for review see [33]). Future analyses of *post mortem* AD tissue and in vivo imaging of AD patients should shed light on the impact of gender on tau phosphorylation.

In addition to the severe increases in tau phosphorylation also the ThioS signal, that includes amyloid plaques, was highly increased in several allo- and isocortical regions. The ThioS signal increase partially already started at Braak stage III/IV and thus earlier than the increase of ptau. It has been shown that ptau residues, Tyr18, Thr231 and Ser199 can be phosphorylated by Aβ via different kinases like Fyn [18, 35, 37, 42, 44, 45], GSK-3β [3, 23, 24] or CDK5 [6, 25]. The activation of tau by Aβ is further demonstrated to be involved in the early formation of neurofibrillary tangles, synaptic loss, neurodegeneration as well as cognitive deficits [2, 3, 11, 25] and thus in the development of the most prominent AD pathologies. These results were derived from AD cell and animal models but our study might suggest a similar effect of Aβ on tau phosphorylation at residues Tyr18, Thr231 and Ser199 in the human disease. Further analyses are needed to validate this hypothesis in humans. A valid tool to analyze such events in vivo might be the use of Pittsburgh compound B (PiB) analysis combined with tau tracer that are currently under development [5, 17, 19, 36].

Isocortical tau pathology is only sparse at early Braak stages I to IV and hence use of higher magnification and modified image analysis parameters may be necessary to reveal subtle changes in isocortical tau pathology at early Braak stages. Additionally, co-labeling of different markers to quantify ptau only in distinct areas or cell populations combined with an increased number of investigated brain regions may be helpful to gain a better understanding of earliest isocortical tau pathology. A tissue microarray method that was developed recently allows to examine over 35 brain regions on one slide could be used for such purposes [43]. However, our study investigated only five brain regions in a limited number of cases and represents therefore proof of principle. Our data need to be confirmed by including more cases and assessing a higher number of brain areas.

Conclusion

We show here for the first time the phosphorylation profile of different tau sites in allocortical and isocortical regions of human brains with tau pathology ranging from Braak stages 0 to VI using quantitative rater-independent immunofluorescent labeling. Our data suggest that the profile of ptau in the isocortex is comparable between all of the analyzed ptau sites while expression levels of ptau at Thr231 and Tyr18 in the transentorhinal region are distinguishable between Braak stages combined with the highest tau phosphorylation relative to controls at these sites and Ser199.

Additional file

Additional file 1: Online Source 9 Double labeling of pSer262 and pSer202/Thr205 tau in the temporal cortex at Braak stage V/VI. Images show different labeling pattern of pSer262 (arrows) and pS202 (white arrowheads) as well as their overlay (yellow arrowheads) (a1) and single fluorescence images (a2,3,4) of case 17. AF: autofluorescence. Scale bar: 20 μm. Online Source 10 Example of measurement procedure of tau pSer262. Objects in the unlabeled autofluorescence channel were detected by thresholding (red in a1). The resulting mask images (a2) were then subtracted from tau pSer262 images to remove autofluorescence (a3). The resulting images were Edge+ filtered (a4) to facilitate threshold-based detection of tau pSer262-positive objects (red outline in a5). These outlines were then loaded onto the raw images to quantify original tau pSer262 signal (red outline in a6). AF: autofluorescence. Scale bar: 20 μm. Online Source 11 Example of detecting ThioS-positive amyloid-β but not NFTs. Image a displays the co-labeling of ThioS (green) and HT7 (red), while images b and c, respectively, show single channel images. ThioS shows intense labeling of plaque-associated β-sheets (b, asterisk) whereas tangles are only weakly labeled (c, arrows) (c). A combination of threshold-based identification of ThioS and size restriction (d', green rectangle) enables quantification of ThioS+ plaque labeling (red highlighted) but not tangles (d). ThioS: ThioflavinS. Scale bar: 20 μm. (PDF 599 kb)

Abbreviations

AD: Alzheimer's disease; ANOVA: Analysis of variance; CiCtx: Cingulate cortex; FrCtx: Frontal cortex; NFT: Neurofibrillary tangle; NT: Neuropil threads; OcCtx: Occipital cortex; ptau: Phosphorylated tau; TeCtx: Temporal cortex; TEntR: Transentorhinal region

Acknowledgements

The authors greatly thank the whole research team of QPS Austria GmbH for their technical support.

Funding

This work was supported by the Austrian Research Promotion Agency (FFG), R&D Projects (FFG# 844453, 851079, 855287).
Tissue for this study was provided by the Newcastle Brain Tissue Resource, which is funded in part by a grant from the UK Medical Research Council (G0400074), by Brains for Dementia research, a joint venture between Alzheimer's Society and Alzheimer's Research UK and by the NIHR Newcastle Biomedical Research Centre awarded to the Newcastle upon Tyne Hospitals NHS Foundation Trust and Newcastle University.

Authors' contributions

JN designed the overall project, analyzed data, interpreted experiments and prepared the manuscript; MT analyzed data and prepared figures; SF interpreted experiments and prepared figures and the manuscript; BK and CH performed experiments and analyzed data; VN edited the manuscript; JA provided neuropathologically diagnoses and edited the manuscript; GD interpreted experiments and edited the manuscript; BHP designed and interpreted experiments and edited the manuscript. All authors read and approved the final manuscript.

Competing interests

JN, MT, SF, BK, CH, TL, VN and BHP are employees of QPS Austria GmbH. The authors declare that they have no other competing interests.

Author details

[1]QPS Austria GmbH, Neuropharmacology, Parkring 12, 8074 Grambach, Austria. [2]Institute for Biochemistry, Graz University of Technology, Graz, Austria. [3]Institute of Zoology, Karl Franzens University, Graz, Austria. [4]Institute of Neuroscience and Newcastle University Institute for Ageing Campus for Ageing and Vitality, Newcastle University, Newcastle upon Tyne NE4 5PL, UK.

References

1. Alafuzoff I, Arzberger T, Al-Sarraj S, Bodi I, Bogdanovic N, Braak H, Bugiani O, Del-Tredici K, Ferrer I, Gelpi E et al (2008) Staging of neurofibrillary pathology in Alzheimer's disease: a study of the BrainNet Europe consortium. Brain Pathol 18: 484–496 doi 10.1111/j.1750-3639.2008.00147.x
2. Amadoro G, Corsetti V, Atlante A, Florenzano F, Capsoni S, Bussani R, Mercanti D, Calissano P (2012) Interaction between NH(2)-tau fragment and Abeta in Alzheimer's disease mitochondria contributes to the synaptic deterioration. Neurobiol Aging 33(833):e831–e825. https://doi.org/10.1016/j.neurobiolaging.2011.08.001
3. Amadoro G, Corsetti V, Ciotti MT, Florenzano F, Capsoni S, Amato G, Calissano P (2011) Endogenous Abeta causes cell death via early tau hyperphosphorylation. Neurobiol Aging 32:969–990. https://doi.org/10.1016/j.neurobiolaging.2009.06.005
4. Andorfer C, Kress Y, Espinoza M, de Silva R, Tucker KL, Barde YA, Duff K, Davies P (2003) Hyperphosphorylation and aggregation of tau in mice expressing normal human tau isoforms. J Neurochem 86:582–590
5. Ariza M, Kolb HC, Moechars D, Rombouts F, Andres JI (2015) Tau positron emission tomography (PET) imaging: past, present, and future. J Med Chem 58:4365–4382. https://doi.org/10.1021/jm5017544
6. Billingsley ML, Kincaid RL (1997) Regulated phosphorylation and dephosphorylation of tau protein: effects on microtubule interaction, intracellular trafficking and neurodegeneration. Biochem J 323(Pt 3):577–591
7. Boekhoorn K, Terwel D, Biemans B, Borghgraef P, Wiegert O, Ramakers GJ, de Vos K, Krugers H, Tomiyama T, Mori H et al (2006) Improved long-term potentiation and memory in young tau-P301L transgenic mice before onset of hyperphosphorylation and tauopathy. J Neurosci 26:3514–3523. https://doi.org/10.1523/JNEUROSCI.5425-05.2006
8. Braak H, Alafuzoff I, Arzberger T, Kretzschmar H, Del Tredici K (2006) Staging of Alzheimer disease-associated neurofibrillary pathology using paraffin sections and immunocytochemistry. Acta Neuropathol 112:389–404. https://doi.org/10.1007/s00401-006-0127-z

9. Braak H, Braak E (1991) Neuropathological stageing of Alzheimer-related changes. Acta Neuropathol 82:239–259
10. Bramblett GT, Goedert M, Jakes R, Merrick SE, Trojanowski JQ, Lee VM (1993) Abnormal tau phosphorylation at Ser396 in Alzheimer's disease recapitulates development and contributes to reduced microtubule binding. Neuron 10:1089–1099
11. Chang KH, de Pablo Y, Lee HP, Lee HG, Smith MA, Shah K (2010) Cdk5 is a major regulator of p38 cascade: relevance to neurotoxicity in Alzheimer's disease. J Neurochem 113:1221–1229. https://doi.org/10.1111/j.1471-4159.2010.06687.x
12. Cho JH, Johnson GV (2004) Primed phosphorylation of tau at Thr231 by glycogen synthase kinase 3beta (GSK3beta) plays a critical role in regulating tau's ability to bind and stabilize microtubules. J Neurochem 88:349–358
13. Cook C, Stankowski JN, Carlomagno Y, Stetler C, Petrucelli L (2014) Acetylation: a new key to unlock tau's role in neurodegeneration. Alzheimers Res Ther 6:29. https://doi.org/10.1186/alzrt259
14. Flunkert S, Hierzer M, Loffler T, Rabl R, Neddens J, Duller S, Schofield EL, Ward MA, Posch M, Jungwirth H et al (2013) Elevated levels of soluble total and hyperphosphorylated tau result in early behavioral deficits and distinct changes in Brain Pathol in a new tau transgenic mouse model. Neurodegener Dis 11:194–205. https://doi.org/10.1159/000338152
15. Goedert M, Jakes R, Crowther RA, Six J, Lubke U, Vandermeeren M, Cras P, Trojanowski JQ, Lee VM (1993) The abnormal phosphorylation of tau protein at Ser-202 in Alzheimer disease recapitulates phosphorylation during development. Proc Natl Acad Sci U S A 90:5066–5070
16. Green KN, Steffan JS, Martinez-Coria H, Sun X, Schreiber SS, Thompson LM, LaFerla FM (2008) Nicotinamide restores cognition in Alzheimer's disease transgenic mice via a mechanism involving sirtuin inhibition and selective reduction of Thr231-phosphotau. J Neurosci 28:11500–11510. https://doi.org/10.1523/JNEUROSCI.3203-08.2008
17. Hashimoto H, Kawamura K, Igarashi N, Takei M, Fujishiro T, Aihara Y, Shiomi S, Muto M, Ito T, Furutsuka K et al (2014) Radiosynthesis, photoisomerization, biodistribution, and metabolite analysis of 11C-PBB3 as a clinically useful PET probe for imaging of tau pathology. J Nucl Med 55: 1532–1538. https://doi.org/10.2967/jnumed.114.139550
18. Ittner LM, Ke YD, Delerue F, Bi M, Gladbach A, van Eersel J, Wolfing H, Chieng BC, Christie MJ, Napier IA et al (2010) Dendritic function of tau mediates amyloid-beta toxicity in Alzheimer's disease mouse models. Cell 142:387–397. https://doi.org/10.1016/j.cell.2010.06.036
19. Kadir A, Nordberg A (2010) Target-specific PET probes for neurodegenerative disorders related to dementia. J Nucl Med 51:1418–1430. https://doi.org/10.2967/jnumed.110.077164
20. Koss DJ, Jones G, Cranston A, Gardner H, Kanaan NM, Platt B (2016) Soluble pre-fibrillar tau and beta-amyloid species emerge in early human Alzheimer's disease and track disease progression and cognitive decline. Acta Neuropathol 132:875–895. https://doi.org/10.1007/s00401-016-1632-3
21. Lauckner J, Frey P, Geula C (2003) Comparative distribution of tau phosphorylated at Ser262 in pre-tangles and tangles. Neurobiol Aging 24:767–776
22. Lee G, Thangavel R, Sharma VM, Litersky JM, Bhaskar K, Fang SM, Do LH, Andreadis A, Van Hoesen G, Ksiezak-Reding H (2004) Phosphorylation of tau by fyn: implications for Alzheimer's disease. J Neurosci 24:2304–2312. https://doi.org/10.1523/JNEUROSCI.4162-03.2004
23. Li T, Hawkes C, Qureshi HY, Kar S, Paudel HK (2006) Cyclin-dependent protein kinase 5 primes microtubule-associated protein tau site-specifically for glycogen synthase kinase 3beta. Biochemistry 45:3134–3145. https://doi.org/10.1021/bi051635j
24. Li T, Paudel HK (2006) Glycogen synthase kinase 3beta phosphorylates Alzheimer's disease-specific Ser396 of microtubule-associated protein tau by a sequential mechanism. Biochemistry 45:3125–3133. https://doi.org/10.1021/bi051634r
25. Liao X, Zhang Y, Wang Y, Wang J (2004) The effect of cdk-5 overexpression on tau phosphorylation and spatial memory of rat. Sci China Ser C Life Sci 47:251–257
26. Lichtenberg-Kraag B, Mandelkow EM, Biernat J, Steiner B, Schroter C, Gustke N, Meyer HE, Mandelkow E (1992) Phosphorylation-dependent epitopes of neurofilament antibodies on tau protein and relationship with Alzheimer tau. Proc Natl Acad Sci U S A 89:5384–5388
27. Luna-Munoz J, Garcia-Sierra F, Falcon V, Menendez I, Chavez-Macias L, Mena R (2005) Regional conformational change involving phosphorylation of tau protein at the Thr231, precedes the structural change detected by Alz-50 antibody in Alzheimer's disease. J Alzheimer's Dis: JAD 8:29–41
28. Lund H, Cowburn RF, Gustafsson E, Stromberg K, Svensson A, Dahllund L, Malinowsky D, Sunnemark D (2013) Tau-tubulin kinase 1 expression, phosphorylation and co-localization with phospho-Ser422 tau in the Alzheimer's disease Brain. Brain Pathol 23:378–389. https://doi.org/10.1111/bpa.12001
29. Luo HB, Xia YY, Shu XJ, Liu ZC, Feng Y, Liu XH, Yu G, Yin G, Xiong YS, Zeng K et al (2014) SUMOylation at K340 inhibits tau degradation through deregulating its phosphorylation and ubiquitination. Proc Natl Acad Sci U S A 111:16586–16591. https://doi.org/10.1073/pnas.1417548111
30. Mondragon-Rodriguez S, Perry G, Zhu X, Moreira PI, Acevedo-Aquino MC, Williams S (2013) Phosphorylation of tau protein as the link between oxidative stress, mitochondrial dysfunction, and connectivity failure: implications for Alzheimer's disease. Oxidative Med Cell Longev 2013: 940603. https://doi.org/10.1155/2013/940603
31. Montine TJ, Phelps CH, Beach TG, Bigio EH, Cairns NJ, Dickson DW, Duyckaerts C, Frosch MP, Masliah E, Mirra SS et al (2012) National Institute on Aging-Alzheimer's Association guidelines for the neuropathologic assessment of Alzheimer's disease: a practical approach. Acta Neuropathologica 123:1–11. https://doi.org/10.1007/s00401-011-0910-3
32. Moszczynski AJ, Gohar M, Volkening K, Leystra-Lantz C, Strong W, Strong MJ (2015) Thr175-phosphorylated tau induces pathologic fibril formation via GSK3beta-mediated phosphorylation of Thr231 in vitro. Neurobiol Aging 36: 1590–1599. https://doi.org/10.1016/j.neurobiolaging.2014.12.001
33. Munoz-Mayorga D, Guerra-Araiza C, Torner L, Morales T (2018) Tau phosphorylation in female neurodegeneration: role of estrogens, progesterone, and prolactin. Front Endocrinol 9:133. https://doi.org/10.3389/fendo.2018.00133
34. Pevalova M, Filipcik P, Novak M, Avila J, Iqbal K (2006) Post-translational modifications of tau protein. Bratislavske Lekarske Listy 107:346–353
35. Roberson ED, Halabisky B, Yoo JW, Yao J, Chin J, Yan F, Wu T, Hamto P, Devidze N, Yu GQ et al (2011) Amyloid-beta/Fyn-induced synaptic, network, and cognitive impairments depend on tau levels in multiple mouse models of Alzheimer's disease. J Neurosci 31:700–711. https://doi.org/10.1523/JNEUROSCI.4152-10.2011
36. Rowe CC, Villemagne VL (2013) Amyloid imaging with PET in early Alzheimer disease diagnosis. Med Clin North Am 97:377–398. https://doi.org/10.1016/j.mcna.2012.12.017
37. Scales TM, Derkinderen P, Leung KY, Byers HL, Ward MA, Price C, Bird IN, Perera T, Kellie S, Williamson R et al (2011) Tyrosine phosphorylation of tau by the SRC family kinases lck and fyn. Mol Neurodegen 6:12. https://doi.org/10.1186/1750-1326-6-12
38. Simic G, Babic Leko M, Wray S, Harrington C, Delalle I, Jovanov-Milosevic N, Bazadona D, Buee L, de Silva R, Di Giovanni G et al (2016) Tau protein hyperphosphorylation and aggregation in Alzheimer's disease and other Tauopathies, and possible neuroprotective strategies. Biomolecules 6: 6 doi https://doi.org/10.3390/biom6010006
39. Su JH, Cummings BJ, Cotman CW (1994) Early phosphorylation of tau in Alzheimer's disease occurs at Ser-202 and is preferentially located within neurites. Neuroreport 5:2358–2362
40. Tenreiro S, Eckermann K, Outeiro TF (2014) Protein phosphorylation in neurodegeneration: friend or foe? Front Mol Neurosci 7:42. https://doi.org/10.3389/fnmol.2014.00042
41. Thal DR, Rub U, Orantes M, Braak H (2002) Phases of a beta-deposition in the human brain and its relevance for the development of AD. Neurology 58:1791–1800
42. Usardi A, Pooler AM, Seereeram A, Reynolds CH, Derkinderen P, Anderton B, Hanger DP, Noble W, Williamson R (2011) Tyrosine phosphorylation of tau regulates its interactions with Fyn SH2 domains, but not SH3 domains, altering the cellular localization of tau. FEBS J 278:2927–2937. https://doi.org/10.1111/j.1742-4658.2011.08218.x
43. Walker L, McAleese KE, Johnson M, Khundakar AA, Erskine D, Thomas AJ, McKeith IG, Attems J (2017) Quantitative neuropathology: an update on

automated methodologies and implications for large scale cohorts. J Neural Transm (Vienna). https://doi.org/10.1007/s00702-017-1702-2
44. Williamson R, Scales T, Clark BR, Gibb G, Reynolds CH, Kellie S, Bird IN, Varndell IM, Sheppard PW, Everall I et al (2002) Rapid tyrosine phosphorylation of neuronal proteins including tau and focal adhesion kinase in response to amyloid-beta peptide exposure: involvement of Src family protein kinases. J Neurosci 22:10–20
45. Williamson R, Usardi A, Hanger DP, Anderton BH (2008) Membrane-bound beta-amyloid oligomers are recruited into lipid rafts by a fyn-dependent mechanism. FASEB J 22:1552–1559. https://doi.org/10.1096/fj.07-9766com
46. Zhou XW, Li X, Bjorkdahl C, Sjogren MJ, Alafuzoff I, Soininen H, Grundke-Iqbal I, Iqbal K, Winblad B, Pei JJ (2006) Assessments of the accumulation severities of amyloid beta-protein and hyperphosphorylated tau in the medial temporal cortex of control and Alzheimer's brains. Neurobiol Dis 22: 657–668. https://doi.org/10.1016/j.nbd.2006.01.006

Permissions

 Every chapter published in this book has been scrutinized by our experts. Their significance has been extensively debated. The topics covered herein carry significant findings which will fuel the growth of the discipline. They may even be implemented as practical applications or may be referred to as a beginning point for another development.

The contributors of this book come from diverse backgrounds, making this book a truly international effort. This book will bring forth new frontiers with its revolutionizing research information and detailed analysis of the nascent developments around the world.

We would like to thank all the contributing authors for lending their expertise to make the book truly unique. They have played a crucial role in the development of this book. Without their invaluable contributions this book wouldn't have been possible. They have made vital efforts to compile up to date information on the varied aspects of this subject to make this book a valuable addition to the collection of many professionals and students.

This book was conceptualized with the vision of imparting up-to-date information and advanced data in this field. To ensure the same, a matchless editorial board was set up. Every individual on the board went through rigorous rounds of assessment to prove their worth. After which they invested a large part of their time researching and compiling the most relevant data for our readers.

The editorial board has been involved in producing this book since its inception. They have spent rigorous hours researching and exploring the diverse topics which have resulted in the successful publishing of this book. They have passed on their knowledge of decades through this book. To expedite this challenging task, the publisher supported the team at every step. A small team of assistant editors was also appointed to further simplify the editing procedure and attain best results for the readers.

Apart from the editorial board, the designing team has also invested a significant amount of their time in understanding the subject and creating the most relevant covers. They scrutinized every image to scout for the most suitable representation of the subject and create an appropriate cover for the book.

The publishing team has been an ardent support to the editorial, designing and production team. Their endless efforts to recruit the best for this project, has resulted in the accomplishment of this book. They are a veteran in the field of academics and their pool of knowledge is as vast as their experience in printing. Their expertise and guidance has proved useful at every step. Their uncompromising quality standards have made this book an exceptional effort. Their encouragement from time to time has been an inspiration for everyone.

The publisher and the editorial board hope that this book will prove to be a valuable piece of knowledge for researchers, students, practitioners and scholars across the globe.

List of Contributors

Brandon F. Maziuk, Daniel J. Apicco, Anna Lourdes Cruz, Lulu Jiang, Peter E. A. Ash and Benjamin Wolozin
Department of Pharmacology and Experimental Therapeutics, Boston University School of Medicine, Boston, MA, USA

Edroaldo Lummertz da Rocha, Cheng Zhang and Hu Li
Mayo Clinic, Rochester, MN, USA

Wai Haung Yu
Department of Pathology and Cell Biology, Taub Institute for Alzheimer's Disease Research, Columbia University Medical Center, New York, NY, USA

John Leszyk
University of Massachusetts Medical Center, Worcester, MA, USA

Jose F. Abisambra
Sanders-Brown Center on Aging, Department of Physiology, Spinal Cord and Brain Injury Research Center, and Epilepsy Center, University of Kentucky, Lexington, KY, USA

Benjamin Wolozin
Department of Neurology, Boston University School of Medicine, Boston, MA, USA
Department of Pharmacology and Neurology Program in Neuroscience, Boston University School of Medicine, 72 East Concord St., R614, Boston, MA 02118-2526, USA

Jenny Lange, Brenda P. Williams and Jonathan D. Cooper
Department of Basic and Clinical Neuroscience, King's College London, Institute of Psychiatry, Psychology & Neuroscience, Maurice Wohl Clinical Neuroscience Institute, 5 Cutcombe Road, London SE5 9RX, UK

Luke J. Haslett and Emyr Lloyd-Evans
School of Biosciences, Cardiff University, Museum Avenue, Cardiff CF10 3AX, UK

Jennifer M. Pocock
Department of Neuroinflammation, Institute of Neurology, University College London, 1 Wakefield Street, London WC1N 1PK, UK

Mark S. Sands
Departments of Medicine and Genetics, Washington University School of Medicine, 660 South Euclid Avenue, St. Louis, MO 63110, USA

Jonathan D. Cooper
Department of Pediatrics, Los Angeles Biomedical Research Institute at Harbor-UCLA Medical Center, and David Geffen School of Medicine UCLA, 1124 West Carson Street, Hanley Hardison Building, Torrance, CA 90502, USA
Department of Pediatrics, Washington University School of Medicine, Campus S Euclid Avenue, St. Louis, MO 63110, USA

Hemant Kumar, Hyemin Choi, Min-Jae Jo, Hari Prasad Joshi, Seil Sohn and Inbo Han
Department of Neurosurgery, CHA University School of Medicine, CHA Bundang Medical Center, Seongnam-si, Gyeonggi-do 13496, Republic of Korea

Manjunatha Muttigi and Soo-Hong Lee
Department of Biomedical Science, CHA University, Seongnam-si, Gyeonggi-do, Republic of Korea

Dario Bonanomi
Molecular Neurobiology Laboratory, Division of Neuroscience, San Raffaele Scientific Institute, Milan, Italy

Sung Bum Kim
Department of Neurosurgery, Kyung Hee University, Dongdaemun-gu, Seoul 02447, Republic of Korea

Eunmi Ban and Aeri Kim
College of Pharmacy, CHA University, Seongnam-si, Gyeonggi-do, Republic of Korea

Kyoung-Tae Kim
Department of Neurosurgery Kyungpook National University Hospital, Kyungpook National University, 130, Dongdeok-ro, Jung-gu, Daegu 41944, Republic of Korea
Department of Neurosurgery, School of Medicine, Kyungpook National University, 130, Dongdeok-ro, Jung-gu, Daegu 41944, Republic of Korea

Xiang Zeng
Department of Histology and Embryology, Zhongshan School of Medicine, Sun Yat-sen University, Guangzhou 510080, Guangdong Province, China

Caterina Masaracchia, Marilena Hnida, Ellen Gerhardt, Tomás Lopes da Fonseca, Anna Villar-Pique and Tiago F. Outeiro
Department of Experimental Neurodegeneration, Center for Biostructural Imaging of Neurodegeneration, Center for Nanoscale Microscopy and Molecular Physiology of the Brain, University Medical Center Goettingen, 37073 Göttingen, Germany

Tiago Branco and Ira Milosevic
European Neuroscience Institute, University Medical Center Göttingen, Göttingen, Germany

Claudio O. Fernández
Max Planck Laboratory for Structural Biology, Chemistry and Molecular Biophysics of Rosario (MPLbioR, UNR-MPIbpC) and Instituto de Investigaciones para el Descubrimiento de Fármacos de Rosario (IIDEFAR, UNR-CONICET), Universidad Nacional de Rosario, Ocampo y Esmeralda, S2002LRK Rosario, Argentina

Markus A. Stahlberg and Camin Dean
Trans-synaptic Signaling Group, European Neuroscience Institute, Grisebachstrasse 5, 37077 Goettingen, Germany

Tiago F. Outeiro
Max Planck Institute for Experimental Medicine, Göttingen, Germany
Institute of Neuroscience, The Medical School, Newcastle University, Framlington Place, Newcastle Upon Tyne NE2 4HH, UK

P. J. van der Spek
Department of Bioinformatics, Erasmus University Medical Center, Wytemaweg 80, 3015, GE, Rotterdam, The Netherlands
Department of Pathology, Erasmus University Medical Center, Wytemaweg 80, 3015, GE, Rotterdam, The Netherlands

T. A. M. Bouwens van der Vlis
Department of Neurosurgery, Maastricht University Medical Center, P. Debyelaan 25, 6229, HX, Maastricht, The Netherlands
Department of Pathology, Erasmus University Medical Center, Wytemaweg 80, 3015, GE, Rotterdam, The Netherlands
Department of Bioinformatics, Erasmus University Medical Center, Wytemaweg 80, 3015, GE, Rotterdam, The Netherlands

J. M. Kros and D. A. M. Mustafa
Department of Pathology, Erasmus University Medical Center, Wytemaweg 80, 3015, GE, Rotterdam, The Netherlands

R. T. A. van Wijck
Department of Internal Medicine, Erasmus University Medical Center, Wytemaweg 80, 3015, GE, Rotterdam, The Netherlands

P. M. van Hagen
Department of Immunology, Erasmus University Medical Center, Wytemaweg 80, 3015, GE, Rotterdam, The Netherlands
Department of Internal Medicine, Erasmus University Medical Center, Wytemaweg 80, 3015, GE, Rotterdam, The Netherlands

Romain Appay and Dominique Figarella-Branger
APHM, Hôpital de la Timone, Service d'Anatomie Pathologique et de Neuropathologie, Marseille, France

Emeline Tabouret
APHM, Hôpital de la Timone, Service de Neurooncologie, Marseille, France
Aix-Marseille Univ, CNRS, INP, Inst Neurophysiopathol, Marseille, France

Romain Appay, Dominique Figarella-Branger and Carole Colin
Aix-Marseille Univ, CNRS, INP, Inst Neurophysiopathol, Marseille, France

Mehdi Touat, Ahmed Idbaih and Caroline Dehais
AP-HP, Hôpitaux Universitaires La Pitié Salpêtrière-Charles Foix, Service de Neurologie 2–Mazarin, Paris, France
Sorbonne Université, Inserm, CNRS, UMR S 1127, Institut du Cerveau et de la Moelle épinière, ICM, AP-HP, Hôpitaux Universitaires La Pitié Salpêtrière - Charles Foix, Service de Neurologie 2-Mazarin, F-75013 Paris, France

Catherine Carpentier and Karima Mokhtari
Sorbonne Université, Inserm, CNRS, UMR S 1127, Institut du Cerveau et de la Moelle épinière, ICM, AP-HP, Hôpitaux Universitaires La Pitié Salpêtrière - Charles Foix, Service de Neurologie 2-Mazarin, F-75013 Paris, France

François Ducray
Hospices Civils de Lyon, Hôpital Pierre Wertheimer, Service de Neuro-oncologie, Bron, France
Department of Cancer Cell Plasticity, Cancer Research Centre of Lyon, Inserm U1052, CNRS UMR5286, Lyon, France

Karima Mokhtari
AP-HP, Hôpitaux Universitaires La Pitié Salpêtrière-Charles Foix, Service de Neuropathologie Raymond Escourolle, Paris, France

Emmanuelle Uro-Coste
CHU Toulouse, Hôpital Rangueil, Service d'Anatomie Pathologique et Histologie-Cytologie, Toulouse, France
Inserm U1037, Centre de Recherche en Cancérologie de Toulouse, Université de Toulouse, Toulouse, France

Daniel Twohig, Kalicharan Patra and Henrietta M. Nielsen
Department of Biochemistry and Biophysics, Stockholm University, Svante Arrhenius väg 16B, 106 91 Stockholm, Sweden

Elena Rodriguez-Vieitez and Agneta Nordberg
Department of Neurobiology, Care Sciences and Society, Karolinska Institutet, Stockholm, Sweden

Ina Møller
Department of Neurology, University Hospital of Trondheim, Trondheim, Norway

Sigrid B. Sando, Geir Bråthen and Linda R. White
Department of Neurology, University Hospital of Trondheim, Trondheim, Norway
Department of Neuroscience, Norwegian University of Science and Technology, Trondheim, Norway

Guro Berge and Gøril R. Grøntvedt
Department of Neuroscience, Norwegian University of Science and Technology, Trondheim, Norway

Guojun Bu
Department of Neuroscience, Mayo Clinic College of Medicine, Jacksonville, FL, USA

Tammie L. S. Benzinger
Department of Radiology, Washington University School of Medicine, St Louis, MO, USA

Celeste M. Karch
Department of Psychiatry, Washington University School of Medicine, St Louis, MO, USA

Anne Fagan, John C. Morris and Randall J. Bateman
Department of Neurology, Washington University School of Medicine, St Louis, MO, USA

Agneta Nordberg
The Aging Research Center, Karolinska University Hospital, Stockholm, Sweden

Carl Julien, Colson Tomberlin, Christine M. Roberts and Christopher D. Link
Department of Integrative Physiology, University of Colorado at Boulder, Boulder, CO, USA

Aumbreen Akram and Michael A. Silverman
Department of Biological Sciences, and the Centre for Cell Biology, Development, and Disease, Simon Fraser University, Burnaby, British Columbia, Canada

Gretchen H. Stein
Department of Molecular, Cellular, and Developmental Biology, University of Colorado, Boulder, Colorado 80309-0347, USA

Carl Julien
Centre de Recherche en Sciences Animales de Deschambault (CRSAD), Deschambault, Quebec G0A 1S0, Canada

Michael A. Silverman
Centre for Cell Biology, Development, and Disease, Simon Fraser University, British Columbia V5A 1S6, Burnaby, Canada

Jarek Wegiel, Wojciech Kaczmarski, Krzysztof Nowicki, Izabela Kuchna and Jerzy Wegiel
Department of Developmental Neurobiology, New York State Institute for Basic Research in Developmental Disabilities, 1050 Forest Hill Road, Staten Island, NY 10314, USA

Michael Flory
Research Design and Analysis Services, New York State Institute for Basic Research in Developmental Disabilities, Staten Island, NY 10314, USA

Veronica Martinez-Cerdeno
Pathology and Laboratory Medicine, Institute for Pediatric Regenerative Medicine, MIND Institute, University of California, Davis, CA, USA

Thomas Wisniewski
Departments of Neurology, Pathology and Psychiatry, NYU Langone Medical Center, New York, NY, USA

Hunter C. Gits, Maia Anderson, Stefanie Stallard, Becky Zon and Carl Koschmann
Department of Pediatrics, Division of Pediatric Hematology-Oncology; Michigan Medicine, Ann Arbor, MI 48109, USA

Drew Pratt
Department of Pathology, Michigan Medicine, Ann Arbor, MI 48109, USA

Tingting Qin
Department of Computational Medicine and Bioinformatics; Michigan Medicine, Ann Arbor, MI 48109, USA

Lili Zhao
Department of Biostatistics, University of Michigan School of Public Health, Ann Arbor, MI 48109, USA

James Leach
Division of Radiology, Cincinnati Children's Hospital, Cincinnati, OH 45229, USA

Brooklyn Chaney
Division of Oncology, Cincinnati Children's Hospital, Cincinnati, OH 45229, USA

Maryam Fouladi
Department of Pediatrics, Division of Oncology, Cincinnati Children's Hospital, Cincinnati, OH 45229, USA

Nicholas G. Gottardo
Michigan Center for Translational Pathology, Michigan Medicine, Ann Arbor, MI 48109, USA
Kids Cancer Centre, Telethon Kids Institute, Subiaco, WA 6008, Australia
Division of Paediatrics, University of Western Australia, Crawley, WA 6009, Australia

Emilie L. Castranio, Cody M. Wolfe, Kyong Nyon Nam, Florent Letronne, Nicholas F. Fitz, Iliya Lefterov and Radosveta Koldamova
Department of Environmental and Occupational Health, University of Pittsburgh, Pittsburgh, PA 15261, USA

Arthur Lionnet, Anne-Gaëlle Corbillé, Alice Prigent, Jacques Gonzales, Emilie Durieu, Malvyne Rolli-Derkinderen, Emmanuel Coron, Emilie Duchalais, Michel Neunlist and Pascal Derkinderen
Inserm, U1235, 1 rue Gaston Veil, F-44035 Nantes, France
Department of Neurology, CHU Nantes, F-44093 Nantes, France
University Nantes, F-44000 Nantes, France

Sébastien Paillusson, Maddalena Tasselli, Diane P. Hanger, Wendy Noble and Matthew A. Wade
King's College London, Institute of Psychiatry, Psychology and Neuroscience, Department of Basic and Clinical Neuroscience, Maurice Wohl Clinical Neuroscience Institute, Rm 1.23, 5 Cutcombe Road, Camberwell, London SE5 9RX, UK

Michael S. Perkinton
Neuroscience, IMED Biotech Unit, AstraZeneca, Cambridge CB21 6GH, UK

Mirta Borin, Erika Lorenzetto, Valeria Pontelli, Anna Avesani, Gianluigi Zanusso and Mario Buffelli
Department of Neurosciences, Biomedicine and Movement Sciences, University of Verona, Strada Le Grazie, 8, 37134 Verona, Italy

Claudia Saraceno, Anna Paterlini, Silvia Fostinelli, Silvia Bolognin, Luisa Benussi and Roberta Ghidoni
Molecular Markers Laboratory, IRCCS Istituto Centro San Giovanni di Dio Fatebenefratelli, Via Pilastroni, 4, 25125 Brescia, Italy

Marcella Catania and Giuseppe Di Fede
Division of Neurology 5 and Neuropathology, IRCCS Foundation - Carlo Besta Neurological Institute, Via Celoria 11, 20133 Milan, Italy

Giuliano Binetti
MAC Memory Center, IRCCS Istituto Centro San Giovanni di Dio Fatebenefratelli, Brescia, Italy

Simone Zorzan
Environmental Research and Innovation (ERIN) Department, Luxembourg Institute of Science and Technology (LIST), Belvaux L-4422, Luxembourg

Silvia Bolognin
Luxembourg Centre for Systems Biomedicine, University of Luxembourg, 6 avenue du Swing, L-4367 Belvaux, Luxembourg

Joerg Neddens, Magdalena Temmel, Stefanie Flunkert, Bianca Kerschbaumer, Christina Hoeller, Tina Loeffler, Vera Niederkofler and Birgit Hutter-Paier
QPS Austria GmbH, Neuropharmacology, Parkring 12, 8074 Grambach, Austria

Bianca Kerschbaumer and Guenther Daum
Institute for Biochemistry, Graz University of Technology, Graz, Austria

Christina Hoeller
Institute of Zoology, Karl Franzens University, Graz, Austria

Johannes Attems
Institute of Neuroscience and Newcastle University Institute for Ageing Campus for Ageing and Vitality, Newcastle University, Newcastle upon Tyne NE4 5PL, UK

Index

P

R

S

T

CPSIA information can be obtained
at www.ICGtesting.com
Printed in the USA
LVHW060859030122
707669LV00002B/261